Advances in Sustainable Biomaterials

Sustainable biomaterials are used as substitutions for traditional materials in aerospace, automotive, civil, mechanical, environmental engineering, medical, and other industries. This book presents the current knowledge and recent developments on the characterization and application of sustainable biomaterials with biomanufacturing 4.0 techniques. The book also describes the unique properties of various classes of sustainable biomaterials, making them highly suitable for many industrial applications.

Advances in Sustainable Biomaterials: Bioprocessing 4.0, Characterizations, and Applications presents key chapters on smart biopolymer composites production and processing methods and provides a wide range of applications in a variety of fields such as medical, food, agriculture, electronics, manufacturing, and chemical engineering. The book features the most recent and detailed information on advancements in biopolymer biomaterials and emphasizes synthesis, characterization, modeling, manufacturing, and testing strategies.

Written to be used as a resource guide on biomaterials and innovations, undergraduate and postgraduate students studying manufacturing and materials science will find this book very useful in addition to those working in mechanical engineering, biomedical engineering, manufacturing of pharmaceuticals, biotechnology, and electronics engineering fields. The book can also be used as additional classroom reading for an advanced course on biomaterials modeling and optimization.

Advancements in Intelligent and Sustainable Technologies and Systems

Series Editor: Ajay Kumar

This book series aims to provide a platform for academicians, researchers, professionals, and individuals to participate and to provide novel systematic theoretical, experimental, computational work in the form of edited text or reference books, monographs in the area of intelligent and sustainable technologies and systems from engineering, management, applied science, healthcare, etc. domains. This book series will educate and inform the readers with a comprehensive overview of advancements in intelligent and sustainable techniques and systems with novel intelligent tools and algorithms to move industries from different domains from a data-centric community to a sustainable world. The book series covers ideas and innovations to help the research community and professionals to understand fundamentals, opportunities, challenges, future outlook, layout, lifecycle, and framework of intelligent and sustainable technologies and systems for different sectors. It serves as a guide for computer science, mechanical, manufacturing, electrical, electronics, civil, automobile, industrial engineering, biomedical, healthcare, and management professionals.

If you are interested in writing or editing a book for the series or would like more information, please contact Cindy Carelli, cindy.carelli@taylorandfrancis.com

5G-Based Smart Hospitals and Healthcare Systems
Evaluation, Integration, and Deployment
Edited by Arun Kumar, Sumit Chakravarty, Aravinda K., and Mohit Kumar Sharma

Handbook of Intelligent and Sustainable Smart Dentistry
Nature and Bio-Inspired Approaches, Processes, Materials, and Manufacturing
Edited by Ajay Kumar, Namrata Dogra, Sarita, Surbhi Bhatia, M. S. Sidhu

Handbook of Intelligent and Sustainable Manufacturing
Tools, Principles, and Strategies
Edited by Ajay Kumar, Parveen Kumar, Yang Liu, and Rakesh Kumar

Advances in Sustainable Biomaterials
Bioprocessing 4.0, Characterizations, and Applications
Edited by Ajay Kumar, D. K. Rajak, Parveen Kumar, and Ashwini Kumar

Advances in Sustainable Biomaterials

Bioprocessing 4.0, Characterizations, and Applications

Edited by Ajay Kumar, D. K. Rajak,
Parveen Kumar, and Ashwini Kumar

CRC Press
Taylor & Francis Group
Boca Raton London New York

CRC Press is an imprint of the
Taylor & Francis Group, an **informa** business

Designed cover image: Shutterstock

First edition published 2025
by CRC Press
2385 NW Executive Center Drive, Suite 320, Boca Raton FL 33431

and by CRC Press
4 Park Square, Milton Park, Abingdon, Oxon, OX14 4RN

CRC Press is an imprint of Taylor & Francis Group, LLC

ISBN: 978-1-032-55687-1 (hbk)
ISBN: 978-1-032-56187-5 (pbk)
ISBN: 978-1-003-43431-3 (ebk)

DOI: 10.1201/9781003434313

Typeset in Times
by codeMantra

Contents

About the Editors

Dr. Ajay Kumar is currently serving as a Professor in the Department of Mechanical Engineering, School of Engineering and Technology, JECRC University, Jaipur, Rajasthan, India. He received his Ph.D. in the field of Advanced Manufacturing and Automation from Guru Jambheshwar University of Science & Technology, Hisar, India after B.Tech. (Hons.) in Mechanical Engineering and M.Tech. (Distinction) in Manufacturing and Automation. His areas of research include biomedical engineering, incremental sheet forming, artificial intelligence, sustainable materials, robotics and automation, additive manufacturing, mechatronics, smart manufacturing, Industry 4.0, industrial engineering systems, waste management, and optimization techniques. He has over 100 publications in international journals of repute including Scopus, Web of Science, and SCI indexed database and refereed international conferences. He has organized various national and international events including an international conference on Mechatronics and Artificial Intelligence (ICMAI-2021) and international conference on Artificial Intelligence, Advanced Materials, and Mechatronics Systems (AIAMMS-2023) as conference chair. He has more than 20 national and international patents to his credit. He has supervised more than 8 M.Tech. and Ph.D. scholars and numerous undergraduate projects/theses. He has a total of 15 years of experience in teaching and research. He is Guest Editor of many reputed journals. He has contributed to many international conferences/symposiums as a session chair, expert speaker, and member of the editorial board. He has won several proficiency awards during the course of his career, including merit awards, best teacher awards, and so on. He has also co-authored, co-edited books and proceedings including:

- Ajay Kumar, Ravi Kant Mittal, *Incremental Sheet Forming Technologies: Principles, Merits, Limitations, and Applications*, CRC Press, Taylor and Francis, ISBN 9780367276744
    https://www.routledge.com/Incremental-Sheet-Forming-Technologies-Principles-Merits-Limitations/jay-Mittal/p/book/9780367276744#
- Ajay, Parveen, Sharif Ahmad, Jyotsna Sharma, Victor Gambhir, *Handbook of Sustainable Materials: Modelling, Characterization, and Optimization*, CRC Press, Taylor and Francis, ISBN: 9781032286327,
    https://www.routledge.com/Handbook-of-Sustainable-Materials-Modelling-Characterization-and-Optimization/Ajay-Parveen-Ahmad-Sharma-Gambhir/p/book/9781032286327
- Ajay, Parveen, Ravi Kant Mittal, Rajesh Goel (2022), *Waste Recovery and Management: An Approach Towards Sustainable Development Goals*, CRC Press, Taylor and Francis, ISBN: 9781032281933
    https://www.taylorfrancis.com/books/edit/10.1201/9781003359784/waste-recovery-management-ajay-kumar-ashwini-kumar-parveen-rathee-rajesh-goel-ravi-kant-mittal
- Ajay, Parveen, Bandar AlMangour, Hari Singh (2023), *Smart Manufacturing: Forecasting the Future of Industry 4.0*, CRC Press,

Taylor and Francis, ISBN: 9781032363431, https://www.routledge. com/Handbook-of-Smart-Manufacturing-Forecasting-the-Future-of-Industry-40/jay-Singh-Parveen-AlMangour/p/book/9781032363431
- *Handbook of Flexible and Smart Sheet Forming Techniques: Industry 4.0 Perspectives*, WILEY Publications, ISBN: 9781119986409, https://doi.org/10.1002/9781119986454
- *3D Printing Technologies: Digital Manufacturing, Artificial Intelligence, Industry 4.0*, De Gruyter, ISBN: 9783111214597, https://doi.org/10.1515/9783111215112
- *Smart Electric and Hybrid Vehicles: Fundamentals, Strategies and Applications*, CRC Press, Taylor and Francis, ISBN: 9781032600369, https://doi.org/10.1201/9781003495574
- *Smart Electric and Hybrid Vehicles: Design, Modeling, and Assessment*, CRC Press, Taylor and Francis, ISBN: 9781032600376, https://doi.org/10.1201/9781003502470
- *Handbook of Intelligent and Sustainable Manufacturing: Tools, Principles, and Strategies*, CRC Press, Taylor and Francis, ISBN: 9781032519838, https://doi.org/10.1201/9781003405870 He has also authored many in-house course notes, lab manuals, monographs, and invited chapters in books. He has organized a series of faculty development programs, international conferences, workshops, and seminars for researchers, Ph.D., UG and PG level students. He is associated with many research, academic, and professional societies in various capacities.

https://www.webofscience.com/wos/author/record/D-5813-2019
Publons Profile: https://publons.com/researcher/1596469/ajay-kumar/
Research Gate: https://www.researchgate.net/profile/Ajay_Kumar349

Dr. D. K. Rajak is a Scientist at CSIR-Advanced Materials and Processes Research Institute, Bhopal, Madhya Pradesh, India. He completed his Ph.D. from the Indian Institute of Technology (ISM), Dhanbad, Jharkhand. He received many prestigious awards and was recognized in the World Top 2% Scientists by Stanford University in 2022 and 2023. He has published over 100 peer-reviewed journal papers and has 3 books, 16 book chapters, 18 patents, 2 designs, and 1 copyright. His recent work involves developing lightweight alloy composite foams and composites for applications requiring low density and high specific energy absorption in automobile, aerospace, marine, noise, and thermal management sectors.

Parveen Kumar is currently serving as an Assistant Professor and Head in the Department of Mechanical Engineering at Rawal Institute of Engineering and Technology, Faridabad, Haryana, India. Currently, his Ph.D. is in final evaluation through the National Institute of Technology, Kurukshetra, Haryana, India. He completed his B.Tech. (Hons.) from Kurukshetra University, Kurukshetra, India and M.Tech. (Distinction) in Manufacturing and Automation from Maharshi Dayanand University, Rohtak, India. His areas of research include intelligent manufacturing systems, materials; mechatronics systems, advanced materials, Industry 4.0, die less forming, design of automotive systems, additive manufacturing, CAD/CAM and artificial intelligence, machine learning and Internet of Things in manufacturing, and multi-objective optimization techniques. He has over 40 publications in international journals of repute including Scopus, Web of Science, and SCI indexed database and refereed international conferences. He has eight national and international patents to his credit. He has supervised two M.Tech. scholars and numerous undergraduate projects/theses. He has a total of 13 years of experience in teaching and research. He has co-authored/co-edited several books including:

- *Waste Recovery and Management: An Approach Towards Sustainable Development Goals*, CRC Press, Taylor and Francis, ISBN: 9781032281933, https://doi.org/10.1201/9781003359784
- *Handbook of Sustainable Materials: Introduction, Modelling, Characterization and Optimization*, CRC Press, Taylor and Francis, ISBN: 9781032295874, https://doi.org/10.1201/9781003297772
- *Handbook of Smart Manufacturing: Forecasting the Future of Industry 4.0*, CRC Press, Taylor and Francis, ISBN: 9781032363431, https://doi.org/10.1201/9781003333760
- *Modeling, Characterization, and Processing of Smart Materials*, IGI Global, EISBN13: 9781668492260
- *Handbook of Flexible and Smart Sheet Forming Techniques: Industry 4.0 Perspectives*, WILEY Publications, ISBN: 9781119986409, https://doi.org/10.1002/9781119986454
- *3D Printing Technologies: Digital Manufacturing, Artificial Intelligence, Industry 4.0*, De Gruyter, ISBN: 9783111214597, https://doi.org/10.1515/9783111215112
- *Smart Electric and Hybrid Vehicles: Fundamentals, Strategies and Applications*, CRC Press, Taylor and Francis, ISBN: 9781032600369, https://doi.org/10.1201/9781003495574
- *Smart Electric and Hybrid Vehicles: Design, Modeling, and Assessment*, CRC Press, Taylor and Francis, ISBN: 9781032600376, https://doi.org/10.1201/9781003502470
- *Handbook of Intelligent and Sustainable Manufacturing: Tools, Principles, and Strategies*, CRC Press, Taylor and Francis, ISBN: 9781032519838, https://doi.org/10.1201/9781003405870

He has organized a series of faculty development programs, workshops, and seminars for researchers and UG level students. He is Editor of the book series, *Industrial and Manufacturing Systems and Technologies: Sustainable and Intelligent Perspectives* CRC Press, Taylor and Francis.

https://scholar.google.com/citations?hl=en&authuser=1&user=ESQ-RnYAAAAJ

Dr. Ashwini Kumar is currently working as an Associate Professor in the Department of Mechanical Engineering, Manav Rachna International Institute of Research and Studies, Faridabad, India. Prior to joining the university, Dr. Kumar gained 13+ years of experience holding different academic and research responsibilities in different organizations. He has completed his research on the optimal thermohydraulic performance of three sides artificially roughened solar air heaters at the National Institute of Technology, Jamshedpur. He received a Doctor of Philosophy from the Department of Mechanical Engineering,National Institute of Technology, Jamshedpur. He has authored several scientific articles in high impact journals and has also written five books. His research contribution includes articles on heat transfer and optimization techniques. He received the Best Session Paper Award in International Conferences, 2016 from NIT Patna. Dr. Kumar is associated with many universities, and along with that, he is an editorial board member of reputed journals and also a reviewer of international journals (SCI & Scopus Index). He also looks forward to guiding many research scholars often developing his own interest in the field in which he is an expediter. He believes that Science, Engineering, and Technology are advancing at a fast pace, and the obsolescence of physical infrastructure, skills, and competence take place rapidly. He is keen on taking steps to improve the existing infrastructure, investment, and intellectual strength wherever they exist and network them so as to utilize them effectively and optimally for meeting the changing needs. Not only does he profess the values of Indian culture but also does he practice many of the fundamental principles of humanity and society. While dedicating himself to the cause of creating world-class infrastructure, he has also undertaken many community services in various parts of India out of his quest to contribute more to society. His research interest includes artificial roughness, solar air heaters, heat transfer, electronic thermal cooling systems, refrigeration and air conditioning, fluid mechanics, and computational fluid dynamics.

Contributors

Juber Akhtar
Department of Pharmacy, Faculty of
    Pharmacy
Integral University
Lucknow, Uttar Pradesh, India

Vaseem A. Ansari
Department of Pharmacy, Faculty of
    Pharmacy
Integral University
Lucknow, Uttar Pradesh, India

Mahalakshmi Suresha Biradar
Department of Pharmaceutical
    Chemistry, Al-Ameen College of
    Pharmacy
Rajiv Gandhi University of Health
    Sciences
Bengaluru, Karnataka, India

G. Boopathi
Department of Agricultural Engineering
Amrita School of Agricultural Sciences
Coimbatore, Tamil Nadu, India

Prafulla B. Choudhari
Department of Pharmaceutical
    Chemistry
Bharati Vidyapeeth College of
    Pharmacy
Kolhapur, India

Ramesh Daryapurkar
LARS Enviro Private Limited
Nagpur, Maharashtra, India

Sanket Ghanti
School of Architecture, KLE
    Technological University
Hubli, Karnataka, India

Syed Misbahul Hasan
Department of Pharmacy, Faculty of
    Pharmacy
Integral University
Lucknow, Uttar Pradesh, India

Mohit Kale
LARS Enviro Private Limited
Nagpur, Maharashtra, India

Sachin N. Kothawade
RSM's N. N. Sattha College of
    Pharmacy
Ahmednagar, Maharashtra, India

Ajay Kumar
Department of Mechanical Engineering,
    School of Engineering and
    Technology
JECRC University
Jaipur, Rajasthan, India

Ashwini Kumar
Mechanical Engineering Research
    Wings
Manav Rachna International Institute of
    Research and Studies
Faridabad, Delhi-NCR, India

Parveen Kumar
Department of Mechanical Engineering
Rawal Institute of Engineering and
    Technology
Faridabad, Haryana, India

D. Praveen Kumar
Department of Agricultural Engineering
Bannari Amman Institute of
    Technology
Sathyamangalam, Erode, Tamil Nadu,
    India

Shubhrat Maheshwari
Faculty of Pharmaceutical Sciences
Rama University
Kanpur, Uttar Pradesh, India

Purba Mandal
Department of Pharmacy, Faculty of
    Pharmacy
Integral University
Lucknow, Uttar Pradesh, India

Gresi D. Mate
Department of Pharmaceutical Quality
    Assurance
Dr. D. Y. Patil Institute of
    Pharmaceutical Sciences and
    Research
Pimpri, Pune, Maharashtra, India

Alok Kumar Mishra
Department of Microbiology, School of
    Bioengineering and Biosciences
Lovely Professional University
Phagwara, Punjab, India

Sanskruti Ajay Mukwane
Yeshwantrao Chavan College of
    Engineering (YCCE)
Nagpur, Maharashtra, India

Sharon Nagpal
Department of Microbiology, School of
    Bioengineering and Biosciences
Lovely Professional University
Phagwara, Punjab, India

Odangowei Inetiminebi Ogidi
Department of Biochemistry, Faculty of
    Basic Medical Sciences
Bayelsa Medical University
Yenagoa Bayelsa State, Nigeria

Kalusing S. Padvi
Department of Chemical Technology
Dr. Babasaheb Ambedkar Marathwada
    University
Aurangabad, Maharashtra, India

Vishal V. Pande
RSM's N. N. Sattha College of
    Pharmacy
Ahmednagar, Maharashtra, India

Aniruddha B. Patil
Department of Chemistry
Maharshi Dayanand College
Parel, Mumbai, Maharashtra, India

Shubhangi P. Patil
Department of Chemistry
The Institute of Science
Mumbai, Maharashtra, India

Bhupendra G. Prajapati
Shree S. K. Patel College of
    Pharmaceutical Education and
    Research
Ganpat University
Gujarat, India

Sanket S. Rathod
Department of Pharmaceutical
    Chemistry
Bharati Vidyapeeth College of
    Pharmacy
Kolhapur, India

Avani S
Department of Microbiology, School of
    Bioengineering and Biosciences
Lovely Professional University
Phagwara, Punjab, India

Pawan Saini
CSB-Central Sericultural Research &
    Training Institute (CSR & TI)
Pampore, Jammu-Kashmir, India

Tejal A. Salunkhe
Department of Quality Assurance
Progressive Education Society's Modern
    College of Pharmacy (For Ladies)
Moshi, Pune, Maharashtra, India

Aniket P. Sarkate
Department of Chemical Technology
Dr. Babasaheb Ambedkar Marathwada
   University
Aurangabad, Maharashtra, India

Divya Sharma
School of Architecture
KLE Technological University
Hubli, Karnataka, India

Isha Sharma
Department of Microbiology, School of
   Bioengineering and Biosciences
Lovely Professional University
Phagwara, Punjab, India

Sonali S. Shinde
Department of Chemical Technology
Dr. Babasaheb Ambedkar Marathwada
   University
Aurangabad, Maharashtra, India

Abhishek Singh
Amity Institute of Pharmacy
Amity University
Lucknow, Uttar Pradesh, India

Aditya Singh
Department of Pharmacy, Faculty of
   Pharmacy
Integral University
Lucknow, Uttar Pradesh, 226026, India

Harsh Vardhan Singh
Department of Microbiology, School of
   Bioengineering and Biosciences
Lovely Professional University
Phagwara, Punjab, India

Sudarshan Singh
Faculty of Pharmacy
Chiang Mai University
Chiang Mai 50200, Thailand

Samiksha P. Surwade
Department of Pharmaceutics
Marathwada Mitra Mandal's College of
   Pharmacy
Thergaon, Pune, Maharashtra, India

Jayprakash Suryawanshi
RSM's N. N. Sattha College of
   Pharmacy
Ahmednagar, Maharashtra, India

Rajendra R. Tayade
Department of Chemistry
The Institute of Science, Institute of
   Science
Nagpur, R. T. Road, Civil Lines,
   Nagpur, India

Shankar Thapa
Department of Pharmacy
Universal College of Medical Sciences
Bhairahawa, Nepal

Aditi M. Thorve
Department of Pharmaceutical Quality
   Assurance
Dr. D. Y. Patil Institute of
   Pharmaceutical Sciences and
   Research
Pimpri, Pune-411018, Maharashtra,
   India

Amita Verma
Bioorganic and Medicinal Chemistry
   Research Laboratory, Department of
   Pharmaceutical Sciences
Sam Higginbottom University of
   Agriculture, Technology and
   Sciences
Prayagraj, Uttar Pradesh, India

V. Karuppasamy Vikraman
Department of Agricultural Engineering
Amrita School of Agricultural Sciences
Coimbatore, Tamil Nadu, India

Harshal Warade
Yeshwantrao Chavan College of
  Engineering (YCCE)
Nagpur, Maharashtra, India

Jagat Pal Yadav
Bioorganic and Medicinal Chemistry
  Research Laboratory, Department of
  Pharmaceutical Sciences
Sam Higginbottom University of
  Agriculture, Technology and
  Sciences
Prayagraj, Uttar Pradesh, India

Seema Yadav
Amity Institute of Pharmacy
Amity University
Lucknow, Uttar Pradesh, India

Bancha Yingngam
Department of Pharmaceutical
  Chemistry and Technology, Faculty
  of Pharmaceutical Sciences
Ubon Ratchathani University
Ubon Ratchathani, Thailand

Preface

Sustainable biomaterials are the materials used throughout our consumer and industrial economy that can be produced in required volumes without depleting non-renewable resources and without disrupting the established steady-state equilibrium of the environment and key natural resource systems. In today's era, there is a high demand for a pollution-free environment, reducing hazardous waste from industries, diverting manufacturing sectors toward green production, minimizing the use of fossil fuels, to reduce carbon contents from atmosphere and to decrease hazardous environmental impact of current engineering materials that are in use to produce customized and cultured products. Sustainable biomaterials are the best option that can meet such requirement across the globe as they have renewability, biodegradability, are cost-effective, have higher chemical and mechanical resistance and biocompatibility for various medical, pharmaceutical, electronics equipment and other engineering applications. Thus, environmentally responsible biomaterials are designed and developed by scientists/engineers to meet the current demand of environment and industries. Sustainable biomaterials are used as substitutions of traditional materials in aerospace, automotive, civil, mechanical, environmental engineering, medical, and other industries. This book provides a one-stop guide for UG/PG students, research scholars, academicians, researchers, and engineers working in the niche area of sustainable biomaterials, biopolymer composites, polymer science, smart biomaterials, materials science, pharmacy, chemical, biomedicine and biomedical engineering. The present book collects the current knowledge and recent developments in biomaterials and application of sustainable biomaterials with bio-manufacturing 4.0 techniques. This volume describes the outstanding properties of various classes of sustainable biomaterials which recommend it for various industrial applications.

Advances in Sustainable Biomaterials: Bioprocessing 4.0, Characterizations, and Applications contains 16 chapters that describe relevant perspectives in this field. Chapter "Biopolymer composites for sustainable and green technology" explores the possibilities of using natural fibers, including hemp fibers, to make renewable biomaterials and talks about the difficulties in combining renewable and petroleum-based materials to get the right performance characteristics. Chapter "A Feasibility study on Characterization of Lignocellulosic Biomass and Control Parameters in the Anaerobic Digestion Process" examines the chemical composition of Napier grass (energy crop) and results are compared between lignin, cellulose, and hemicellulose content of raw Napier grass for different harvesting ages of 15, 30, 45, and 60 days. Additionally, the comparison between the chemical compositions of raw Napier grass for different harvesting days from various literature studies was evaluated. The biogas and methane yield from anaerobic digestion of Napier grass from various literatures was also evaluated. Chapter "Lignocellulose-Based Sustainable Biomaterials" discusses notable properties of lignocellulose like its ability to extend down to nanoscale, strength, biocompatibility, versatility, biodegradability, renewability, etc. which make it a potential compound to hold a future in helping restore, improve, and repair human health. Chapter "Thermo Bi-Metal. A Sustainable Biomaterial for Breathing Façade" aims at the application of thermo-bimetal as a façade material that can self-manage building systems for light and ventilation. Chapter "Lignocellulose-Based Sustainable Biomaterials in the Fabrication of Pharmaceutical Products" explores the diverse applications of lignocelluloses biomass and lignin derivatives, highlighting their role in addressing environmental challenges, advancing sustainable technologies, and fostering innovation across various industries. Chapter "Biodegradable and Non-Biodegradable Sustainable Biomaterials" explores the intricate landscape of sustainable biomaterials, delineating the significance of biodegradability and the role of non-biodegradable alternatives in environmental sustainability. Chapter "Smart Sustainable Biomaterials and their Recent Advancement for Targeting the Nervous System" covers the both natural and artificial neuro-regenerative biomaterials for neural tissue engineering. And application of 0D, 1D, 2D, and 3D categorized nanostructures of biomaterials is also covered. Chapter "Processing of Sustainable Biomaterials by Additive Manufacturing Methods" explores the impact of nanoparticles on the mechanical and biological properties of 3D printed tissue scaffolds and discusses the difficulties of today and what lies in the future for ecologically friendly polymeric substances made using additive manufacturing. Chapter "Comprehensive Characterization of Biochar from Coconut Shell Pyrolysis: Effect of Temperature" emphasizes the critical role of pyrolysis temperature in controlling biochar properties for specific applications. Chapter "Applications of Sustainable Biomaterials in the Pharmaceutical and Medical Field" discusses a variety of biomaterials, such as natural polymers, synthetic polymers, metals, and recyclable polymers, as well as their applications in drug delivery, medical devices, and tissue engineering. Chapter "The Utilization of Biomaterials in the Diagnostic, Preventive, and Therapeutic Approaches for the Management of Covid-19" covers the etiology, epidemiology, and morphology of COVID-19, as well as the mechanism of infection and transmission, a general overview of biomaterials, their significance, applications, the benefits and drawbacks of using biomaterials for COVID-19 diagnosis and therapy, current challenges, and future prospects. Chapter "An Intriguing, Versatile, and Sustainable Biomaterial: Nanocellulose" goes over different forms of nanocellulose, extraction techniques, and sources of cellulose and focuses on modifying nanocellulose in a revised way to make it better at meeting specified requirements based on its use. Chapter "Sustainable Biomaterials for Pharmaceutical and Medical Applications" offers a comprehensive overview of the various types of biopolymers and bioprocessing techniques used in biomaterial production, such as extraction, purification, and modification methods. Chapter "Polymeric Biomaterials for Potential Pharmaceutical and Biomedical Applications" underscores the diverse pharmaceutical applications of lignin-derived compounds and the importance of ongoing research and innovation in harnessing lignin's full potential for the benefit of healthcare and beyond. Chapter "The Advent of Biopolymers for Potential Applications in the Medical and Pharmaceutical Sector" led to the search for materials that were tough, and versatile but also non-toxic, biocompatible, and bioresorbable. Chapter "Applications of Biopolymers in the Agriculture and Food Industry" emphasizes to improve the characteristics of biopolymers, addresses limitations like heat resistance and opacity, and explores novel uses in food packaging.

This book is for researchers, academicians, industrialists, and scientists working in the field of medical, biomedical, healthcare, materials science, polymer science, biotechnology, manufacturing, production of materials, and smart materials.

The editors acknowledge the professional support received from CRC Press and express their gratitude for this opportunity.

Reader's observations, suggestions, and queries are welcome.

Acknowledgments

The editors are grateful to the CRC Press for showing their interest to publish this book in the active research area of Advances in Sustainable Materials. The editors express their personal adulation and gratitude to Ms. Cindy Renee Carelli (Executive Editor) CRC Press, for giving consent to publish our work. She undoubtedly imparted the great and adept experience in terms of systematic and methodical staff who have helped the editors compile and finalize the manuscript. The editors also extend their gratitude to Ms. Kaitlyn Fisher, CRC Press, for supporting during her tenure.

The editors wish to thank all the chapter authors to contribute their valuable research and experience to compile this volume. The chapter authors, corresponding author in particular, deserve special acknowledgments for bearing with the editors, who persistently kept bothering them for deadlines, and with their remarks.

Dr. Ajay also wishes to express his gratitude to his parents, Shri. Jagdish and Smt. Kamla, and his loving brother Shri. Parveen for their true and endless support. They have made him walk tall before the world regardless of sacrificing their happiness and living in a small village. He cannot close these prefatory remarks without expressing his deep sense of gratitude and reverence to his life partner Mrs. Sarita Rathee for her understanding, care, support, and encouragement for having kept his morale high all the time. No magnitude of words can ever quantify the love and gratitude the he feels in thanking his daughters, Sejal Rathee and Mahi Rathee, and son Kushal Rathee who are the world's best children.

Finally, the editors obligate this work to the divine creator and express their indebtedness to the Almighty for gifting them power to yield their ideas and concepts into substantial manifestation. The editors believe that this book would enlighten the readers about each feature and characteristics of sustainable biomaterials.

Ajay Kumar
D. K. Rajak
Parveen Kumar
Ashwini Kumar

1 Biopolymer Composites for Sustainable and Green Technology

Aditya Singh
Integral University

Shubhrat Maheshwari
Rama University and Sam Higginbottom, University
of Agriculture, Technology and Sciences

Amita Verma
Sam Higginbottom, University of Agriculture,
Technology and Sciences

Vaseem A. Ansari
Integral University

Bhupendra G. Prajapati
Ganpat University

Sudarshan Singh
Chiang Mai University

1 INTRODUCTION

Schmidt and colleagues have documented that the transportation of plastics from land to oceans occurs through rivers (van Emmerik and Schwarz 2020). When plastics waste finds its way into the environment, it leads to severe problems such as the obstruction of waterways, resulting in stagnation (Kumar et al. 2021). Plastics also pose significant challenges to block their respiratory pathways, leading to their demise (Santhoskumar et al. 2010). According to Jambeck and colleagues, marine litter is projected to harm nearly 600 species by 2050, with 90% of seabirds being threatened by plastics ingestion, and approximately 15% of marine species falling under endangered categories due to plastics ingestion and entanglement. The degradation of large plastics occurs through various physical, chemical, and biological processes when they interact with the natural environment. Consequently, these small-sized particles can be found in various environmental systems, such as soil, sub-surface systems, groundwater, atmosphere, wetlands, rivers, and marine environments (Kumar et al.

DOI: 10.1201/9781003434313-1

"

2021). Bionanocomposites offer numerous benefits across various fields, including medicine, pharmaceutics, cosmetics, and food packaging (Cheung et al. 2007). The polylactic acid (PLA), which is widely recognized, is aliphatic polyester that can be derived from agricultural resources like corn and through the ring-opening polymerization of the lactides (Inkinen et al. 2011). The biocomposites that consist of PLA-based matrices are considered to be highly intriguing due to their combination of favorable mechanical and physical properties and their strong commitment to sustainability. With this in mind, a recent report has presented a novel approach to producing block copolymers of PLA by incorporating multiple "inifer" groups, such as tetraphenylethane (TPE) (Lin et al. 2010; Murariu et al. 2022). The biopolymers exhibit numerous interesting characteristics, including biocompatibility, biodegradability, and antibacterial activity (Stephen et al. 2023; Varghese et al. 2020). This transistor effectively imitates biological synapses and is expected to serve as a fundamental component in future bio-friendly neuromorphic computing systems (Cesca et al. 2010). This material is currently gaining attention not only in traditional applications like packaging, but also in the production of textile fibers (Zhao et al. 2010). PLA is finding higher value in durable/technical and biomedical applications. However, when it comes to durable applications, PLA faces limitations due to its poor thermal resistance, low heat distortion temperature, and rate of crystallization, as well as its high sensitivity to hydrolysis. These shortcomings prevent the material from meeting specific end-use requirements (Taib et al. 2023). PLA has already been melt-mixed with various mineral fillers, such as $CaCO_3$, talc, kaolin, $BaSO_4$, and others, commonly used in the polymer composites industry (MacArthur, Waughray, and Stuchtey 2016). It should be noted that composites of polylactic acid (PLA) with calcium sulfate $CaSO_4$ is considered an uncommon biocompatible material that is completely absorbed after being implanted (Tokiwa and Calabia 2007).

2   RENEWABLE PLASTICS

Plastics play a crucial role in various aspects of sustainability, such as improving fuel efficiency in transportation (Sousa and Silvestre 2022), increasing energy savings through insulation, and extending the shelf life of food, thereby reducing food waste (Walker and Rothman 2020). The production of plastic has reached over 380 million tons annually and continues to increase at a rate of 4% per year (Chen 2019; Gandini et al. 2016; Yuntawattana et al. 2021). It is important to note that some biodegradable plastics derived from fossil resources are also labeled as bioplastics, but this terminology is considered misleading and discouraged (Miller 2020). Currently, 100% bio-based bioplastics are being produced at a scale of approximately 2 million tons per year (Mrowiec 2018). Major international organizations, such as the UN, World Economic Forum, World Health Organization, and European Union, have made environmental plastic pollution a top concern (Fadeeva and Van Berkel 2021; MacArthur 2017). In 2015, the commercial sectors with the highest global plastic volumes were packaging (35.9%), building (16.0%), textiles (14.5%), and consumer goods (10.3%) (Foschi and Bonoli 2019; Geyer, Jambeck, and Law 2017) (for instance, 10.1%, 6.2%, and 3.4% (Foschi and Bonoli 2019). Roughly 80% of plastic waste found in the ocean originates on land, usually from poorly maintained landfills and street corners that are ravaged by wind and sea waves. Every year, almost 2 million tons of plastic garbage seep into rivers. This happens in both industrialized countries like the United States and China and in poor nations like Bangladesh, which lack the necessary infrastructure for waste collection and treatment (Jambeck et al. 2015; Schmidt, Krauth, and Wagner 2017). "External dumping," or the shipping of plastic from more affluent countries to those with less developed waste management systems and laws, exacerbates this problem (Ahmed 2019; Messerli et al. 2019). With over 300 new industrial projects (worth over US$200 billion) petrochemical plastic capacity are increasing. Approximately half of these projects are either in the construction or completion stages (Ziccardi et al. 2016). By 2050, the manufacture and incineration of plastic might result in 2.8 trillion tons of CO_2 emissions due to this trend (Messerli et al. 2019). Microplastic debris may have an impact on maritime species' ability to absorb CO_2, in addition to carbon emissions linked to plastic (Ten Brink et al. 2018). "Biorefineries," like an oil refinery, are facilities that turn renewable feedstocks derived from plants into valuable compounds. The resource biomass may be replenished rather quickly and is commonly separated into feedstocks of the first and second generations (Coates and Getzler 2020). The former usually refers to easily fermentable sugars from edible vegetable oils and sources of edible polysaccharides, including maize and sugarcane (Pudack, Stepanski, and Fässler 2020). While some research indicates that biomass can be sustainably co-produced for food and fuel (consequently, for bio-based goods) (Chu and Majumdar 2012). First-generation biomass is still debatable because of moral worries about possible rivalry with food supplies, particularly in rural areas. Only 0.02% of the world's agricultural land is now used to grow the predecessors of bioplastics (Mülhaupt 2013). It is doubtful that biomass will completely replace fossil fuels as a source of plastics, which emphasizes the need for better recycling and consumption reduction. An

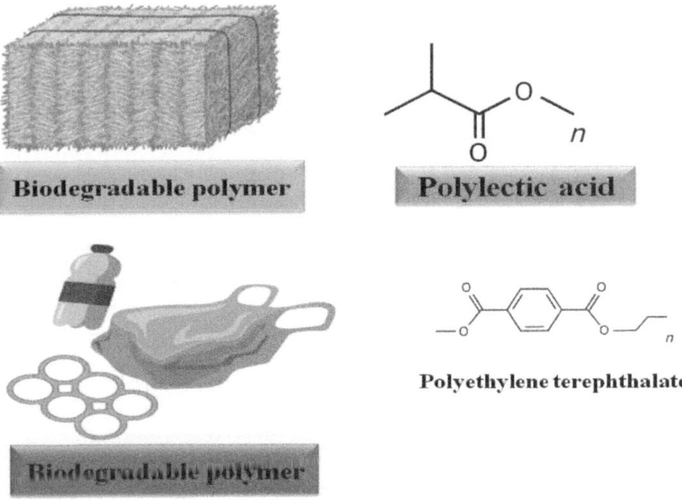

FIGURE 1.1 Synthesis methods of biodegradable plastics.

TABLE 1.1

Compares the average costs and environmental characteristics of a few commercially significant synthetic polymers derived from fossil fuels and biofuels

Polymer	Biodegradation (industrial)	AP cradle-to-gate (kg SO$_2$ eq per ton polymer)	Biodegradation (ocean)	GWP cradle-to-gate (ton CO$_2$ eq. per ton polymer)	Price (US$ per kg)	Reference
LDPE	NA	27	NA	1.9–3.1	1.36	(Wohlleben et al. 2023)
PET	NA	10–18	NA	2.4–5	1.2–1.4	(Penczek et al. 2003)
PVC	NA	3	NA	1.5–2.2	1.9	(Labet and Thielemans 2009)
PCL	4–6 weeks	NA	6 weeks	NA	NA	(Conix 1958; Labet and Thielemans 2009; Woodruff and Hutmacher 2010)
PBAT	2–3 months	NA	>1 year	NA	4.1	(Burgstaller et al. 2018; Wohlleben et al. 2023)
bioPE	NA	30	NA	0.68	1.8–2.4	(Koch and Mihalyi 2018)
bioPBS	>3 months	75	>1 year	2.2	NA	(Burgstaller et al. 2018)
PGA	2–3 months	NA	1–2 months	NA	NA	(Jem and Tan 2020; Lamberti, Román-Ramírez, and Wood 2020)
P4HB	4–6 weeks	NA	1–6 months	NA	NA	(Burgstaller et al. 2018; Lamberti, Román-Ramírez, and Wood 2020)
PEF	9 months	NA	NA	2.1	NA	(Rosenboom et al. 2018)
PP	NA	49	NA	1.5–3.6	1.1	(Harding et al. 2007; Lamberti, Román-Ramírez, and Wood 2020)

entire transition to bioplastics from the 170 million tons of packaging plastics produced annually worldwide is predicted to require 54% of current maize production and 60% more than Europe's yearly withdrawal of freshwater (Brizga, Hubacek, and Feng 2020). Second-generation biomass refers to a variety of inedible biowastes that provide a more generally accessible, commercially viable, and morally sound feedstock, albeit being more complex. For instance, each year, more than 1 billion tons of food and agricultural waste are produced worldwide, and food waste makes up around 20% of household waste (Lin et al. 2013; Varsha Vijay et al. 2016; Zheng and Suh 2019). The goal of research into potential bio refineries is to develop methods for converting lignocellulosic biomass, like sugarcane biogases and wheat straw. These agricultural wastes are generally cheap, but in order to release the sugars that can be fermented, cellulose and hemicellulose, from the protective network of cross-linked, phenolic lignin polymers, further pre-treatment procedures are needed. Brown and red algae, which are classified as seaweed, are another source of polysaccharides (Schmidt, Krauth, and Wagner 2017). Alginates are the most prevalent polysaccharides in brown algae, making up as much as 40% of their dry weight (Grignard et al. 2019). Access to different monomers is also made possible by vegetable and plant oils. Edible oils, like sugars, cause issues with food rivalry and deforestation, whereas waste oils or non-edible oils are more morally and environmentally sound. Triglycerides with unsaturated fatty acids found in vegetable oils can be epoxidized to create epoxy resins. When combined with efficient catalysts, CO_2 provides a low-energy, abundant feedstock that can be utilized straight for polymer synthesis. With this method, CO_2 from the air can be trapped in a polymer until it is eventually released through composting or burning (Kucherov et al. 2018). CO_2 (sold at less than US\$70 a ton) makes up approximately 97% and 15% of the exhaust gases from coal power plants and NH_3 factories, respectively. Therefore, CO_2 can be obtained from these emission-intensive businesses and used as a raw material to synthesize polycarbonates, polyurethanes, and other compounds. The majority of conventional durable polymers are made from crude oil, but it is also a key component of many biodegradable polymers, which are covered later. Crude oil is regarded as a non-sustainable feedstock because of the environmental damage caused by the extraction and burning of fossil fuels (Harmsen, Hackmann, and Bos 2014). Nowadays, biomass can provide practically all of the monomers needed to produce drop-in polymers, which are chemically identical substitutes for polymers derived from fossil fuels (Sajid, Zhao, and Liu 2018). There has been a thorough analysis of the processes used to convert biomass to produce rubbers, amides, alcohols, vinyl monomers, and carboxylic acids (Sajid, Zhao, and Liu 2018). The synthesis methods are illustrated in Figure 1.1, and Table 1.1, presents a comparison and characteristics of the bioplastics (Chafran et al. 2019; Kucherov et al. 2018).

3 STARCH-BASED PLASTICS

This bioplastic was first synthesized in the 1970s and is currently produced globally by businesses, including Futerro, Biome, and Biotec. Carbohydrates are also known by the terms starch, sugars, saccharides, and polysaccharides. When CO_2 and water combine during photosynthesis, they are created (Costa et al. 2023). The general representation of the chemical symbol is $C_x(H_2O)_y$, where x and y are integers between 3 and 12. One kind of polysaccharide that comes from floral sources is starch (Shafqat et al. 2020). Plasticizers (glycerol, sorbitol) or a sizable amount of water are needed to produce polymeric film from starch. They are widely employed in many thermal and mechanical applications around the world as a PS alternative. Though there are many diverse potential sources of starch, the primary ones with the intrinsic biodegradability, massive availability, and yearly renewability of starch make it one of the most promising natural polymers. Given their cheap material cost and suitability for processing with standard plastic processing machinery, starches present a very appealing low cost basis for novel biodegradable polymers. Following the widely acknowledged problems of the oil scarcity and the increasing desire to lessen the environmental impact resulting from the widespread usage of polymers generated from petrochemicals, research and development of biodegradable starch-based materials have gained more and more attention. Nowadays, limitations and outright bans on disposable plastics are being implemented in an increasing number of nations. This polymer is produced by the majority of green plants as an energy reserve (Sanyang et al. 2016). This same carbohydrate is found in large quantities in fundamental foods, including rice, cassava, corn (maize), wheat, and potatoes, and it is also a part of human diets. Out of all of them, cassava starch is the most significant because its dry mass comprises over 80% starch (Lii, Shao, and Tseng 1995). Glycosidic linkages hold a large number of glucose units together in starch, making it a type of carbohydrate. Cassava starch ranks third among the most important food sources for people living in tropical areas (Imamura et al. 2006; Sriroth et al. 2000). Using several surface treatments, a biodegradable polymer was created from cassava starch for a range of uses (Sriroth et al. 2000). The different mechanical, thermal, and physical qualities were discussed. For packaging purposes, researchers created bioplastic film based on sugar starch with a variety of reinforcements (Sultan and Johari 2017). White is the color of pure starch (Rittenauer et al. 2016). The starch powder has no distinct flavor or aroma (Klucinec and Thompson 1999; Salo and Salmi 1968; Vamadevan and Bertoft 2020). The chemical amylopectin is far more superior to amylose (Bello-Perez et al. 1996). Popular biocompatible and biodegradable materials like polycaprolactone are employed in implantable medicine delivery devices and sutures. In humans, polycaprolactone hydrolyzes non-enzymatically in years, while in seawater; it is biodegraded by bacteria and fungi in weeks (Jem and Tan 2020). For an aliphatic ester, polyglycolic acid is the most basic (Sun et al. 2017). It is a compelling option for plastic packaging due to its quick rates of industrial and marine deterioration and high gas barrier. From a commodity perspective, polyglycolic acid is produced in small quantities, despite having a significant financial market share in the biomedical industry. Polylactide-co-glycolide (PLGA), is frequently utilized in biomedical applications since it degrades more quickly than polycaprolactone and is biocompatible (Danhier et al. 2012). The process of thermally processing starch involves a variety of chemical and physical events, such as water diffusion, granule expansion, gelatinization, breakdown, melting, and crystallization, due to its unique microstructures, which include high molecular weight and hydrogen bonds (Chen et al. 2006), wherein the gelatinization is distinct and holds the most significance due to its strong relationship with the other processes. The temperature

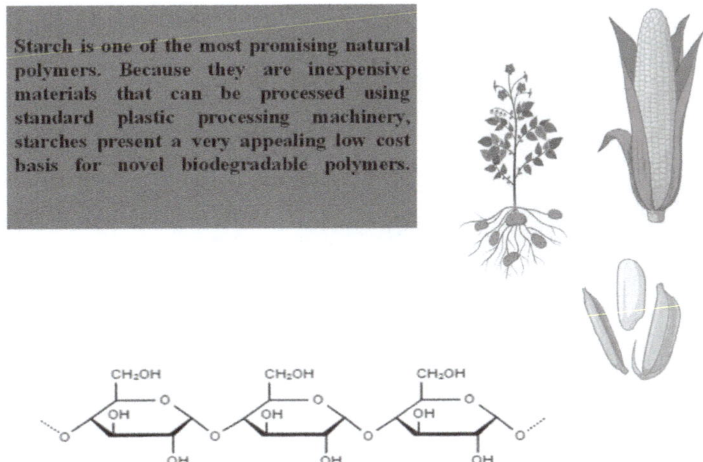

FIGURE 1.2 Schematic representation of natural starch.

at which starch breaks down before gelatinization is lower than the temperature at which it melts. Thus, starch-based polymers (shown in Figure 1.2) cannot be thermally processed with traditional plastic equipment, especially not an extruder (Lan Cao et al. 2010). One of the most widely used methods for processing polymeric materials, mostly involving melting and solidification, is extrusion. Starch granules, however, experience a number of intricate phase changes throughout the extrusion process, such as starch swelling, loss of birefringence, melting, and solubilization. The thermal process of gelatinization is primarily governed by temperature and water content in the absence of shear stress (Xie et al. 2006). The starch granular structure maintains its stability below 50°C (Lim, Wu, and Reid 2000).

4 TYPES OF BIOPLASTICS FOR SUSTAINABLE AND GREEN TECHNOLOGY

Certain kinds of polymeric materials derived from bioresources are known as "bioplastics," and they are thought to be a financially and environmentally sound alternative to traditional petrochemical plastics. "Bio-based" are industrial or commercial goods (apart from food and feed) that contain a significant portion of biological materials (Mekonnen et al. 2013). The primary component of bio-based bioplastics is renewable material produced by living things (Bátori et al. 2018; Bowens et al. 2021). As shown in Figure 1.3, a paradigm shift from conventional plastics to bioplastics might be made based on environmental sustainability (Bezirhan Arikan et al. 2021; Lee, Nagalingam, and Yeo 2021). Note that not all plastics that are biodegradable are also bio-based, nor are all plastics that are bio-based biodegradable. Polyethylene terephthalate (de P. Daubeny, Bunn, and Brown 1954), for instance, can be made using bioresources and derivatives of fossil fuels. Without a controlled composting environment, bio-based polyethylene terephthalate, like its petrochemical equivalent, can remain in the environment. Conversely, polylactic acid derived from plants

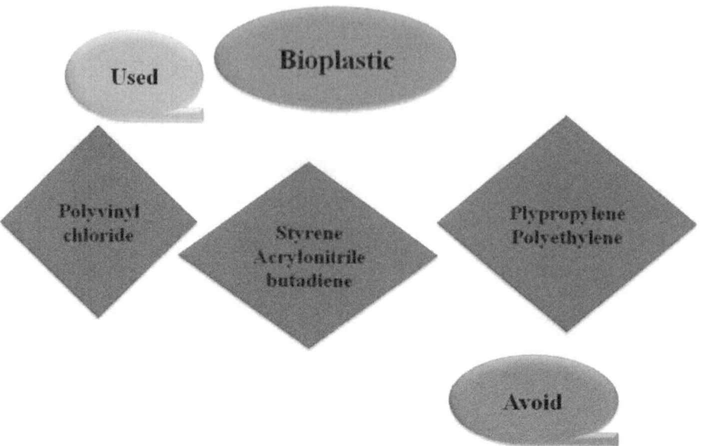

FIGURE 1.3 Most popular materials are bioplastics, which are followed by acrylonitrile butadiene styrene, polyethylene terephthalate, polyurethane, and polystyrene.

is compostable, biodegradable, and recyclable. On the other hand, non-biodegradable yet bio-based materials include bio-based poly (trimethylene terephthalate), bio-based polyethylene, and bio-based polyamide (Nisticò 2020). Biodegradable polymers, also known as biopolymers, have garnered significant attention in recent years due to their eco-friendly nature and various applications in both medical and industrial fields. One notable class of biodegradable polymers is polyhydroxyalkanoate (PHA), elastomeric or thermoplastic polyester synthesized biologically by bacteria (Gross and Kalra 2002; Philip, Keshavarz, and Roy 2007). PLA's environmental impact is noteworthy, releasing CO_2, water, and decomposed organic matter upon biodegradation. The closed-loop system of PLA production from renewable agricultural materials enhances its environmental benefits (Merino, Zych, and Athanassiou 2022) as drug encapsulation and agricultural uses like fertilizer encapsulation (Mukherjee et al. 2019). Polyamide, a bio-based polymer derived from castor oil, stands out for its high chemical and mechanical stability. Its non-biodegradable nature makes it suitable for applications requiring longevity (Abe et al. 2021). Polyhydroxyurethanes, emerging as an alternative to traditional polyurethanes, address environmental concerns related to isocyanate usage. The use of CO_2 in polyhydroxyurethane synthesis has shown promise in reducing isocyanate dependency and associated CO_2 emissions. Cellulose-based biopolymers are derived from natural sources like agricultural and forestry biomass (Simon et al. 1998). Starch-based biopolymers, derived from food crops like corn, cassava, wheat, rice, and potatoes, are gaining popularity due to their renewability, biodegradability, and affordability. However, concerns about competition with food crops for large-scale production raise environmental considerations (Wang, Yang, and Wang 2003). Protein-based biopolymers, including soy protein zein, casein, offer advantages with mechanical properties and are widely used in edible films as well as the incorporation of keratin protein into synthetic elastomers (Vaz et al. 2003).

5　ENVIRONMENTAL IMPACT

The impact of biodegradable bioplastics on the environment is a subject of ongoing debate, marked by a complex interplay of positive and negative factors. These materials, designed to degrade naturally and reduce environmental harm, present a nuanced scenario with potential benefits and drawbacks (Haider et al. 2019). One of the key advantages touted for biodegradable bioplastics is their ability to break down into natural materials through microbial mechanisms. In ideal conditions, such as composting, these materials decompose, blending harmlessly into the soil. This process is facilitated by water and oxygen, allowing the material to transform into smaller fragments digestible by bacteria. However, the effectiveness of this decomposition varies among different types of biodegradable bioplastics, with some requiring specific conditions like high temperatures or treatment in specialized composting facilities. One significant environmental concern arises from the decomposition process itself, particularly during composting, which can result in the release of methane gas. Methane is a potent greenhouse gas, contributing to global warming. This unintended consequence raises questions about the overall environmental benefit of certain biodegradable bioplastics, especially when considering their end-of-life disposal. The production of bioplastics from agricultural crops introduces another layer of environmental impact. The repurposing of land, particularly for crops like corn and maize used in bioplastic production, can lead to land-use conflicts (Touchaleaume et al. 2016), potentially impacting food prices and affecting vulnerable communities. Additionally, the cultivation of crops for bioplastics involves the use of fertilizers and pesticides, contributing to environmental pollution. However, some bioplastics, such as those made from polylactic acid (PLA), present eco-friendly characteristics. PLA production is energy-efficient, saving two-thirds of the energy required for traditional plastics. During PLA bioplastics' biodegradation, there is no net increase in carbon dioxide, as the plants absorb the same amount of carbon dioxide during cultivation as is released during their breakdown. This suggests that certain bioplastics can contribute to a reduction in greenhouse gas emissions and utilize renewable energy sources. Polyhydroxyalkanoate (PHA) stands out for its biodegradability, making it a favorable choice in minimizing plastic waste. As PHA undergoes degradation, it releases CO_2, water, and decomposed organic matter. This process facilitates the utilization of these byproducts by green plants, contributing to a more sustainable and closed-loop system. Polylactic acid (PLA) further enhances the environmental profile of biodegradable polymers. Its production from renewable agricultural materials ensures a lower carbon footprint compared with traditional synthetic polymers. PLA's biodegradability, coupled with its relatively lower greenhouse gas emission rate, aligns with environmental goals, providing a more sustainable option for various applications. Poly-3-hydroxybutyrate (PHB) showcases environmental benefits through its microbial fermentation process, utilizing renewable carbohydrate precursors. As a biodegradable polymer, PHB contributes to the reduction of persistent plastic waste and demonstrates promise in applications ranging from surgical implants to agricultural uses. Polyamide 11, derived from castor oil, exhibits both chemical and mechanical stability, contributing to its longevity in applications such as natural gas piping and electrical cables. While not biodegradable, its recyclability and potential use in wood-plastic composites contribute to a more sustainable approach in various industries. Polyhydroxyurethanes offer an environmentally friendly alternative to traditional polyurethanes, addressing concerns related to isocyanate usage. The synthesis routes, particularly those involving CO_2, contribute to reducing dependence on toxic isocyanates and associated CO_2 emissions, aligning with the broader goal of mitigating environmental impact. Cellulose-based biopolymers provide a green solution due to their biodegradability, high durability, and strength. Blending cellulose with other polysaccharides enhances stability, promoting longer-lasting applications while maintaining eco-friendly characteristics. Starch-based biopolymers, despite concerns about food crop competition, present an environmentally sound option due to renewability and biodegradability. Their mechanical properties and transparency, akin to conventional plastics, make them a promising candidate for various applications. Protein-based biopolymers, with their use in edible films and enhanced mechanical properties, contribute positively to the environment. These materials, derived from abundant sources like wheat gluten and soy protein, offer sustainable alternatives with biodegradability and reduced reliance on traditional plastics (Avérous and Pollet 2012).

6 CONCLUSIONS

In conclusion, the exploration of biopolymer composites for sustainable and green technology represents a promising avenue toward addressing environmental challenges associated with conventional materials. The utilization of biopolymers derived from renewable resources, combined with reinforcing elements or additives, has demonstrated significant potential in various applications. This multifaceted approach contributes to the development of eco-friendly alternatives in diverse sectors, from packaging materials to construction and electronics. The integration of biopolymer composites aligns with the principles of sustainability by reducing reliance on finite resources, minimizing environmental impact, and promoting circular economy practices. The use of renewable feedstocks, such as plant-based polymers or biowastes, not only mitigates the carbon footprint but also addresses concerns related to the depletion of fossil fuels and accumulation of non-degradable waste. Bioplastics, often considered environmentally friendly alternatives to conventional plastics, have become a focal point in the quest for sustainable materials. However, their impact on the environment is multifaceted, and a comprehensive understanding involves evaluating their production processes, applications, recyclability, and end-of-life considerations. Bio-based replacements for fossil-based applications exist, albeit often in small quantities and at higher costs. Evaluating the sustainability and environmental impact of bioplastics requires robust tools such as Life Cycle Assessments. However, standardizing methodology standards for Life Cycle Assessments is essential for transparency, consistency, and comparability. Existing bioplastic labels must be revised to convey globally recognized standards and indicate the end-of-life possibilities, contributing to consumer education and informed decision-making. Various types of bioplastics, from well-known ones like polyhydroxybutyrate, polycaprolactone, and polylactic acid to newer additions like mycelium-based and chitin-based biopolymers, address environmental concerns. However, challenges such as limited recycling options, gradual decomposition in landfills, and the potential release of methane gas need careful consideration.

REFERENCES

Abe, Mateus Manabu, Júlia Ribeiro Martins, Paula Bertolino Sanvezzo, João Vitor Macedo, Marcia Cristina Branciforti, Peter Halley, Vagner Roberto Botaro, and Michel Brienzo. 2021. "Advantages and Disadvantages of Bioplastics Production from Starch and Lignocellulosic Components." *Polymers* 13 (15): 2484.

Ahmed, Ishtiaque. 2019. "The Basel Convention on the Control of Transboundary Movements of Hazardous Wastes and Their Disposal: A Legal Misfit in Global Ship Recycling Jurisprudence." *Washington International Law Journal* 29: 411.

Avérous, Luc, and Eric Pollet (Eds). 2012. "Biodegradable Polymers." In *Environmental Silicate Nano-biocomposites*, 13–39. Springer.

Bátori, Veronika, Dan Åkesson, Akram Zamani, Mohammad J. Taherzadeh, and Ilona Sárvári Horváth. 2018. "Anaerobic Degradation of Bioplastics: A Review." *Waste Management* 80: 406–413.

Bello-Perez, Luis Arturo, Octavio Paredes-Lopez, Philippe Roger, and Paul Colonna. 1996. "Amylopectin—Properties and Fine Structure." *Food Chemistry* 56 (2): 171–176.

Bezirhan Arikan, Ezgi, Esma Mahfouf Bouchareb, Raouf Bouchareb, Nevin Yağcı, and Nadir Dizge. 2021. "Innovative Technologies Adopted for the Production of Bioplastics at Industrial Level." *Bioplastics for Sustainable Development* 83–102.

Bowens, Rory, Michael R. Meyer, Christian Delacroix, Olivier Absil, Roy van Boekel, Sascha Patrick Quanz, M. Shinde, Matthew Kenworthy, Brunella Carlomagno, and Gilles Orban de Xivry. 2021. "Exoplanets with ELT-METIS-I. Estimating the Direct Imaging Exoplanet Yield around Stars within 6.5 Parsecs." *Astronomy & Astrophysics* 653: A8.

Brizga, Janis, Klaus Hubacek, and Kuishuang Feng. 2020. "The Unintended Side Effects of Bioplastics: Carbon, Land, and Water Footprints." *One Earth* 3 (1): 45–53.

Burgstaller, Maria, Alexander Potrykus, Jakob Weißenbacher, Stephan Kabasci, Ute Merrettig-Bruns, and Bettina Sayder. 2018. "Gutachten zur Behandlung biologisch abbaubarer Kunststoffe."

Cesca, Fabrizia, Pietro Baldelli, Flavia Valtorta, and Fabio Benfenati. 2010. "The Synapsins: Key Actors of Synapse Function and Plasticity." *Progress in Neurobiology* 91 (4): 313–348.

Chafran, Liana S., Mateus F. Paiva, Juliene O.C. França, Maria José A. Sales, Sílvia C.L. Dias, and José A. Dias. 2019. "Preparation of PLA Blends by Polycondensation of D, L-lactic Acid Using Supported 12-Tungstophosphoric Acid as a Heterogeneous Catalyst." *Heliyon* 5 (5).

Chen, J. 2019. *Global Markets and Technologies for Bioplastics*. BCC Research.

Chen, Pei, Long Yu, Ling Chen, and Xiaoxi Li. 2006. "Morphology and Microstructure of Maize Starches with Different Amylose/Amylopectin Content." *Starch-Stärke* 58 (12): 611–615.

Cheung, Hoi-Yan, Kin-Tak Lau, Tung-Po Lu, and David Hui. 2007. "A Critical Review on Polymer-based Bio-Engineered Materials for Scaffold Development." *Composites Part B: Engineering* 38 (3): 291–300.

Chu, Steven, and Arun Majumdar. 2012. "Opportunities and Challenges for a Sustainable Energy Future." *Nature* 488 (7411): 294–303.

Coates, Geoffrey W., and Yutan D.Y.L. Getzler. 2020. "Chemical Recycling to Monomer for an Ideal, Circular Polymer Economy." *Nature Reviews Materials* 5 (7): 501–516.

Conix, Andre. 1958. "Aromatic Polyanhydrides, a New Class of High Melting Fiber-forming Polymers." *Journal of Polymer Science* 29 (120). 343–353.

Costa, Ana, Telma Encarnação, Rafael Tavares, Tiago Todo Bom, and Artur Mateus. 2023. "Bioplastics: Innovation for Green Transition." *Polymers* 15 (3): 517.

Danhier, Fabienne, Eduardo Ansorena, Joana M. Silva, Régis Coco, Aude Le Breton, and Véronique Préat. 2012. "PLGA-based Nanoparticles: An Overview of Biomedical Applications." *Journal of Controlled Release* 161 (2): 505–522.

de P. Daubeny R., Charles William Bunn, and C.J. Brown. 1954. "The Crystal Structure of Polyethylene Terephthalate." *Proceedings of the Royal Society of London. Series A. Mathematical and Physical Sciences* 226 (1167): 531–542.

Fadeeva, Zinaida, and Rene Van Berkel. 2021. "Unlocking Circular Economy for Prevention of Marine Plastic Pollution: An Exploration of G20 Policy and Initiatives." *Journal of Environmental Management* 277: 111457.

Foschi, Eleonora, and Alessandra Bonoli. 2019. "The Commitment of Packaging Industry in the Framework of the European Strategy for Plastics in a Circular Economy." *Administrative Sciences* 9 (1): 18.

Gandini, Alessandro, Talita M. Lacerda, Antonio J.F. Carvalho, and Eliane Trovatti. 2016. "Progress of Polymers from Renewable Resources: Furans, Vegetable Oils, and Polysaccharides." *Chemical Reviews* 116 (3): 1637–1669.

Geyer, Roland, Jenna R. Jambeck, and Kara Lavender Law. 2017. "Production, Use, and Fate of All Plastics Ever Made." *Science Advances* 3 (7): e1700782.

Grignard, Bruno, Sandro Gennen, Christine Jérôme, Arjan W. Kleij, and Christophe Detrembleur. 2019. "Advances in the Use of CO_2 as a Renewable Feedstock for the Synthesis of Polymers." *Chemical Society Reviews* 48 (16): 4466–4514.

Gross, Richard A., and Bhanu Kalra. 2002. "Biodegradable Polymers for the Environment." *Science* 297 (5582): 803–807.

Haider, Tobias P., Carolin Völker, Johanna Kramm, Katharina Landfester, and Frederik R. Wurm. 2019. "Plastics of the Future? The Impact of Biodegradable Polymers on the Environment and on Society." *Angewandte Chemie International Edition* 58 (1): 50–62.

Harding, Kevin Graham, J.S. Dennis, Harro Von Blottnitz, and Susan T.L. Harrison. 2007. "Environmental Analysis of Plastic Production Processes: Comparing Petroleum-based Polypropylene and Polyethylene with Biologically-based Poly-β-hydroxybutyric Acid Using Life Cycle Analysis." *Journal of Biotechnology* 130 (1): 57–66.

Harmsen, Paulien F.H., Martijn M. Hackmann, and Harriëtte L. Bos. 2014. "Green Building Blocks for Bio-based Plastics." *Biofuels, Bioproducts and Biorefining* 8 (3): 306–324.

Imamura, Koreyoshi, Keisuke Sakaura, Ken-ichi Ohyama, Atsushi Fukushima, Hiroyuki Imanaka, Takaharu Sakiyama, and Kazuhiro Nakanishi. 2006. "Temperature Scanning FTIR Analysis of Hydrogen Bonding States of Various Saccharides in Amorphous Matrixes below and above their Glass Transition Temperatures." *The Journal of Physical Chemistry B* 110 (31): 15094–15099.

Inkinen, Saara, Minna Hakkarainen, Ann-Christine Albertsson, and Anders Södergård. 2011. "From Lactic Acid to Poly (lactic acid)(PLA): Characterization and Analysis of PLA and Its Precursors." *Biomacromolecules* 12 (3): 523–532.

Jambeck, Jenna R., Roland Geyer, Chris Wilcox, Theodore R. Siegler, Miriam Perryman, Anthony Andrady, Ramani Narayan, and Kara Lavender Law. 2015. "Plastic Waste Inputs from Land into the Ocean." *Science* 347 (6223): 768–771.

Jem, K. Jim, and Bowen Tan. 2020. "The Development and Challenges of Poly (Lactic Acid) and Poly (Glycolic Acid)." *Advanced Industrial and Engineering Polymer Research* 3 (2): 60–70.

Klucinec, Jeffrey D., and Donald B. Thompson. 1999. "Amylose and Amylopectin Interact in Retrogradation of Dispersed High-amylose Starches." *Cereal Chemistry* 76 (2): 282–291.

Koch, Daniel, and Bettina Mihalyi. 2018. "Assessing the Change in Environmental Impact Categories When Replacing Conventional Plastic with Bioplastic in Chosen Application Fields." *Chemical Engineering Transactions* 70: 853–858.

Kucherov, Fedor A., Leonid V. Romashov, Konstantin I. Galkin, and Valentine P. Ananikov. 2018. "Chemical Transformations of Biomass-derived C6-Furanic Platform Chemicals for Sustainable Energy Research, Materials Science, and Synthetic Building Blocks." *ACS Sustainable Chemistry & Engineering* 6 (7): 8064–8092.

Kumar, Rakesh, Anurag Verma, Arkajyoti Shome, Rama Sinha, Srishti Sinha, Prakash Kumar Jha, Ritesh Kumar, Pawan Kumar, Shubham, and Shreyas Das. 2021. "Impacts of Plastic Pollution on Ecosystem Services, Sustainable Development Goals, and Need to Focus on Circular Economy and Policy Interventions." *Sustainability* 13 (17): 9963.

Labet, Marianne, and Wim Thielemans. 2009. "Synthesis of Polycaprolactone: A Review." *Chemical Society Reviews* 38 (12): 3484–3504.

Lamberti, Fabio M., Luis A. Román-Ramírez, and Joseph Wood. 2020. "Recycling of Bioplastics: Routes and Benefits." *Journal of Polymers and the Environment* 28 (10): 2551–2571.

Lan, Cao, Hongsheng Liu, Pei Chen, Long Yu, Ling Chen, Xiaoxi Li, and Ximei Zhang. 2010. "Gelatinization and Retrogradation of Hydroxypropylated Cornstarch." *International Journal of Food Engineering* 6 (4).

Lee, Jian-Yuan, Arun Prasanth Nagalingam, and S.H. Yeo. 2021. "A Review on the State-of-the-Art of Surface Finishing Processes and Related ISO/ASTM Standards for Metal Additive Manufactured Components." *Virtual and Physical Prototyping* 16 (1): 68–96.

Lii, C., Yi-Yuan Shao, and Kuo-Hsuen Tseng. 1995. "Gelation Mechanism and Rheological Properties of Rice Starch."

Lim, Miang H., Hongbing Wu, and David S. Reid. 2000. "The Effect of Starch Gelatinization and Solute Concentrations on T g′ of Starch Model System." *Journal of the Science of Food and Agriculture* 80 (12): 1757–1762.

Lin, Carol Sze Ki, Lucie A. Pfaltzgraff, Lorenzo Herrero-Davila, Egid B. Mubofu, Solhy Abderrahim, James H. Clark, Apostolis A. Koutinas, Nikolaos Kopsahelis, Katerina Stamatelatou, and Fiona Dickson. 2013. "Food Waste as a Valuable Resource for the Production of Chemicals, Materials and Fuels. Current Situation and Global Perspective." *Energy & Environmental Science* 6 (2): 426–464.

Lin, Chien-Ting, Shiao-Wei Kuo, Chih-Feng Huang, and Feng-Chih Chang. 2010. "Glass Transition Temperature Enhancement of PMMA through Copolymerization with PMAAM and PTCM Mediated by Hydrogen Bonding." *Polymer* 51 (4): 883–889.

MacArthur, Dame Ellen, Dominic Waughray, and Martin R. Stuchtey. 2016. *The New Plastics Economy, Rethinking the Future of Plastics.* World Economic Forum.

MacArthur, Ellen. 2017. *Beyond Plastic Waste.* American Association for the Advancement of Science.

Mekonnen, Tizazu, Paolo Mussone, Hamdy Khalil, and David Bressler. 2013. "Progress in Bio-based Plastics and Plasticizing Modifications." *Journal of Materials Chemistry A* 1 (43): 13379–13398.

Merino, Danila, Arkadiusz Zych, and Athanassia Athanassiou. 2022. "Biodegradable and Biobased Mulch Films: Highly Stretchable PLA Composites with Different Industrial Vegetable Waste." *ACS Applied Materials & Interfaces* 14 (41): 46920–46931.

Messerli, Peter, Endah Murniningtyas, Parfait Eloundou-Enyegue, Ernest G. Foli, Eeva Furman, Amanda Glassman, Gonzalo Hernández Licona, Eun Mee Kim, Wolfgang Lutz, and J.-P. Moatti. 2019. "Global Sustainable Development Report 2019: The Future Is Now–Science for Achieving Sustainable Development."

Miller, Shelie A. 2020. "Five Misperceptions Surrounding the Environmental Impacts of Single-Use Plastic." *Environmental Science & Technology* 54 (22): 14143–14151.

Mrowiec, Bozena. 2018. "Plastics in the Circular Economy (CE)." *Environmental Protection and Natural Resources* 29 (4): 16–19

Mukherjee, Agneev, Simon Knoch, Gérald Chouinard, Jason R. Tavares, and Marie-Josée Dumont. 2019. "Use of Bio-based Polymers in Agricultural Exclusion Nets: A Perspective." *Biosystems Engineering* 180: 121–145.

Mülhaupt, Rolf. 2013. "Green Polymer Chemistry and Bio-based Plastics: Dreams and Reality." *Macromolecular Chemistry and Physics* 214 (2): 159–174.

Murariu, Marius, Yoann Paint, Oltea Murariu, Fouad Laoutid, and Philippe Dubois. 2022. "Recent Advances in Production of Ecofriendly Polylactide (PLA)–Calcium Sulfate (Anhydrite II) Composites: From the Evidence of Filler Stability to the Effects of PLA Matrix and Filling on Key Properties." *Polymers* 14 (12): 2360.

Nisticò, Roberto. 2020. "Polyethylene Terephthalate (PET) in the Packaging Industry." *Polymer Testing* 90: 106707.

Penczek, Stanislaw, Ryszard Szymanski, Andrzej Duda, and Jolanta Baran. 2003. "Living Polymerization of Cyclic Esters–A Route to (Bio) Degradable Polymers. Influence of Chain Transfer to Polymer on Livingness." *Macromolecular Symposia* 201 (1): 261–270.

Philip, Stephen, Tajalli Keshavarz, and Ipsita Roy. 2007. "Polyhydroxyalkanoates: Biodegradable Polymers with a Range of Applications." *Journal of Chemical Technology & Biotechnology: International Research in Process, Environmental & Clean Technology* 82 (3): 233–247.

Pudack, Claudia, Manfred Stepanski, and Peter Fässler. 2020. "PET Recycling–Contributions of Crystallization to Sustainability." *Chemie Ingenieur Technik* 92 (4): 452–458.

Rittenauer, Michael, L. Kolesnik, Martina Gastl, and T. Becker. 2016. "From Native Malt to Pure Starch–Development and Characterization of a Purification Procedure for Modified Starch." *Food Hydrocolloids* 56: 50–57.

Rosenboom, Jan-Georg, Diana Kay Hohl, Peter Fleckenstein, Giuseppe Storti, and Massimo Morbidelli. 2018. "Bottle-grade Polyethylene Furanoate from Ring-Opening Polymerisation of Cyclic Oligomers." *Nature Communications* 9 (1): 2701.

Sajid, Muhammad, Xuebing Zhao, and Dehua Liu. 2018. "Production of 2, 5-furandicarboxylic acid (FDCA) from 5-hydroxymethylfurfural (HMF): Recent Progress Focusing on the Chemical-catalytic Routes." *Green Chemistry* 20 (24): 5427–5453.

Salo, Maija-Liisa, and Marja Salmi. 1968. "Determination of Starch by the Amyloglucosidase Method." *Agricultural and Food Science* 40 (1): 38–45.

Santhoskumar, Anniyyappa Umapathi, Komaragounder Palanivelu Komaragounder Palanivelu, S. K. Sharma, and S. K. Nayak. 2010. "Comparison of Biological Activity Transistion Metal 12 Hydroxy Oleate on Photodegradation of Plastics." (2010): 1000109.

Sanyang, Muhammad Lamin, S.M. Sapuan, Mohammad Jawaid, Mohamad Ridzwan Ishak, and J. Sahari. 2016. "Development and Characterization of Sugar Palm Starch and Poly (Lactic Acid) Bilayer Films." *Carbohydrate Polymers* 146: 36–45.

Schmidt, Christian, Tobias Krauth, and Stephan Wagner. 2017. "Export of Plastic Debris by Rivers into the Sea." *Environmental Science & Technology* 51 (21): 12246–12253.

Shafqat, Arifa, Arifa Tahir, Adeel Mahmood, Amtul Bari Tabinda, Abdullah Yasar, and Arivalagan Pugazhendhi. 2020. "A Review on Environmental Significance Carbon Foot Prints of Starch Based Bio-plastic: A Substitute of Conventional Plastics." *Biocatalysis and Agricultural Biotechnology* 27: 101540.

Simon, J., H.P. Müller, R. Koch, and V. Müller. 1998. "Thermoplastic and Biodegradable Polymers of Cellulose." *Polymer Degradation and Stability* 59 (1–3): 107–115.

Sousa, Andreia F., and Armando J.D. Silvestre. 2022. "Plastics from Renewable Sources as Green and Sustainable Alternatives." *Current Opinion in Green and Sustainable Chemistry* 33: 100557.

Sriroth, Klanarong, Kuakoon Piyachomkwan, Sittichoke Wanlapatit, and Christopher G. Oates. 2000. "Cassava Starch Technology: The Thai Experience." *Starch-Stärke* 52 (12): 439–449.

Stephen, Meera, Ali Nawaz, Sang Yeon Lee, Prashant Sonar, and Wei Lin Leong. 2023. "Biodegradable Materials for Transient Organic Transistors." *Advanced Functional Materials* 33 (6): 2208521.

Sultan, Noor Fatimah Kader, and Wan Lutfi Wan Johari. 2017. "The Development of Banana Peel/Corn Starch Bioplastic Film: A Preliminary Study." *Bioremediation Science and Technology Research* 5 (1): 12–17.

Sun, Xiaoyu, Chun Xu, Gang Wu, Qingsong Ye, and Changning Wang. 2017. "Poly (lactic-co-glycolic acid): Applications and Future Prospects for Periodontal Tissue Regeneration." *Polymers* 9 (6): 189.

Taib, Nur-Azzah Afifah Binti, Md Rezaur Rahman, Durul Huda, Kuok King Kuok, Sinin Hamdan, Muhammad Khusairy Bin Bakri, Muhammad Rafiq Mirza Bin Julaihi, and Afrasyab Khan. 2023. "A Review on Poly Lactic Acid (PLA) as a Biodegradable Polymer." *Polymer Bulletin* 80 (2): 1179–1213.

Ten Brink, Patrick, Jean-Pierre Schweitzer, Emma Watkins, Charlotte Janssens, Michiel De Smet, Heather Leslie, and François Galgani. 2018. *Circular Economy Measures to Keep Plastics and their Value in the Economy, Avoid Waste and Reduce Marine Litter.* Economics Discussion Papers.

Tokiwa, Yutaka, and Buenaventurada P. Calabia. 2007. "Biodegradability and Biodegradation of Polyesters." *Journal of Polymers and the Environment* 15: 259–267.

Touchaleaume, François, Lluís Martin-Closas, Helene Angellier-Coussy, Anne Chevillard, Guy Cesar, Nathalie Gontard, and Emmanuelle Gastaldi. 2016. "Performance and Environmental Impact of Biodegradable Polymers as Agricultural Mulching Films." *Chemosphere* 144: 433–439.

Vamadevan, Varatharajan, and Eric Bertoft. 2020. "Observations on the Impact of Amylopectin and Amylose Structure on the Swelling of Starch Granules." *Food Hydrocolloids* 103: 105663.

van Emmerik, Tim, and Anna Schwarz. 2020. "Plastic Debris in Rivers." *Wiley Interdisciplinary Reviews: Water* 7 (1): e1398.

Varghese, Sandhya Alice, Sanjay Mavinkere Rangappa, Suchart Siengchin, and Jyotishkumar Parameswaranpillai. 2020. "Natural Polymers and the Hydrogels Prepared from Them." In Chen, Yu (Ed), *Hydrogels Based on Natural Polymers*, 17–47. Elsevier.

Vijay, Varsha, Stuart L. Pimm, Clinton N. Jenkins, and Sharon J. Smith. 2016. The Impacts of Oil Palm on Recent Deforestation and Biodiversity Loss." *PloS One* 11 (7): e0159668.

Vaz, Claudia M., M. Fossen, Robert F. Van Tuil, Leontine A. de Graaf, Rui L. Reis, and Adam M. Cunha. 2003. "Casein and Soybean Protein-based Thermoplastics and Composites as Alternative Biodegradable Polymers for Biomedical Applications." *Journal of Biomedical Materials Research Part A* 65 (1): 60–70.

Walker, Stuart, and Rachel Rothman. 2020. "Life Cycle Assessment of Bio-based and Fossil-based Plastic: A Review." *Journal of Cleaner Production* 261: 121158.

Wang, Xiu-Li, Ke-Ke Yang, and Yu-Zhong Wang. 2003. "Properties of Starch Blends with Biodegradable Polymers." *Journal of Macromolecular Science, Part C: Polymer Reviews* 43 (3): 385–409.

Wohlleben, Wendel, Markus Rückel, Lars Meyer, Patrizia Pfohl, Glauco Battagliarin, Thorsten Hüffer, Michael Zumstein, and Thilo Hofmann. 2023. "Fragmentation and Mineralization of a Compostable Aromatic–Aliphatic Polyester during Industrial Composting." *Environmental Science & Technology Letters* 10 (8): 698–704.

Woodruff, Maria Ann, and Dietmar Werner Hutmacher. 2010. "The Return of a Forgotten Polymer—Polycaprolactone in the 21st Century." *Progress in Polymer Science* 35 (10): 1217–1256.

Xie, Fengwei, Hongshen Liu, Pei Chen, Tao Xue, Ling Chen, Long Yu, and Penny Corrigan. 2007. "Starch Gelatinization under Shearless and Shear Conditions." *International Journal of Food Engineering* 2 (5).

Yuntawattana, Nattawut, Georgina L. Gregory, Leticia Peña Carrodeguas, and Charlotte K. Williams 2021. "Switchable Polymerization Catalysis Using a Tin (II) Catalyst and Commercial Monomers to Toughen Poly (L-lactide)." *ACS Macro Letters* 10 (7): 774–779.

Zhao, Peng, Wanqiang Liu, Qingsheng Wu, and Jie Ren. 2010. "Preparation, Mechanical, and Thermal Properties of Biodegradable Polyesters/Poly (Lactic Acid) Blends." *Journal of Nanomaterials* 2010: 1–6.

Zheng, Jiajia, and Sangwon Suh. 2019. "Strategies to Reduce the Global Carbon Footprint of Plastics." *Nature Climate Change* 9 (5): 374–378.

Ziccardi, Linda M., Aaron Edgington, Karyn Hentz, Konrad J Kulacki, and Susan Kane Driscoll. 2016. "Microplastics as Vectors for Bioaccumulation of Hydrophobic Organic Chemicals in the Marine Environment: A State-of-the-Science Review." *Environmental Toxicology and Chemistry* 35 (7): 1667–1676.

2 A Feasibility Study on Characterization of Lignocellulosic Biomass and Control Parameters in the Anaerobic Digestion Process

Sanskruti Ajay Mukwane and Harshal Warade
Yeshwantrao Chavan College of Engineering (YCCE)

Mohit Kale and Ramesh Daryapurkar
LARS Enviro Private Limited

1   INTRODUCTION

In the past decades, an increasing number of nations have used AD process with great success for tasks including waste management and energy production [1]. AD is a microbial process that produces carbon dioxide and methane when intricate organic substances including lipids, carbohydrates, and proteins are degraded in the absence of oxygen [2]. Figure 2.1 illustrates the four stages of AD wherein first stage, fermentative bacteria carry out the hydrolysis and acidogenesis of the complex molecule, converting the intermediate products into acetic acid and other volatile fatty acids (VFA). Second, the first group's volatile fatty acid can be converted by the hydrogen-producing bacteria into hydrogen and acetate by acetogenesis. Methane-producing bacteria convert carbon dioxide, hydrogen, and acetate into methane in the final stage known as methanogenesis [3].

The only alternate organically produced carbon source that has been identified to be significant enough to replace fossil fuels is biomass [4]. When energy crops are taken into account as a biomass source, the overall potential energy of biomass for the production of energy might be significantly higher than the potential energy of biomass wastes [5]. India's vast agricultural sector makes it ideal for growing a variety of crops, particularly yearly ones that might be used as energy crops for bio-methanation process [6]. Lignocellulose is one of the major promising renewable substances in land and accounts for the bulk of biomass or around 50% of the plant

DOI: 10.1201/9781003434313-2

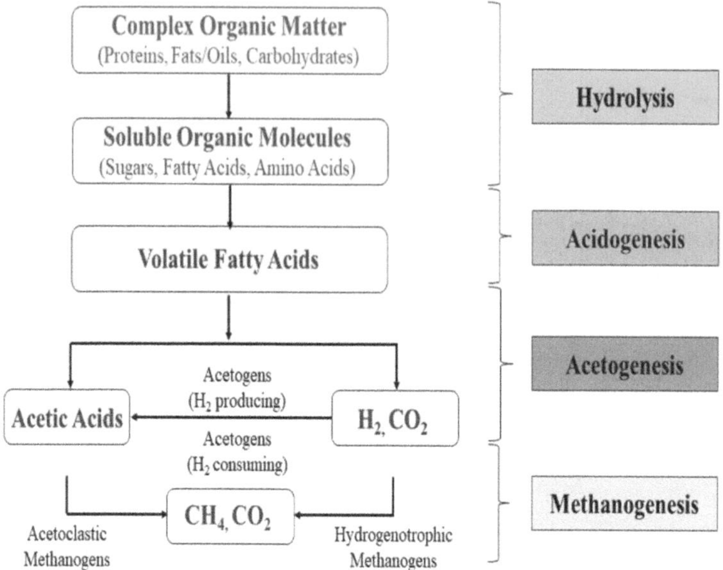

FIGURE 2.1   Anaerobic digestion process.

matter generated due to photosynthesis [7]. Grasses are effective in producing cellulosic biomass. Because they store more carbon, require less tillage, and use fewer pesticides and fertilizers, grasses are more beneficial [8]. In latest years, a significant improvement in the use of grass lands for energy generation, especially for the generation of methane and digestate as a fertilizer [9]. Another aspect that influences the generation of biogas is the type of grass utilized; the composition of various kinds of grass results in varying substrates for the process of AD [10].

To evaluate the chemical composition of Napier grass is the major objective of current study and to achieve this fiber analysis is performed. The comparison between lignin, hemicellulose, and cellulose content (in percentage) of raw Napier grass for different harvesting days of 15, 30, 45, and 60 days is assessed and tabulated and a graphical representation is illustrated for the same. Additionally, another graphical representation showing the comparison between the chemical composition of raw Napier grass for different harvesting days from various literature studies is evaluated.

2 MATERIALS AND METHODS

2.1 MATERIALS

One species that is cultivated mainly for grazing is Napier grass (*Pennisetum purpureum*) as shown in Figure 2.2. Napier grass production is about 150 MT/acrc/year. To stop stem-boring insects from ruining the agricultural produce, it has, however, lately been included into a pest control plan by being grown around the edge of the farm. Napier grass additionally avoids erosion of soil in arid areas and increases the fertility of the soil [1, 2, 11].

FIGURE 2.2 Napier grass.

Napier grass (*Pennisetum purpureum* × *Pennisetum americanum*) has a high content of nitrogen and residual LCB, and thus it doesn't require additional co-digestion, nutrition, or pretreatment for the AD process. The crude protein level of Napier grass is approximately 10.2%, whereas the crude fiber content is 30.5% [3, 12]. The broad, big, green leaves of the Napier grass have softer sheaths and are not as fibrous and readily digested. Every 45–60 days, harvesting may be required. It can withstand protracted periods of extreme drought and regenerate itself when it rains [13, 14]. Growing Napier grass costs around 50% below than the growing crops that are harvested seasonally as compared with typical grasses. Moreover, Napier grass generates around double the quantity per hectare or acre [1, 13]. Once harvested, Napier grass can be cultivated up to seven years. It has a strong growth rate, can resist drought well, and recovers quickly from rainfall. It also adapts effectively to tropical and subtropical regions [15]. Because Napier grass emits photons during the extraction of organic fuels like alcohols, methane, and pyrolytic oils, it is considered the energy crop of the future. For the purpose of producing biofuels and bio-based goods, Napier grass is specifically kept. This is because of its higher content of cellulose matter (34–40%) and maximum production of green fodder per acre, also resistance to drought [16–18]. Researchers looking into Napier grass' potential for biogas generation discovered that it could potentially be grown as a feedstock for this purpose and that the greatest kinetics for the generation of biogas could be reached at 5% total solids [19]

In this study, Fresh Napier grass (Pakchong-1) was cut at different harvesting age of 15, 30, 45, and 60 days and was chaff cut to a size of 1 cm. The chaff cut grass was then oven dried at 100°C for about 8 hours and further grinded and sieved to obtain fine powder for the fiber analysis.

2.2 PROCEDURE

The pulverized Napier grass was processed for structural characterization like cellulose, hemicellulose, and lignin using the Van Soest technique, as shown in Figure 2.3.

In 1963 and 1967, Van Soest devised a method for analyzing silage residue in terms of wall structure and cell elements using different detergents. This method defines cell components that are soluble in neutral detergents as soluble proteins, starch, organic acids, sugars, nitrogenous substances, and other water-soluble materials. On the other hand, heat-damaged proteins, silica, keratins, lignin, cellulose, hemicellulose, and lignified nitrogenous molecules are among the components of the insoluble material known as neutral detergent fiber (NDF) [20, 21]. Acid detergent process releases hemicellulose, certain nitrogenous compounds from the cell wall, and acid soluble ash from NDF levels, leaving the portion that is insoluble known as acid detergent fiber (ADF) or lignocellulose with the remaining cell wall components.

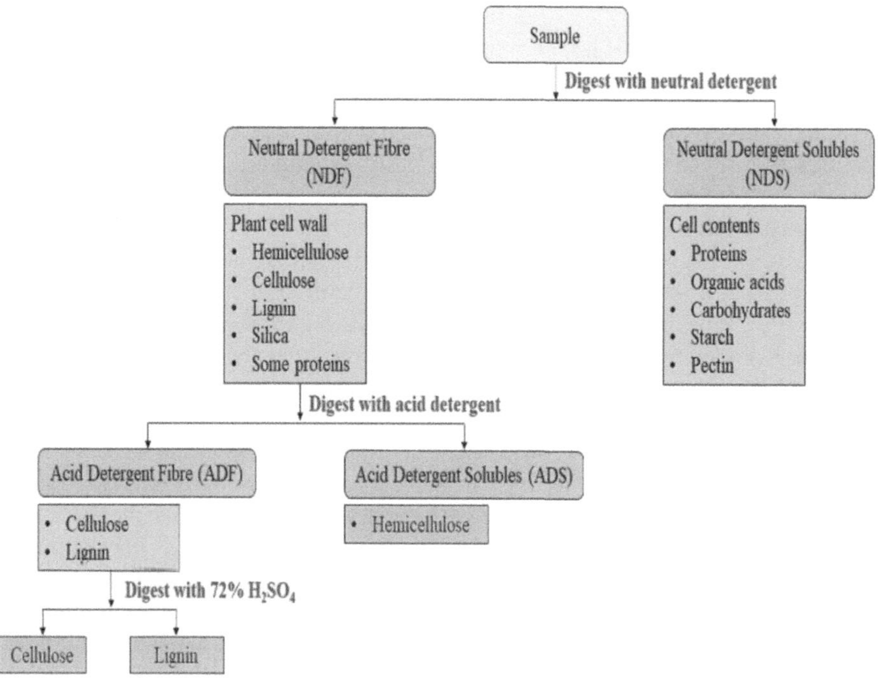

FIGURE 2.3 Van Soest method for fiber analysis.

FIGURE 2.4 Fiber analysis equipment.

2.3 EXPERIMENTAL SETUP

The pulverized Napier grass was fed in the fiber analysis equipment to determine its chemical composition by following the Van Soest method. The fiber analysis equipment is as shown in Figure 2.4. The apparatus required for this experiment included refluxing apparatus and sintered glass crucibles with coarse porosity (grade) of about 50 ml capacity.

The chemicals required for this experiment included neutral detergent solution, decahydronaphthalenes, acetone, sodium sulfite (anhydrous), acid detergent solution (prepared by adding acetyl trimethyl ammonium bromide to 1 N H_2SO_4), n-hexane and H_2SO_4 72% by weight. EDTA and sodium borate decahydrate ($Na_2B_4O_7 10 H_2O$) were combined with distilled water in a big beaker, and the mixture was boiled until the ingredients were entirely dissolved, to create the neutral detergent solution. Next, two ethoxy ethanol and sodium lauryl sulfate were added to the same beaker. In addition, some of the distilled water was combined with disodium hydrogen phosphate (Na_2HPO_4) and boiled in a separate beaker until it was dissolved. Ultimately, the two solutions were appropriately blended [22–24].

3 RESULTS AND DISCUSSIONS

3.1 CHARACTERIZATION OF NAPIER GRASS

Lignocellulose is globally abundant in materials such as energy crops (maize, Napier grass), forest residue (fuel wood, branches, sawdust), agricultural residue (rice husk, wheat straw), industrial residue (pulp and paper), food Digestry (citrus crops), municipal residue (grass, newspaper, cardboard) etc., consisting high proportion of carbohydrates that may be utilized for AD [25]. The main structural element of plants, LCB is often found in the leaves, root systems, and stalks. LCB is mostly composed of three kinds of polymers that are connected to one another in a hetero-matrix: lignin (10–25%), cellulose (30–60%), and hemicelluloses (20–40%) as shown in Figure 2.5.

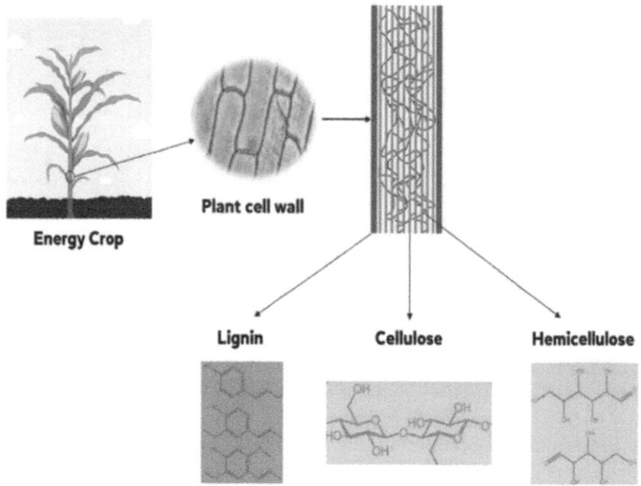

FIGURE 2.5   Structure of lignocellulosic biomass.

The method of determining the mentioned characteristics is the Van Soest method of fiber analysis as explained in Sections 2.2 and 2.3. In LCB, lignin, hemicellulose, and cellulose comprise around 90% of the dry matter; the remaining portion is made up of ashes and extract [26–30]. Variations in the genetic makeup of the plant and its surroundings impact the structure of LCB [26, 27, 29, 31, 32].

3.1.1   Lignin

An extensive three-dimensional network arrangement made up of phenyl propane units makes up the aromatic polymer lignin [7, 27]. In LCB, lignin functions as adhesive, filling in the spaces surrounding and between cellulose and hemicelluloses to firmly bind these together [33]. Lignin is an amorphous heteropolymer that prevents oxidative and bacterial damage and renders the cell membrane impermeable [34]. Hardwoods often contain more holocellulose than softwoods, but softwoods typically have more lignin [26, 27]. In present experimentation, the lignin content of Napier grass of 15th, 30th, 45th, and 60th day was assessed and it was found to be 16.5%, 26%, 28.5%, and 32% respectively as shown in Table 2.1. The lignin content increased by 57.58% from 15th day to 30th day, 9.62% from 30th day to 45th day, and 12.28% from 45th day to 60th day Napier grass.

3.1.2   Cellulose

The homopolysaccharide cellulose ($C_6H_{10}O_5$) consists of β-D-glucose units arranged in a linear network joined by β-1, four glycosidic bonds. These structural elements are united with rigid bonds of hydrogen, ensuring the crystalline structure of cellulose and forms microfibrils [35, 36]. Bundling of microfibrils produces cellulosic fibers. Amorphous (unorganized) and crystalline (organized) regions make up cellulose, with the former being more resistant to deterioration than the latter [37]. An amorphous framework made of lignin, pectin, and hemicelluloses envelops cellulose fibers. In both, the middle lamellae and the cellulose microfibrils present in cell walls with lignin and hemicellulose are dispersed [26, 38, 39].

3.1.3   Hemicellulose

L-arabinose and D-xylose are pentose sugars and D-mannose, D-galactose, and D-glucose are hexose sugars that combine to form hemicelluloses, including xylose as particularly dominant [26, 40, 41]. The primary heteropolymers of hemicelluloses are galactan, mannan, xylan, and arabinan [42]. After cellulose, xylan is the next largest renewable polysaccharide that occurs naturally and serves as the main structural element of plant hemicelluloses [43]. Roughly 1/3rd of the carbon that is renewable on Earth is found in xylan [26, 44]. Xylan is an intricate polysaccharide

TABLE 2.1

Harvesting age of Napier grass and variation in chemical composition

Particulars	Harvesting age of grass in days			
	15	30	45	60
Moisture content	81	75.65	69.32	64.16
Lignin	16.5	26	28.5	32
Crude protein	14.87	11	11.81	8.75
Ether extract	3	3.3	3	3.3
Nitrogen free extract	53.63	47.67	43.69	43.95
Total ash	12	12	13	12
Total	100	100	100	100

(Excluding moisture content)

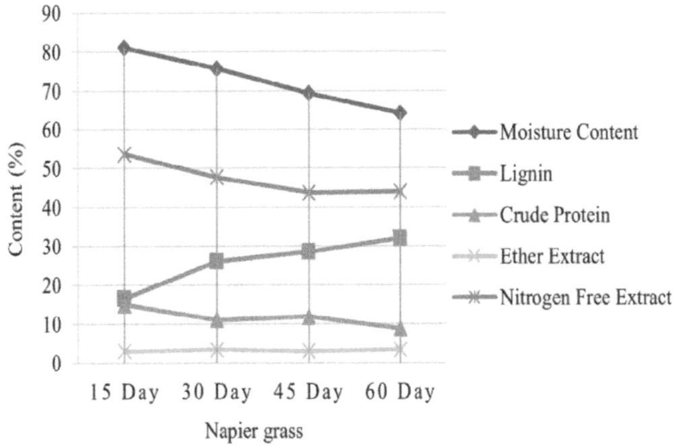

FIGURE 2.6   Comparison between chemical composition of Napier grass of different harvesting age.

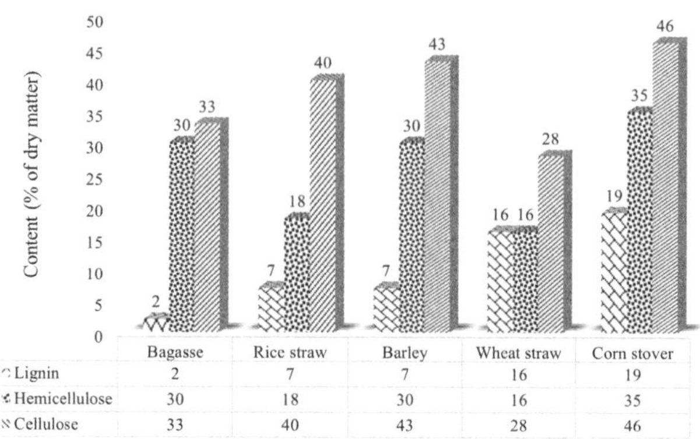

	Bagasse	Rice straw	Barley	Wheat straw	Corn stover
Lignin	2	7	7	16	19
Hemicellulose	30	18	30	16	35
Cellulose	33	40	43	28	46

FIGURE 2.7 Lignocellulose contents of different crop residues.

made up of trace amounts of L-arabinose and an underlying structure of xylose residues joined by β-1, 4-glycosidic linkage [42, 45]. With its non-covalent bond with cellulose and covalent connection with lignin, the xylan layer might be crucial for preserving the structural integrity of cellulose in its natural environment and shielding cellulosic fibers from cellulase destruction [42, 46].

A graphical representation for the comparison between chemical composition of Napier grass of different harvesting age is illustrated in Figure 2.6.

The lignocellulose contents of different crop residues were also evaluated from various literature studies. Figure 2.7 indicates lignocellulose contents of different crop residues [47].

Table 2.2 below shows the characterization of raw Napier grass that was studied and evaluated from a variety of literature evaluations.

4 CONTROL PARAMETERS OF AD PROCESS

Total Solids (TS) and Volatile Solids (VS): The decomposable component of TS that exists in the substrates is known as VS. The concentration of volatile solids in organic substrates aids in assessing the substrates' ability for degradation, the generation of biogas, the rate of organic load, and the C/N ratio [13, 56, 57]. VS are the most significant solids present in the substrates because they clearly indicate the degree of energy conversion [58]. Additionally, the conversion of volatile solids indicates how much methane is present in the biogas [13, 59].

Volatile Fatty Acids (VFA): AD process always affected by VFAs. It includes dissolvable organic acids such as valeric, acetic, butyric, and propionic acids. The AD process' fermentation is harmed by an increase in the VFA level [13, 60, 61]. The AD process is greatly impacted by the excess VFA generated in the reactor, which lowers pH [62]. There has been a detrimental effect on the AD, if volatile fatty acids are produced in large amounts, which eventually has an impact on biogas production [63]. The production of methane can be inhibited by VFA. Significant VFA density lowers pH and indirectly interferes with fermentation. As was already explained, the AD processes require a specific pH rate to function. It has been shown that the

TABLE 2.2

Chemical composition of raw Napier grass from literature

Harvesting days	Total solids (%)	Volatile solids (%)	Lignin (%)	Cellulose (%)	Hemicellulose (%)	Total carbon C (%)	Total nitrogen N (%)	C/N ratio	References
45	19.11	10.75	11.08	84.56	4.36	11.37	0.44	25.84	[3]
–	–	–	–	–	–	43.9	2	21.95	[4]
–	–	–	–	–	–	44.2	2	22.1	[4]
—	—	—	—	—	—	44	1.9	23.16	[4]
—	—	—	24	22	24	—	—	—	[26]
30–45	—	—	10–33	25–43	15–53	49.93	2.05	24.35	[11]
—	—	—	32.04	34.25	17.36	49.93	2.02	24.72	[48]
—	28	25	30.23	29.48	17.32	45.36	0.78	58.15	[49]
45	12.19	8.17	30.40	36.34	34.12	44.19	2.00	22.10	[50]
—	25.67	21.97	21.67	32.26	19.73	—	—	—	[51]
—	31.60	29.84	—	—	—	10	2.26	17.64	[52]
—	31.6	29.6	—	—	—	10	2	5	[53]
—	—	—	21.6	47.1	31.2	—	—	—	[54]
60	—	—	8.27	36.81	26.16	24.3	2.67	9.10	[55]

highest levels of VFA more than 4 g/l inhibit glucose fermentation [64]. During AD, acetic acid is frequently detected in more quantum than the rest of the fatty acids [65].

Temperature: Temperature serves as one of the most crucial factors in AD; also it is usually achievable in both ranges of temperature. Mesophilic temperature operates up to 35°C whereas thermophilic temperature operates up to 55°C [66]. The duration of time needed for retention in the reactor is reduced by a greater thermophilic temperature range. Additionally, the range of thermophilic temperatures is more vulnerable to hazardous substances and changes in operational conditions that result in higher methane generation levels [13, 67]. According to a 1992 National Renewable Energy Laboratory report, an increased thermophilic range of temperature reduces HRT maintained in the digester. The HRT increases and the bio-methanation process is slowed when the digester is maintained at the mesophilic range [68]. It was discovered that operating a digester at a lower thermophilic temperature can both lessen the amount of time needed for retention and provide better quality of gas. In addition to focusing on bacterial growth, the extreme temperatures in the AD system also affected mass transfer characteristics as well as physical indicators like surface tension and viscosity. In order to shorten the duration of process of digestion, investigations on the AD of natural solid waste part in semi-dry thermophilic temperature were carried out at large-scale reactors at Verona-Cadel Blue in Italy [69, 70].

pH: While the optimum pH range is 6.5–7.5, the ideal pH range for anaerobic bacteria, such as those that generate methane, is 6.8–7.6. It is possible to reduce the rate of generation of methane when pH is between 6.3 and 7.8 [71]. An elevated concentration of acidogenesis and hydrolysis with a pH of around 7.0 was obtained by AD of household wastes with 86% of total organic carbon and 82% of COD. The pH of the digester is raised to 6.4 by adding lime, and then bicarbonate or carbonate salt should be added to reach the ideal pH [13, 72]. The ideal pH range for AD was discovered to be in the range of about 5.5 and 8.5 in 1998. Methanogenesis required a pH of about 7.0. Because acid-consuming bacteria predominate over acid-producing bacteria, this could be the result of deflective fluctuations in the digester.

VFA to Alkalinity Ratio: The optimal ratio of VFA to alkalinity is 0.05–0.25. As VFA rise, there are grounds for severe worries if the ratio exceeds 0.15. It requires enough buffering, such as the addition of lime or more water, in order to be prohibited. The quantity of methane that governs the generation of biogas is inversely correlated with the VFA accumulation. When VFA increases in comparison with alkalinity, a ratio greater than 0.25 poses serious issues [12, 63]. If VFA/alkalinity ratio in the range of 0.4, AD system worked well and there was no risk of acidification [73].

Carbon to Nitrogen (C/N) Ratio: The C/N balance is one of the major variables influencing the effective digestion of substrate in reactor. For AD, the optimal C/N ratio is between roughly 20:1 and 30:1 [74]. Optimizing production of methane requires an appropriate C/N ratio. Insufficient nitrogen for cells to operate properly is indicated by a high C/N ratio. This restricts microbial development, which lowers the amount of biogas produced. Reduced VFA and ammonia production is another benefit of a higher C/N ratio [75]. The feedstock's nitrogen helps to make nucleic acids, proteins, and amino acids, and the carbon serves as both a structural component and an important source of energy for microorganisms [76]. It is possible to code the high C/N ratio found in rich organic waste substrates, such as vegetable and livestock dung, to maintain the digester content's optimal C/N values [77]. The few researches have indicated that the average C/N ratios for methane fermentation are in the range of 25–30 [78, 79], it is possible for operating circumstances, such as temperature, to have an inhibiting influence on the depletion of both carbon and nitrogen [80].

Alkalinity: Alkalinity is a definite observation for the bases within water that are capable of neutralizing acids, or the water's capacity to buffer acids. Low alkalinity in water samples is more sensitive to variations in pH; water with higher alkalinity is more resistant to these changes and has a larger buffering capacity [81]. The quantity of acid required to bring a water sample's pH down to a predetermined level is used to measure alkalinity; the results are often expressed in standardized units, such as milligrams of $CaCO_3$ per liter. Agdag and Sponza used a controlled deposition bioreactor to study how alkalinity affected the digestion process of organic solid waste. Alkalinity reduced the quantity of pollutants as well as the amount of organic matter and the time it took for biodegradation [82].

Moisture Content: Moisture content is the total water content into any substrate which evaporates with processes of sunlight, burning, etc. The primary types of anaerobic bio-methanation reactors or digesters can be identified based on the feedstock material's moisture content (dry or wet digestion) and the processing mode (single stage, continuous mode, batch type, and two stage). Napier grass offers a special chance for fragmentation into solid and liquid streams due to its higher moisture content (80% w/w). The juice that is high in nutrients can be utilized as a substrate for the development of co-products, and the formed cellulosic fibers may be utilized as a substrate for the generation of biofuel [83]. Since the drying process changed the forages' nutritional quality, moisture content is a crucial element that should be determined in the sample of forage. The sample can be dried until a constant weight is reached in order to measure it. The study's findings indicate that the moisture content of several Napier grass cultivars is significantly impacted ($p < 0.05$) by the harvesting ages. As the harvesting ages grew, the moisture content of the Napier grasses in Zanzibar, Uganda, Red, and India reduced, with values of 6.11%, 10.53%, 3.51%, and 9.63%, respectively, dropping from 45 days to 75 days [84]. Similar results were documented by Ansah et al., who discovered that the percentage of moisture increased when harvested at 60 days as opposed to 120 days, recording 52.13% and 48.89%, respectively. With a moisture percentage of 92.99%, the India Napier cultivar has the greatest moisture content [85, 86].

Hydraulic Retention Time (HRT) & Organic Loading Rate (OLR): The OLR implies quantity of organic material to be processed in certain period with certain quantity of feedstock material which is monitored and calculated by volatile solids or COD per volume of substrate while HRT for a certain feedstock is a measure of average retention time in liquid slurry tank of digester. Processes need a lower OLR in their beginning phases, while a higher OLR can be dealt by mature and well-functioning processes. About 4–5 kg VS/m³/day can be treated by thermophilic processes, whereas mesophilic processes usually operate at 2–3 kg VS/m³/day. The anaerobic system will collapse if the HRT is shorter due to an extraction of the slowest growing anaerobic micro-organisms from the reactor. The HRT of anaerobic digesters handling municipal solid wastes ranges from 3 to 55 days, based on characteristics of substrates,

TABLE 2.3

Outcomes of control parameters of the anaerobic digestion process

No.	Parameters	Description	Outcomes		References
			Effect on AD	Gas production	
1	Total solids	The dissolved combined content of all inorganic and organic materials is measured as total solids.	The morphology of the microorganisms in the system will alter as the total solids concentration varies.	Research revealed that there was a tendency for hydrolysis and methane generation to increase initially and then decrease as TS increased.	[89, 90]
2	Volatile solids	The decomposable component of TS that exists in the substrates is known as VS.	The concentration of VS in organic substrates aids in assessing the substrates' ability for degradation, the generation of biogas, the rate of organic load, and the C/N ratio.	The conversion of volatile solids indicates how much methane is present in the biogas.	[13, 57]
3	Volatile fatty acids	The secondary substances created during the AD process are known as volatile fatty acids.	The AD process's fermentation is harmed by increase in VFA level.	If VFA are generated in large amounts during AD, it has a detrimental effect on the anaerobic process, which eventually has an impact on bio methanation.	[13, 60, 61]
4	Temperature	AD is usually achievable in two temperature phases. Mesophilic condition maintained at approximately 35°C and thermophilic condition maintained at approximately up to 55°C.	The HRT increases if the reactor operates at mesophilic condition. Extreme temperatures in AD system also affect mass transfer characteristics as well as physical indicators.	The bio-methanation process slows down when the digester operates at mesophilic temperatures.	[66, 68]
5	pH	The negative logarithm of the concentration of hydrogen ions is known as pH.	The optimal pH range for AD was discovered to be in the range of about 5.5 and 8.5. Methanogenesis required a pH of about 7.0.	The optimal pH range for anaerobic bacteria, such as those that generate methane, is 6.8 to 7.6.	[13, 71]

No.	Parameters	Description	Outcomes		References
			Effect on AD	Gas production	
6	VFA to alkalinity ratio	The optimal range of VFA to alkalinity is 0.05 to 0.25.	If VFA/alkalinity ratio was in range of 0.4, AD system worked well and there was no risk of acidification.	The quantity of methane that governs the generation of biogas is inversely correlated with the VFA accumulation.	[12, 73]
7	(C/N) ratio	The C/N ratio is one of the major variables influencing the effective digestion of substrate and the production of biogas.	For AD, the optimal C/N ratio is between roughly 20:1 and 30:1.	Higher C/N ratio lowers the amount of biogas produced.	[74]
8	Moisture content	Moisture content is the total water content into any substrate which evaporates with processes of sunlight, burning, etc.	The primary types of anaerobic bio-methanation reactors or digesters can be identified based on the feedstock material's moisture content (dry or wet digestion) and the processing mode.	According to Fujishima, when the moisture level on the AD decreased from 97% to 89%, the generated methane yield decreased from 340 to 275 mL/g-VSS.	[91]
9	Hydraulic retention time (HRT) and organic loading rate (OLR)	The OLR implies quantity of organic material to be processed in certain period with certain quantity of feedstock material while HRT for a certain feedstock is a measure of average retention time in liquid slurry tank of digester.	The anaerobic system will collapse if the HRT is shorter due to an extraction of the slowest-growing anaerobic micro-organisms from the reactor.	Studies showed that the optimized biogas output is in range of 0.51 and 0.56 m^3/kg VS up to an OLR of 0.8 kg VS/m^3 d by considering HRT of 45 to 100 days.	[57]

atmospheric temperature, hydraulic design of digester, and time required for completion of stages in degradation process. The OLR is an extremely important parameter for control in continuous feed reactor [87]. The majority of anaerobic process reactors were found to have failed as a result of the high substrate overload. The HRT of degradation process varies from 25 to 50 days, according to the type of feedstock material, average pH range, and digester temperature during anaerobic process. Overall, when the OLR is high, a longer HRT is needed to avoid too little degradation. The solid as well as hydraulic retention time of substrates must be investigated during the AD process. HRT and SRT indicate the amount of time required to degrade the organic matter. A decline in SRT is correlated with a reduction in reaction rate [70].

Mixing Condition: The various study reports suggest that the blending of substrates offers sufficient interaction between the inlet slurry of homogeneously mixed substrates and the feasible bacterial population, also preventing thermal inequality and accumulation of surface layer of scum in an anaerobic reactor [88]. The mixing of substrates could be carried out via a number of methods, including mechanical blenders, slurry recirculation (application of digestate sludge), and provision of pressured biogas in slurry. On the basis of various reports that the limited stirring is preferred in digesters even when several types of microbial accelerators is being used to stop the destruction of effective microbial biomass. However, the appropriate blending method has always remained the subject of discussion.

The results of the aforementioned AD process and their control parameters are displayed in Table 2.3.

5 ENERGY GENERATION

Numerous plant materials and crops have demonstrated the ability to produce methane. Methane production has been shown to be possible in a variety of crop combinations, such as sunflowers and cotton, maize and cereals, grass clover, and several other crops. Even with miscanthus, cabbage, corn, onions, green beans, beets, potato, flax, and rhubarb, these trials are successfully carried out. Table 2.4 presents a comparatively wide diversity of biogas and methane contributions produced by AD of different crops obtained from various literatures. These contributions mostly depend on different circumstances.

The average values of biogas production are measured in m^3/kg VS added and also in m^3/MT substrate as shown in Table 2.4. A graphical representation for the same is also illustrated in Figure 2.8.

The biomass is more resistant to chemical and biological degradation the more lignin it has. A significant obstacle to the bioconversion processes that use LCB is lignin [105]. In simple words, higher lignin content leads to lower biogas yields and thus for the purpose of producing biogas, energy crops with a reduced lignin content are typically used.

6 INTERPRETATIONS

From Table 2.2, a graph was computed as illustrated in Figure 2.9 showing the comparison between lignin, hemicellulose, and cellulose content (in percentage)

TABLE 2.4

Review of biogas/methane yield from various energy crops

No.	Substrates	Biogas yield (m³/MT substrate)	Methane yield (m³/MT substrate)	Biogas yield (m³/ Kg VS)	References
1	Cane grass	87.70	62.64	0.52	[92]
2	Napier grass	91.85	65.61	0.54	[92]
3	Napier grass	74.34	53.1	0.44	[92]
4	Napier grass	26.86	16.11	0.158	[1]
5	Miscanthus grass	77.46	55.33	0.46	[92] [93]
6	Sorghum	142.31	101.65	0.84	[94]
7	Rye grass	120.40	86	0.71	[95] [96]
8	Rye grass	144.06	102.9	0.85	[97]
9	Ryegrass mix	119.22	85.16	0.70	[92] [98]
10	Mix grass	116.73	83.38	0.69	[99]
11	Clover and Alfalfa mix grass	102.82	73.44	0.60	[100] [101]
12	Alfalfa mix grass	102.82	73.44	0.60	[99]
13	Tall wheatgrass	91.74	65.53	0.54	[102]
14	Country mallow	85.68	61.2	0.50	[100] [103]
15	Giant Cane	89.74	64.1	0.53	[104]
16	Corn	85.82	61.3	0.50	[104]
	Average	94.22	67.306	0.60	—

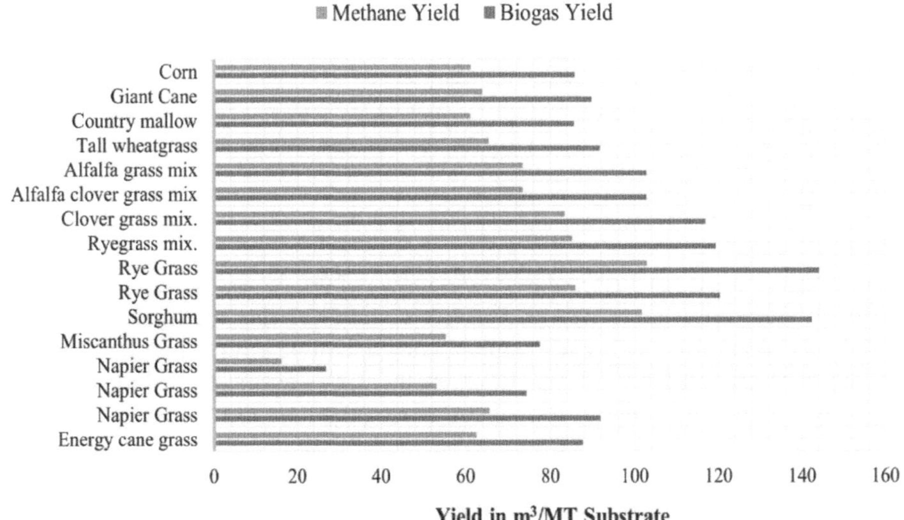

FIGURE 2.8 Review of biogas/methane yield from various energy crops.

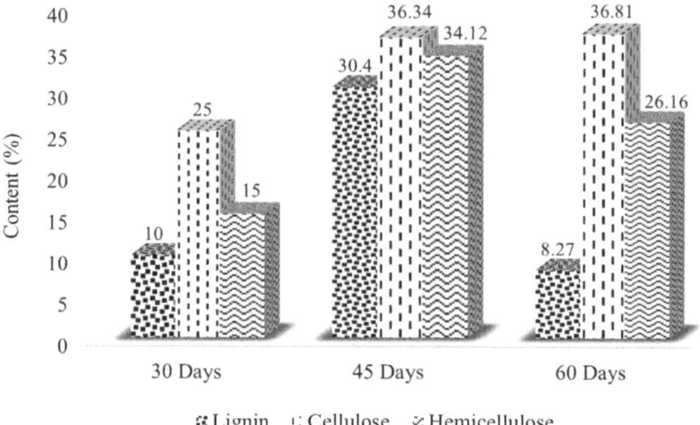

FIGURE 2.9 Comparison between lignin, cellulose, and hemicellulose content for different harvesting days.

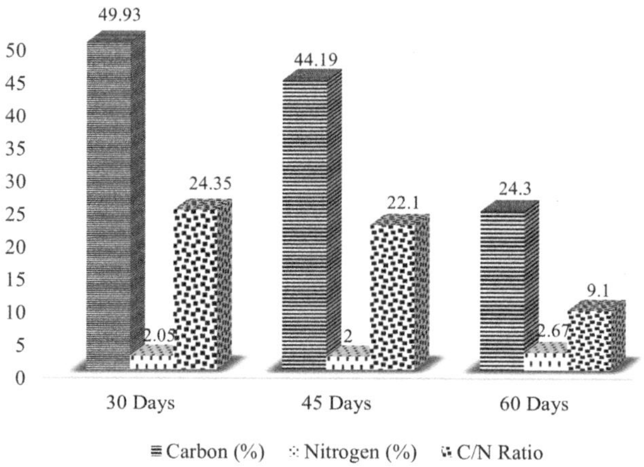

FIGURE 2.10 Comparison between carbon and nitrogen content for different harvesting days.

of raw Napier grass for different harvesting days of 30, 45, and 60 days from three different literature studies [42, 45, 50]. It was observed that the lignin and hemicellulose content was the highest at harvesting age of 45 days, while the cellulose content was highest for 60 days. However, there was not much difference in the cellulose content of Napier grass at harvesting age of 45 and 60 days as depicted Figure 2.9.

In a similar vein, Figure 2.10 compares the carbon and nitrogen contents (in % values) of raw Napier grass at harvesting ages of 30, 45, and 60 days. It is seen that the maximum percentage of carbon was present at harvesting age of 30 days. Similarly, at 30 days of harvesting age, Napier grass' nitrogen content (in%) peaked.

7 CONCLUSION AND FUTURE PERSPECTIVES

In perspective of the access of feedstock, the technical maturity of various processes and economic support networks, it is necessary to contemplate the feasibility of biomass conversion. The abundance of lignocellulose in the world provides higher prospects to yield lignocellulosic biogas. This chapter analyzed the chemical composition of Napier grass, an energy crop and to achieve this, fiber analysis was performed. The comparison between lignin, hemicellulose, and cellulose content (in percentage) of raw Napier grass for different harvesting days of 15, 30, 45, and 60 days was tabulated. The results showed that the lignin content increased by 93.94% from the 15th day to the 60th day Napier grass whereas the moisture content decreased by 20.79% from the 15th day to the 60th day Napier grass. The lignin content increased by 57.58% from the 15th day to the 30th day, 9.62% from the 30th day to the 45th day and 12.28% from the 45th day to the 60th day Napier grass. The moisture content decreased by 6.6% from the 15th day to the 30th day, 8.37% from the 30th day to the 45th day and 7.44% from the 45th day to the 60th day of Napier grass. Additionally, it was observed from various studies that the lignin content of Napier grass ranges from 20% to 30%, cellulose content ranges from 20% to 40%, whereas the hemicellulose content ranges from 15% to 35%. It was also noted that the C/N ratio of Napier grass ranges from 20% to 25%. The rate at which the bacteria grow in number is essential for the AD process. Therefore, the digester's operating parameters are controlled to amplify the microbial activity and maximize productivity. Different factors, like form of feedstock used, as well as numerous operational variables i.e., organic loading rate, mixing conditions, energy generation, hydraulic retention time, etc. influence biogas yield process. It was proved that a number of properties of raw Napier grass, including pH, temperature, volatile and total solids, and C/N range, significantly influenced bio-methanation during the AD process. Future prospects for energy crops may involve effective and low-cost pretreatment of energy crops, the production of bio-refineries, atmospheric carbon "scrubbing" and an increasing small-scale trend, the transmission of energy systems that could potentially contribute to the hydrogen economy. The recalcitrance of biomass, the high % of lignin and increased cellulose limits microbial degradation, thus pretreatment measures are beneficial. Pretreatment tends to unveil the structure and enhance the production of biogas subsequently.

REFERENCES

[1] N. RekhaB. and A. B. Pandit, 'Performance Enhancement of Batch Anaerobic Digestion of Napier Grass by Alkali Pre-Treatment', *Int. J. ChemTech Res.*, vol. 5, no. 2, pp. 558–564, 2013, [Online]. https://api.semanticscholar.org/CorpusID:43883129.

[2] V. Souvannasouk, M. Shen, M. Trejo, and P. Bhuyar, 'Biogas Production from Napier Grass and Cattle Slurry Using a Green Energy Technology', *Int. J. Innov. Res. Sci. Stud.*, vol. 4, no. 3, pp. 174–180, Jun. 2021, https://doi.org/10.53894/ijirss.v4i3.74.

[3] A. Janejadkarn and O. Chavalparit, 'Biogas Production from Napier Grass (Pak Chong 1) (Pennisetum Purpureum × Pennisetum Americanum)', *Adv. Mater. Res.*, vol. 856, pp. 327–332, Dec. 2013, https://doi.org/10.4028/www.scientific.net/AMR.856.327.

[4] S. Kasulla and S. J. Malik, 'Potential of Biogas Generation from Hybrid Napier Grass', *Int. J. Trend Sci. Res. Dev.*, vol. 6, no. 2, pp. 277–281, 2022, https://doi.org/10.5281/zenodo.5886088.

[5] P. McKendry, 'Energy Production from Biomass (Part 1): Overview of Biomass', *Bioresour. Technol.*, vol. 83, no. 1, pp. 37–46, May 2002, https://doi.org/10.1016/S0960-8524(01)00118-3.

[6] M. J. Sukhesh and P. V. Rao, 'Anaerobic Digestion of Crop Residues: Technological Developments and Environmental Impact in the Indian Context', *Biocatal. Agric. Biotechnol.*, vol. 16, pp. 513–528, Oct. 2018, https://doi.org/10.1016/j.bcab.2018.08.007.

[7] C. Sánchez, 'Lignocellulosic Residues: Biodegradation and Bioconversion by Fungi', *Biotechnol. Adv.*, vol. 27, no. 2, pp. 185–194, Mar. 2009, https://doi.org/10.1016/j.biotechadv.2008.11.001.

[8] A. Haryanto, U. Hasanudin, C. Afrian, and I. Zulkarnaen, 'Biogas Production from Anaerobic Codigestion of Cowdung and Elephant Grass (Pennisetum Purpureum) Using Batch Digester', *IOP Conf. Ser. Earth Environ. Sci.*, vol. 141, p. 012011, Mar. 2018, https://doi.org/10.1088/1755-1315/141/1/012011.

[9] C. Rodriguez, A. Alaswad, K. Y. Benyounis, and A. G. Olabi, 'Pretreatment Techniques Used in Biogas Production from Grass', *Renew. Sustain. Energy Rev.*, vol. 68, pp. 1193–1204, Feb. 2017, https://doi.org/10.1016/j.rser.2016.02.022.

[10] L. Carlier, I. Rotar, V. Mariana, and R. Vidican, 'Importance and Functions of Grasslands', *Not. Bot. Horti Agrobot. Cluj-Napoca*, vol. 37, 2009.

[11] H. Warade, R. Daryapurkar, and P. B. Nagarnaik, 'Assessment of Biogas Production from Energy Crop Using Animal Manure as Co-substrate Through Portable Reactor', in *Smart Technologies for Energy, Environment and Sustainable Development*, 2019, pp. 165–174. https://doi.org/10.1007/978-981-13-6148-7_18.

[12] J. Lindmark, N. Leksell, A. Schnürer, and E. Thorin, 'Effects of Mechanical Pre-treatment on the Biogas Yield from Ley Crop Silage', *Appl. Energy*, vol. 97, pp. 498–502, Sep. 2012, https://doi.org/10.1016/j.apenergy.2011.12.066.

[13] H. Warade et al., 'Optimizing the Grass Bio Methanation in Lab Scale Reactor Utilizing Response Surface Methodology', *Biofuels*, vol. 14, no. 7, pp. 721–732, Aug. 2023, https://doi.org/10.1080/17597269.2023.2170034.

[14] X. Zhao, K. Cheng, and D. Liu, 'Organosolv Pretreatment of Lignocellulosic Biomass for Enzymatic Hydrolysis', *Appl. Microbiol. Biotechnol.*, vol. 82, no. 5, pp. 815–827, Apr. 2009, https://doi.org/10.1007/s00253-009-1883-1.

[15] A. Chiluwal, H. P. Singh, K. Sahoo, R. Paudel, W. F. Whithead, and B. P. Singh, 'Napiergrass Has Dual Use as Biofuel Feedstock and Animal Fodder', *Agron. J.*, vol. 111, no. 4, pp. 1752–1759, Jul. 2019, https://doi.org/10.2134/agronj2018.09.0601.

[16] A. Gudimetla, P. Kumar, S. S. Prasad, S. Geeri, and V. V. N. Sarath, 'Towards Smart Materials: Enhancing the Efficiency of the Materials', in *Modeling, Characterization, and Processing of Smart Materials*, 2023, pp. 1–30.

[17] A. Kumar, P. Kumar, R. K. Mittal, and V. Gambhir, 'Materials Processed by Additive Manufacturing Techniques', in *Advances in Additive Manufacturing*, A. Kumar, R. K. Mittal, and A. Haleem, Eds., Elsevier, 2023, pp. 217–233, https://doi.org/10.1016/B978-0-323-91834-3.00014-4.

[18] V. G. Ajay Kumar, P. Kumar, and A. K. Srivastava, *Modeling, Characterization, and Processing of Smart Materials*, in *Advances in Chemical and Materials Engineering*, IGI Global, 2023.

[19] B. Aanjna, A. K. Gour, and V. Singh, 'The Future of Napier Grass as a Biofuel', *Just Agric.*, vol. 2, no. 5, pp. 3–6, 2022.

[20] Ajay, Parveen, S. Ahmad, J. Sharma, and V. Gambhir, *Handbook of Sustainable Materials: Modelling, Characterization, and Optimization*. Boca Raton: CRC Press, 2023. https://doi.org/10.1201/9781003297772.

[21] A. Kumar, H. Singh, P. Kumar, and B. AlMangour, *Handbook of Smart Manufacturing: Forecasting the Future of Industry 4.0*. Boca Raton: CRC Press, 2023, https://doi.org/10.1201/9781003333760.

[22] S. Boopathi and P. Kumar, '5 Advanced Bioprinting Processes Using Additive Manufacturing Technologies: Revolutionizing Tissue Engineering', in *3D Printing Technologies*, De Gruyter, 2024, pp. 95–118, https://doi.org/10.1515/9783111215112-005.

[23] H. K. Tiwari, A. K. Srivastava, P. Kumar, M. K. Singh, H. Kumar, and A. Bhadauria, 'Materials and Technology for Implant Manufacturing: Challenges and Opportunity', in *Modeling, Characterization, and Processing of Smart Materials*, 2023, pp. 83–106.

[24] A. K. Srivastava, A. Kumar, P. Kumar, P. Gautam, and N. Dogra, 'Research Progress in Metal Additive Manufacturing: Challenges and Opportunities', *Int. J. Interact. Des. Manuf.*, pp. 1–17, Dec. 2023, https://doi.org/10.1007/s12008-023-01661-6.

[25] D. G. Cirne, A. Lehtomäki, L. Björnsson, and L. L. Blackall, 'Hydrolysis and Microbial Community Analyses in Two-stage Anaerobic Digestion of Energy Crops', *J. Appl. Microbiol.*, vol. 103, no. 3, pp. 516–527, Sep. 2007, https://doi.org/10.1111/j.1365-2672.2006.03270.x.

[26] A. Kumar, A. Gautam, and D. Dutt, 'Biotechnological Transformation of Lignocellulosic Biomass in to Industrial Products: An Overview', *Adv. Biosci. Biotechnol.*, vol. 7, no. 3, pp. 149–168, 2016, https://doi.org/10.4236/abb.2016.73014.

[27] S. Nanda, J. Mohammad, S. N. Reddy, J. A. Kozinski, and A. K. Dalai, 'Pathways of Lignocellulosic Biomass Conversion to Renewable Fuels', *Biomass Convers. Biorefinery*, vol. 4, no. 2, pp. 157–191, Jun. 2014, https://doi.org/10.1007/s13399-013-0097-z.

[28] S. J. Horn, G. Vaaje-Kolstad, B. Westereng, and V. Eijsink, 'Novel Enzymes for the Degradation of Cellulose', *Biotechnol. Biofuels*, vol. 5, no. 1, p. 45, Dec. 2012, https://doi.org/10.1186/1754-6834-5-45.

[29] S. Nanda, P. Mohanty, K. K. Pant, S. Naik, J. A. Kozinski, and A. K. Dalai, 'Characterization of North American Lignocellulosic Biomass and Biochars in Terms of their Candidacy for Alternate Renewable Fuels', *BioEnergy Res.*, vol. 6, no. 2, pp. 663–677, Jun. 2013, https://doi.org/10.1007/s12155-012-9281-4.

[30] R. Estela and J. Luis, 'Hydrolysis of Biomass Mediated by Cellulases for the Production of Sugars', in *Sustainable Degradation of Lignocellulosic Biomass - Techniques, Applications and Commercialization*, InTech, 2013. https://doi.org/10.5772/53719.

[31] S. Malherbe and T. E. Cloete, 'Lignocellulose Biodegradation: Fundamentals and Applications', *Rev. Environ. Sci. Bio/Technology*, vol. 1, no. 2, pp. 105–114, Jun. 2002, https://doi.org/10.1023/A:1020858910646.

[32] S. Soltanian et al., 'A Critical Review of the Effects of Pretreatment Methods on the Exergetic Aspects of Lignocellulosic Biofuels', *Energy Convers. Manag.*, vol. 212, p. 112792, May 2020, https://doi.org/10.1016/j.enconman.2020.112792.

[33] T. Shahzadi et al., 'Advances in Lignocellulosic Biotechnology: A Brief Review on Lignocellulosic Biomass and Cellulases', *Adv. Biosci. Biotechnol.*, vol. 5, no. 3, pp. 246–251, 2014, https://doi.org/10.4236/abb.2014.53031.

[34] S. Mussatto and J. Teixeira, 'Lignocellulose as Raw Material in Fermentation Processes', *Formatex Research Center* vol. 2, 2010, pp. 897–907, 2010, ISBN-13: 978-84-614-6195-0.

[35] A. T. W. M. Hendriks and G. Zeeman, 'Pretreatments to Enhance the Digestibility of Lignocellulosic Biomass', *Bioresour. Technol.*, vol. 100, no. 1, pp. 10–18, Jan. 2009, https://doi.org/10.1016/j.biortech.2008.05.027.

[36] R. R. Singhania, R. K. Sukumaran, A. K. Patel, C. Larroche, and A. Pandey, 'Advancement and Comparative Profiles in the Production Technologies Using Solid-State and Submerged Fermentation for Microbial Cellulases', *Enzyme Microb. Technol.*, vol. 46, no. 7, pp. 541–549, Jun. 2010, https://doi.org/10.1016/j.enzmictec.2010.03.010.

[37] V. B. Agbor, N. Cicek, R. Sparling, A. Berlin, and D. B. Levin, 'Biomass Pretreatment: Fundamentals toward Application', *Biotechnol. Adv.*, vol. 29, no. 6, pp. 675–685, Nov. 2011, https://doi.org/10.1016/j.biotechadv.2011.05.005.

[38] J. S. Van Dyk and B. I. Pletschke, 'A Review of Lignocellulose Bioconversion Using Enzymatic Hydrolysis and Synergistic Cooperation between Enzymes—Factors Affecting Enzymes, Conversion and Synergy', *Biotechnol. Adv.*, vol. 30, no. 6, pp. 1458–1480, Nov. 2012, https://doi.org/10.1016/j.biotechadv.2012.03.002.

[39] K.-E. L. Eriksson and H. Bermek, 'Lignin, Lignocellulose, Ligninase', in *Encyclopedia of Microbiology*, Elsevier, 2009, pp. 373–384, https://doi.org/10.1016/B978-012373944-5.00152-8.

[40] R. Kumar, S. Singh, and O. V. Singh, 'Bioconversion of Lignocellulosic Biomass: Biochemical and Molecular Perspectives', *J. Ind. Microbiol. Biotechnol.*, vol. 35, no. 5, pp. 377–391, May 2008, https://doi.org/10.1007/s10295-008-0327-8.

[41] V. Juturu and J. C. Wu, 'Microbial Xylanases: Engineering, Production and Industrial Applications', *Biotechnol. Adv.*, vol. 30, no. 6, pp. 1219–1227, Nov. 2012, https://doi.org/10.1016/j.biotechadv.2011.11.006.

[42] Q. K. Beg, M. Kapoor, L. Mahajan, and G. S. Hoondal, 'Microbial Xylanases and their Industrial Applications: A Review', *Appl. Microbiol. Biotechnol.*, vol. 56, no. 3–4, pp. 326–338, Aug. 2001, https://doi.org/10.1007/s002530100704.

[43] T. Collins, C. Gerday, and G. Feller, 'Xylanases, Xylanase Families and Extremophilic Xylanases', *FEMS Microbiol. Rev.*, vol. 29, no. 1, pp. 3–23, Jan. 2005, https://doi.org/10.1016/j.femsre.2004.06.005.

[44] R. A. Prade, 'Xylanases: From Biology to BioTechnology', *Biotechnol. Genet. Eng. Rev.*, vol. 13, no. 1, pp. 101–132, Dec. 1996, https://doi.org/10.1080/02648725.1996.10647925.

[45] K. B. Bastawde, 'Xylan Structure, Microbial Xylanases, and their Mode of Action', *World J. Microbiol. Biotechnol.*, vol. 8, no. 4, pp. 353–368, Jul. 1992, https://doi.org/10.1007/BF01198746.

[46] R. L. Uffen, 'Xylan Degradation: A Glimpse at Microbial Diversity', *J. Ind. Microbiol. Biotechnol.*, vol. 19, no. 1, pp. 1–6, Jul. 1997, https://doi.org/10.1038/sj.jim.2900417.

[47] D. P. Ho, H. H. Ngo, and W. Guo, 'A Mini Review on Renewable Sources for Biofuel', *Bioresour. Technol.*, vol. 169, pp. 742–749, Oct. 2014, https://doi.org/10.1016/j.biortech.2014.07.022.

[48] S. Sittijunda, 'Biogas Production from Hydrolysate Napier Grass by Co-digestion with Slaughterhouse Wastewater using Anaerobic Mixed Cultures', *KKU Res. J.*, vol. 20, no. 3, pp. 323–336, 2015.

[49] W. Prapinagsorn, S. Sittijunda, and A. Reungsang, 'Co-Digestion of Napier Grass and Its Silage with Cow Dung for Bio-Hydrogen and Methane Production by Two-Stage Anaerobic Digestion Process', *Energies*, vol. 11, no. 1, p. 47, Dec. 2017, https://doi.org/10.3390/en11010047.

[50] N. Dussadee, R. Ramaraj, and T. Cheunbarn, 'Biotechnological Application of Sustainable Biogas Production through Dry Anaerobic Digestion of Napier Grass', *3 Biotech*, vol. 7, no. 1, p. 47, May 2017, https://doi.org/10.1007/s13205-017-0646-4.

[51] L. Lianhua et al., 'Low-cost Additive Improved Silage Quality and Anaerobic Digestion Performance of Napiergrass', *Bioresour. Technol.*, vol. 173, pp. 439–442, Dec. 2014, https://doi.org/10.1016/j.biortech.2014.09.011.

[52] M. M. Victor, A. C. Igboanugo, U. J. Alukwe, and M. Okwu, 'Enhanced Biogas Production from Fresh Elephant Grasses, Using Liquid Extract from Plantain Pseudo Stem', *Int. J. Sci. Technol. Res.*, vol. 10, no. 3, pp. 123–127, 2021.

[53] V. M. Mbachu, I. Anthony C., and U. Alukwe, 'Development of Enhanced Substrate from Fresh Elephant Grass for Biogas Production', *Int. J. Sci. Technol. Res.*, vol. 8, pp. 44–49, 2019.

[54] K. O. Reddy et al., 'Extraction and Characterization of Cellulose Single Fibers from Native African Napier Grass', *Carbohydr. Polym.*, vol. 188, pp. 85–91, May 2018, https://doi.org/10.1016/j.carbpol.2018.01.110.

[55] P. Kullavanijaya and O. Chavalparit, 'The Production of Volatile Fatty Acids from Napier Grass via an Anaerobic Leach Bed Process: The Influence of Leachate Dilution, Inoculum, Recirculation, and Buffering Agent Addition', *J. Environ. Chem. Eng.*, vol. 7, no. 6, p. 103458, Dec. 2019, https://doi.org/10.1016/j.jece.2019.103458.

[56] S. Zahmatkesh et al., 'Machine Learning Modeling of Polycarbonate Ultrafiltration Membranes at Different Temperatures, Al2O3 Nanoparticle Volumes, and Water Ratios', *Chemosphere*, vol. 313, p. 137424, Feb. 2023, https://doi.org/10.1016/j.chemosphere.2022.137424.

[57] E. Salminen and J. Rintala, 'Anaerobic Digestion of Organic Solid Poultry Slaughterhouse Waste – A Review', *Bioresour. Technol.*, vol. 83, no. 1, pp. 13–26, May 2002, https://doi.org/10.1016/S0960-8524(01)00199-7.

[58] A. D. Ibrahim, 'Production of Biogas Using Abattoir Waste at Different Retention Time', *Sci. World J.*, vol. 5, pp. 23–26, 2010.

[59] I. M. Buendía, F. J. Fernández, J. Villaseñor, and L. Rodríguez, 'Feasibility of Anaerobic Co-digestion as a Treatment Option of Meat Industry Wastes', *Bioresour. Technol.*, vol. 100, no. 6, pp. 1903–1909, Mar. 2009, https://doi.org/10.1016/j.biortech.2008.10.013.

[60] W. Parawira, J. S. Read, B. Mattiasson, and L. Björnsson, 'Energy Production from Agricultural Residues: High Methane Yields in Pilot-Scale Two-Stage Anaerobic Digestion', *Biomass and Bioenergy*, vol. 32, no. 1, pp. 44–50, Jan. 2008, https://doi.org/10.1016/j.biombioe.2007.06.003.

[61] T. S. O. Souza, A. Carvajal, A. Donoso-Bravo, M. Peña, and F. Fdz-Polanco, 'ADM1 Calibration Using BMP Tests for Modeling the Effect of Autohydrolysis Pretreatment on the Performance of Continuous Sludge Digesters', *Water Res.*, vol. 47, no. 9, pp. 3244–3254, Jun. 2013, https://doi.org/10.1016/j.watres.2013.03.041.

[62] S. Tedesco, K. Y. Benyounis, and A. G. Olabi, 'Mechanical Pretreatment Effects on Macroalgae-Derived Biogas Production in Co-digestion with Sludge in Ireland', *Energy*, vol. 61, pp. 27–33, Nov. 2013, https://doi.org/10.1016/j.energy.2013.01.071.

[63] H. Warade, R. Daryapurkar, and P. B. Nagarnaik, 'Review of Biogas Production from Various Substrates and Co-substrates through Different Anaerobic Reactor', *Int J Emerg Technol*, vol. 10, no. 2, pp. 235–242, 2019.

[64] Y. Wang, Y. Zhang, J. Wang, and L. Meng, 'Effects of Volatile Fatty Acid Concentrations on Methane Yield and Methanogenic Bacteria', *Biomass and Bioenergy*, vol. 33, no. 5, pp. 848–853, May 2009, https://doi.org/10.1016/j.biombioe.2009.01.007.

[65] Q. Wang, M. Kuninobu, H. I. Ogawa, and Y. Kato, 'Degradation of Volatile Fatty Acids in Highly Efficient Anaerobic Digestion', *Biomass and Bioenergy*, vol. 16, no. 6, pp. 407–416, Jun. 1999, https://doi.org/10.1016/S0961-9534(99)00016-1.

[66] A. Babaee, J. Shayegan, and A. Roshani, 'Anaerobic Slurry Co-digestion of Poultry Manure and Straw: Effect of Organic Loading and Temperature', *J. Environ. Heal. Sci. Eng.*, vol. 11, no. 1, p. 15, Dec. 2013, https://doi.org/10.1186/2052-336X-11-15.

[67] Y. Li, S. Y. Park, and J. Zhu, 'Solid-State Anaerobic Digestion for Methane Production from Organic Waste', *Renew. Sustain. Energy Rev.*, vol. 15, no. 1, pp. 821–826, Jan. 2011, https://doi.org/10.1016/j.rser.2010.07.042.

[68] N. B. Soni, 'Food Waste Based Kitchen Gas Plant-A Sustainable Solution Forfood Waste Management', *Volatiles Essent. Oils*, vol. 8, no. 4, pp. 12932–12939, 2021.

[69] A. Kumar, P. Kumar, N. Dogra, and A. Jaglan, 'Application of Incremental Sheet Forming (ISF) toward Biomedical and Medical Implants', in *Handbook of Flexible and Smart Sheet Forming Techniques*, Wiley, 2023, pp. 247–263. https://doi.org/10.1002/9781119986454.ch13.

[70] A. Kumar, P. Kumar, N. Sharma, and A. K. Srivastava, *3D Printing Technologies: Digital Manufacturing, Artificial Intelligence, Industry 4.0*, De Gruyter, 2024. https://doi.org/10.1515/9783111215112.

[71] Y. Chen, J. J. Cheng, and K. S. Creamer, 'Inhibition of Anaerobic Digestion Process: A Review', *Bioresour. Technol.*, vol. 99, no. 10, pp. 4044–4064, Jul. 2008, https://doi.org/10.1016/j.biortech.2007.01.057.

[72] L. Yang, F. Xu, X. Ge, and Y. Li, 'Challenges and Strategies for Solid-State Anaerobic Digestion of Lignocellulosic Biomass', *Renew. Sustain. Energy Rev.*, vol. 44, pp. 824–834, Apr. 2015, https://doi.org/10.1016/j.rser.2015.01.002.

[73] J. L. De Boever, A. De Smet, D. L. De Brabander, and C. V. Boucque, 'Evaluation of Physical Structure. 1. Grass Silage', *J. Dairy Sci.*, vol. 76, no. 1, pp. 140–153, Jan. 1993, https://doi.org/10.3168/jds.S0022-0302(93)77333-6.

[74] Rajlakshmi et al., 'Co-digestion Processes of Waste: Status and Perspective', in *Bio-Based Materials and Waste for Energy Generation and Resource Management*, Elsevier, 2023, pp. 207–241, https://doi.org/10.1016/B978-0-323-91149-8.00010-7.

[75] I. TG, I. Haq, and A. S. Kalamdhad, 'Factors Affecting Anaerobic Digestion for Biogas Production: a Review', in *Advanced Organic Waste Management*, Elsevier, 2022, pp. 223–233, https://doi.org/10.1016/B978-0-323-85792-5.00020-4.

[76] S. K. Khanal, T. G. Tirta Nindhia, and S. Nitayavardhana, 'Biogas from Wastes', in *Sustainable Resource Recovery and Zero Waste Approaches*, Elsevier, 2019, pp. 165–174, https://doi.org/10.1016/B978-0-444-64200-4.00011-6.

[77] U. Zaher, R. Li, U. Jeppsson, J.-P. Steyer, and S. Chen, 'GISCOD: General Integrated Solid Waste Co-Digestion Model', *Water Res.*, vol. 43, no. 10, pp. 2717–2727, Jun. 2009, https://doi.org/10.1016/j.watres.2009.03.018.

[78] U. Marchaim and C. Krause, 'Propionic to Acetic Acid Ratios in Overloaded Anaerobic Digestion', *Bioresour. Technol.*, vol. 43, no. 3, pp. 195–203, Jan. 1993, https://doi.org/10.1016/0960-8524(93)90031-6.

[79] H. Yen and D. Brune, 'Anaerobic Co-Digestion of Algal Sludge and Waste Paper to Produce Methane', *Bioresour. Technol.*, vol. 98, no. 1, pp. 130–134, Jan. 2007, https://doi.org/10.1016/j.biortech.2005.11.010.

[80] X. Wang, X. Lu, F. Li, and G. Yang, 'Effects of Temperature and Carbon-Nitrogen (C/N) Ratio on the Performance of Anaerobic Co-Digestion of Dairy Manure, Chicken Manure and Rice Straw: Focusing on Ammonia Inhibition', *PLoS One*, vol. 9, no. 5, p. e97265, May 2014, https://doi.org/10.1371/journal.pone.0097265.

[81] V. Martínez-Alvarez, M. J. González-Ortega, B. Martin-Gorriz, M. Soto-García, and J. F. Maestre-Valero, 'Seawater Desalination for Crop Irrigation—Current Status and Perspectives', in *Emerging Technologies for Sustainable Desalination Handbook*, Elsevier, 2018, pp. 461–492, https://doi.org/10.1016/B978-0-12-815818-0.00014-X.

[82] O. N. Ağdağ and D. T. Sponza, 'Anaerobic/aerobic Treatment of Municipal Landfill Leachate in Sequential Two-Stage Up-flow Anaerobic Sludge Blanket Reactor (UASB)/ Completely Stirred Tank Reactor (CSTR) Systems', *Process Biochem.*, vol. 40, no. 2, pp. 895–902, Feb. 2005, https://doi.org/10.1016/j.procbio.2004.02.021.

[83] D. Takara and S. K. Khanal, 'Characterizing Compositional Changes of Napier Grass at Different Stages of Growth for Biofuel and Biobased Products Potential', *Bioresour. Technol.*, vol. 188, pp. 103–108, Jul. 2015, https://doi.org/10.1016/j.biortech.2015.01.114.

[84] J. Y. Ahn, D. Y. Kil, C. Kong, and B. G. Kim, 'Comparison of Oven-drying Methods for Determination of Moisture Content in Feed Ingredients', *Asian-Australasian J. Anim. Sci.*, vol. 27, no. 11, pp. 1615–1622, Oct. 2014, https://doi.org/10.5713/ajas.2014.14305.

[85] T. Ansah, E. Osafo, and H. Hansen, 'Herbage Yield and Chemical Composition of Four Varieties of Napier (Pennisetum purpureum) Grass Harvested at Three Different Days after Planting', *Agric. Biol. J. North Am.*, vol. 1, no. 5, pp. 923–929, Sep. 2010, https://doi.org/10.5251/abjna.2010.1.5.923.929.

[86] N. A. Kamaruddin, N. Ahmad, N. Zubaidah, and A. Rahman, 'Chemical Composition and Dry Matter Yield of four Napier Grass Cultivars Influenced by Harvesting Ages', *Biosci. Res.*, vol. 18, no. 2, pp. 225–235, 2021, [Online]. www.isisn.org.

[87] P. Yadav, P. Kumar, M. Saxena, Y. Dwivedi, and A. Dubey, '15 Application of Three-Dimensional Printing in Medical, Agriculture, Engineering, and Other Sectors', in *3D Printing Technologies*, De Gruyter, 2024, pp. 311–334, https://doi.org/10.1515/9783111215112-015.

[88] Ajay, Parveen, A. Kumar, R. Goel, and R. K. Mittal, *Waste Recovery and Management: An Approach Toward Sustainable Development Goals*, Boca Raton: CRC Press, 2023, https://doi.org/10.1201/9781003359784.

[89] J. Yi, B. Dong, J. Jin, and X. Dai, 'Effect of Increasing Total Solids Contents on Anaerobic Digestion of Food Waste under Mesophilic Conditions: Performance and Microbial Characteristics Analysis', *PLoS One*, vol. 9, no. 7, p. e102548, Jul. 2014, https://doi.org/10.1371/journal.pone.0102548.

[90] X.-F. Zhao et al., 'Effect of Total Solids Contents on the Performance of Anaerobic Digester Treating Food Waste and Kinetics Evaluation', *E3S Web Conf.*, vol. 272, p. 8, Jun. 2021, https://doi.org/10.1051/e3sconf/202127201026.

[91] S. Fujishima, T. Miyahara, and T. Noike, 'Effect of Moisture Content on Anaerobic Digestion of Dewatered Sludge: Ammonia Inhibition to Carbohydrate Removal and Methane Production', *Water Sci. Technol.*, vol. 41, no. 3, pp. 119–127, Feb. 2000, https://doi.org/10.2166/wst.2000.0063.

[92] K. C. Surendra et al., 'Anaerobic Digestion of High-yielding Tropical Energy Crops for Biomethane Production: Effects of Crop Types, Locations and Plant Parts', *Bioresour. Technol.*, vol. 262, pp. 194–202, 2018, https://doi.org/10.1016/j.biortech.2018.04.062.

[93] R. Wahid et al., 'Methane Production Potential from Miscanthus sp.: Effect of Harvesting Time, Genotypes and Plant Fractions', *Biosyst. Eng.*, vol. 133, pp. 71–80, 2015, https://doi.org/10.1016/j.biosystemseng.2015.03.005.

[94] A. Mahmood and B. Honermeier, 'Chemical Composition and Methane Yield of Sorghum Cultivars with Contrasting Row Spacing', *F. Crop. Res.*, vol. 128, pp. 27–33, 2012, https://doi.org/10.1016/j.fcr.2011.12.010.

[95] M. Hübner et al., 'Impact of Genotype, Harvest Time and Chemical Composition on the Methane Yield of Winter Rye for Biogas Production', *Biomass and Bioenergy*, vol. 35, no. 10, pp. 4316–4323, 2011, https://doi.org/10.1016/j.biombioe.2011.07.021.

[96] P. Weiland, 'Production and Energetic Use of Biogas from Energy Crops and Wastes in Germany', *Appl. Biochem. Biotechnol.*, vol. 109, pp. 263–274, 2003, https://doi.org/10.1385/ABAB:109:1-3:263.

[97] M. Seppälä, T. Paavola, A. Lehtomaki, and J. Rintala, 'Biogas Production from boreal herbaceous Grasses – Specific Methane Yield and Methane Yield per Hectare', *Bioresour. Technol.*, vol. 100, pp. 2952–2958, 2009, https://doi.org/10.1016/j.biortech.2009.01.044.

[98] R. Gao et al., 'Methane Yield through Anaerobic Digestion for Various Maize Varieties in China', *Bioresour. Technol.*, vol. 118, pp. 611–614, 2012, https://doi.org/10.1016/j.biortech.2012.05.051.

[99] B. Molinuevo-Salces, R. Fernández-Varela, and H. Uellendahl, 'Key Factors Influencing the Potential of Catch Crops for Methane Production', *Environ. Technol.*, vol. 35, no. 13–16, pp. 1685–1694, Aug. 2014, https://doi.org/10.1080/09593330.2014.880515.

[100] V. Dandikas, H. Heuwinkel, F. Lichti, J. E. Drewes, and K. Koch, 'Correlation between Biogas Yield and Chemical Composition of Energy Crops', *Bioresour. Technol.*, vol. 174, pp. 316–320, 2014, https://doi.org/10.1016/j.biortech.2014.10.019.

[101] C. Gissén et al., 'Comparing Energy Crops for Biogas Production – Yields, Energy Input and Costs in Cultivation using Digestate and Mineral Fertilisation', *Biomass and Bioenergy*, vol. 64, pp. 199–210, 2014, https://doi.org/10.1016/j.biombioe.2014.03.061.

[102] T. Amon et al., 'Methane Production through Anaerobic Digestion of Various Energy Crops Grown in Sustainable Crop Rotations', *Bioresour. Technol.*, vol. 98, no. 17, pp. 3204–3212, 2007, https://doi.org/10.1016/j.biortech.2006.07.007.

[103] V. N. Gunaseelan, 'Predicting Ultimate Methane Yields of Jatropha Curcus and Morus Indica from their Chemical Composition', *Bioresour. Technol.*, vol. 100, no. 13, pp. 3426–3429, Jul. 2009, https://doi.org/10.1016/j.biortech.2009.02.005.

[104] L. Corno, R. Pilu, F. Tambone, B. Scaglia, and F. Adani, 'New Energy Crop Giant Cane (Arundo donax L.) Can Substitute Traditional Energy Crops Increasing Biogas Yield and Reducing Costs', *Bioresour. Technol.*, vol. 191, pp. 197–204, 2015, https://doi.org/10.1016/j.biortech.2015.05.015.

[105] Y. Zheng, J. Zhao, F. Xu, and Y. Li, 'Pretreatment of Lignocellulosic Biomass for Enhanced Biogas Production', *Prog. Energy Combust. Sci.*, vol. 42, pp. 35–53, 2014, https://doi.org/10.1016/j.pecs.2014.01.001.

3 Lignocellulose-Based Sustainable Biomaterials

Sonali S. Shinde and Aniket P. Sarkate
Dr. Babasaheb Ambedkar Marathwada University

Parveen Kumar
Department of Mechanical Engineering, Rawal Institute of
Engineering and Technology, Faridabad, Haryana, India

Gresi D. Mate and Aditi M. Thorve
Dr. D. Y. Patil Institute of Pharmaceutical
Sciences and Research

LIST OF ABBREVIATIONS

Abbreviation	Word
Ca–P	Calcium–Phosphorous
LBP	Lignocellulose biopolymers
XNP	Xylan nanoparticles
LNP	Lignin nanoparticles
CNC	Cellulose nanocrystals
SCP	Single-cell protein
CO2	Carbon dioxide
CNF	Cellulose nanofibers
NaOH	Sodium hydroxide
DES	Deep eutectic solvents
IL	Ionic liquids
3D	Three dimensional

1 INTRODUCTION

1.1 BIOMATERIALS

Biomaterials, which are materials that can interact with living things and be used for medical and diagnostic purposes, have been around for a while. There's been a steady increase in people wanting to use foreign substances to help heal and fix their bodies [1]. The utilization of biomaterials has a long history, when aseptic surgical

DOI: 10.1201/9781003434313-3

techniques and plates were employed for internal bone fixation. Subsequently, biomaterials were incorporated into 3D printing and the development of nanoparticles. Later it was discovered that biomaterials could be used in more than just implanted devices.

The concept of a biomaterial was first defined by the European Society for Biomaterials as a non-biological material incorporated into a medical device for the purpose of interacting with biological systems and further sustainable biomaterials are referred to as the development of innovative biomaterials using building blocks obtained from natural, renewable resources [2].

At the time, the preferred definition was that of a material designed to interact with biological systems in order to assess, cure, augment, or substitute any organ, tissue, or process of the human body. A substance's ability to stimulate the host to respond appropriately in a particular situation known as biocompatibility distinguishes a biomaterial from other materials. Additionally, it was demonstrated that biological responses to a biomaterial vary among tissues and that tissues themselves influence the biocompatibility of a material. In order to fully recognize the advantages, it is crucial to not only focus on the development of biomaterials but also to incorporate effective engineering design and proper characterization of materials and devices [3].

In the past, material selection for prosthetics was primarily based on availability and the skills of the individuals creating and using them. However, synthetic materials such as metal alloys, ceramics, and polymers began to take the place of naturally generated materials around the beginning of the 20th century [4]. This shift resulted in improved performance, enhanced functionality, and greater consistency compared with their natural counterparts. As a result of these advancements, the application of biomaterials expanded significantly, leading to the development of life-saving or life-enhancing devices, such as vascular stents, dental restorations, contact lenses, and artificial hips, benefiting millions of people.

Biomaterials is a topic that touches on a lot of different topics, from chemistry to chemical engineering to materials science to bioengineering to mechanics to surface science biology and medicine. Ethicists, government regulators, and even entrepreneurs have all had their say on the topic. Biomaterials have transformed the landscape of modern medicine and continue to push the boundaries of what is possible in healthcare. Their unique ability to interact with living systems and provide tailored solutions to medical challenges makes them indispensable in addressing complex clinical needs [5]. By understanding the principles of biomaterial design, their applications, and the challenges they present, we can unlock their full potential and pave the way for groundbreaking innovations in patient care.

1.2 TYPES OF BIOMATERIALS

Biomaterials are a group of materials that are compatible with living systems and possess specific functional qualities suitable for their intended applications. The diagnosis and treatment of a variety of human ailments now frequently includes the use of biomaterials. The traditional classification of biomaterials has four main types: (i) ceramic, (ii) metallic, (iii) composites, and (iv) polymeric [6].

1.2.1 Metallic Biomaterial

Metal biomaterials are mainly used in the manufacturing of medical implants for hard tissue replacement, such as hip implants, bone implants, and dental implants, as they are highly reliable from a mechanical point of view [7]. Materials implanted in vivo are initially exposed to extracellular body fluids such as blood and interstitial fluid. Body fluids contain proteins and amino acids, which have a tendency to promote corrosion. Depending upon the isoelectric points of biomolecules like proteins, the pH of bodily fluids close to those materials may fluctuate. When metallic materials are corroded by body fluid, when metal ions are released into the fluid over a long period of time, or when the ions interact with biomolecules such as proteins and enzymes, toxicity and allergies occur in vivo. Therefore, the ability of metallic biomaterials to resist corrosion is crucial. To enhance biocompatibility and minimize adverse reactions in metallic biomaterials, employ strategies such as selecting biocompatible metals, applying surface treatments and coatings, optimizing surface roughness, controlling alloy composition, and conducting rigorous biocompatibility testing. Additionally, consider design modifications, proper sterilization, and patient-specific approaches to tailor implants, all while ensuring long-term monitoring and education for healthcare professionals. These measures collectively improve the safety and effectiveness of metallic biomaterials in medical applications. Stainless steels, titanium, cobalt–chromium–molybdenum alloys and commercially pure titanium alloys are among the biomaterials that are now in use [8,9].

1.2.2 Ceramic Biomaterial

Ceramics encompass a variety of materials, including metal and metalloid oxides, nitrides, sulfides, and carbides. Bioceramics, specifically, are characterized by their hardness, elastic modulus, fragility, and low fracture toughness compared with bone. In dentistry, implants, coatings, orthopedics, calcified tissues, scientific sensors, and diverse different fields, artificial bioceramics like zirconia, titania, Ca–P primarily based totally porous materials, alumina, bioactive glass–ceramics locate applications. More recently, we've been using natural materials like agro waste/ashes from rice husks and sugarcane leaves, as well as corn cob as a great way to get silica. These materials can be turned into cost-effective, bio-based nanoparticles that add value and can be used to make glasses and glass-like ceramics. Similarly, egg shell powder and membrane incorporate calcium and peels of banana incorporates potassium. Additionally, polyphenolic compounds and important amino acids are found in meals waste, which may be applied for the manufacturing of biocatalytic bio glasses. These glasses are an alternative to artificial bio glasses utilized in prostheses and scientific therapies [10].

Ceramic materials are increasingly important because they are biocompatible, corrosion-resistant and mainly because mineral phases are a significant component of bone. Ceramics are therefore used as bone substitutes to stimulate bone regeneration [11].

1.2.3 Polymeric Biomaterial

The most adaptable class of biomaterials is still polymers, which are widely used in biotechnology, medicine, the food, and cosmetics sectors. It is possible to synthesize polymers with the necessary chemical, physical, interfacial, and biomimetic properties for a variety of specialized applications [12]. A prosthesis is intended to replace a function or organ that has completely failed and is no longer necessary for healing to occur naturally. Polymeric biomaterials contribute significantly to the development of permanent prostheses. These polymers are utilized as matrixes in tissue engineering, implants, and artificial organs, aiming to replicate the properties found in natural organs' tissues.

1.2.4 Composites

Biological composites, which consist of biopolymers and minerals, are commonly found in natural materials. One of the best features of these composites is their ability to perform many different functions [13]. The term "composite" refers to materials that combine various polymers, whether they are naturally occurring or synthetic, and may also incorporate inorganic components. Additionally, there are biomaterials that combine polymers with biological factors to achieve intelligent functionalities. These smart composite biomaterials are designed to be used in a wide range of target tissues, including hard tissues like bones and teeth, as well as soft tissues that require specific composite formulations [14] (Figure 3.1).

This flowchart offers a structured classification of biomaterials widely employed in the medical and biotechnological sectors. It provides a thorough understanding of the two primary categories of biomaterials: natural biomaterials derived from plants, animals, or microorganisms, and synthetic biomaterials created through artificial means. The flowchart emphasizes the broad spectrum of materials within each category.

1.3 Lignocellulose as a Biomaterial

The term "lignocellulosic materials" refers to the organic matter extracted from trees, agricultural crops, and shrubs. These materials are a key source of raw materials for

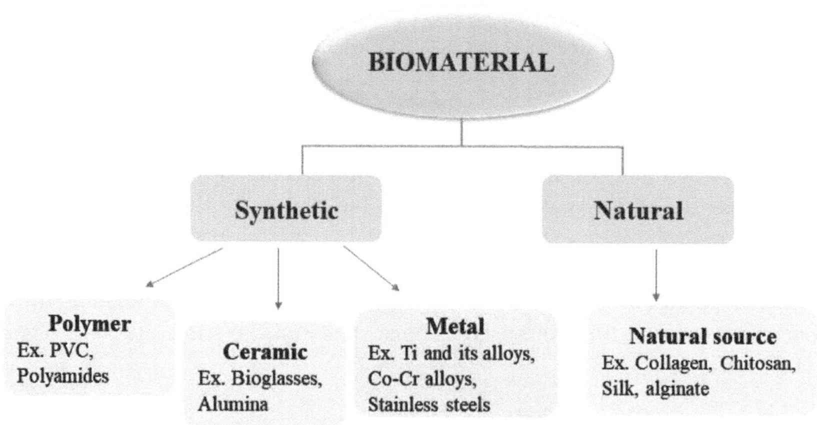

FIGURE 3.1 Types of biomaterials.

the pulp and paper business, the composites industry, and packaging materials. Plant species have the greatest influence on the composition of lignocellulose. Generally, lignocellulose contains 29–47 wt.% cellulose, 25–35 wt.% hemicelluloses, and 16–31 wt.% lignin. The strength and stiffness of lignocellulose materials are sufficient; however, due to their fibrous structure, they are difficult to utilize in load-bearing applications on their own.

1.4 BENEFITS OF LIGNOCELLULOSE AS SUSTAINABLE BIOMATERIALS

The global consumption of fossil fuels and their derivatives, including synthetic plastics, is increasing, resulting in considerable amounts of waste that has a detrimental effect on the natural environment. Plant biomass has the potential to replace fossil fuels and their associated by-products. Plant biomass comes from a variety of sources, including agro-industrial waste, natural fibers, and forestry biological products and crops. Lignocellulosic materials are a great way to help reduce your carbon footprint because they're renewable. Utilizing lignocelluloses as a biomaterial can help you battle climate change by lowering your greenhouse gas emissions. Lignocellulosic biomass can be converted into various valuable products, including biofuels, bioplastics, biochemicals, and biomaterials. It serves as a versatile feedstock for numerous industries, promoting the development of a sustainable and bio-based economy.

1.5 SOURCES OF LIGNOCELLULOSE

Lignocellulose is a renewable resource that is abundant and widely available. It can be found in a variety of sources, including agricultural residues, forestry residues, and waste materials. Agricultural residue is the residual plant material that remains after the harvesting of a crop. Common examples of agricultural residue include corn stover residue, wheat straw residue, rice straw residue, and sugar cane bagasse residue. Forestry residues are plant material left over after a tree is cut down. Sawdust, bark, and chips are examples of forestry residues. Lignocelluloses can be found in different types of waste materials, including paper sludge, city solid waste, and food waste [15].

Agricultural residues are by far the most widespread source of lignocellulose in the world, while forestry residues are also an important source, although they are more prevalent in certain regions. The extent to which lignocellulose is available in a particular area will depend on the location of the source. Energy crops are a newer source of lignocellulose; they are plants that are grown specifically in order to generate energy which include switchgrass, miscanthus, and sweet sorghum. They are becoming more popular as a way to produce renewable energy [16]. Waste is a potential raw material for lignocellulose. However, it is often contaminated by other raw materials, making it more challenging to process. Lignocellulose is a valuable resource that can be used to produce a variety of products, including biofuels, bioproducts, and chemicals [17]. The development of new technologies to process lignocellulose is ongoing, and it is likely future developments will see a rise in the utilization of lignocellulose as a feedstock.

2 PHARMACEUTICAL APPLICATIONS OF LIGNOCELLULOSE BIOMATERIALS

2.1 3D PRINTING

Exploring the potential of lignocellulosic materials for 3D bio manufacturing has gained popularity in recent years. Lignocellulose biopolymers (LFBs) and their derivatives are a vast, largely undiscovered source of materials that offer tremendous opportunities for light-based printing [18]. Biopolymers that are naturally derived as 3D printable feedstocks, such as cellulose, hemicellulose, and lignin monomers, starch monomers, alginates and chitosan, and their derivatives, not only fulfill the need for sustainability but also greatly cut down on the possibility of adverse impacts linked with certain synthetic polymers used in biomedical applications, such as degradable, recyclable, toxic breakdown products, release of additives, and reduced cell adhesion [19]. Medical and pharmaceutical applications can benefit greatly from 3D printing because it can be used to create custom-made, patient-specific medical items, tools, and drugs. These 3D printing technologies would primarily include extrusion-based, particle fused-based, inkjet-based and light-assisted 3D printing technologies [20]. The future and ongoing research in the pharmaceutical products and 3D printing of biomedical products mainly include: (i) manufacturing of organs and tissues; personalization of implants, prosthetics, and anatomically accurate representations; (ii) pharmaceutical research involves creating personalized dosage forms, delivering drugs, testing drugs, and discovering new drugs.

2.2 DRUG DELIVERY SYSTEM

Drug delivery systems play a critical role in medical treatment. A good drug delivery system aids in boosting efficiency and effectiveness of a treatment for a disease so that it has a significant impact on the treatment of a disease. Drug delivery systems based on nanoparticles use a variety of synthetic and natural nanoparticles to efficiently store and transport a variety of drugs. Lignocellulosic nanoparticles are made from lignin, which is a type of natural biomass made up of three main components. These components are 5–30% of lignin, 20–35% hemicellulose, and 35–50% cellulose. There are several kinds of lignin-based nanoparticles, which have been extensively studied as drug delivery systems such as xylan nanoparticles (XNPs), lignin nanoparticles (LNPs), and cellulose nanocrystals (CNCs) [21]. LNPs are used as drug transporters for a variety of drugs, including curcumin and ovalbumin. Other drugs that LNPs can be used as transporters include the following: Resveratrol, Sorafenib, Doxorubicin, Irinotecan XNPs have excellent properties similar to other naturally derived nanoparticles, including biocompatible, non-toxic, non-immunogenetic, anti-tumor properties [22]. XNPs are better suited for colon-specific drug delivery because they can only be broken down by the colon's microflora, which produces multiple enzymes, like β-xylosidase, β-glucuronidase, α-arabinosidase, β-galactosidase, nitro reductase, Az reductase, deaminase, and urea dehydroxylase [23]. Lignocellulosic hydrogels are three-dimensional networks that can absorb and retain large amounts of water. They offer a high degree of swelling and gel strength which make them suitable for use in drug delivery. The use of lignocellulose hydrogels allows for the encapsulation of drugs and the gradual release of the drug, resulting in a sustained release profile for long-term therapeutic effects. Hydrogels are widely used in contact lens production, as they have been found to be softer than hydrophobic contact lenses. Hydrogels are used for a variety of functions, like rebuilding and regenerating cartilage, making artificial organs, creating drug systems, and giving wound dressings the moisture they need to heal wounds [24].

2.3 BIOENGINEERING

The application of lignocellulose in bioengineering encompasses different areas where lignocellulosic biomaterials are utilized for various purposes. Cellulose is a material used in the production of conductive nano-fiber scaffolds for the engineering and repair of nerve tissue. The structure of a network of interconnected cellulose ribbons extending to the nanoscale is capable of reproducing the structure of collagen, which is why it is being employed in tissue engineering [25].

2.4 SINGLE-CELL PROTEIN

The term single-cell protein is used to describe dehydrated microbial cells or the overall protein obtained from fungi, bacteria, algae, and yeast. These microorganisms are cultivated in large-scale fermentation systems and are utilized as protein sources in animal feed or human food, either as supplements or directly incorporated into the diet [26]. Utilizing biodegradable agricultural and industrial waste as a source of vital nutrients for the development of microorganisms is crucial to reduce the cost of single-cell protein (SCP) production. Lignocellulosic biomass is often considered non-food biomass, meaning it does not compete directly with the food supply chain. The use of lignocellulose cellulose in SCP production facilitates the transformation of biomass which would otherwise be disregarded or classified as waste, including crop residue, forest residue, and certain industrial waste. This helps to maximize resource efficiency and minimize waste generation. Lignocellulosic biomass contains a significant amount of sugars that can be readily fermented. It is rich in fiber but has a comparatively low protein content. An example of this type of lignocellulosic biomass is the cladodes of the prickly pear cactus, *Opuntia ficus-indica*. The utilization of cactus pear cladodes for the bioconversion into SCP presents a novel, efficient, and cost-effective approach compared with traditional methods of utilizing this waste raw material. The process of SCP is based on the utilization of a variety of hydrocarbon and nitrogenous compounds, as well as polysaccharides, and agricultural wastes, including hemicellulose and cellulose waste derived from plants, as well as fibrous proteins derived from animal sources, such as horns, feathers, nails, and hair [27].

2.5 BIOMEDICAL IMPLANTS

Lignocellulose is a natural, biodegradable material that has been studied for a variety of pharmaceutical applications, including biomedical implants. Biomedical implants are medical devices that are implanted in the body to replace or repair damaged tissue. Biomedical implants that are derived from lignocelluloses have the potential to provide a range of benefits over conventional biomedical implants, such as biocompatibility, biodegradability, mechanical robustness, and cost-effectiveness [28]. Lignocellulosic biological materials are naturally biocompatible which means that they are well tolerated by living tissue without causing any side effects or toxicity. This property is crucial for implant materials to ensure compatibility with the surrounding biological environment and promote proper tissue integration [29]. Biomedical implants made of lignocellulose can have a variety of mechanical qualities designed into them, making them suitable for a variety of applications. Due to its cost-effectiveness, the use of lignocellulose as a material for biomedical implants may be more cost-effective than the use of traditional biomedical implants. The development of lignocellulose-based biomedical implants is an active area of research, and there is potential for this technology to revolutionize the field of biomedical implants (Figure 3.2).

3 OTHER APPLICATIONS OF LIGNOCELLULOSE

3.1 BIOCHEMICALS

Recently, an extensive variety of appealing and opportunity substances that may update artificial fibers together with lignocellulosic fibers are an increasing number

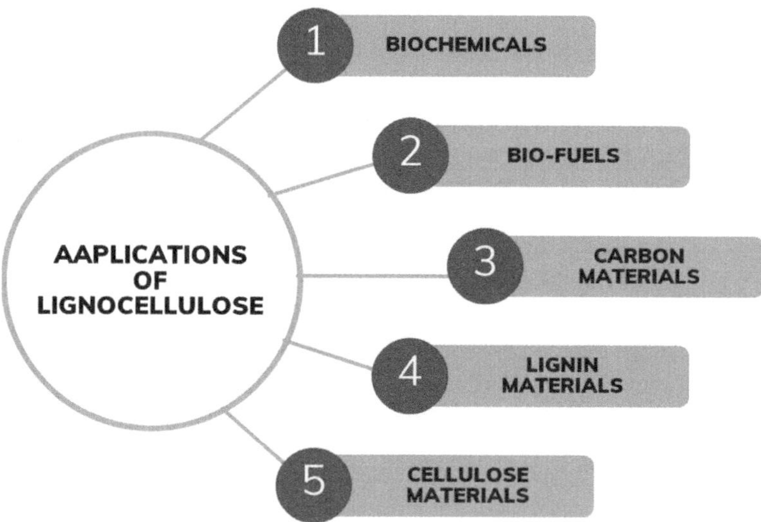

FIGURE 3.2 Flowchart of lignocellulose biomaterial applications. This flowchart offers depiction of the extensive applications of lignocellulose biomaterial. It showcases how lignocellulose is utilized in diverse industries and sectors, including biochemicals, biofuels, carbon materials, lignin materials, and cellulose materials. This demonstrates the versatility and importance of lignocellulose in various contexts, underscoring its wide-ranging significance.

being applied as environmentally pleasant substances that may lessen the full-size dependence on fossil fuels and beautify the surroundings and economic system simultaneously. In contrast to artificial fibers, lignocellulosic fibers are biodegradable, renewable, and extensively available; moreover, they are of low density, aggressive particular mechanical residences, and an enormously low cost [30].

3.2 BIOFUELS

One of the most plentiful natural resources is lignocellulosic biomass, which has attracted significant research and development attention as a promising renewable resource for the generation of various biofuels. Lignocellulose feedstock can be used to make a variety of liquid fuels using either a biochemical or a thermochemical process. The primary source of liquid biofuel for second-generation applications is derived primarily from agricultural and forestry residues, as well as the specialized energy crops [31]. Lignocellulosic biomass is a valuable feedstock for the production of biofuels and bioenergy. It can be enzymatically or chemically converted into bioethanol, biobutanol, and other biofuels through processes such as pretreatment, hydrolysis, and fermentation. If you're looking for a way to cut down on your carbon footprint and reduce your impact on the environment, you should consider using lignocellulosic materials. These materials are made from things like agricultural residues, wood from forests, and renewable energy crops. They're a sustainable, renewable way to reduce your reliance on fossil fuels. Agricultural waste and forest trash, which are examples of lignocellulosic biomass, can be used to create biochar [32].

Biofuels, also referred to as biomass-based fuels, possess a wide range of benefits over petroleum-based fuels: (i) biomass-based biofuels are readily available; (ii) they depict a CO_2-cycle in combustion; (iii) biofuels have a considerable environment. Biofuels have the potential to reduce greenhouse gas emissions, generate clean and sustainable energy, and improve the agricultural income of rural poor people in developing countries. Lignocellulosic materials can produce as much as 442 billion liters of bioethanol annually. Rice straw is one of the world's most abundant lignin-based waste materials that could certainly contribute to bioethanol production. The process of saccharization and fermentation is the primary method of biochemistry for the production of biomass-based bioethanol [33].

3.3 CARBON MATERIALS

Lignocellulosic biomasses derived from agricultural by-products have been identified as a potential raw material for the production of activated carbon, particularly due to their affordability. Lignin-based biomass is the most plentiful renewable carbon source on the planet, second only to cellulose. It's produced around the world at around 40–50 million tons a year. Since lignin has a lot of carbon, it's a good choice for being used as a precursor for making activated carbon. Widespread methods of pollutant removal employ activated carbon made from lignocellulosic biomass. Given that lignocellulose is the most prevalent biologically generated polymer in nature, it may be an excellent target for microbes seeking low-cost sources of carbon and energy. The celluloses must be separated from the lignin before being hydrolyzed with an acid or an enzyme to produce simple monosaccharides, which may then be used to extract the fermentable sugars from lignocellulose. Glycoside hydrolases hydrolyze the 1,4-glycosidic bonds in cellulose, converting it into glucose units through enzymatic techniques. Lignocellulose is being used in the preparation of nanotubules and nanofibers which are rich in carbon. It is possible to extract cellulose nanofibers (CNFs) through the removal of lignin hemicellulose and breaking the hydrogen bonds between cellulose microfibrils. The pseudo-stem of *Musa basjoo* can be used as a raw material to create lignocellulose nanotubes using a straightforward one-step treatment of NaOH [34].

3.4 LIGNIN MATERIALS

Lignin, a type of aromatic polymer, plays a critical role in the exchange of hydrogen and ester bonds between cellulose molecules and hemicellulos polymers. Lignin represents approximately 15–35% of the lignin-like cellulose biomass derived from the operations of the wood and paper industries. The majority of the lignin is used as lignosulfonates that are used as building additives, while the other remaining is mostly used as low-quality fuel or simply disposed of as waste [35]. Historical research has concentrated primarily on hydrothermal, acidic, alkaline, ammonia fiber explosion, organosolv, wet oxidation, and ionic liquid pretreatment methods for extracting lignin and cellulose from various sources. The most popular technique for lignin extraction and degradation involves preconditioning lignocellulose using deep eutectic solvents, which are physicochemical comparable to ionic liquids but overcome the drawbacks of ILs. Lignocellulose can be transformed into nanocellulose, which is useful as a resin reinforcement agent. According to one study, ferulic acid derived from lignocellulose and vegetal oil components like fatty acids and glycerol can be converted into reusable epoxy-amine resins with adjustable thermo-metric properties, water permeability, and corrosion resistance using highly selective lipase-catalyzed trans-esterification. For the manufacturing of plasticizer for polyvinyl alcohol films or starch-based biodegradable thermoplastic compositions, lignocellulosic biomass can be employed as a sustainable platform. In submerged fermentation, lignocellulose can serve as a source of exo-polysaccharide with antioxidant action [36].

3.5 CELLULOSE MATERIAL

Cellulose is one of the most widely used renewable polymer materials out there, and it's seen as an almost limitless source of raw materials for the growing need for eco-friendly and bio-friendly products. Cellulose has been utilized as a source of energy for thousands of years through the use of wood and vegetable fibers in construction, paper, textile, and apparel. The utilization of cellulose in composite materials is a significant area of study, as it can be used to reinforce engineering polymer systems [37]. Lignocellulose-based hydrogels can be prepared using two methods, one by the lignin-containing cellulose nanofibers derived from woody plants' physical or chemical crosslinking and another way the cellulose lignocellulosic hydrogel is created by the lignocellulose's dissolution process. Lignocellulose is a raw fiber source for animal production that is mycotoxin-free and hygienically perfect. It has been utilized as a high-quality dietary fiber source in animal nutrition for a few years and is made of fresh wood [38].

4 FUTURE PROSPECTS

In recent years, greater focus has been placed on lignocellulose, a promising sustainable biomaterial. It can be transformed into industrial products in a biorefinery and is a sustainable, renewable resource. Despite all this advancement in use of lignocellulose in the field of medicine, it had low impact on the patients. The use of lignocellulosic biomaterials for improvement of human health with minimum adverse effects and biosustainability is still a challenge in front of the researchers [39]. Nevertheless, it is expected, based on the properties of lignocellulose that it will have a great contribution in the field of pharmacy and medicine. Due to its sustainability and numerous uses, lignocellulose has a bright future as a versatile biomaterial that can be converted into biodegradable plastics, biofuels, building materials, and biomedical products, providing eco-friendly alternatives across numerous industries [40]. It is projected to play a significant part in the growth of the bioeconomy and has strong future prospects as a sustainable biomaterial.

5 CONCLUSION

Biomaterials are materials that interact with living biological tissues and have been used for therapeutic and diagnostic purposes since the 1860s. Synthetic polymers, ceramics, and metal alloys replaced natural materials in the early 20th century, resulting in improved performance and enhanced functionality. There are four main types: metallic, ceramic, polymeric, and composites. Lignocellulose is a sustainable biomaterial that reduces greenhouse gas emissions and promotes a sustainable economy. It is a renewable resource found in various industries and can be used to produce biofuels, bioproducts, and chemicals. Its potential for 3D bio fabrication is increasing, offering significant advantages for pharmaceutical and medical applications. The text discusses the use of lignocellulosic-derived nanoparticles, hydrogels, and biomass in various industries. SCP is obtained from microbial cells in large-scale fermentation systems. To reduce the production cost, agro-industrial by-products and waste which is biodegradable are used as a source of nutrients. Lignocellulosic biomass, such as the cladodes of *Opuntia ficus-indica*, is one example of such biomass. Nanoparticle-based drug delivery systems combine nanoparticles from synthetic and natural sources to improve the efficacy and treatment of diseases. Lignocellulosic-derived nanoparticles, such as LNPs and XNPs, are used as drug carriers. These materials offer advantages such as biocompatibility, biodegradability, mechanical strength, and cost-effectiveness. They are well tolerated by living tissues and can be engineered to have a broad range of mechanical properties. They can also be converted into biofuels and bioenergy. Lignin is a polymer used in wood and paper processing industries. Cellulose is a renewable resource used for building materials, textiles, and clothing. Lignocellulose plays a crucial role in medicine and pharmacy as a sustainable biomaterial due to its biocompatibility and biodegradability. It is utilized in tissue engineering for regenerating damaged tissues, drug delivery systems for controlled release, and eco-friendly wound care products. Being derived from renewable sources, lignocellulose reduces the environmental impact of medical and pharmaceutical applications, aligning with sustainability goals in healthcare. Its potential for innovative drug formulations and personalized medicine further underscores its relevance in these fields. This shows the sustainability and abundance potential of lignocellulose as a biomaterial.

REFERENCES

1. Amato, Stephen F., and Robert M. Ezzell (Eds.). *Regulatory affairs for biomaterials and medical devices.* Elsevier, 2014.
2. Festas, A. J., A. Ramos, and J. P. Davim. "Medical devices biomaterials–A review." *Proceedings of the Institution of Mechanical Engineers, Part L: Journal of Materials: Design and Applications* 234, no. 1 (2020): 218–228.
3. Burg, Karen J.L., Scott Porter, and James F. Kellam. "Biomaterial developments for bone tissue engineering." *Biomaterials* 21, no. 23 (2000): 2347–2359.
4. Huebsch, Nathaniel, and David J. Mooney. "Inspiration and application in the evolution of biomaterials." *Nature* 462, no. 7272 (2009): 426–432. https://doi.org/10.1038/nature08601
5. Mitragotri, Samir, and Joerg Lahann. "Physical approaches to biomaterial design." *Nature Materials* 8, no. 1 (2009): 15–23. https://doi.org/10.1038/nmat2344
6. Arjunan, Arun, Ahmad Baroutaji, Ayyappan S. Praveen, John Robinson, and Chang Wang. "Classification of biomaterial functionality." (2020). http://doi.org/10.1016/b978-0-12-815732-9.00027-9
7. Niinomi, Mitsuo. "Metallic biomaterials." *Journal of Artificial Organs* 11 (2008): 105–110. https://doi.org/10.1007/s10047-008-0422-7
8. Hanawa, Takao. "In vivo metallic biomaterials and surface modification." *Materials Science and Engineering: A* 267, no. 2 (1999): 260–266. https://doi.org/10.1016/S0921-5093(99)00101-X
9. Park, Joon B., and Young Kon Kim. "Metallic biomaterials." In Joyce Y. Wong and Joseph D. Bronzino (Eds.), *Biomaterials*, pp. 1–21. CRC Press, 2007.
10. Punj, Shivani, Jashandeep Singh, and K. Singh. "Ceramic biomaterials: Properties, state of the art and future prospectives." *Ceramics International* 47, no. 20 (2021): 28059–28074. https://doi.org/10.1016/j.ceramint.2021.06.238
11. Sáenz, Alejandro, E. Rivera, Witold Brostow, and Victor M. Castaño. "Ceramic biomaterials: An introductory overview." *Journal of Materials Education* 21, no. 5/6 (1999): 267–276.
12. Angelova, Nela, and David Hunkeler. "Rationalizing the design of polymeric biomaterials." *Trends in Biotechnology* 17, no. 10 (1999): 409–421. https://doi.org/10.1016/S0167-7799(99)01356-6
13. Dunlop, John W.C., and Peter Fratzl. "Biological Composites." *Annual Review of Materials Research* 40 (2010): 1–24. https://doi.org/10.1146/annurev-matsci-070909-104421
14. Perez, Roman A., Jong-Eun Won, Jonathan C. Knowles, and Hae-Won Kim. "Naturally and synthetic smart composite biomaterials for tissue regeneration." *Advanced Drug Delivery Reviews* 65, no. 4 (2013): 471–496. https://doi.org/10.1016/j.addr.2012.03.009
15. Searle, Stephanie, and Chris Malins. *Availability of cellulosic residues and wastes in the EU.* International Council on Clean Transportation, 2013.
16. Koçar, Günnur, and Nilgün Civaş. "An overview of biofuels from energy crops: Current status and future prospects." *Renewable and Sustainable Energy Reviews* 28 (2013): 900–916. https://doi.org/10.1016/j.rser.2013.08.022

17. Lange, Jean-Paul. "Lignocellulose conversion: An introduction to chemistry, process and economics." *Biofuels, Bioproducts and Biorefining: Innovation for a Sustainable Economy* 1, no. 1 (2007): 39–48. https://doi.org/10.1002/bbb.7

18. Melilli, Giuseppe, Irene Carmagnola, Chiara Tonda-Turo, Fabrizio Pirri, Gianluca Ciardelli, Marco Sangermano, Minna Hakkarainen, and Annalisa Chiappone. "DLP 3D printing meets lignocellulosic biopolymers: Carboxymethyl cellulose inks for 3D biocompatible hydrogels." *Polymers* 12, no. 8 (2020): 1655. https://doi.org/10.3390/polym12081655

19. Shavandi, Amin, Soraya Hosseini, Oseweuba Valentine Okoro, Lei Nie, Farahnaz Eghbali Babadi, and Ferry Melchels. "3D bioprinting of lignocellulosic biomaterials." *Advanced Healthcare Materials* 9, no. 24 (2020): 2001472. https://doi.org/10.1002/adhm.202001472

20. Tai, Chia, Soukaina Bouissil, Enkhtuul Gantumur, Mary Stephanie Carranza, Ayano Yoshii, Shinji Sakai, Guillaume Pierre, Philippe Michaud, and Cédric Delattre. "Use of anionic polysaccharides in the development of 3D bioprinting technology." *Applied Sciences* 9, no. 13 (2019): 2596. https://doi.org/10.3390/app9132596

21. Gudimetla, Avinash, Parveen Kumar, S. Sambhu Prasad, Satish Geeri, and V. V. N. Sarath. "Towards smart materials: Enhancing the efficiency of the materials." In *Modeling, characterization, and processing of smart materials*, pp. 1–30. IGI Global, 2023.

22. Ghorbani, Fereshteh Mohammad, Babak Kaffashi, Parvin Shokrollahi, Ehsan Seyedjafari, and Abdolreza Ardeshirylajimi. "PCL/chitosan/Zn-doped nHAelectrospun nanocomposite scaffold promotes adipose derived stem cells adhesion and proliferation." *Carbohydrate Polymers* 118 (2015): 133–142. https://doi.org/10.1016/j.carbpol.2014.10.071

23. Kumar, Samit, and Yuvraj Singh Negi. "Corn cob xylan-based nanoparticles: Ester prodrug of 5-aminosalicylic acid for possible targeted delivery of drug." *Journal of Pharmaceutical Sciences and Research* 4, no. 12 (2012): 1995.

24. Mankoo, Ramandeep Kaur, Jaswinder Kaur, Isha Dudeja, and Sristhi Kapil. "Recent approaches to the synthesis of hydrogels from lignocellulosic biomass: A review." *Cellulose Chemistry and Technology* 56 (2022): 891–906.

25. Kuzmenko, Volodymyr, Sanna Sämfors, Daniel Hägg, and Paul Gatenholm. "Universal method for protein bioconjugation with nanocellulose scaffolds for increased cell adhesion." *Materials Science and Engineering: C* 33, no. 8 (2013): 4599–4607. https://doi.org/10.1016/j.msec.2013.07.031

26. Bajpai, Pratima. *Single cell protein production from lignocellulosic biomass.* Springer, 2017. https://doi.org/10.1007/978-981-10-5873-8

27. Suman, Gour, Mathur Nupur, Singh Anuradha, and Bhatnagar Pradeep. "Single cell protein production: A review." *International Journal of Current Microbiology and Applied Sciences* 4, no. 9 (2015): 251–262.

28. Tiwari, Himanshu Kumar, Ashish Kumar Srivastava, Parveen Kumar, Manish Kumar Singh, Hritik Kumar, and Akshit Bhadauria. "Materials and Technology for Implant Manufacturing: Challenges and Opportunity." In *Modeling, Characterization, and Processing of Smart Materials*, pp. 83–106. IGI Global, 2023.

29. Kumar, Ajay, Parveen Kumar, Namrata Dogra, and Archana Jaglan. "Application of Incremental Sheet Forming (ISF) Toward Biomedical and Medical Implants." In *Handbook of Flexible and Smart Sheet Forming Techniques: Industry 4.0 Approaches,* 2023, p. 247. Wiley.

30. Tan, Boon Khoon, Yern Chee Ching, Sin Chew Poh, Luqman Chuah Abdullah, and Seng Neon Gan. "A review of natural fiber reinforced poly (vinyl alcohol) based composites: Application and opportunity." *Polymers* 7, no. 11 (2015): 2205–2222. https://doi.org/10.3390/polym7111509

31. Naik, Satya Narayan, Vaibhav V. Goud, Prasant K. Rout, and Ajay K. Dalai. "Production of first and second generation biofuels: A comprehensive review." *Renewable and Sustainable Energy Reviews* 14, no. 2 (2010): 578–597. https://doi.org/10.1016/j.rser.2009.10.003

32. Kumar, Ajay, Ravi Kant Mittal, and Abid Haleem, eds. *Advances in additive manufacturing: Artificial intelligence, nature-inspired, and biomanufacturing.* Elsevier, 2022. https://doi.org/10.1016/C2020-0-03877-6

33. Balat, Mustafa. "Production of bioethanol from lignocellulosic materials via the biochemical pathway: A review." *Energy Conversion and Management* 52, no. 2 (2011): 858–875.

34. Fukugaichi, Satoru, Erna Mayasari, Erni Johan, and Naoto Matsue. "One-step preparation of lignocellulose nanofibers from Musa basjoo pseudo-stem." *Chemical Papers* (2023): 1–9. https://doi.org/10.1007/s11696-023-02724-4

35. Sugiarto, Sigit, Yihao Leow, Chong Li Tan, Guan Wang, and Dan Kai. "How far is Lignin from being a biomedical material." *Bioactive Materials* 8 (2022): 71–94. https://doi.org/10.1016/j.bioactmat.2021.06.023

36. Kumar, Ajay, Parveen Kumar, Ravi Kant Mittal, and Victor Gambhir. "Materials processed by additive manufacturing techniques." In *Advances in additive manufacturing artificial intelligence, nature-inspired, and biomanufacturing*, pp. 217–233. Elsevier, 2023. https://doi.org/10.1016/B978-0-323-91834-3.00014-4

37. Brinchi, Lucia, Franco Cotana, Elena Fortunati, and Jose Maria Kenny. "Production of nanocrystalline cellulose from lignocellulosic biomass: Technology and applications." *Carbohydrate Polymers* 94, no. 1 (2013): 154–169. https://doi.org/10.1016/j.carbpol.2013.01.033

38. Kumar, Ajay, et al. (Eds.). *Modeling, Characterization, and Processing of Smart Materials.* IGI Global, 2023. https://doi.org/10.4018/978-1-6684-9224-6

39. Ajay, Parveen, Sharif Ahmad, Jyotsna Sharma, and Victor Gambhir (Eds.). *Handbook of sustainable materials: Modelling, characterization, and optimization* (1st ed.). CRC Press, 2023. https://doi.org/10.1201/9781003297772

40. Kumar, Ajay, Hari Singh, Parveen Kumar, and Bandar AlMangour (Eds.). *Handbook of smart manufacturing: Forecasting the future of industry 4.0.* CRC Press, 2023. https://doi.org/10.1201/9781003333760

4 Thermo Bi-Metal
A Sustainable Biomaterial for Breathing Façade

Divya Sharma and Sanket Ghanti
KLE Technological University

1   INTRODUCTION

Researchers have undertaken various studies related to the consumption of energy in built-form. Currently, 40% of the total energy is procured by the building sector, 30% of natural resources are exploited by this sector and, parallelly, it exploits 70% of the electricity produced. In developed countries, the construction precinct emits greenhouse gases approximately by 30% [1]. Globally by 2035, the consumption of energy is probably to rise by 50% juxtaposing to the echelon in the past era [2]. The building sector procures approximately part of global steel, glass, brick, and concrete, and yet, their use is envisioned to multiply eventually. As a result of organic evolution in the building sector, human beings are repeatedly implicating the habitat amidst essentially unfavorable bias. To compel Mother Earth with a healthy environment, one of the key role models can be availed with sustainable materials. To attain the objective of decarbonization, productive and sturdy building envelope retrofit resolution is to be implied desperately. Conventional building materials being steel and concrete are yet cast-off in the building sector [16]. After all, depending exclusively on such materials can result in serious environmental threats under their means of making. A couple of such materials need a bulk of energy for production and frequently they incorporate toxic chemicals that influence human health when discharged into Nature. Complications in the domain comprise intense weather, unrivaled global warming due to carbon emissions, and additional toxic discharge with higher consumption of energy which enforces universal evolution and strategies to encounter sustainable objectives. The consequence of this would be a tremendous chunk of carbon emissions. Enhancing the thermal parameters in a built-form can predominantly demote energy consumption from the use of air conditioning, ventilation, well-lit places, and cooling facilities, thus rejuvenating inner occupants comfortably. This is why it was determined to scrutinize the various materials and the possible ways in which we can degrade carbon emissions, and since the traditional concept of using concrete embodies a quantum of energy, architects should contemplate sequential materials and sustainable design strategies where bioinspired principles and smart materials calibrate for zero-energy building. Materials that are pro-earth curtail heat deficit by managing built-forms to remain cool in summer and

DOI: 10.1201/9781003434313-4

warm in winter and that can assist in bringing down energy costs. The bioinspired design explores regional climatic parameters to cut down energy needs. Building envelope employs Nature as an exemplar that intensifies sustainability and accomplishes a regenerative approach [3]. It aims for building design to be well-ventilated with sufficient day lighting, cutting down glare, and improvising the orientation of the building thus enhancing shading. Adapting a building envelope to intensify solar gain will pare with a seasonal stipulation, enhancing natural solar light and cutting down the use of artificial lighting. Special focus should be given to the attributes of built-form which would reduce the consumption of energy [4]. The building sector has a vital hazardous effect on Nature. It is imperative to lessen carbon footprint due to built-forms right out of resource manufacturing until discarding extravagance. Bioinspired strategies will benefit the diminishing of carbon emissions and safeguard Mother Earth's resources. The current chapter scrutinizes the paradigm of breathing skin with thermo bi-metal as an enveloping material in tune with bioinspired principles. Thermo bi-metal is composed of various metals which possess variant thermal expansion coefficients. As sunrays hit the metal skin, the bilayer expands at different speeds allowing the building skin to breathe. Design methodology imitates bio mimicry as a precedent. The morphological and adaptive properties of the *Oxalis oregana* plant were examined which traps solar energy by photoreceptors. The remnant of the chapter is aligned as follows: Once the adapted design methodology is portrayed in Section 2, the author illustrates the importance of adaptive façade in Section 3 and continues with an overview of bioinspired principles in Section 4. Energy efficiency can be gained by retrofitting the building envelope.

Further, different buildings are examined with bioinspired strategies and materials that reduce energy consumption. Bioinspired architecture is a determinate engineering domain, where designer's essence on precise elements to mimic flourishing strategies adverse to the explicative trail are required for innovative realms [5]. Most of the built-forms are adaptable up to a certain extent depending upon their surroundings wherein material and technical systems play a vital role in determining the adaptive parameters in building envelopes. In view of gearing up energy efficiency, smart materials are adapted. Biomimicry principles are considered guidelines to achieve an adaptive façade. The author tries to mimic the physical properties of the *Oxalis oregana* plant which tends to capture solar rays through photoreceptors. The leaves of the plant tend to wrap when the sun's rays hit and resume their original shape in the shade which traps solar energy and intensifies the incident light. The author tries to mimic the morphological and physical behavior of the leaf on the façade with thermo bi-metal as façade material on pseudo skin which can orient itself as per solar angles. The design methodology for the retrofitted façade is mentioned further. Active adaptive façade is retrofitted as a pseudo façade which possesses to capture solar energy, reducing glare and monitoring exterior vision. The morphological property of the plant is mimicked on a retrofitted façade where thermo bi-metal folds in the *x* axis and folds horizontally. Finally, the proposed retrofit prototype is performed with approved thermal modeling software which can rate the thermal performance of the building. The software further works with plugin software which gives clarity of energy analysis, net carbon emission, water efficacy, and daylight analysis giving a comparison of existing glass façade, green building parameters, and proposed retrofitted thermo bi-metal façade, reducing glare and allowing required daylight. It also provides material efficient analysis to procure the smart and efficient material required for a sustainable green building which reduces energy loads increasing daylight and shade on building envelopes. Materials when efficiently used increase the life of prevailing material applied in building and reduce the quantity of material required by replacing them with adequate carbon reduction components. In terms of manufacturing materials, it invokes to curtail the quantity of raw materials implied to fabricate the component, to accomplish slight leftovers for each material thus promoting waste management. Substantial material efficiency can be attained through numerous parameters for instance design for environment, energy, and water efficiency. Along with this, water efficiency parameters are also considered. The latest strategies in design which limit the use of water have become more noteworthy than before and it clinches that designers protect the current source of water which is the need of the hour to avert water crunch subsequently.

2 ADAPTIVE SKIN

Buildings that adapt to their corresponding environment and maintain the inner temperature cool and keep occupants comfortable are termed adaptive architecture and along with this when adaptive sustainable materials are applied to the building envelope, the building is termed an adaptive façade or adaptive skin. Assimilating the distinct design methodology along with biology and its neighborhood can be contemplated as one of the vital strategies for adaptive skin [6]. Adaptive facades constitute a promising approach for reducing the energy and resource consumption of buildings while improving occupant's comfort [15]. Adaptive façade uses the common description according to which it comprises a technology-oriented intensely revamped practice in which the tangible barrier amidst indoor as well as outdoor habitat can alter related function, attributes, and conduct eventually in reflex to provisional execution requisite and confined limitations in addition to upgrading the comprehensive built-form pursuance [7]. Creatures have consistent prototypes to modifications since they emerged, exploiting utmost necessities by rebuilding firm innermost exemptions for survival [7]. Researchers have identified numerous adaptive facades which include smart, kinetic, intelligent, bionic, responsive, or active façade, and many more. Irrespective of the façade working mechanism, every adaptive built envelope is calibrated with a nature-oriented work phase preferably a digital adaptation. Building envelopes should be altered from passive construction techniques to adaptive envelopes which can generate energy for itself and revamp internal comfort constraints. Disparate adaptive design strategies have evolved which would intensify futuristic resolution to the global challenge of energy consumption. Currently, built-forms are substantially worn out from economizing energy. They cannot either alter to reform the context of the habitat or the modified requirements of the occupants. Comprehensive, smart, adaptive, and responsive facades can be evaluated as the succeeding state-of-the-art or advanced breakthrough for building envelopes. Adaptive facades can sustain collaboration with the external surroundings and the inner occupant by responding to their etiquette. The façade wraps when required, bring out energy feasibly and shades as per the stipulation of intrinsic conditions.

The performance of adaptive built-form foremost depends on the criteria of adaptive materials, design methodology, and consequences of adaptation. Climate, external parameters, and inner occupancy also play a vital role in fine-tuning the working mechanism for the building envelope. Since building envelopes have a strong connection between inner space and the external environment, adaptation experiments on an envelope with materials on a building work. The adaptive facades can regulate heat forfeiture and furnish ingress of natural light by constantly modifying to diverse atmospheric circumstances [9]. Adaptive façade works on a concept mechanism that either self-adjusts to building environmental changes or works with a solar screen or panels that expand and contract over time. Irrespective of the type of adaptive façade, as a result, it cools the building by reflecting heat and diffuses sunlight; it reduces solar gain, reduces energy need, and controls daylight eventually reducing carbon emission. The author's concern in this research chapter is to apply an active façade along with bioinspired principles to rule out an energy crisis. Energy efficiency can be gained by retrofitting building envelopes.

Further, the chapter explains bioinspired principles and various types of façade. In addition to this, the author aims at thermo bi-metal properties and mechanisms applied to retrofit the façade. Then the prototype is simulated with appropriate software to examine various analyses. Authors try to apply a responsive façade that truncates the procuring of cosmic heat such that the imbibing of energy in built space results in less depletion by perceptible relaxation of occupants. Due to the adaptive façade and bioinspired design strategies, a tremendous change in glare reduction is observed which reduces the heat load of the building. Numerous analyses helps to understand the reduction of energy consumption in buildings, reducing carbon footprint, and providing a sustainable green zero-energy building.

2.1 BIOINSPIRED PRINCIPLES

Employing processes from Nature's prototype and mimicking them to resolve individual complications can be specified as bioinspired design, bionic, bio mimicry, and biomimetic. These terminologies are off times availed but they are not similar. The most used phrase is a bioinspired design which is similar to biomimetic and gravitates fronting to the process of design. Bio mimicry connotes the ideology and integrative design accession holding Nature as an exemplary to attain the provocation of global challenges from a sustainable perspective. Bionics embodies the vocational restraint which pursues to stimulate, boost, or reconstitute their natural activity by their responsive or adaptive substitute [10]. Bioinspired principles speak about Nature's prototype and mimic these strategies to resolve individual complications. Bioinspired design can be defined as the implementation of the wisdom of biological structure to construe practical complications and flourish high-tech revolution in design processes. It is merely a provocation since it demands skills in both biological fields with technical setups. The anomaly of bioinspired architecture being multidisciplinary stimulates a revolution in divergent extents like medicine, nanotechnology, mechanical agriculture, and many more. Any building can be termed as energy-efficient if it utilizes the least energy compared with a building that is inefficient while performing a common function. To make any building efficient, along with the selection of materials, the construction process, and design strategies should be considered. Bioinspired principles can be one of the approaches at the design stage for passive ventilation, maximum daylight, less glare, and utilizing the captured energy that buildings need, making it energy-efficient. Furthermore, these buildings reduce inner air pollution since they furnish virtuous combustion with superior ventilation compared with regular built-forms. Bioinspired architecture has three levels which comprise the organism level being the first, the behavior level in the middle, and lastly, the ecosystem level [9]. While considering the organism level, the building mimics any particular organism. At this level, the integration with its external habitat is not considered. In the second level, the building mimics the way an organism behaves with its external habitat and interacts with Nature. At the ecosystem level, the building imitates the processes in a larger context. A subsidiary among this is adaptive architecture which encloses the façade or roof. It is the juncture between the built environment and the inner atmosphere. Hence it's perfect chemistry to combine biology and natural processes to accomplish sustainability. Hence practices that are pro-Earth should be followed along with the motive of adaptive façade, passive design principles, and energy efficiency with smart materials. Hence bionic design practices can consequently be avenued adopting various design criteria susceptible to the threat and prevailing scope.

3   SMART BUILDING MATERIALS

As per researchers, it is initiated that the materials which tend to transform their properties in a determined way, at any time are disclosed to a fragment of outward stimulants and are termed smart materials [11]. Urbanization is a dominant complication globally be it in interurban or metropolitan areas. As a result of immense population growth and requisite for occupancy, there is inflation observed in building needs, resulting in carbon dioxide emissions. It's neither energy consumption nor carbon dioxide emissions, but even the inappropriate application of the material in built-form gradually affects the resources and aftermath of which has accounted for environmental changes [19]. A sustainable city seeks to accomplish sustainable buildings with smart materials. Adapting energy-efficient strategies in the proposed built-form can be the initial step toward sustainability, but it is equally essential to ratify and explore sustainable adaptive strategies in actual existing built-forms also. Contiguous to smart materials, design practices should ensure in subduing the consumption of energy in buildings and cutting down carbon dioxide emissions. Smart materials comprise substances that behave rapidly in the habitat where they are positioned. One of its specific features which is mutual among all smart materials is the efficiency to thoroughly modify multiple aspects while performing to achieve an inner restrained habitat [14]. Each one of the materials reciprocates one way or the other to the described incentive and there is an absolutely precise gap that differentiates smart materials from alternate materials [18]. Feasibly, it is practical to determine the adapting conjointly notorious as "smart performance" that transpires when a material undergoes stimulation in distinction to its neighborhood environment and counters in a conducive and sustainable manner. Such overview is observed only when materials experience a sensual notion with their external surroundings.

Distinct subfields of smart materials have developed among which shape-changing materials, self-healing components, and sensors and communication equipment are widely explored. Shape-changing components have been advanced to supersede automatic actuators that interpret climatic variation with zero-energy consumption [20]. In addition to this, natural polymers are also one of the responsive materials that are in trend because of their biodegradable and bioactive properties. All substantiate adaptability regarding varying natural conditions while pledging comfort and a low carbon footprint and thermo bi-metal is one among them.

3.1 Scope of Thermo Bi-Metal on Façade

Redeveloping the façade to attain enriched energy levels in built-form is an ideal methodology in the changeover to a zero-carbon world. To retrofit the façade, thermo bi-metal can be used to function as an automatic self-managing system for built-forms irrespective of sensors. This material has the unique character of expanding and wrapping back with changes in solar temperature. It also helps to reduce overheating since it can change its façade arrangement across the day when exposed to sunlight. Thermo bi-metal is a combination of dual metal (especially steel and copper) with varied thermal expansion coefficients. As solar rays heat the metal façade, both the metals expand in distinct directions and dimensions as well. The greater the distinctness in the material's thermal coefficients, the greater the movement of metals will be. The thermo bi-metal façade comprises various, single-cut, plates and is placed with gaps in between which permits for deviation in expansion, dimension, and curvature. The pseudo metal façade is unitized by a steel substructure for strength. As the material tends to expand and wrap back, it is designated as a breathing façade. Since the material is self-sustaining and automatic, it is primitively put on with a shading mechanism on the façade and allows enough day lighting to inner occupants allowing the inner occupants to be at a comfortable level. The material works efficiently in any weather circumstances and also can be amended for specific climates. Research is going on for sandwiching thermo bi-metal with glazing sheets for the application in window openings which will be the future application of the material. It is widely adapted for adequate ventilation to move hot air at higher levels and exhaust it through gaps. Smart façade that authorizes for thermal regulation where the built-form restrains its climate, thermo bi-metal is the perfect material. Further research is in process which might come up with different uses of the material in the construction sector. As an outcome of this advancement and further analysis, digital sustainable architecture is momentarily transforming into actuality. Adaptive façade can allocate diverse thermal insulation depending on the season, harvest energy, and fraternize in conferring with the movement of the sun, humidity control, and many more so that sustainable parameters can be achieved in the built-form [8]. The eventual objective is to attain a "smart envelope" that can perform better technically when compared with stereotype building pod [12].

In pursuance to escalate the practicality of biopolymers, supplementary multiple structures of inorganic utilitarian and practical elements are popularized by assimilating titanium dioxide, graphene, tin oxide, and nanoparticles along with biopolymers [21]. Progress in green innovation has made it further sensible to employ materials

TABLE 4.1

Built-form with different material and bioinspired design strategies [17]

No.	Built-form	Material	Mimicked by	Methodology	Inference
1	Arab World Institute, Paris	Metallic screen	Human eye	Thermal comfort with natural lighting	Regulates light-enhancing energy conservation
2	One Ocean, Thematic Pavilion, South Korea	Glass fiber-reinforced polymers	Characteristics of fish	The moving mechanism helps in the entry of natural light	Natural ventilation and daylighting
3	Heliotrope, Freiburg	Photovoltaic tracking panels	Leaves absorbing sunlight	To track sunlight	Plus energy, geothermal heat exchanger
4	Las Palmas water theater, Spain	ETFE high-strength polymer	African Beetle	Water condensation to capture vapor	Optimizes fresh water and consigns surplus water to nearby buildings and landscapes
5	Esplanade Theater	Aluminum louvers	Durian fruit	Responsive shading with self-adjustable louvers	Natural light and dramatic effect of shadow
6	Beijing National stadium	ETFE panels	Birds nest	Acoustic insulation and wind protection	Reduces dead load on the roof and acoustic insulation
7	Olympiastadion, Munchen	Acrylic glass and steel fiber polymer	Spider web	Cable structure	Provides resilient strength
8	Eden Project, England	Three layers of ETFE hexagonal frames	Soap bubbles and natural stone	Greenhouse cycle with self-cleaning	Geodesics acts as a thermal blanket and traps air reducing energy consumption

that are in harmony with Nature. In association with this, activators are accustomed to strengthening the capacity of material, consistency, and resilience. Various extrinsic elements also leverage the working of smart materials. These materials play a vital role in detecting structural faults in buildings. Illustration of smart materials applied in smart urbanite involves ethyl tetro fluoro ethylene, fiber optics, metallic screens of polymeric metals, hyperbolic reinforced concrete, electrorheological fluids, and many more. Diverse materials are also explored in building envelopes which are selected as per the need for the inner comfort of occupants. Several facts have found numerous materials with divergent design methodologies resulting in the energy efficacy of a built-form. Following is a study associated with adaptive façade with biomimicry strategies in the design along with smart biomaterials (Table 4.1).

The above study gives a clear notion of adapting sustainable bimetals along with bioinspired design strategies that have immense optimistic speculation possibilities. A responsive façade authorizes buildings to envelop and modify their pattern, configuration, and color compatibly. This envelope allows substantial adaptation swapping with the external neighborhood through determined building components. A responsive façade tends to bring down the consumption of energy and carbon dioxide emissions in Nature, susceptible to the temperature, orientation, and climatic condition of a particular built-form [13]. The extent of daylight penetrated a building bank on the local climate. Building envelopes influence the indoor thermal comfort of the inner occupancy. To facilitate comfortable indoor air quality and temperature in all respects of time, double skin façade or pseudo façade plays a vital role along with sustainable materials. Along with the façade, extrusion for overhangs, fins, and integrated glass with film, coatings on materials with opening patterns also help in enhancing energy saving in the building. Based on the case studies, it is observed that the adaptive envelop, sustainable materials, and bionic design principles restrain the flow of energy and even adapt to a series which suffices to synthesize energy, heat, and ultimately dispense with Nature.

4   DESIGN METHODOLOGY

In this section, we examine the interlacing between biology and adaptive architecture. Further, the author confers the design methodology for the proposed adaptive bioinspired built envelope mimicked by the *Oxalis oregano* leaf (Figure 4.1). Additionally, the morphological and adaptive properties of the leaf are traversed and are imitated on the kinematics and responsive mechanism of the proposed façade. Nature revamps subsequently interchanging, enclosed within its terrain; likewise, the building envelope also can be practicable by mimicking the practice followed by Nature. The integrant mechanism of the leaf is implemented on the proposed façade which acts as a pseudo skin for the envelop and simulation is performed through EDGE software which gives the graphical data related to material for the analysis of energy, net carbon emission, and water efficiency with base case and improved cases. The radical intention of choosing the *Oxalis* leaf is the rapid light response property the leaf possesses. It photosynthesizes at a comparably bottom-most level of diffusive sunlight.

The leaves fold downward when direct sunlight strikes on them and the leaves tend to reopen during shade. This morphological movement of leaves to wrap themselves and open again occurs in smaller duration but is observed by the human naked eye. The proposed module has been derived from the leaf shape at an organism level which is the intersection of three inclined lines of the leaflets (Figure 4.2). The proposed module of the pseudo façade is a rectangular-shaped shading device (Figure 4.3). The kinetic movement of the proposed pseudo façade emerges from the functioning of the leaf to capture sunlight through a photoreceptor in a couple of minutes and orient its form depending on the sunlight intensity at the behavior level

FIGURE 4.1 *Oxalis oregano* leaf which tends to track sunlight.

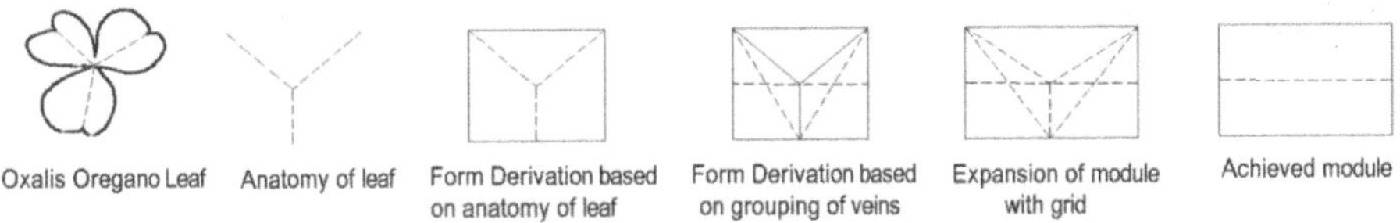

FIGURE 4.2 Derivation of module based on the anatomy of leaf structure.

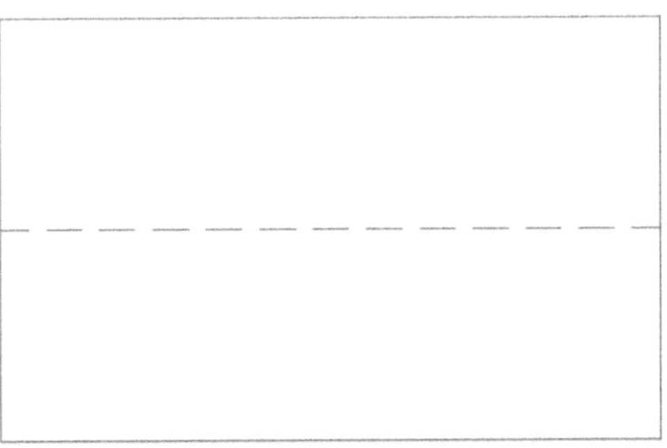

FIGURE 4.3 Rectangular module derived from the anatomy of leaf structure.

of bioinspired architecture. Depending on the movement of the sun, the module on the façade partially folds (Figure 4.4) and wraps completely in a horizontal direction (Figure 4.5).

As the leaf of the plant tends to determine light and alienate itself bilaterally, the shading device of the pseudo façade tries to imitate this property and orient itself

 Advances in Sustainable Biomaterials

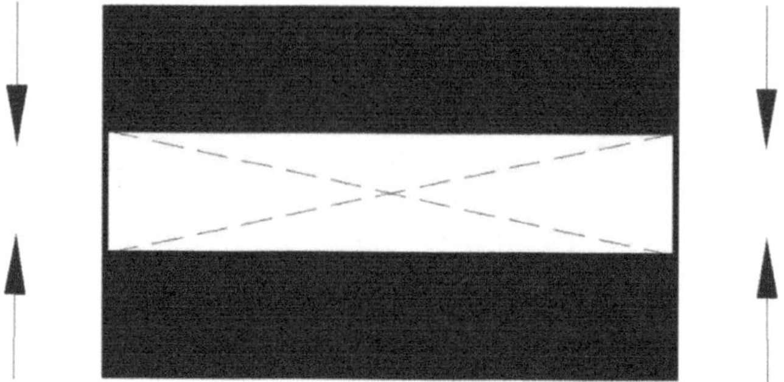

FIGURE 4.4 Derived rectangular module partially open to capture sunlight and allow daylight.

FIGURE 4.5 Derived rectangular module completely closed in X axis to allow daylight and shade.

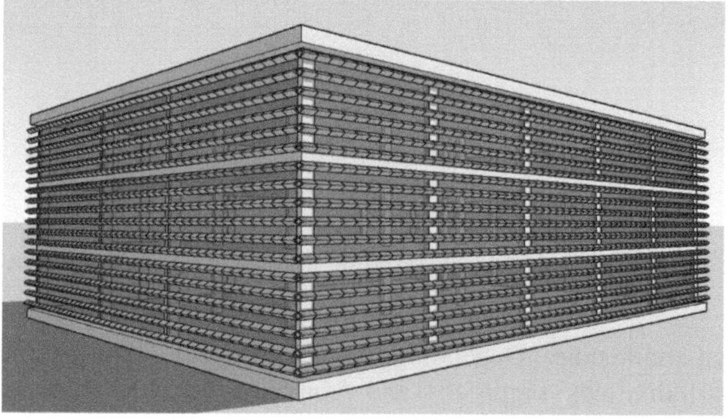

FIGURE 4.6 View of the retrofitted façade with module closed to allow daylight.

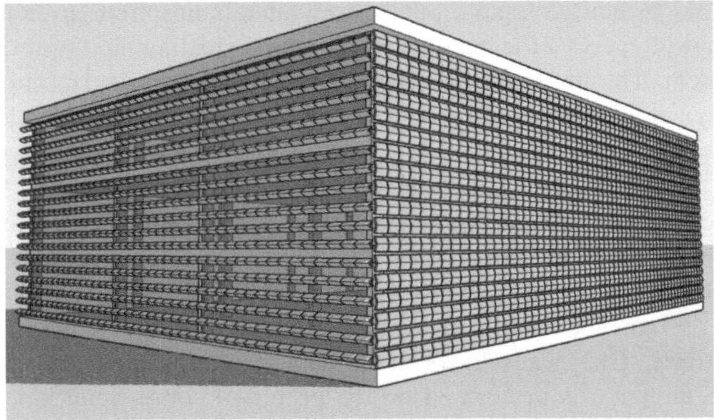

FIGURE 4.7 View of the retrofitted pseudo façade with a few modules partially enclosed.

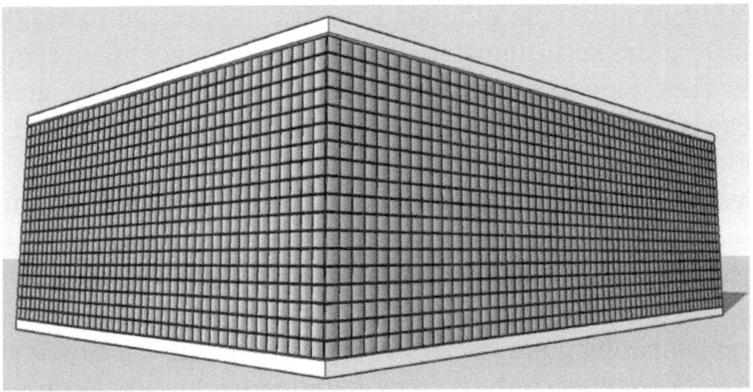

FIGURE 4.8 View of the retrofitted pseudo façade with modules open to capture sunlight.

depending upon the movement and intensity of the sun. When the façade experiences shade, the proposed retrofitted façade tends to close itself and allow daylight cutting down glare and making its inner occupants comfortable (Figure 4.6). As the change in solar intensity is experienced, the module tends to wrap itself partially (Figure 4.7) and the module opens completely (Figure 4.8) when it experiences the harsh sun's rays and tends to capture solar energy and store it by means of a grid network of sensors and actuators.

4.1 FUNCTIONING OF THE RETROFITTED ADAPTIVE FAÇADE

The radical component of the pseudo façade is a rectangular-shaped shading device of thermo bi-metal. The proposed rectangular module move horizontally on the x axis by the aluminum grid frame which is further bridged to the façade by means of guide rails controlled by a technical mechanism. It is further connected to sensors that tend to analyze the disparity of a peripheral stimulant that channels messages to the actuator. The proposed module uses shape memory polymers as sensors and experiences horizontal wrapping. Such sustainable bioinspired envelop offers multiple leverages of energy efficiency, daylight factor, shading, and water efficiency to a certain extent. The proposed envelope as integrated with thermo bi-metal shape memory polymer and photovoltaic on external façade reduces cooling loads.

A hypothetical commercial block is scrutinized for the simulation process. The commercial block is basement ground plus two structures at Nagpur, Maharashtra. Nagpur experiences a tropical climate with less rainfall in winter and hot summers. Since it is situated nearer to the equator, summers are severely troublesome. The proposed prototype is a commercial block with an internal area of 1980 sqm and floor height being 3.5 m. The south and west façade are exposed to sunlight by 30% and 22% respectively. The façade looks monotonous; built envelope experiences harsh solar rays. Simulation is performed on the proposed commercial block with glass as cladding material using EDGE software. EDGE stands for Excellence in Design for Greater Efficiencies by International Finance Corporation (IFC). It is a free open-source software and also an international green building certification system. The IFC is a member of the World Bank Group that focuses on private sector investment and development in emerging markets. IFC is involved in various initiatives and programs to promote sustainable development, climate finance, and innovative technologies. They support projects related to renewable energy, green buildings, financial inclusion, and other areas of sustainable investment. It is a green building certification software aiming to make built-forms more energy-efficient. EDGE empowers designers to analyze cost-effective design strategies to minimize energy and water use, and exemplified energy in materials. The applied strategies integrated into the project design are authenticated by experts in their respective fields. Firstly simulation is performed on the conventional commercial block with glass as the base case and on green building parameters as an improved case. The second simulation is performed with a comparison of green building parameters as an improved case and thermo bi-metal as an improved case 1. The simulation includes the base case as a conventional building with glass as façade material without using any green building strategies and the improved case using different green building strategies and improved case 1 by the application of thermo bi-metal to achieve net zero-energy consumption. This application gives overall data in regard to three components: energy, water, and material efficiency. It assesses the above-mentioned components consumed during construction, operation, and maintenance. While performing the simulation, numerous strategies like climate data, wind speed, wall window ratio, solar reflectance index, the orientation of built-form, exterior wall reflectance index, relative humidity, external shading devices, and roof insulation are considered to calculate the analysis on built-form to calculate the energy consumption and carbon emission. The above-mentioned strategies are considered in all three cases while performing the simulation. Simulation is performed through EDGE software for the base case and improved case for energy efficiency (Figure 4.9), net carbon emissions (Figure 4.10), water efficiency measures (Figure 4.11), and material efficiency analysis (Figure 4.12). Further simulation for the second instance is performed on the improved case with green building material and improved case 1 with thermo bi-metal on energy efficiency (Figure 4.13), net carbon emissions (Figure 4.14), water efficiency measures (Figure 4.15), and material efficiency analysis (Figure 4.16).

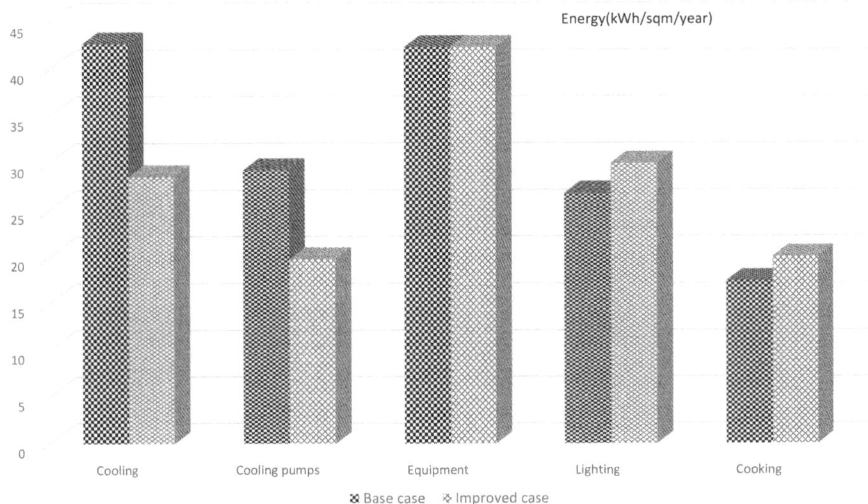

FIGURE 4.9 Energy efficiency measures for a base case with glass as a façade material.

The dual simulation clarifies the comparison with respect to energy, material, and water efficiency which finally concludes with the efficient design among the materials explored in the cases.

5 NUMERICAL RESULTS

In this section, we review in detail the correlative analysis of the working of the proposed built envelope before and after applying the pseudo façade. Energy analysis, net carbon emission, and water efficiency simulation are executed on the hypothetical prototype with glass as base material, green building strategies as the improved case material, and thermo bi-metal as pseudo material on the retrofitted façade for improved case 1. Various parameters are considered while performing the simulation in which climate data, relative humidity, wind speed, solar angles thermal performance, and orientation of the building play a dynamic function in analyzing the energy efficiency and material efficacy [21]. These parameters are the key factors to embed as data in EDGE software to perform simulation. Through this software, simulation is performed on the hypothetical commercial block with glass cladding as material and then retrofitted with a pseudo façade and thermo bi-metal being the material. Table 4.2 gives a comparative analysis of the cases mentioned.

The increase in energy consumption due to changes in the external fenestration indicates the impact of fenestration on the internal heat gain due to sunlight falling on different facades, mainly the west and southern façade. As the bilayer expands at different speeds, it allows the façade to capture and store energy for its further need and improves the thermal performance of the building. It also helps to monitor vision and allows required sufficient daylight to cut down glare. Due to heat gain, the load on the district cooling systems is increased and hence increases in energy and water consumption to maintain thermal comfort making the building self-breathable. Another

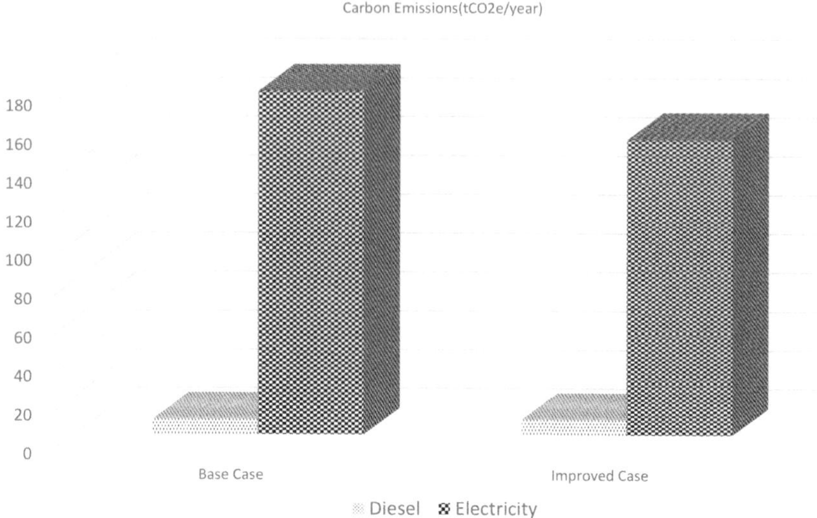

FIGURE 4.10   Net carbon emissions analysis for a base case with glass as a façade material.

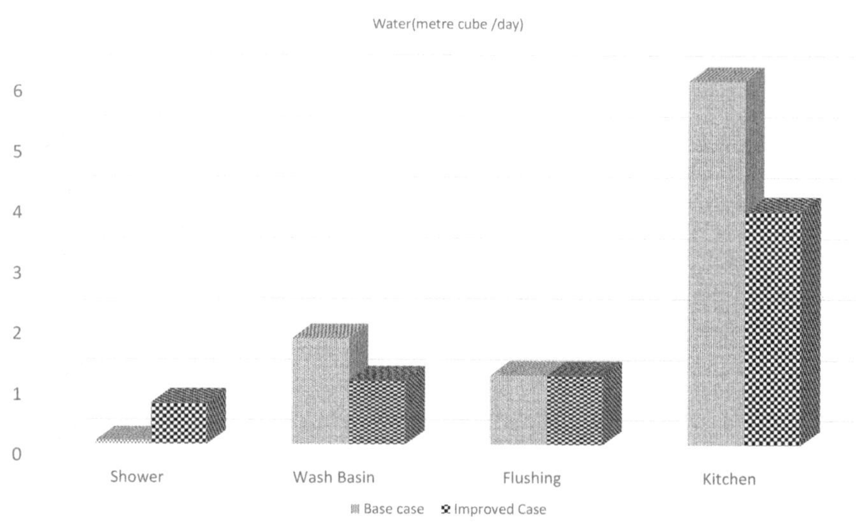

FIGURE 4.11   Water efficiency measures for a base case with glass as a façade material.

factor is due to proper utilization of daylight less energy is consumed resulting in the built-form being a self-manageable building. Energy efficiency on the retrofitted façade solely quantifies to dodge energy stipulation in zero-emission building, with the correlated compute of detectable swap in electrification, material efficiency, and digital responsive technology. Energy-efficient strategies moderate the rise in energy requirements and enact a crucial aspect to narrow down the utilization of fossil fuel and carbon emissions to Mother Earth. This transition is added on with integrated

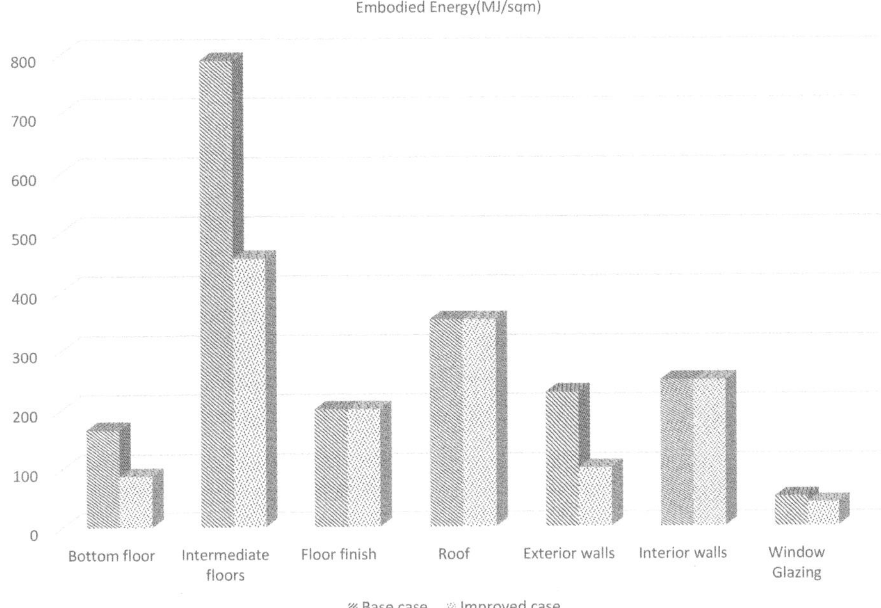

FIGURE 4.12 Material efficiency measures for a base case with glass as a façade material.

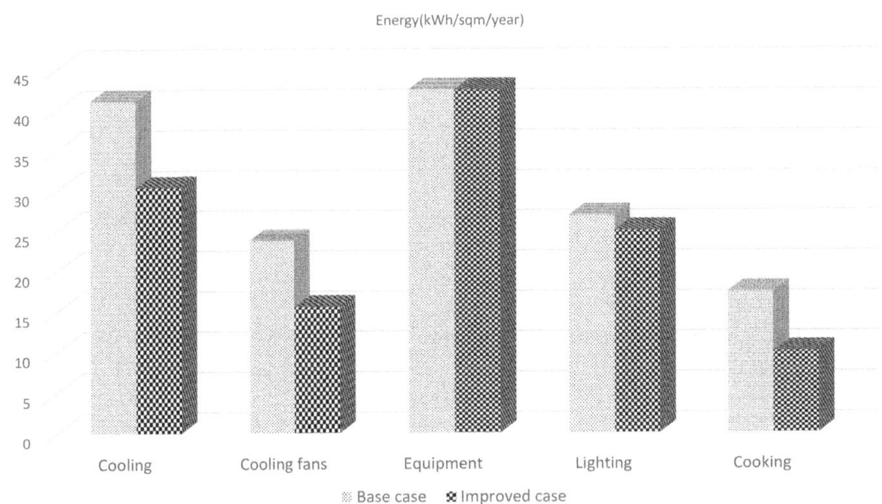

FIGURE 4.13 Energy efficiency measures for improved case with thermo bi-metal as a façade material.

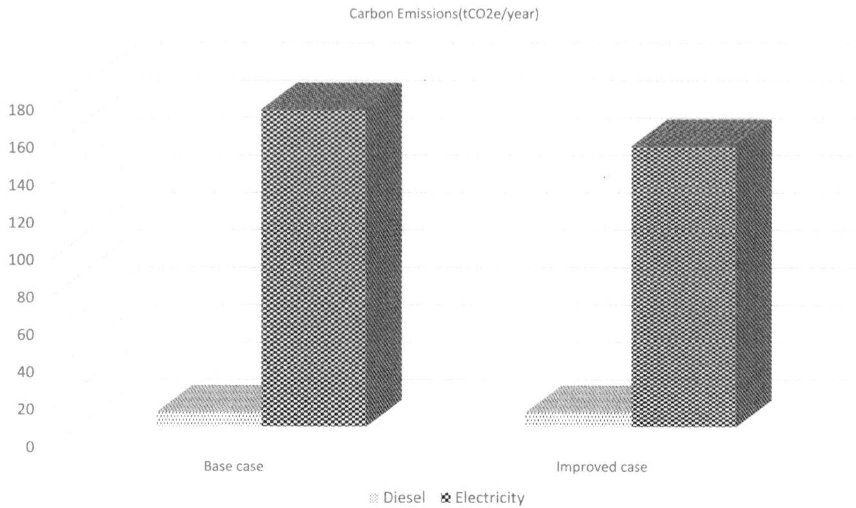

FIGURE 4.14 Net carbon emissions for improved case with thermo bi-metal as a façade material.

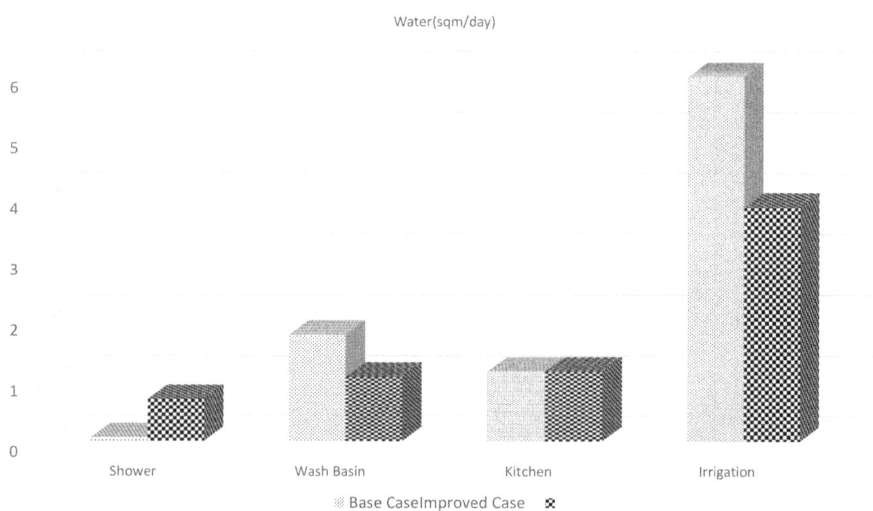

FIGURE 4.15 Water efficiency analysis for improved case with thermo bi-metal as a façade material.

grid technologies, actuators, and sensors to consume and save energy which helps to attain decarbonization. Further, the simulation on material efficacy gives clarification in choosing eco-friendly materials which reduce energy loads and carbon emissions. Genuinely material and energy efficiency together bond in perfect chemistry and assist in reducing manufacturing costs but also shrink carbon emissions and wastes.

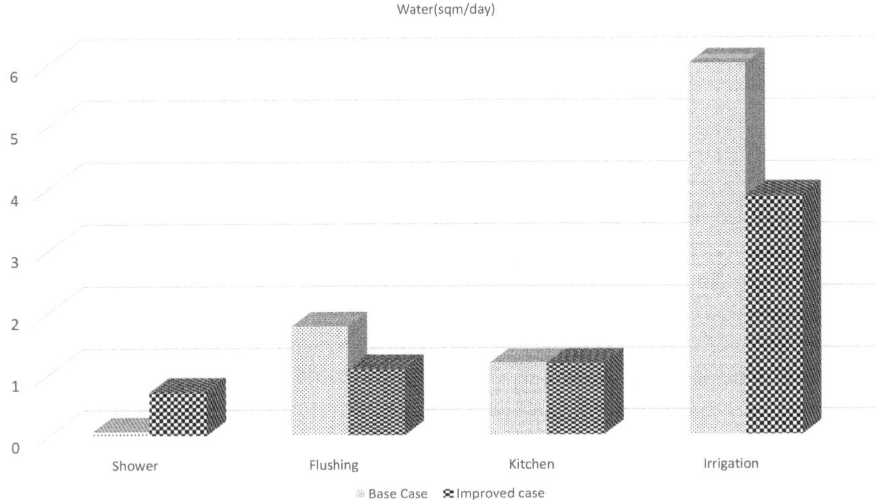

FIGURE 4.16 Materials efficiency analysis for improved case with thermo bi-metal as a façade material.

TABLE 4.2

Comparative analysis of simulation performed on a base case, improved case, and Improved case 1 through EDGE software

Strategies	Conventional building	Green building	Green building with thermo bi-metal	Inference
	Base case	Improved case	Improved case 1	
Net Carbon emissions. (tCO2e/year)	185.7	162.3	161	Reduced carbon emissions by 1.3 tCO2e/year
EPI (kWh/m²/year)	164	143	141	Reduced EPI of 2 kWh/m²/year
District cooling demand (kWh/month)	27,860.18	20,209.07	19,504.29	Reduced cooling demand of 704.78 kWh/month
Embodied energy savings (GJ)	154.7	987.05	1,136.2	Increased embodied energy savings 149.15GJ
Energy savings (MWh/year)	0	42.97	45.14	Increased energy savings of 2.17 MWh/year

6   CONCLUSION

The current research emphasizes retrofitting of a bioinspired adaptive façade with thermo bi-metal on built envelop with base material or structural glazing. The author aims to explore with responsive adaptive façade which curtails heat gain and increases required daylight reducing glare and carbon emission. Simulation executed on built envelop prescribes that the building's significant energy load can be demoted by 45% annually by retrofitting the façade. Based on the simulation graph, net carbon emissions performed on the south and west façade for weekdays and whole day duration show data where carbon emissions are reduced by 1.3 te/year for thermo bi-metal whereas when compared with base material, the carbon emissions are high by 42%. The energy consumption is reduced by 2 kWh/m²/year for retrofitted thermo bi-metal façade. Cooling demand is reduced by 704.78 kWh/month for pseudo façade and increased embodied energy savings 149.15 GJ and 2.17 MWh/year for retrofitted façade. This drop in the building after retrofying the façade can assist us to downscale consumption of energy by the application of biomaterial. Adding to this, material efficiency lessens the effect kindred to material consumption. All the mentioned strategies lay down a standard criterion for the selection of materials, openings, and orientation of the built-form minimizing carbon footprint, allocating a sustainable green design expedient for buildings with glazed façade preferably in a hotter climatic zone. As erecting an interlaced photovoltaic grid is blended with a bioinspired adaptive façade, the superstructure is accomplished to generate its electrical needs along with energy consumption. Numerous responsive adaptive facades can be explored eventually to attain an eco-friendly sustainable built-form with the aim of zero-energy consumption with subordinate loss of raw materials excelling ascent in view of the potential application of biomaterial in the field of construction.

ACKNOWLEDGMENTS

The authors acknowledge Ar. Priyanka for her sincere efforts in contributing an excellent retrofitted view of the hypothetical built-form.

REFERENCES

1. Dixit, Manish K., Jose L. Fernádez-Solís, Sarel Lavy, and Charles H. Culp: "Need for an Embodied Energy Measurement Protocol for Buildings: A Review Paper." Volume 16, Issue 6, August 2012, pp. 4–25 (https://doi.org/10.1016/j.rser.2012.03.021).
2. Zhang, Xi, Hao Zhang, Yuyan Wang, and Xuepeng Shi: "Adaptive Facades: Review of Designs, Performance Evaluation and Control Systems." Volume 12, Issue 12, December 2022, pp. 25–38 (https://doi.org/10.3390/buildings12122112).
3. MDPI: Bioinspired Architecture and Climatisation, pp. 20–43 (https://doi..org/10.3390/biomimetics10070114).
4. Sheikh, Wajiha Tariq, and Quratulian Asghar: "Frontiers of Architectural Research: Adaptive Biomimetic Facades: Enhancing Energy Efficiency of Highly Glazed Buildings." Volume 8, Issue 3, September 2019, pp. 1–25.
5. Sedky, Rania Raouf: "The Interrelationship between Kinetic Architecture and Human Bodily Movement: A Kinematic Experimental Analysis of Adaptive and Responsive Structural Systems." Volume 11, Issue 2, 2021, pp. 15–30 (doi: 10.5923/j.arch.20211102.02).

6. Dugue, Antoine, Tingting Vogt Wu, Fabienne Aujard, and Denis Bruneau MDPI: "An Adaptive Building Skin Concept Resulting from a New Bio Inspiration Process: Design, Prototyping, and Characterization." Volume 15, Issue 3, 2022, pp. 20–25 (https://doi.org/10.3390/en15030891).

7. Sheikh, Wajiha Tariq, and Quratulain Asghar: "Adaptive Biomimetic Façades: Enhancing Energy Efficiency of Highly Glazed Buildings." Volume 8, Issue 3, September 2019, pp. 20–34.

8. Zari, Maibritt Pedersen: "Biomimetic Approaches to Architectural Design for Increased Sustainability." Sustainable Building Conference Auckland, 2007.

9. Mukherjee, Abhilash, Deepmala, Prateek Srivastava and Jasminder Kaur Sandhu: "Applications of Smart Materials in Civil Engineering : A review." 2021 (https://doi.org/10.1016/j.matpr.2021.03.304).

10. Soliman, Samar Awad Abdelhamed, Ibrahim Elsayed Maarouf Ibrahim, and Mai Mohamed Abdo Ibrahim: "Smart Materials and Adaptive Building Envelopes as an Approach for Reducing Energy Consumption in Egypt: A Literature Review." Volume 260, 2021 (https://doi.org/10.2495/SC220021).

11. Loonen, Roel C.G.M., Jose Miguel Rico-Martinez, Fabio Favoino, Brezeicki Marcin, Christophe Ménézo, Giuseppe La Ferla, and Laura Aelenei: "Façade Design Adaptability-Towards a Unified and Systematic Characterization." Proc. 10th Energy Forum-Adv. Building Skin, Bern, Switzerland: October 2015, pp. 1274–1284.

12. Erratum Regarding Missing Declaration of Competing Interest Statements in Previously Published Articles-Frontiers of Architectural Research, 2021.

13. Voigt, Michael P., Daniel Roth and Hansgeorg Binz: "Challenges with Adaptive Facades – A Life Cycle Perspective." University of Stuttgart Germany. October 2021.

14. Rana El-Dabaa, Sherif Abdelmohsen: "Hygroscopy and Adaptive Architectural Facades: An Overview." 2022 (https://doi.org/10.1007/s00226-023-01464-8).

15. Ahmed, Noha, Mohamed Adbel-Hamid, Mahmoud M. Abdl El-Razik, and Karim M. El-Dash: "Impact of Sustainable Design in the Construction Sector on Climate Change." *Ain Shams Engineering Journal.* Volume 12, Issue 2, June 2021, p. 4 (https://doi.org/10.1016/j.asej.2020.11.002).

16. Jamei, Elmira, and Zora Vrcelj: "Bio Mimicry and the Built Environment, Learning from Nature's Solutions." *Applied Sciences.* Volume 11, Issue 16, 2021 (https://doi.org/10.3390/app11167514).

17. Ajay, Parveen, Sharif Ahmad, Jyotsna Sharma, and Victor Gambhir. (Eds.): *Handbook of Sustainable Materials: Modelling, Characterization, and Optimization* (1st ed.). CRC Press, 2023 (https://doi.org/10.1201/9781003297772).

18. Kumar, Ajay: *Modeling, Characterization, and Processing of Smart Materials.* IGI Global, 2023.

19. Gudimetla, Avinash, Parveen Kumar, S. Sambhu Prasad, Satish Geeri, and V. V. N. Sarath: "Towards Smart Materials: Enhancing the Efficiency of the Materials." In *Modeling, Characterization, and Processing of Smart Materials*, pp. 1–30. IGI Global, 2023.

20. Kumar, Ajay, Parveen Kumar, Ravi Kant Mittal, and Victor Gambhir: "Materials Processed by Additive Manufacturing Techniques." In *Advances in Additive Manufacturing Artificial Intelligence, Nature-Inspired, and Biomanufacturing*, pp. 217–233. Elsevier, 2023 (https://doi.org/10.1016/B978-0-323-91834-3.00014-4).

21. Kumar, Ajay, Parveen Kumar, Namrata Dogra, and Archana Jaglan: "Application of Incremental Sheet Forming (ISF) Toward Biomedical and Medical Implants." In Vishal Gulati (Ed.), *Handbook of Flexible and Smart Sheet Forming Techniques: Industry 4.0 Approaches*, p. 247. Wiley, 2023.

5 Lignocellulose-Based Sustainable Biomaterials in the Fabrication of Pharmaceutical Products

Aditya Singh
Integral University

Shubhrat Maheshwari
Rama University and Sam Higginbottom University
of Agriculture, Technology and Sciences

Amita Verma
Sam Higginbottom University of Agriculture,
Technology and Sciences

Syed Misbahul Hasan
Integral University

Bhupendra G. Prajapati
Ganpat University

Sudarshan Singh
Chiang Mai University

1   INTRODUCTION

Lignocellulose derived from plant cells serves as a crucial raw material and poses a significant challenge in terms of converting it into fermentable sugars, particularly when it comes to using renewable resources (Ashokkumar et al. 2022). As a result of the exploitation of forests and agriculture, this material is widely available (Haldar and Purkait 2020). Its composition varies considerably from one location to another, comprising pectin, lignin, hemicellulose, and cellulose, in varying arrangement proportions (Menon and Rao 2012). This material offers the potential to obtain various valuable products including sugars that ferment such as arabinose, galactose, glucose, mannose, and xylose, which can be fermented into biofuel ethanol at different rates (Rastogi and Shrivastava 2017). Additionally, methane production can be achieved through the collaborative action of anaerobic

                              DOI: 10.1201/9781003434313-5

microorganisms (Rastogi and Shrivastava 2017). Numerous physicochemical and structural elements, such as the amount of lignin present, the degree of polymerization, the surface area, the crystallinity of cellulose, and the strength of the connection between cellulose and hemicellulose microfibrils, affect enzymatic hydrolysis (Vasić, Knez, and Leitgeb 2021). Consequently, pretreating the raw material is essential (Vohra et al. 2014). The main goals of this pre-treatment are to break the structure of the crystals of cellulose and the lignin seal, which is an amorphous heteropolymer made of phenylpropanoid units that are dimethoxylated, monomethoxylated, and nonmethoxylated and bound together by different kinds of linkages (Zabed et al. 2016). The key characteristic of interest in the utilization of extremophiles is the utilization of high temperatures either in combination with or without catalysts. It should be noted, On the other hand, it is possible that applying high temperatures (which can occasionally exceed 200°C) in the presence in water will cause lignin to become soluble, produce acetic acid by hydrolyzing the acetyl groups linked to hemicellulose, and produce other acids (Turner, Mamo, and Karlsson 2007). In addition to enhancing the disruptive power of heat, these effects also result in secondary reactions and compounds that have the potential to impede the future activity of microbes and enzymes. Because of its availability, affordability, and potential to reduce greenhouse gas emissions, lignocellulosic biomass, which is formed from plant cell walls, has attracted a lot of interest as a potentially renewable energy source (Hendriks and Zeeman 2009; Mosier et al. 2005). Extensive research has been conducted on the effective conversion of various forms of biomass, which have the potential to be utilized as raw materials for the manufacture of bioethanol. This is due to the growing need for renewable as well as carbon-neutral energy solutions (Klein-Marcuschamer et al. 2012). These biomass sources may be divided into three primary types based on their chemical makeup, which is mostly composed of carbs. The first group consists of raw materials that include sugar, including molasses, whey, sugarcane, sugar beet, and sweet sorghum. The second category is made up of starch-containing (Ibarra-Gonzalez and Rong 2019) feedstocks like corn, wheat, and cassava. Finally, lignocellulosic biomass, which comprises wood wastes, agricultural waste, and straw, makes up the third group (Cheah et al. 2020). However, the use of sugar- and starch-containing feedstocks (known as first generation) for bioethanol production competes with their utilization as food or feed, thus affecting their availability (Sasmal and Mohanty 2018).

2 LIGNOCELLULOSE AS RENEWABLE RAW MATERIALS

The bulk of biomass is lignocellulosic matter, which is composed of lignin, ashes, sugars, and other extractives (LB) (Das et al. 2021). Lignin and carbohydrates remain the two basic components of LB; however, their exact proportion changes according to the species of plant, age, and development stage (da Costa Lopes et al. 2013). In minor levels, cellulose, hemicellulose, and pectin are among the carbohydrates found in LB. LB is mostly composed of cellulose, with 39–54% coming from hardwood species, 41–50% coming from softwood, and 32–47% coming from agricultural leftovers, such as rice and wheat straw. About 20–50%

of LB is composed of hemicellulose, with the remaining 1–5% coming from extractives such as waxes, terpenes, steroids, lipids, and phenolic components (de Paula Protásio et al. 2022). Cellulose is made up of repeating units called β-1,4 glycosidic linkages connecting d-glucose molecules, which polymerize to produce cellulose microfibrils (Heinze, El Seoud, and Koschella 2018). Together, these microfibrils form cellulose's crystalline ultrastructure, which is extremely resistant to enzymatic and chemical breakdown. Furthermore, the presence of lignin and hemicellulose shields cellulose fibers from hydrolytic components, decreasing their accessibility. Hexoses (d-glucose, d-mannose, and d-galactose) and pentoses (d-xylose and l-arabinose) are among the heteropolymers that make up hemicellulose (Mochochoko et al. 2016). It is susceptible to thermochemical treatments, and hemicellulose sugars may decompose into compounds like furfural and acetic acid during pre-treatment or hydrolysis operations. These chemicals have the potential to obstruct further hydrolysis steps or microbial fermentation (Lehrhofer et al. 2022). The third most prevalent component, lignin, is an unstructured heteropolymer that gives materials mechanical stiffness, recalcitrance, and impermeability. Although terpenes and rosin acids are good raw materials for biobased goods, lignocellulose—the world's most abundant source of biomass—has attracted a lot of attention (Chen 2014). Isolated from a range of waste streams, including agriculture, forestry, and paper industry, it has the potential to serve as a sustainable carbon source to progressively replace hydrocarbons (Lange 2007). The cell wall of a plant cell is made up of various layers and different biopolymer compositions and layouts, giving it a complicated structure. Lignocellulose is made up of cellulose, hemicellulose (Fatma et al. 2018), and lignin, which are crosslinked and non-covalently bonded together to form a stiff structure (Østby and Várnai 2023). About 15–20 weight percent of lignocellulose is made up of the complex aromatic macromolecule known as lignin (Østby and Várnai 2023). It has an irregular, non-crystalline three-dimensional structure that is kept together by both carbon–carbon and ether bonds, while typical lignin is primarily linked by the former (Lange 2018). There is still room for additional degradation of the polymer's basic aromatic ring structures, such as the methoxy, phenol, and guaiacyl subunits. The kind of biomass used determines the lignin composition; lignin from hardwood has a lower G unit content than lignin from softwood. By 2035, bioresources might account for 10% of global energy production, according to IAEA projections (Jeenkeawpieam et al. 2012). By 2050, it is possible to replace roughly 27% of petroleum-based petrol with renewable liquid fuels like ethanol. Bioplastics can also be made from renewable chemicals produced by various biorefinery processes, such as lactic acid, succinic acid, and ethanol. Bioplastics enhance environmental benefits and are inherently compliant with green chemistry principles, being biodegradable and eco-friendly. They might be non-biodegradable like biobased polyethylene terephthalate (Bio-PET) and biobased polylactic acid (PLA), or biodegradable like polyhydroxyalkanoates (PHA), polybutylene succinate (PBS), and polylactic acid (PLA). Because bioplastics share traits with synthetic plastics, they have the potential to completely replace polymers derived from fossil fuels on a global scale (Bontaş et al. 2023).

3 PRE-TREATMENT OF LIGNOCELLULOSIC BIOMASS

In the process of converting lignocellulosic biomass, pre-treatment is crucial. In order to prepare lignocellulosic biomass for enzymatic saccharification, it must undergo structural modification. Among the various chemical pre-treatment techniques that have been thoroughly studied is alkali pre-treatment (Rizal et al. 2018). Sodium hydroxide, potassium hydroxide, calcium hydroxide, ammonium hydroxide, and aqueous ammonia are the commonly utilized agents for alkaline pre-treatment. In delignification, alkali pre-treatment is crucial (Honarmandrad, Kucharska, and Gębicki 2022; Pendse, Deshmukh, and Pande 2023), and so are their mechanisms of action, circumstances of alkali pre-treatment, prospects and consequences, and commercialization considerations. Understanding the basic components of any type of biomass is essential to appreciating the structural intricacies and obstinate nature of a typical LCB. Generally speaking, cellulose (35–50%), lignin (5–30%), and hemicellulose (20–35%) are the three primary ingredients of LCB (Devi et al. 2022). Depending on where the biomass feedstock comes from, the percentage mix of the three distinct components can change considerably (Table 5.1) (Brar et al. 2024). The cellulosic fraction is present in the majority of LCBs at the highest proportion. To effectively break down lignocellulosic biomass (LCB) and turn it into bioenergy, pre-treatment is an essential first step. Energy usage and the economics of the pre-treatment process are two factors that affect its industrial feasibility.

It is crucial that the LCB components' natural structures are not materially changed by the pre-treatment procedure (Yan et al. 2022). For instance, delignification procedures may change lignin's natural structure and increase its value. Furthermore, pre-treatment can decrease the crystallinity of the cellulose fraction, which can impact the enzymes' accessibility to it. Cellulose is a polysaccharide consisting of linearly connected glucose monomer units by β-1,4 glycosidic connections, with a complex structure that contains intra- with intermolecularly hydrogen-bound

TABLE 5.1

Lignocellulosic biomass (LCB) composition derived from various sources

LCB Type	Cellulose (%)	Hemicellulose (%)	Lignin (%)	References
Corn stalks	50	20	30	(Christopher et al. 2017; Ma et al. 2020)
Rice husk	37.1	29.4	24.1	(Kalita et al. 2015)
Bamboo	41.8	18	29.3	(Ma et al. 2020)
Sugarcane bagasse	35	35.8	16.1	(Kalita et al. 2015; Sasaki, Adschiri, and Arai 2003)
Rice straw	35.8	21.5	24.4	(Imman et al. 2015)
Chili post-harvest residue	39.9	17.8	25.3	(Sindhu et al. 2017)
Spruce wood	43	29.4	27.6	(Demirbaş 2005)

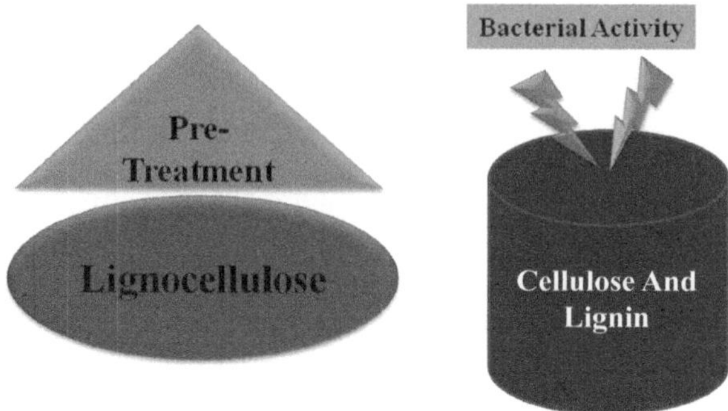

FIGURE 5.1 Pre-treatment of lignocellulosic biomass is a crucial step in the production of biofuels and other value-added products.

hydroxyl groups and a high rate of polymerization (up to ≈10,000 units). (Devi et al. 2022). These attributes add to the complex and uncooperative nature of lignocellulosic biomass (LCB). As opposed to cellulose, hemicellulose, another polysaccharide found in LCB, is hydrophilic, amorphous, and has a lower degree of polymerization (Brar et al. 2024). Hemicellulose is made up of sugars with five and six carbons joined together by β-1,4 glycosidic linkages. It becomes a binder in between the lignin and cellulose fractions, increasing the stiffness of the biomass matrix. Because cellulose and hemicellulose both contain reducing sugars, they are significant sources of chemicals of economic value. Lignocellulosic biomass (LCB) pre-treatment is a crucial initial step in the effective conversion of biomass into biofuels. The breakdown of cellulose, hemicellulose, and lignin—the three main components of LCB—is necessary to enable their subsequent conversion. However, it's critical to consider factors like energy consumption, budget, and preserving the first structure for LCB components during pre-treatment. To get above these obstacles, a variety of pre-treatment (Figure 5.1) methods have been devised, including physicochemical, physical, and chemical ones (Ghimire, Bakke, and Bergland 2021; Rizal et al. 2018).

3.1 PHYSICAL PRE-TREATMENT

These methods aim to reduce the particle size of the biomass to maximize the surface area accessible for enzymatic hydrolysis. One popular physical pre-treatment method is milling, which is reducing the particle size of biomass by grinding, milling, chipping, or shredding it. This procedure aids in mass transfer, enzymatic hydrolysis efficiency, and cellulose crystallinity reduction. But it frequently uses a lot of energy and might not be able to completely eliminate lignin (Gallego-García et al. 2023).

3.2 EXTRUSION PRE-TREATMENT

Breaks down the structure of biomass by using high pressure and temperature produced by extruder screws. Defibrillation, fiber shortening, and chemical structural alteration are the outcomes, which enhance enzyme accessibility and surface area.

To maximize sugar recovery, screw speed, temperature, and moisture content must all be optimized. Extrusion can be used in conjunction with other pre-treatment techniques to increase biomass-derived sugar yields (Zheng and Rehmann 2014).

3.3 MICROWAVE PRE-TREATMENT

By subjecting biomass to non-ionizing electromagnetic radiation, microwave pre-treatment produces quick and even heating. The structural fragmentation of biomass components brought about by this process enhances the effectiveness of enzymatic hydrolysis. The application of microwave-assisted pre-treatment techniques has yielded encouraging outcomes in terms of elevated sugar yields from diverse biomass feedstocks. However, there are certain drawbacks to this approach, including significant capital costs and energy consumption (Krakowska-Sieprawska et al. 2022).

3.4 ULTRASOUND PRE-TREATMENT

By creating shockwaves and acoustic cavitation, ultrasound pre-treatment uses acoustic energy waves to break down the structure of biomass components. By making biomass more accessible to enzymes, this method improves the efficiency of enzymatic hydrolysis. High sugar recovery from various biomass sources has been successfully attained with ultrasound-assisted pre-treatment techniques. For effective pre-treatment, nevertheless, variables including residence duration, reactor architecture, and ultrasonic frequency must be optimized (Pandiselvam et al. 2023).

3.4.1 Hemicellulosic Biomass Undergoes Acid Pre-Treatment

It extracts hemicellulose mostly using phosphoric acid, acetate, or sulfuric acid. Enzymatic hydrolysis, as well as fermentation, are facilitated by the greater accessibility of enzymes to cellulose. Concentrated acidic pre-treatment at temperatures below 50°C along with diluted acid pre-treatment at temperatures above 200°C are two common techniques. Dilution of acid pre-treatment requires more energy due to higher temperatures, but it can produce fermentation inhibitors including furfural and 5-hydroxymethylfurfural which may hinder microbial activity (Rizal et al. 2018).

3.4.2 Alkali Pre-Treatment

By employing hydroxides like potassium, sodium, or calcium, alkali pre-treatment attempts to improve enzyme accessibility to cellulose by dissolving lignin–carbohydrate complexes (LCC) by saponification and solvation processes. Additionally, it eliminates hemicellulose's acetyl and uronic acid groups. NaOH, or sodium hydroxide, is widely utilized because of its efficiency (Pol Segura et al. 2023).

3.4.3 Organosolv Pre-Treatment

In order to improve cellulose accessibility for enzymatic hydrolysis, organosolv pre-treatment uses organic solvents or solvent combinations to solubilize lignin and break lignin–hemicellulose linkages. Common solvents include acetone, ethanol, and methanol. As catalysts, organic acids or bases may also be used in the process. Due to difficulties with solvent recovery, organosolv pre-treatment can be expensive, volatile, and energy-intensive even though it is successful (de Milliano et al. 2017).

3.4.4 Enhanced Lignocellulosic Fractionation with a Co-Solvent

Lignocellulosic fractionation with a co-solvent combines the advantages of organic solvents and water to pre-treat biomass while concurrently delignifying and converting cellulose into fuel precursors for example furfural and levulinic acid. It is economically feasible due to its high lignin removal and glucose yields, particularly when combined with simultaneous saccharification and fermentation procedures. Pre-treatment with AFEX entails treating biomass at moderate pressures and temperatures with anhydrous liquid ammonia in order to destroy lignin and hemicellulose structures while maintaining the integrity of cellulose. Benefits include reduced inhibitor production, brief residence periods, and enhanced enzyme accessibility. However, disadvantages of high-lignin biomass include high equipment costs and limited effectiveness. SCFs provide an effective and environmentally friendly pre-treatment solution for biomass, much like supercritical CO_2 and supercritical water. They efficiently break up lignin and promote enzymatic hydrolysis by penetrating biomass. Large-scale industrial applications have difficulties when using SCF pre-treatment because of its high temperature and pressure requirements, notwithstanding its effectiveness. Hydrogen bond acceptors and donors make up DESs (Table 5.2), which have become popular environmentally acceptable solvents for processing of biomass. They provide a gentle and effective pre-treatment alternative by removing lignin alone while leaving cellulose intact. But there are still issues like high cost and constrained scalability (Jaffur, Jeetah, and Kumar 2021).

TABLE 5.2

Soluble cellulose and lignin in various DESs

DES (HBD:HBA)	Molar ratio (HBD:HBA)	Temp. (°C)	Lignin (wt.%)	Cellulose (wt.%)
LA: Proline	3.3:1	60	9	<1 (Lynam, Kumar, and Wong 2017)
LA: DBU	3:1	90	37.3	Not reported (Liu, Mou, et al. 2019)
LA: ChCl	5:1	60	7.7	0 (Francisco, Van Den Bruinhorst, and Kroon 2012)
Malic acid: Proline	1:3	100	14.9	0.7 (Francisco, Van Den Bruinhorst, and Kroon 2012)
Acetic acid: ChCl	2:1	60	12	<1 (Lynam, Kumar, and Wong 2017)
LA: ATMAC	2:1	70	48.5	Not reported (Liu, Zhao, et al. 2019)
LA: BTMAC	2:1	70	48.6	Not reported (Liu, Zhao, et al. 2019)

3.5 LIGNIN-BASED HYDROGELS

The hydrophilicity of lignin is facilitated by certain functional groups in its structure as a chemical, and the extraction process also modifies its structure, which results in its chemical reactivity (Liu et al. 2022). Lignin is capable of undergoing the standard chemical reactions that result in crosslinked structures, such as hydroxymethylation using formaldehyde especially epoxidation with epichlorohydrin, because of the large number of phenolic along with aliphatic groups of hydroxyl on its side chain. Formaldehyde and lignin combine to produce hydroxymethylation, which adds OH groups. Similar to the manufacture of phenol–formaldehyde resins, lignin-based polymerized networks are generated following reaction with phenol moieties (Komisarz, Majka, and Pielichowski 2022). Formaldehyde is electrophilically substituted with hydroxymethyl groups at the reactive sites in lignin. Because of this, lignin has been used as a whole or partially renewable phenol substitute in phenol–formaldehyde-based resins (Rico-García et al. 2020a). Nevertheless, steric hindrances limit lignin's reactivity with formaldehyde, which frequently results in phase separation for large lignin/phenol ratios. Reactivity is enhanced by chemically altering lignin by methods such as acid hydrolysis, acetosolv, and organosolv. An epoxidation reaction with epichlorohydrin in alkaline circumstances produces lignin-based epoxy resins, providing a bio-replacement for bisphenol-A in epoxy resin synthesis. By mixing lignin with petroleum-based epoxy resin and allowing the lignin to react with the epoxy resin during curing, these resins can be created (Rico-García et al. 2020a). Poly (ethylene) glycol diglycidyl ether (PEGDGE) is one flexible crosslinker agent that may be immediately copolymerized with lignin to produce highly swellable lignin derivatives. To create the crosslinking process, phenolic OH groups in PEGDGE and lignin etherify. These lignin–epoxy networks show dimensional stability, mechanical resilience, and flexibility (de Haro et al. 2021). To create cellulose–lignin hydrogels, Ciolacu et al. made use of lignin's and cellulose's reactivity with epichlorohydrin. These hydrogels were easily created by dissolving lignin and cellulose in an alkaline solution, followed by covalent crosslinking with epichlorohydrin (Bashir et al. 2020). It has been reported that lignin-grafted copolymers within polymerized networks, generating semi-interpenetrating polymeric networks, can be produced by polymerization of free radicals with several monomers, particularly *N*-isopropyl acrylamide (NIPAAm) and acrylamide (AAm). These copolymers have a high water absorption capacity and swelling that is pH-sensitive (Barbero 2023). Radical polymerization can take place with or without the use of external crosslinking agents; in the former scenario, covalent bonds are produced between the crosslinker and polymers, and in the latter, hydrogen bonds are formed between the graft-copolymer and monomers (Rico-García et al. 2020a). NIPAAm, or *N*-isopropyl acrylamide, has been used to polymerize phenolated alkali lignin. The pre-activation of lignin increased reaction sites, affecting the lower critical solution temperature and crosslinking density of hydrogels. This led to the production of lignin hydrogels with high water-absorbing capabilities, using GRANTEZ, a non-lignin-based copolymer (Parvathy et al. 2021). This one-step, solid-state procedure offers an easy-to-use, environmentally friendly way for industrial manufacturing because it doesn't require any additional reagents like initiators or crosslinkers or organic solvents. The esterification process between the carboxylic acids of maleic acid units and the hydroxyl groups of lignin is the basis for the strategy. The qualities of the hydrogels created with organosolv, kraft, and straw soda, three different forms of lignin, to remove cationic dyes in wastewater were studied by Espino-Pérez et al. (2014). The ability of poly(ethylene glycol) to esterify (Yousefiasl et al. 2023) PEG was added to the mixture containing poly(methyl vinyl ether co-maleic acid) to further improve network crosslinking (Rico-García et al. 2020a).

3.6　Lignin-Based Composite/Bio-Composite in the Management of Wound Healing

Alkali lignin and chitosan were used to create friendly physic hydrogels, as described by Ravishankar and colleagues (Rico-García et al. 2020b). The cytotoxicity of these hydrogels was examined in vitro and in vivo, respectively, against mesenchymal stem cells and zebrafish. The hydrogels were safe, according to the findings of the two trials. The scientists also note a good potential relevance for wound healing since these hydrogels show good NIH 3T3 murine fibroblast cell attachment and proliferation. Lin and Dufresne (2014) gave an additional illustration of lignin-based hydrogels that may be used in wound healing applications. Utilizing coniferyl alcohol (DHP) as a model chemical for lignin, hydrogels containing bacterial cellulose were created (Lin and Dufresne 2014). To examine the antibacterial activity of these hydrogels, a variety of microorganisms, including *S. aureus* and *P. aeruginosa*, were evaluated. These gels' antibacterial activity is connected to the release of the antibacterial chemical concentrations were observed to remain constant over 72 hours while using DHP oligomers (Wang, Wang, and Xu 2020). The release of oligomers may be very helpful in the healing of chronic wounds. Other authors also assessed the cytotoxicity of hydrogels based on lignin (Sagadevan and Oh 2023). Recent investigations indicate the utilization of lignin-based hydrogels as well for highly antibacterial medication delivery; the lignin derivatives ratio formed in the material is regarded as non-toxic as it does not significantly impair cell viability. Hydrogels prepared using lignin, Gantrez S-97 (poly[methyl vinyl ether-co-maleic acid]), and poly(ethylene glycol) or glycerol were shown to have a logarithmic reduction in the adhesion of *Staphylococcus aureus* and *Proteus mirabilis*. These hydrogels have swelling ratios between 24% and 40%, which makes them more effective as drug delivery methods. Furthermore, curcumin was discovered for a sustained delivery in pharmacological models for as long as four days. Furthermore, it has been claimed that lignin-based hydrogels can be used to distribute active substances, such as fragrance compounds or insecticides, in different forms than pharmaceuticals. Raschip et al. produced hydrogels based on lignin/xanthan gum for the release of vanillin, an active element in aromas. However, lignin and acrylic acid-based hydrogels have been effectively employed to release insecticides. The following was the order in which the three pesticides were loaded: paraquat > cyhalofop-butyl > cyfluthrin. Cyfluthrin and paraquat both had greater cumulative releases. Hydrogels based on lignin have also shown promise in water treatment applications. Because heavy metal ions in polluted water are very harmful, there is a growing demand for effective remediation materials. This reactivity and swelling properties of lignin-based hydrogels enhance their ability to absorb both inorganic ions and organic pollutants like colors. For example, hydrogels consisting of bentonite/sodium lignosulfonate grafted with maleic anhydride and acrylamide demonstrate a notable ability to absorb lead (II). The removal of heavy metal ions from water has also demonstrated promise for other hydrogels, such as those based on lignin/acrylic acid or lignosulfonate/graphene. Furthermore, lignocatechol-based hydrogels have proven to be selective for specific metal ions, which makes them useful for absorbing metal ions (Vojtová et al. 2021). Water-purification potential of lignin-based hydrogels is demonstrated by their ability to effectively adsorb colors such as methylene blue and malachite green (Camargo et al. 2010) from organic contaminants. These results demonstrate how adaptable and effective lignin-based hydrogels are at solving problems related to water pollution. Additionally, lignin-based materials have been investigated for wearable electronics and flexible supercapacitors (FSC). The rise of portable consumer gadgets has led to a surge in demand for FSC, and lignin offers an exciting renewable source for these kinds of applications. Although cellulose-based FSC has received a lot of attention in research, lignin-based materials are starting to attract interest. Research has indicated that the integration of lignin into FSC electrodes and electrolytes is practicable, leading to the production of FSC that exhibit elevated specific capacitance, energy density, and power density. For instance, FSC with remarkable electrochemical performance has been produced using lignosulfonate/single-walled carbon nanotube hydrogel electrodes in conjunction with cellulose hydrogel electrolyte separators. Similar to this, FSC has proven promising capacitance, energy, along with power qualities employing lignin-based electrolytes as well as electrodes, such as lignin/polyacrylonitrile and lignin crosslinked by poly(ethylene glycol) diglycidyl ether (Niknam et al. 2022).

3.7 LIGNIN-BASED NANOPARTICLES

Despite being widely utilized, cancer chemotherapy has many drawbacks, most notably the inability to distribute drugs in a tailored manner (Tang et al. 2020a). The non-specific toxicity of chemotherapy medications, which affects both malignant and healthy cells, is one of the main obstacles. Researchers are looking to cancer nanotechnology, which has the potential to improve medication efficacy while lowering toxicity, to get over this obstacle (Tang et al. 2020b). Because they can achieve steady-state medication concentrations over extended periods of time and lower toxicity levels, nanocarriers like nanoparticles have the potential to increase the therapeutic index of pharmaceuticals. By offering more potent and focused treatments, this strategy is transforming the way cancer is treated. Furthermore, developments in the fields of material science and chemistry have made lignin suitable for use in nanotechnology, where it can be employed as a drug carrier or as a medicinal excipient (Wang et al. 2020). Natural polymer lignin, which is widely distributed in nature, has drawn interest because of its possible application in cancer nanotechnology (Liu, Luo, and Zheng 2018). Concerns about the sustainability of the economy and environment have led researchers to look into using lignin as a precursor for the production of nanoparticles. Because lignin has a variety of chemical functional groups, it can operate as a surfactant, stabilizing matrix, reducing agent, and capping agent. This makes it a good candidate for the creation of effective nanosystems for the treatment of cancer. For example, to increase the bioavailability of medications such as resveratrol, lignin has been added to innovative self-nanoemulsifying drug delivery systems. Supplementing medications with lignin as a surfactant has enhanced their absorption, stability, and therapeutic efficacy. Likewise, lignin has been employed as a green surfactant in the manufacturing of microemulsions loaded with oil in water. Furthermore, the pharmacological excipient lignin has been investigated for the creation of nanoparticles, including silver nanoparticles (L-AgNPs). The special benefits of lignin nanoparticles (LNPs) include safety, biodegradability, and the ability to improve absorption. To create LNPs, a variety of techniques have been used, such as phase separation and solvent exchange by dialysis. The pharmacokinetics and pharmacodynamics of anticancer medications have been demonstrated to be improved by LNPs, as well as to improve their bioavailability, cellular absorption, and oral bioavailability, curcumin, a medication with low water solubility and limited bioavailability, has been encapsulated in organosolv lignin-based nanoparticles. In comparison with the free drug suspension, these optimized nanoparticles showed prolonged release and enhanced intestinal permeability, which raised plasma concentration and oral bioavailability. In addition, lignin-based nanosystems have been created to deliver powerful anticancer drugs to tumor microenvironments specifically. Tumor-associated macrophages have been more effectively targeted by LNPs modified with certain peptides, resulting in decreased tumor volume and increased antitumor efficacy. Hydrogels are three-dimensional networks of hydrophilic molecules that have been crosslinked. They find use in several biomedical domains, including wound healing, tissue engineering scaffolds, controlled drug release equipment, and contact lenses. They are perfect for regulated medication delivery systems because of their ability to swell in response to certain trigger events. Due to their biocompatibility, biodegradability, and physiologically recognized qualities, renewable natural polymers-based hydrogels derived from sources such as gelatin, hyaluronate, lignin, protein, sodium alginate, and starch are being investigated more and more for biomedical applications. Lignin, a natural polymer abundant in biomass, has emerged as a promising candidate for hydrogel fabrication due to its unique properties (Figueiredo et al. 2017). Hydrogel preparation uses lignin, a polymer having hydrophilic functional groups, because of its potential applications and adaptability. It is a biodegradable and sustainable substance because of its abundance of phenolic and aliphatic groups, which allow for considerable chemical modification. Hydrogels based on lignin have the potential to be used in pharmaceutical and medical fields because of their sustainability, biocompatibility, non-toxicity, antioxidant, and antibacterial qualities. To make composite hydrogels, they can be mixed with other polymers such as chitosan, alginate, PVA, and cellulose. Drug release kinetics, swelling capacity, and water absorption are all impacted by the porous architectures of lignin-based hydrogels (Morena et al. 2022). Lignin can increase the porosity of hydrogel, creating a network that is more relaxed and has a greater ability to swell. As a result, the structure becomes more homogeneous, enhancing the mechanical and drug release characteristics. Because they enable the regulated release of hydrophobic along with hydrophilic molecules and provide sustained distribution over extended periods, lignin-based hydrogels have demonstrated promising outcomes in wound healing, tissue engineering, and drug delivery systems. Additionally, lignin-containing hydrogels have been employed as wound dressings, promoting tissue regeneration and accelerating wound healing in animal models (Getya and Gitsov 2023). Moreover, lignin-based hydrogels are being investigated for use in tissue engineering, offering support for the attachment, growth, and differentiation of cells. Two interesting manufacturing strategies for these materials are electrospinning and 3D printing. High porosity along with surface area nanofibers or nanofibrous scaffolds are produced via electrospinning, which makes them ideal for drug delivery and tissue engineering applications. Precise shape and porosity control made possible by 3D printing present prospects for regenerative treatments and customized medicine. Hydrogels based on lignin have a lot of promise for use in pharmaceutical and medical fields. With further research and development, these sustainable and biocompatible materials could pave the way for innovative solutions in healthcare and contribute to advancements in tissue engineering, drug delivery, and regenerative medicine. 3D printing, the development of approaches utilizing lignin-based materials presents a promising frontier that holds immense potential for medical and pharmaceutical applications. Bioprinting for tissue engineering applications, lignin-based hydrogels show promise as innovative bioinks. Various bioprinting techniques, such as microextrusion, inkjet, and laser-assisted bioprinting, offer different approaches to depositing bioinks and constructing three-dimensional tissue scaffolds. Laser-assisted bioprinting leverages the optical properties of the bio-ink or the wavelength of the laser to direct ink ejection, offering precise control over the printing process (Silva et al. 2021). Despite the immense potential of 3D printing in the medical field, several challenges remain to be addressed. Regulatory frameworks concerning the manufacture of 3D-printed medical products need to be established to ensure safety and efficacy. Furthermore, while bioprinting has shown promise in laboratory settings for fabricating various tissues and organs, translating these advancements into clinical and commercial applications requires further research and development. Nonetheless, 3D printing technology holds significant advantages and offers endless possibilities for revolutionizing medical and pharmaceutical applications, particularly with the integration of lignin-based materials. Reaching the full potential of 3D printing in healthcare will need sustained innovation and cross-disciplinary cooperation to overcome current obstacles (Kai et al. 2016).

3.8 CONCLUSION

The utilization of lignocellulose-based sustainable biomaterials presents a promising avenue for innovation in the fabrication of pharmaceutical products. Throughout this exploration, we have witnessed the remarkable potential of lignocellulosic biomass, comprising cellulose, hemicellulose, and lignin, in revolutionizing various aspects of pharmaceutical development and production. Lignocellulose offers unique properties that can be harnessed across a range of pharmaceutical applications, including wound healing, drug delivery systems, nanotechnology, and tissue engineering. Its abundance, renewability, and biocompatibility make it an attractive alternative to conventional materials, aligning with the growing emphasis on sustainability within the pharmaceutical industry. The chemical composition and structural characteristics of lignocellulose components play pivotal roles in shaping their applications.

Cellulose, with its fibrous structure, provides mechanical strength and stability, while hemicellulose offers flexibility and adhesive properties. Lignin, as a natural polymer abundant in biomass, serves as a versatile building block for the fabrication of hydrogels and nanoparticles, enabling controlled drug release and targeted delivery. Lignin-based hydrogels, in particular, exhibit promising features such as sustainability, biodegradability, and biocompatibility. These hydrogels have demonstrated efficacy in facilitating wound healing, promoting tissue regeneration, and delivering therapeutic agents with precision. Moreover, lignin nanoparticles show potential as effective carriers for drug delivery, particularly in cancer therapy, where targeted delivery can enhance therapeutic outcomes while minimizing side effects. The integration of lignocellulose-based materials into advanced pharmaceutical formulations underscores the importance of interdisciplinary collaboration and ongoing research efforts. By leveraging the unique properties of lignocellulose, researchers and pharmaceutical developers can address key challenges in tissue engineering, drug delivery, and regenerative medicine. The considerable progress made in harnessing lignocellulose-based biomaterials, challenges such as scalability, cost-effectiveness, and regulatory considerations remain to be addressed. Continued innovation and investment in development and research are essential to unlocking the full potential of lignocellulose in pharmaceutical applications and realizing its transformative impact on healthcare.

REFERENCES

Ashokkumar, Veeramuthu, Radhakrishnan Venkatkarthick, Shanmugam Jayashree, Santi Chuetor, Selvakumar Dharmaraj, Gopalakrishnan Kumar, Wei-Hsin Chen, and Chawalit Ngamcharussrivichai. 2022. "Recent Advances in Lignocellulosic Biomass for Biofuels and Value-Added Bioproducts-A Critical Review." *Bioresource Technology* 344: 126195.

Barbero, Cesar A. 2023. "Functional Materials Made by Combining Hydrogels (Cross-Linked Polyacrylamides) and Conducting Polymers (Polyanilines)-A Critical Review." *Polymers (Basel)* 15 (10). https://doi.org/10.3390/polym15102240.

Bashir, Shahid, Maryam Hina, Javed Iqbal, A.H. Rajpar, M.A. Mujtaba, N.A. Alghamdi, S. Wageh, K. Ramesh, and S. Ramesh. 2020. "Fundamental Concepts of Hydrogels: Synthesis, Properties, and Their Applications." *Polymers (Basel)* 12 (11). https://doi.org/10.3390/polym12112702.

Bontaş, Marius Gabriel, Aurel Diacon, Ioan Călinescu, and Edina Rusen. 2023. "Lignocellulose Biomass Liquefaction: Process and Applications Development as Polyurethane Foams." *Polymers (Basel)* 15 (3). https://doi.org/10.3390/polym15030563.

Brar, K.K., Y. Raheja, B.S. Chadha, S. Magdouli, S.K. Brar, Y.H. Yang, S.K. Bhatia, and A. Koubaa. 2024. "A Paradigm Shift towards Production of Sustainable Bioenergy and Advanced Products from Cannabis/hemp Biomass in Canada." *Biomass Convers Biorefin* 14: 3161–31822. https://doi.org/10.1007/s13399-022-02570-6.

Camargo, T.M., V.S. Nazato, M.G. Silva, J.C. Cogo, F.C. Groppo, and Y. Oshima-Franco. 2010. "Bothrops Jararacussu Venom-Induced Neuromuscular Blockade Inhibited by Casearia Gossypiosperma Briquet Hydroalcoholic Extract." *Journal of Venomous Animals and Toxins Including Tropical Diseases* 16: 432–441.

Cheah, Wai Yan, Revathy Sankaran, Pau Loke Show, Tg Ibrahim, Tg Nilam Baizura, Kit Wayne Chew, Alvin Culaba, and Jo-Shu Chang. 2020. "Pretreatment Methods for Lignocellulosic Biofuels Production: Current Advances, Challenges and Future Prospects." *Biofuel Research Journal* 7 (1): 1115–1127.

Chen, Hongzhang. 2014. *Biotechnology of Lignocellulose: Theory and Practice.* Chemical Industry Press and Springer.

Christopher, Meera, Anil K. Mathew, M. Kiran Kumar, Ashok Pandey, and Rajeev K. Sukumaran. 2017. "A Biorefinery-based Approach for the Production of Ethanol from Enzymatically Hydrolysed Cotton Stalks." *Bioresource Technology* 242: 178–183.

da Costa Lopes, Andre M., Karen G. João, Ana Rita C. Morais, Ewa Bogel-Łukasik, and Rafał Bogel-Łukasik. 2013. "Ionic Liquids as a Tool for Lignocellulosic Biomass Fractionation." *Sustainable Chemical Processes* 1: 1–31.

Das, Nisha, Pradip Kumar Jena, Diptymayee Padhi, Mahendra Kumar Mohanty, and Gyanaranjan Sahoo. 2021. "A Comprehensive Review of Characterization, Pretreatment and Its Applications on Different Lignocellulosic Biomass for Bioethanol Production." *Biomass Conversion and Biorefinery*: 1–25.

de Haro, Juan Carlos, Elisavet Tatsi, Lucia Fagiolari, Matteo Bonomo, Claudia Barolo, Stefano Turri, Federico Bella, and Gianmarco Griffini. 2021. "Lignin-Based Polymer Electrolyte Membranes for Sustainable Aqueous Dye-Sensitized Solar Cells." *ACS Sustainable Chemistry & Engineering* 9 (25): 8550–8560. https://doi.org/10.1021/acssuschemeng.1c01882.

de Milliano, Inge, Moniek Twisk, Johannes C. Ket, Judith A. Huirne, and Wouter J. Hehenkamp. 2017. "Pre-treatment with GnRHa or Ulipristal Acetate Prior to Laparoscopic and Laparotomic Myomectomy: A Systematic Review and Meta-analysis." *PLoS One* 12 (10): e0186158. https://doi.org/10.1371/journal.pone.0186158.

de Paula Protásio, T., J.S. da Costa, M.V. Scatolino, M.D.R. Lima, M.R. de Assis, M.G. da Silva, L. Bufalino, A.F. Dias Junior, and P.F. Trugilho. 2022. "Revealing the Influence of Chemical Compounds on the Pyrolysis of Lignocellulosic Wastes from the Amazonian Production Chains." *International Journal of Environmental Science and Technology* 19 (5): 4491–4508.

Demirbaş, Ayhan. 2005. "Thermochemical Conversion of Biomass to Liquid Products in the Aqueous Medium." *Energy Sources* 27 (13): 1235–1243.

Devi, Arti, Somvir Bajar, Havleen Kour, Richa Kothari, Deepak Pant, and Anita Singh. 2022. "Lignocellulosic Biomass Valorization for Bioethanol Production: A Circular Bioeconomy Approach." *Bioenergy Research* 15 (4): 1820–1841. https://doi.org/10.1007/s12155-022-10401-9.

Espino-Pérez, Etzael, Sandra Domenek, Naceur Belgacem, Cécile Sillard, and Julien Bras. 2014. "Green Process for Chemical Functionalization of Nanocellulose with Carboxylic Acids." *Biomacromolecules* 15 (12): 4551–4560. https://doi.org/10.1021/bm5013458.

Fatma, Shabih, Amir Hameed, Muhammad Noman, Temoor Ahmed, Muhammad Shahid, Mohsin Tariq, Imran Sohail, and Romana Tabassum. 2018. "Lignocellulosic Biomass: A Sustainable Bioenergy Source for the Future." *Protein and Peptide Letters* 25 (2): 148–163.

Figueiredo, Patricia, Kalle Lintinen, Alexandros Kiriazis, Ville Hynninen, Zehua Liu, Tomas Bauleth-Ramos, Antti Rahikkala, Alexandra Correia, Tomáš Kohout, and Bruno Sarmento. 2017. "In vitro Evaluation of Biodegradable Lignin-based Nanoparticles for Drug Delivery and Enhanced Antiproliferation Effect in Cancer Cells." *Biomaterials* 121: 97–108.

Francisco, María, Adriaan Van Den Bruinhorst, and Maaike C. Kroon. 2012. "New Natural and Renewable Low Transition Temperature Mixtures (LTTMs): Screening as Solvents for Lignocellulosic Biomass Processing." *Green Chemistry* 14 (8): 2153–2157.

Gallego-García, María, Antonio D. Moreno, Paloma Manzanares, María José Negro, and Aleta Duque. 2023. "Recent Advances on Physical Technologies for the Pretreatment of Food Waste and Lignocellulosic Residues." *Bioresource Technolonogy* 369: 128397. https://doi.org/10.1016/j.biortech.2022.128397.

Getya, Dariya, and Ivan Gitsov. 2023. "Synthesis and Applications of Hybrid Polymer Networks Based on Renewable Natural Macromolecules." *Molecules* 28 (16). https://doi.org/10.3390/molecules28166030.

Ghimire, Nirmal, Rune Bakke, and Wenche Hennie Bergland. 2021. "Liquefaction of Lignocellulosic Biomass for Methane Production: A Review." *Bioresource Technology* 332: 125068. https://doi.org/10.1016/j.biortech.2021.125068.

Haldar, Dibyajyoti, and Mihir Kumar Purkait. 2020. "Lignocellulosic Conversion into Value-Added Products: A Review." *Process Biochemistry* 89: 110–133.

Heinze, Thomas, Omar A. El Seoud, and Andreas Koschella. 2018. *Cellulose Derivatives: Synthesis, Structure, and Properties*. Springer.

Hendriks, A.T.W.M., and G. Zeeman. 2009. "Pretreatments to Enhance the Digestibility of Lignocellulosic Biomass." *Bioresource Technology* 100 (1): 10–18.

Honarmandrad, Zhila, Karolina Kucharska, and Jacek Gębicki. 2022. "Processing of Biomass Prior to Hydrogen Fermentation and Post-Fermentative Broth Management." *Molecules* 27 (21). https://doi.org/10.3390/molecules27217658.

Ibarra-Gonzalez, Paola, and Ben-Guang Rong. 2019. "A Review of the Current State of Biofuels Production from Lignocellulosic Biomass Using Thermochemical Conversion Routes." *Chinese Journal of Chemical Engineering* 27(7): 1523–1535.

Imman, Saksit, Jantima Arnthong, Vorakan Burapatana, Verawat Champreda, and Navadol Laosiripojana. 2015. "Fractionation of Rice Straw by a Single-Step Solvothermal Process: Effects of Solvents, Acid Promoters, and Microwave Treatment." *Renewable Energy* 83: 663–673.

Jaffur, N., P. Jeetah, and G. Kumar. 2021. "A Review on Enzymes and Pathways for Manufacturing Polyhydroxybutyrate from Lignocellulosic Materials." *3 Biotech* 11 (11): 483. https://doi.org/10.1007/s13205-021-03009-x.

Jeenkeawpieam, Juthatip, Souwalak Phongpaichit, Vatcharin Rukachaisirikul, and Jariya Sakayaroj. 2012. "Antifungal Activity and Molecular Identification of Endophytic Fungi from the Angiosperm Rhodomyrtus Tomentosa." *African Journal of Biotechnology* 11 (75): 14007–14016.

Kai, Dan, Mein Jin Tan, Pei Lin Chee, Yun Khim Chua, Yong Liang Yap, and Xian Jun Loh. 2016. "Towards Lignin-based Functional Materials in a Sustainable World." *Green Chemistry* 18 (5): 1175–1200.

Kalita, E., B.K. Nath, P. Deb, F. Agan, Md R. Islam, and K. Saikia. 2015. "High Quality Fluorescent Cellulose Nanofibers from Endemic Rice Husk: Isolation and Characterization." *Carbohydrate Polymers* 122: 308–313.

Klein-Marcuschamer, Daniel, Piotr Oleskowicz-Popiel, Blake A. Simmons, and Harvey W. Blanch. 2012. "The Challenge of Enzyme Cost in the Production of Lignocellulosic Biofuels." *Biotechnology and Bioengineering* 109 (4): 1083–1087.

Komisarz, Karolina, Tomasz M. Majka, and Krzysztof Pielichowski. 2022. "Chemical and Physical Modification of Lignin for Green Polymeric Composite Materials." *Materials (Basel)* 16 (1). https://doi.org/10.3390/ma16010016.

Krakowska-Sieprawska, Aneta, Anna Kiełbasa, Katarzyna Rafińska, Magdalena Ligor, and Boguslaw Buszewski. 2022. "Modern Methods of Pre-Treatment of Plant Material for the Extraction of Bioactive Compounds." *Molecules* 27 (3). https://doi.org/10.3390/molecules27030730.

Lange, Jean-Paul. 2007. "Lignocellulose Conversion: An Introduction to Chemistry, Process and Economics." *Biofuels, Bioproducts and Biorefining: Innovation for a Sustainable Economy* 1 (1): 39–48.

Lange, Jean-Paul. 2018. "Lignocellulose Liquefaction to Biocrude: A Tutorial Review." *ChemSusChem* 11 (6): 997–1014. https://doi.org/10.1002/cssc.201702362.

Lehrhofer, Anna F., Takaaki Goto, Toshinari Kawada, Thomas Rosenau, and Hubert Hettegger. 2022. "The In Vitro Synthesis of Cellulose–A Mini-Review." *Carbohydrate Polymers* 285: 119222.

Lin, Ning, and Alain Dufresne. 2014. "Nanocellulose in Biomedicine: Current Status and Future Prospect." *European Polymer Journal* 59: 302–325.

Liu, Chao, Yu Li, Jingshun Zhuang, Zhouyang Xiang, Weikun Jiang, Shuaiming He, and Huining Xiao. 2022. "Conductive Hydrogels Based on Industrial Lignin: Opportunities and Challenges." *Polymers (Basel)* 14 (18). https://doi.org/10.3390/polym14183739.

Liu, Qiaoling, Hongyu Mou, Wenjun Chen, Xinhui Zhao, Haitao Yu, Zhimin Xue, and Tiancheng Mu. 2019. "Highly Efficient Dissolution of Lignin by Eutectic Molecular Liquids." *Industrial & Engineering Chemistry Research* 58 (51): 23438–23444.

Liu, Qiaoling, Xinhui Zhao, Dongkun Yu, Haitao Yu, Yibin Zhang, Zhimin Xue, and Tiancheng Mu. 2019. "Novel Deep Eutectic Solvents with Different Functional Groups towards Highly Efficient Dissolution of Lignin." *Green Chemistry* 21 (19): 5291–5297.

Liu, Qingquan, Le Luo, and Luqing Zheng. 2018. "Lignins: Biosynthesis and Biological Functions in Plants." *International Journal of Molecular Sciences* 19 (2). https://doi.org/10.3390/ijms19020335.

Lynam, Joan G., Narendra Kumar, and Mark J. Wong. 2017. "Deep Eutectic Solvents' Ability to Solubilize Lignin, Cellulose, and Hemicellulose; Thermal Stability; and Density." *Bioresource Technology* 238: 684–689.

Ma, Yan, Weihong Tan, Jingxin Wang, Junming Xu, Kui Wang, and Jianchun Jiang. 2020. "Liquefaction of Bamboo Biomass and Production of Three Fractions Containing Aromatic Compounds." *Journal of Bioresources and Bioproducts* 5 (2): 114–123.

Menon, Vishnu, and Mala Rao. 2012. "Trends in Bioconversion of Lignocellulose: Biofuels, Platform Chemicals & Biorefinery Concept." *Progress in Energy and Combustion Science* 38 (4): 522–550.

Mochochoko, Tanki, Oluwatobis Oluwafemi, Olufemi O. Adeyemi, Denis N. Jumbam, and Sandile P. Songca. 2016. "Natural Cellulose Fibers: Sources, Isolation, Properties and Applications." *Micro-and Nanostructured Polymer Systems: From Synthesis to Applications* 25.

Morena, Angela Gala, Arnau Bassegoda, Michal Natan, Gila Jacobi, Ehud Banin, and Tzanko Tzanov. 2022. "Antibacterial Properties and Mechanisms of Action of Sonoenzymatically Synthesized Lignin-based Nanoparticles." *ACS Applied Materials & Interfaces* 14 (33): 37270–37279.

Mosier, Nathan, Charles Wyman, Bruce Dale, Richard Elander, Y.Y. Lee, Mark Holtzapple, and Michael Ladisch. 2005. "Features of Promising Technologies for Pretreatment of Lignocellulosic Biomass." *Bioresource Technology* 96 (6): 673–686.

Niknam, Zahra, Faezeh Hosseinzadeh, Forough Shams, Leyla Fath-Bayati, Ghader Nuoroozi, Leila Mohammadi Amirabad, Fariba Mohebichamkhorami, Sahar Khakpour Naeimi, Soudeh Ghafouri-Fard, and Hakimeh Zali. 2022. "Recent Advances and Challenges in Graphene-based Nanocomposite Scaffolds for Tissue Engineering Application." *Journal of Biomedical Materials Research Part A* 110 (10): 1695–1721.

Østby, Heidi, and Anikó Várnai. 2023. "Hemicellulolytic Enzymes in Lignocellulose Processing." *Essays in Biochemistry* 67 (3): 533–550. https://doi.org/10.1042/ebc20220154.

Pandiselvam, R., Alev Y. Aydar, Naciye Kutlu, Raouf Aslam, Prashant Sahni, Swati Mitharwal, Mohsen Gavahian, Manoj Kumar, António Raposo, Sunghoon Yoo, Heesup Han, and Anjineyulu Kothakota. 2023. "Individual and Interactive Effect of Ultrasound Pre-treatment on Drying Kinetics and Biochemical Qualities of Food: A Critical Review." *Ultrason Sonochem* 92: 106261. https://doi.org/10.1016/j.ultsonch.2022.106261.

Parvathy, P.A., A.V. Ayobami, A.M. Raichur, and S.K. Sahoo. 2021. "Methacrylated Alkali Lignin Grafted P(Nipam-Co-AAc) Copolymeric Hydrogels: Tuning the Mechanical and Stimuli-responsive Properties." *International Journal of Biological Macromolecules* 192: 180–196.

Pendse, Dhanashri S., Minal Deshmukh, and Ashwini Pande. 2023. "Different Pre-treatments and Kinetic Models for Bioethanol Production from Lignocellulosic Biomass: A Review." *Heliyon* 9 (6): e16604. https://doi.org/10.1016/j.heliyon.2023.e16604.

Pol Segura, Isabel, Navid Ranjbar, Anne Juul Damø, Lars Skaarup Jensen, Mariana Canut, and Peter Arendt Jensen. 2023. "A Review: Alkali-Activated Cement and Concrete Production Technologies Available in the Industry." *Heliyon* 9 (5): e15718. https://doi.org/10.1016/j.heliyon.2023.e15718.

Rastogi, Meenal, and Smriti Shrivastava. 2017. "Recent Advances in Second Generation Bioethanol Production: An Insight to Pretreatment, Saccharification and Fermentation Processes." *Renewable and Sustainable Energy Reviews* 80: 330–340.

Rico-García, Diana, Leire Ruiz-Rubio, Leyre Pérez-Alvarez, Saira L. Hernández-Olmos, Guillermo L. Guerrero-Ramírez, and José L. Vilas-Vilela. 2020a. "Lignin-Based Hydrogels: Synthesis and Applications." *Polymers (Basel)* 12 (1). https://doi.org/10.3390/polym12010081.

Rico-García, Diana, Leire Ruiz-Rubio, Leyre Pérez-Alvarez, Saira L. Hernández-Olmos, Guillermo L. Guerrero-Ramírez, and José Luis Vilas-Vilela. 2020b. "Lignin-based Hydrogels: Synthesis and Applications." *Polymers* 12 (1): 81.

Rizal, Nur Fatin Athirah Ahmad, Mohamad Faizal Ibrahim, Mohd Rafein Zakaria, Suraini Abd-Aziz, Phang Lai Yee, and Mohd Ali Hassan. 2018. "Pre-treatment of Oil Palm Biomass for Fermentable Sugars Production." *Molecules* 23 (6). https://doi.org/10.3390/molecules23061381.

Sagadevan, Suresh, and Won-Chun Oh. 2023. "Comprehensive Utilization and Biomedical Application of MXenes-A Systematic Review of Cytotoxicity and Biocompatibility." *Journal of Drug Delivery Science and Technology* 85: 104569.

Sasaki, Mitsuru, Tadafumi Adschiri, and Kunio Arai. 2003. "Fractionation of Sugarcane Bagasse by Hydrothermal Treatment." *Bioresource Technology* 86 (3): 301–304.

Sasmal, Soumya, and Kaustubha Mohanty. 2018. "Pretreatment of Lignocellulosic Biomass toward Biofuel Production." In Sani, R. K. (Ed.), *Biorefining of Biomass to Biofuels: Opportunities and Perception*, 203–221. Springer.

Silva, Jéssica P., Alonso R.P. Ticona, Pedro R.V. Hamann, Betania F. Quirino, and Eliane F. Noronha. 2021. "Deconstruction of Lignin: From Enzymes to Microorganisms." *Molecules* 26 (8). https://doi.org/10.3390/molecules26082299.

Sindhu, Raveendran, Parameswaran Binod, Anil Kuruvilla Mathew, Amith Abraham, Edgard Gnansounou, Sabeela Beevi Ummalyma, Leya Thomas, and Ashok Pandey. 2017. "Development of a Novel Ultrasound-Assisted Alkali Pretreatment Strategy for the Production of Bioethanol and Xylanases from Chili Post Harvest Residue." *Bioresource Technology* 242: 146–151.

Tang, Qianqian, Yong Qian, Dongjie Yang, Xueqing Qiu, Yanlin Qin, and Mingsong Zhou. 2020a. "Lignin-Based Nanoparticles: A Review on Their Preparations and Applications." *Polymers (Basel)* 12 (11). https://doi.org/10.3390/polym12112471.

Tang, Qianqian, Yong Qian, Dongjie Yang, Xueqing Qiu, Yanlin Qin, and Mingsong Zhou. 2020b. "Lignin-based Nanoparticles: A Review on their Preparations and Applications." *Polymers* 12 (11): 2471.

Turner, Pernilla, Gashaw Mamo, and Eva Nordberg Karlsson. 2007. "Potential and Utilization of Thermophiles and Thermostable Enzymes in Biorefining." *Microbial Cell Factories* 6: 9. https://doi.org/10.1186/1475-2859-6-9.

Vasić, Katja, Željko Knez, and Maja Leitgeb. 2021. "Bioethanol Production by Enzymatic Hydrolysis from Different Lignocellulosic Sources." *Molecules* 26 (3). https://doi.org/10.3390/molecules26030753.

Vohra, Mustafa, Jagdish Manwar, Rahul Manmode, Satish Padgilwar, and Sanjay Patil. 2014. "Bioethanol Production: Feedstock and Current Technologies." *Journal of Environmental Chemical Engineering* 2 (1): 573–584.

Vojtová, Lucy, Veronika Pavliňáková, Johana Muchová, Katarína Kacvinská, Jana Brtníková, Martin Knoz, Břetislav Lipový, Martin Faldyna, Eduard Göpfert, and Jakub Holoubek. 2021. "Healing and Angiogenic Properties of Collagen/Chitosan Scaffolds Enriched with Hyperstable FGF2-STAB® Protein: In Vitro, Ex Ovo and In Vivo Comprehensive Evaluation." *Biomedicines* 9 (6): 590.

Wang, Xiaoju, Qingbo Wang, and Chunlin Xu. 2020. "Nanocellulose-based Inks for 3d Bioprinting: Key Aspects in Research Development and Challenging Perspectives in Applications—A Mini Review." *Bioengineering* 7 (2): 40.

Wang, Yun-Yan, Xianzhi Meng, Yunqiao Pu, and Arthur J. Ragauskas. 2020. "Recent Advances in the Application of Functionalized Lignin in Value-Added Polymeric Materials." *Polymers (Basel)* 12 (10). https://doi.org/10.3390/polym12102277.

Yan, Feng, Shuangqi Tian, Ke Du, Xing'ao Xue, Peng Gao, and Zhicheng Chen. 2022. "Preparation and Nutritional Properties of Xylooligosaccharide from Agricultural and Forestry Byproducts: A Comprehensive Review." *Frontiers in Nutrition* 9: 977548. https://doi.org/10.3389/fnut.2022.977548.

Yousefiasl, Satar, Hamed Manoochehri, Pooyan Makvandi, Saeid Afshar, Erfan Salahinejad, Pegah Khosraviyan, Massoud Saidijam, Sara Soleimani Asl, and Esmaeel Sharifi. 2023. "Chitosan/Alginate Bionanocomposites Adorned with Mesoporous Silica Nanoparticles for Bone Tissue Engineering." *Journal of Nanostructure in Chemistry* 13 (3): 389–403. https://doi.org/10.1007/s40097-022-00507-z.

Zabed, H., Jaya Narayan Sahu, Amru Nasrulhaq Boyce, and Golam Faruq. 2016. "Fuel Ethanol Production from Lignocellulosic Biomass: An Overview on Feedstocks and Technological Approaches." *Renewable and Sustainable Energy Reviews* 66: 751–774.

Zheng, Jun, and Lars Rehmann. 2014. "Extrusion Pretreatment of Lignocellulosic Biomass: A Review." *International Journal of Molecular Sciences* 15 (10): 18967–18984. https://doi.org/10.3390/ijms151018967.

6 Biodegradable and Non-Biodegradable Sustainable Biomaterials

Sachin N. Kothawade, Vishal V. Pande,
and Jayprakash Suryawanshi
RSM's N. N. Sattha College of Pharmacy

Ashwini Kumar
Manav Rachna International Institute
of Research and Studies

Parveen Kumar
Department of Mechanical Engineering, Rawal Institute of
Engineering and Technology, Faridabad, Haryana, India

Ajay Kumar
Department of Mechanical Engineering,
School of Engineering and Technology, JECRC
University, Jaipur, Rajasthan, India

1 INTRODUCTION

Biomaterials are specifically engineered substances intended to interact with living systems. These compounds are capable of rapidly integrating into human tissue and have bioactive properties [1]. Their breakdown is excellent. They are frequently employed in the production of pharmaceuticals, tissue engineering, and the creation of human body parts. The advancement of eco-friendly materials and the progress of biomaterials are mutually reinforced, along with the implementation of state-of-the-art techniques like bioprinting, nanotechnology, utilization of biodegradable materials, integration of bioactive substances, and surface alterations [2–4]. Sustainable biomaterials are created using environmentally sustainable technology or derived from diverse sources of life. Several biomaterials have been discovered and fabricated as potential substitutes for traditional substances, and many of them have been successfully utilized in several biomedical domains. Key uses encompass augmentation of breasts, stem cell therapy for nerve regeneration, restoration of ligaments and tendons, orthopedic procedures, wound healing, and the development of contact lenses for ophthalmic purposes. In addition, they are utilized in a diverse

DOI: 10.1201/9781003434313-6

range of nonbiomedical applications. Nonbiomedical applications encompass a range of uses, including but not limited to filter and membrane technology, energy-related products, packaging components, fabrics, building supplies, food containers, sporting equipment, items for personal care, and cosmetics [5–7].

The utilization of specific biomaterials in the field of biomedicine has experienced significant increase in the past two decades, mostly driven by the expansion of the pharmaceutical sector and the focus on improving patient adherence. A biomaterial is a substance that is mostly or entirely made up of living matter, according to its definition. An instance of such a case might involve a polymer structure that has been infused with cells [8]. These materials can be utilized in medical devices as a way to improve or substitute biological tissue for therapeutic uses. The rejuvenation of cartilage tissue, heart valves, bones, and various other organs can be facilitated by utilizing a combination of synthetic and natural substances derived from stem cells. This is achieved by replacing or attempting to restore the damaged component and utilizing materials that are resistant to immunological rejection due to their genetic similarity to the patient. The bulk of biomaterials consist of metals, ceramics, and polymeric components. Although ceramics and metals are mainly used to substitute hard tissue, polymeric materials can be utilized to temporarily substitute either hard or soft tissues in the field of orthopedics. Chemistry, biology, medicine, materials science, and tissue engineering are subfields of biomaterial science. Initially, cranial deformities were addressed by employing biomaterials composed of gold and ivory [9–12].

Polymethyl methacrylate, also referred to as PMMA is the initial polymer used following World War II. During the 1980s and 1990s, bioactive compounds were used to elicit specific reactions in cells at the boundary of substances, resulting in a shift in the application of biomaterials toward non-reactive substances [13–15]. Any substance, regardless of its origin, can be classified as a biological material. It is an inherent part of a biological structure that operates spontaneously through the display, sudden increase, and change in activity. It has the potential to serve as a cardiac valve. Furthermore, it can be utilized for interactive applications such as hydroxy-apatite-(HA)-coated hip replacements. Biomaterials have practical applications in medicine delivery and dental surgery in our daily lives. Furthermore, they have applications in contact lenses, cochlear implants, bone plates, skin healing instruments, bone cement, blood vessel prostheses, synthetic ligaments, and tendons. Tissues are classified into three kinds based on their response. The first category consists of bio-inert materials that maintain continuous contact with the adjacent bone tissue. There will be no chemical interaction between the tissue and implant. Bioactive compounds refer to biotolerant compounds that are physically separated from bone-related cells by an outer layer of fibrous tissue. These materials demonstrate bone formation, which refers to the capacity to establish chemical connections with bone tissue [16–19].

The different additives in the recipe should be compatible with the biological materials. The human being should exhibit no adverse reaction to the biomaterial, and conversely, the biomaterial should not elicit any negative response from the body [20, 21]. It should not have carcinogenic effects on the body. It should be innocuous. In order to serve as a substitute for or improvement upon bodily tissues, it is necessary for it to possess sufficient physical and mechanical characteristics. It will be

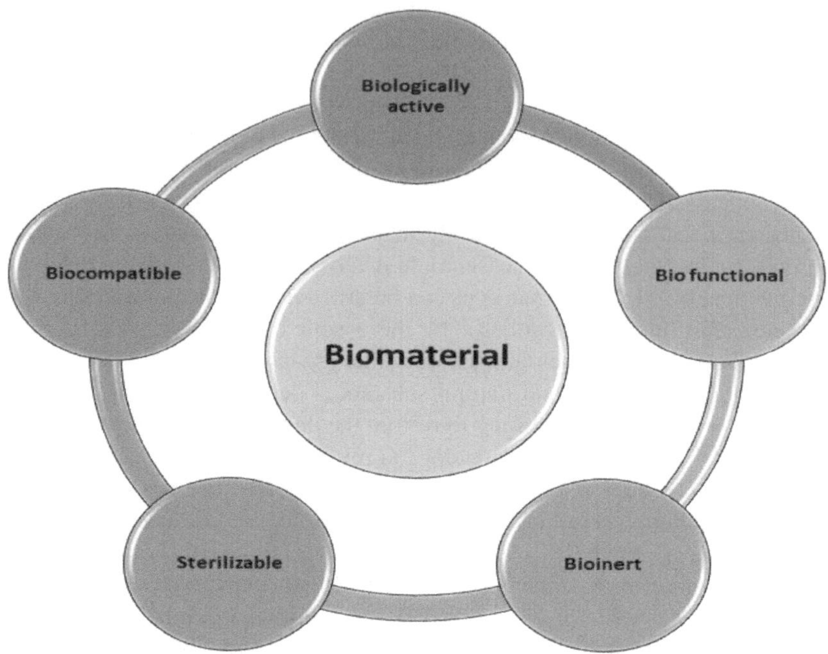

FIGURE 6.1	Properties of biomaterials.

appropriate for commercial utilization, allowing for versatile shaping. Biomaterials should be affordable and readily obtainable. It should possess a unique protective mechanism to prevent deterioration, such as a defense against corrosion for metals or a defense against biological breakdown for polymers. In order to minimize the production of wear particles, it is important for the material to possess a high level of resistance to wear. In order to decrease bone reabsorption, it is necessary for it to possess a lower elastic modulus. The ideal biomaterial should possess the characteristics illustrated in Figure 6.1. Prior research has examined several modes of degradation in composite materials based on biopolymers, as well as parameters related to their sustainability. The authors of this chapter discuss several biomaterials, such as metallic, polymeric, ceramic, and composite substances, along with their properties and applications. This chapter also explores the topic of 3D printing and alternative techniques for producing biomaterials that are environmentally friendly. Various sustainable biomaterials, such as protein-based, cellulose, chitin, and chitosan composite-based, hydroxyapatite-starch-based, and carbonaceous biomaterials, are examined in detail, including their characteristics and applications [22–28].

## 1.1	DIFFERENT TYPES OF BIOMATERIALS

Biomaterials are either artificial or naturally occurring elements that are used to fabricate implants or additional components capable of replacing biological components that have been eliminated or damaged, thereby restoring both function and form. Cellulose, gelatin, alginate, chitosan, and collagen are among the naturally occurring biomaterials. Metal, composites, ceramics, polymers, and hydrogels are instances of biomaterials that are produced using synthetic means. Biomaterials have significantly enhanced the quality of life and lifetime of humans, leading to rapid advancements in the sector to cater to the needs of an aging population. Biomaterials are utilized in many bodily parts such as artificial heart valves, shoulder replacement implants, knee joints, elbow joints, ears, orodental structures, and hip joints. Implants have been made from different materials, chosen based on the specific needs of each application. Metals, alloys, polymers, ceramics, and composite are the biomaterials that are most frequently used. The subsequent sections will cover several categories of biomaterials [19, 29–31].

1.1.1 Metallic Biomaterials

Commonly used metallic biomolecules include chromium–cobalt alloy, stainless steel, gold–silver–copper–palladium alloys, magnesium and its alloys, and nitinol (nickel–titanium). Titanium alloys have become widely used as implant materials in various medical applications for many years. These materials possess a high degree of resisting corrosion with excellent mechanical properties, which have enabled their use [32, 33]. An inherent advantage of the titanium implanted, as initially mentioned, is its ability to seamlessly fuse with the bone in the jaw. However, in recent times, the word "tight apposition" or "mechanical fit" has been more correctly applied to describe this connection, instead of the term "real bonding" [34].

1.1.2 Polymeric Biomaterials

Biopolymers are among the most promising materials that have the potential to be used in several industries, particularly in the field of biomedicine. Additionally, they possess ecological friendliness, chemical adaptability, sustainability, biocompatibility, biodegradability, and natural functionality. High-density polyethylene (HDPE), polytetrafluoroethylene (PTFE), polymethylmethacrylate (PMMA), and other biopolymers are frequently used in applications in medicine due to their outstanding moldability, outstanding biocompatibility, wide availability, and cost. Additional polymer biomaterials include acrylic, polyamide, polyester, polyethylenes, polysiloxanes, and polyurethanes. Polymeric biomaterials find several applications in the medical field, including replacements of joints, synthetic skin, heart rate monitors, soft-tissue substitutes, encapsulations, prosthetic blood vessels, and sutures. Additional uses encompass the use of artificial organs such as the pancreas, hearts, livers, kidneys, and bladders [35–39].

1.1.3 Ceramic Biomaterials

Over the past few years, extensive research has been conducted to develop bioactive composites as substitutes for bone. This involves integrating bioactive HAp ceramic fragments with a bioinert polyethylene with a high-density matrix. The commonly used ceramic implant materials include graphite, calcium phosphate, apatites, and aluminum oxide. Furthermore, glasses were developed for application in the field of healthcare. Ceramics were chosen for their non-toxicity to humans, excellent wear qualities in certain situations, ability to be molded into various shapes and porosities, and inertness within the body. Some applications of ceramics include the use of heart valves, hip prosthesis, artificial knees, bone transplants, and numerous orthopedic, dental, and other applications relating to tissue in-growth. Ceramics frequently exhibit inadequate mechanical characteristics when subjected to load-bearing and stress applications. For implant devices that must endure significant tensile stresses, meticulous development and manufacturing are crucial to ensure the safe use of ceramics [40–45].

1.1.4 Composite Biomaterials

Prosthesis developers are now employing composite biomaterials in numerous uses, following their widespread application in dentistry. Typically, carbon fibers are employed to strengthen an ultrahigh-molecular-weight polyethylene (UHMWPE) matrix. To enhance the Young's modulus and tensile strength, carbon fibers are produced by pyrolyzing acrylic fibers, resulting in the formation of a graphitic structure. The carbon fibers exhibit a random alignment inside the matrix and have a diameter ranging from 6 to 15 nm. While the greater Young's modulus of the reinforcing fibers enhances the strength of the matrix, it is crucial for the manufacturing process to establish a strong interfacial connection between the fiber and matrix. Subsequently, by utilizing this fiber-reinforced composite material, a variety of medical implants, such as intramedullary rods and prosthetic joints, can be manufactured. Due to the mechanical qualities of these composites and the quantity of carbon fibers they contain, it is feasible to modify the adaptability of the material's configuration to suit the final shape of a prosthesis. Composites possess distinct characteristics and frequently exhibit greater strength compared with the individual homogeneous substances they are composed of. Experts in the field have utilized this attribute to address various complex problems that necessitate tissue growth. Examples comprise the deposition of Al_2O_3 on carbon, the combination of carbon and PTFE, the combination of Al_2O_3 and PTFE, and the coating of carbon fibers on polylactic acid (PLA) [46–50].

1.1.5 Nanocellulose

Nanocellulose is widely regarded as an innovative biological material due to its unique features and ability to interact well with living organisms. Cellulose, predominantly present in the cell walls of plants, is the most prevalent natural polymer on our planet [51]. The word "nanocellulose" is used to refer to cellulose that has been reduced to the nanometer and recovered as nanofibers of cellulose or nanocrystals. Scientists and material researchers are currently focused on developing new biomedical materials made from naturally occurring polymers for usage in medical and everyday uses. The term "bio cellulose" has been employed in certain research to denote details about nanocellulose and its various applications [52–54]. This is due to the material's unique properties and its potential for application in the study of diverse biomedical materials. Nanocellulose, an exceptional material, can be produced by mechanically, enzymatically, or chemically breaking down bacteria, cell walls of plants, or cotton linters into nanoscaled cellulose fibrils and nanocrystals. Nanocellulose is derived from natural cellulose fibers and typically has a width of approximately 100 nm and a length of a few micrometers. The classification of cellulose can be divided into three primary categories: nanofibrillated cellulose (NFC), bacterial nanocellulose (BNC), and cellulose nanocrystals (CNC). Nanocellulose possesses remarkable characteristics such as a high elastic modulus (110–220 GPa), tensile strength (7.5–7.7 GPa), customizable aspect ratios, a large specific surface area, easy surface performance, adaptable crystallization, an essential quantity of polymerization, and excellent chemical resistance. Nanocellulose sustainable biomaterials have grown significantly in several fields such as aerospace, automotive, packaging, energy devices, and other transdisciplinary domains due to their distinct physicochemical features. It is imperative to transition into these sustainable nanomaterials in the next years. The engineering, biomaterials, and high-end applications have recently discovered the importance and promise of diverse types of cellulose, especially nanocellulose. These materials are commonly used in veterinary medicine, medical services, and water filtration. Understanding the measurements of nanocellulose fibers obtained from different plants and nanofibrils (NFs) is vital since the effectiveness of nanocellulose-based products relies on achieving precise nanocellulose sizes [55–58].

2 NON-BIODEGRADABLE SUSTAINABLE BIOMATERIALS

2.1 DEFINITION

Non-biodegradable sustainable biomaterials refer to materials derived from renewable resources that possess durability and resilience, yet do not degrade into harmful substances in the environment over a short period. These materials are designed to fulfill specific functional requirements while minimizing their impact on ecosystems and human health. Unlike traditional non-biodegradable materials like conventional plastics, which persist in the environment for hundreds to thousands of years, sustainable biomaterials are engineered to degrade or undergo recycling processes at the end of their useful life cycle, thereby reducing their environmental footprint [35, 59, 60].

2.2 CHARACTERISTICS OF NON-BIODEGRADABLE SUSTAINABLE BIOMATERIALS [61–67]

a. Renewable Sourcing:

Non-biodegradable sustainable biomaterials are sourced from renewable resources, such as plants, algae, fungi, or animal by-products. These resources are replenishable and contribute to reducing dependence on finite fossil fuel-derived materials, making them more sustainable in the long term [61–67].

b. Durability:

One of the key characteristics of non-biodegradable sustainable biomaterials is their durability. Unlike biodegradable materials that break down relatively quickly, these materials maintain their structural integrity and functionality over an extended period, making them suitable for various applications requiring longevity and robustness [61–67].

c. Recyclability:

Sustainable biomaterials are designed to be recyclable, meaning they can be processed and transformed into new products or materials at the end of their lifecycle. This characteristic promotes a closed-loop system, where materials are continuously reused, reducing the demand for virgin resources and minimizing waste generation [61–67].

  d. Upgradability:

Another important aspect of non-biodegradable sustainable biomaterials is their potential for upcycling. Upcycling involves converting waste or by-products into materials of higher value or quality. Sustainable biomaterials can be engineered to undergo upcycling processes, where they are transformed into new materials with improved properties or functionality, extending their lifespan and reducing environmental impact [61–67].

  e. Reduced Environmental Impact:

Non-biodegradable sustainable biomaterials are developed with a focus on minimizing their environmental impact throughout their lifecycle. This includes considering factors such as resource extraction, manufacturing processes, usage, and end-of-life disposal or recycling options. By optimizing these aspects, sustainable biomaterials aim to reduce greenhouse gas emissions, energy consumption, and pollution associated with traditional non-renewable materials [61–67].

  f. Compatibility with Circular Economy Principles:

Sustainable biomaterials align with the principles of the circular economy, which aims to maximize resource efficiency and minimize waste generation. These materials contribute to closing the loop of material flow by being reusable, recyclable, or upgradable, thereby promoting a more sustainable and resilient economy [61–67].

  g. Biocompatibility and Safety:

Non-biodegradable sustainable biomaterials are designed to be biocompatible and safe for use in various applications, including food packaging, medical devices, and consumer goods. They undergo rigorous testing to ensure that they do not pose harm to human health or the environment, meeting regulatory standards and requirements [61–67].

  h. Versatility:

Sustainable biomaterials exhibit versatility in terms of their applications across different industries and sectors. From packaging materials to construction products, textiles, and electronics, these materials offer a wide range of functionalities and properties, making them suitable for diverse end uses [61–67].

2.3 TYPES OF NON-BIODEGRADABLE BIOMATERIALS

Non-biodegradable biomaterials play a crucial role in the development of sustainable solutions to address environmental challenges. Among these biomaterials, two prominent types stand out: recyclable polymers and upcycled materials. This section explores these categories in detail, including their characteristics, processes, and applications.

TABLE 6.1

Recyclable polymers, their properties, and recycling processes

Polymer	Properties	Recycling process
Polyethylene (PE)	High durability, flexibility, and chemical resistance	Mechanical recycling: sorting, shredding, melting, and molding
Polypropylene (PP)	High strength, heat resistance, and recyclability	Chemical recycling: pyrolysis, depolymerization, and conversion to monomers or fuel
Polyethylene terephthalate (PET)	Transparency, strength, and barrier properties	Advanced recycling: chemical depolymerization to produce virgin-quality PET
Polystyrene (PS)	Lightweight, insulating, and versatile	Mechanical recycling: grinding, washing, and extrusion
Polyvinyl chloride (PVC)	Durability, chemical resistance, and versatility	Mechanical recycling: grinding and remelting

2.3.1   Recyclable Polymers

Recyclable polymers are a subset of non-biodegradable biomaterials that are designed to be recycled multiple times without significant degradation in their properties. These polymers offer a sustainable alternative to conventional plastics, which often end up in landfills or oceans, causing environmental pollution. Recyclable polymers can be processed through various recycling methods, including mechanical and chemical recycling, to recover their raw materials for reuse [63, 68, 69]. The following Table 6.1 provides an overview of some common recyclable polymers, their properties, and recycling processes:

Recyclable polymers are essential components of sustainable biomaterials, offering versatility, durability, and recyclability. Among the most commonly used recyclable polymers are polyethylene (PE), polypropylene (PP), polyethylene terephthalate (PET), polystyrene (PS), and polyvinyl chloride (PVC). Each polymer has distinct properties, advantages, and disadvantages that make them suitable for various applications [70].

1. Polyethylene (PE)

     Polyethylene is one of the most widely used thermoplastics, known for its versatility, durability, and chemical resistance. It is categorized into several types based on its density, including high-density polyethylene (HDPE) and low-density polyethylene (LDPE).

Properties:
- High durability and toughness
- Excellent chemical resistance
- Good electrical insulation properties
- Flexible and lightweight
- Transparency (for LDPE)
- Resistance to moisture and environmental stress cracking

Advantages:
- Widely available and cost-effective
- Recyclable through mechanical recycling processes
- Versatile applications in packaging, construction, agriculture, and automotive industries
- Lightweight nature reduces transportation costs and energy consumption

Disadvantages:
- Susceptible to degradation under UV radiation
- Limited resistance to high temperatures
- Difficulties in achieving high-quality recycled materials due to contamination issues
- Limited compatibility with certain chemicals and solvents

2. Polypropylene (PP)

   Polypropylene is a versatile thermoplastic known for its high strength, heat resistance, and recyclability. It is commonly used in packaging, automotive components, textiles, and medical devices.

Properties:
- High strength-to-weight ratio
- Excellent chemical resistance
- Resistance to fatigue and stress cracking
- Heat resistance up to 100°C (depending on grade)
- Low density and lightweight
- Good moisture barrier properties

Advantages:
- Recyclable through both mechanical and chemical recycling processes
- Versatile applications in packaging, automotive, textiles, and consumer goods
- Lightweight nature reduces material usage and transportation costs
- Long-term durability and stability

Disadvantages:
- Susceptible to degradation under prolonged exposure to UV radiation
- Limited resistance to high temperatures compared with other polymers
- Relatively higher production costs compared with some other polymers
- Limited compatibility with certain chemicals and solvents

3. Polyethylene terephthalate (PET)

   Polyethylene terephthalate, commonly known as PET, is a transparent thermoplastic polymer with excellent barrier properties, making it ideal for packaging applications, particularly in the food and beverage industry.

Properties:
- Excellent barrier properties to gases, water, and aromas
- Transparency and clarity
- Lightweight and durable
- Good mechanical strength and stiffness
- Resistance to grease and oils
- Recyclable through both mechanical and chemical recycling processes

Advantages:
- Widely recycled worldwide, with established recycling infrastructure
- Versatile applications in packaging, textiles, and automotive industries
- Lightweight nature reduces transportation costs and environmental impact
- High strength-to-weight ratio

Disadvantages:
- Susceptible to degradation under prolonged exposure to UV radiation
- Limited resistance to high temperatures, limiting its use in certain applications
- Potential for contamination issues during recycling processes
- Concerns regarding chemical migration in food and beverage packaging applications

4. Polystyrene (PS)

Polystyrene is a lightweight and versatile thermoplastic known for its insulation properties, making it suitable for packaging, construction, and consumer goods applications.

Properties:
- Lightweight and rigid
- Excellent insulation properties (thermal and electrical)
- Transparency and clarity (for certain grades)
- Resistance to moisture and chemicals
- Low cost and ease of processing
- Recyclable through mechanical recycling processes

Advantages:
- Versatile applications in packaging, insulation, and consumer goods
- Cost-effective solution for lightweight and disposable products
- Excellent insulation properties make it suitable for food packaging and construction applications
- Recyclable through existing infrastructure, particularly in foam and rigid forms

Disadvantages:
- Susceptible to degradation under prolonged exposure to UV radiation
- Limited resistance to high temperatures, leading to softening and deformation

- Concerns regarding environmental impact, particularly in marine environments
- Challenges in recycling due to contamination and limited market demand for recycled PS

5. Polyvinyl chloride (PVC)

Polyvinyl chloride, commonly known as PVC, is a versatile thermoplastic known for its durability, chemical resistance, and electrical insulation properties. It is widely used in construction, healthcare, automotive, and consumer goods industries.

Properties:
- High durability and toughness
- Excellent chemical resistance
- Flame retardant properties
- Electrical insulation properties
- Versatile processing methods (extrusion, injection molding, etc.)
- Recyclable through both mechanical and chemical recycling processes

Advantages:
- Versatile applications in construction, healthcare, automotive, and consumer goods
- Excellent durability and weather resistance make it suitable for outdoor applications
- Cost-effective solution for a wide range of products and applications
- Recyclable through existing infrastructure, with established recycling programs

Disadvantages:
- Concerns regarding environmental impact, particularly during production and disposal
- Release of toxic gases (hydrochloric acid) during incineration
- Limited compatibility with certain polymers, leading to challenges in recycling
- Susceptible to degradation under prolonged exposure to UV radiation and heat [71–73]

2.3.2 Recycling Processes

I. Mechanical recycling:

Mechanical recycling is the most common method for recycling polymers, particularly thermoplastics like polyethylene, polypropylene, and polystyrene. The process involves several steps:

1. Sorting: Plastic waste is sorted based on polymer type, color, and quality. Automated sorting systems utilize technologies such as near-infrared (NIR) spectroscopy and optical sensors to separate different types of plastics.

2. Shredding: The sorted plastic waste is shredded into smaller pieces or flakes using shredding equipment. This increases the surface area and facilitates subsequent processing steps.
3. Washing: The shredded plastic flakes are washed to remove contaminants, such as dirt, labels, adhesives, and residual contents. Water-based washing systems and detergents are commonly used for this purpose.
4. Melting and molding: The clean plastic flakes are melted and formed into pellets or granules through extrusion or injection molding processes. These pellets can then be used as raw materials for manufacturing new plastic products [70].

II. Chemical recycling:

Chemical recycling, also known as advanced recycling or depolymerization, involves breaking down polymers into their constituent monomers or other valuable chemicals. This process offers a promising solution for recycling plastics that are difficult to mechanically recycle or contain multiple layers or additives. The main methods of chemical recycling include:

1. Pyrolysis: Pyrolysis is a thermal decomposition process that converts plastic waste into oil, gas, and char in the absence of oxygen. The generated oil can be further refined into fuels or chemical feedstocks, while the gas can be used for energy generation.
2. Depolymerization: Depolymerization involves breaking down polymer chains into monomers or oligomers through chemical reactions. These monomers can then be purified and used to produce new polymers with virgin-quality properties [73].

3 UPCYCLED MATERIALS

Upcycled materials represent another category of non-biodegradable biomaterials that are derived from waste or by-products and transformed into products of higher value or quality. Unlike traditional recycling, which often involves downcycling or converting materials into lower-grade products, upcycling aims to retain or enhance the inherent properties of the original materials. Upcycled materials offer a sustainable alternative to conventional materials, reducing waste generation and resource depletion [74, 75]. Table 6.2 highlights some examples of upcycled materials and their characteristics:

Upcycling processes vary depending on the type of material and its intended application. However, they generally involve several common steps:

1. Sorting and cleaning: Waste materials are sorted based on their composition, size, and quality. Contaminants such as dirt, labels, and adhesives are removed through cleaning and washing processes.
2. Processing: The sorted materials undergo mechanical or chemical processing to transform them into usable forms. This may include shredding, grinding, melting, or chemical treatments to break down complex structures and impurities.

TABLE 6.2

Upcycled materials and their characteristics

Material	Source	Upcycling process
Recycled plastic lumber	Post-consumer plastic waste	Sorting, cleaning, melting, and extrusion into lumber boards
Upcycled textiles	Textile waste and scraps	Sorting, shredding, and spinning into yarns for new textiles
Upcycled glass	Post-consumer glass bottles	Crushing, melting, and forming into new glass products
Upcycled metal alloys	Scrap metal and industrial waste	Sorting, melting, and alloying to produce high-quality metal alloys
Upcycled wood composites	Demolition wood and sawmill waste	Grinding, blending with binders, and pressing into composite boards

3. Transformation: The processed materials are transformed into new products or materials through shaping, molding, casting, or extrusion processes. This step may involve combining upcycled materials with other additives or binders to enhance their properties and performance.
4. Quality control: Upcycled products undergo quality control measures to ensure that they meet relevant standards and specifications. This may include testing for strength, durability, chemical resistance, and other performance criteria [76, 77].

3.1 Environmental Impact and End-of-Life Options

Understanding the environmental impact of non-biodegradable sustainable biomaterials is essential for evaluating their sustainability and guiding end-of-life management strategies. This section delves into the environmental implications of these materials and explores various options for their disposal or recycling.

3.1.1 Environmental Impact

1. **Resource Depletion:** Non-biodegradable sustainable biomaterials are often derived from renewable resources, such as plants, but their production may still contribute to resource depletion if not managed sustainably. Extraction of raw materials, energy consumption, and water usage during manufacturing processes can impact ecosystems and biodiversity.
2. **Greenhouse Gas Emissions:** The production and processing of non-biodegradable sustainable biomaterials may generate greenhouse gas emissions, contributing to climate change. Factors such as energy sources, transportation, and chemical processes influence the carbon footprint of these materials.
3. **Pollution:** The disposal of non-biodegradable sustainable biomaterials at the end of their lifecycle can result in pollution of land, water, and air. Improper waste management practices, such as landfilling or incineration, can lead to the release of harmful chemicals and pollutants into the environment.

4. **Microplastics:** Some non-biodegradable biomaterials, particularly plastics, can degrade into microplastics over time, posing risks to aquatic and terrestrial ecosystems. Microplastics can accumulate in the food chain, potentially causing harm to marine life and human health.
5. **End-of-Life Challenges:** Recycling infrastructure, technology, and consumer behavior play significant roles in determining the environmental impact of non-biodegradable sustainable biomaterials. Limited recycling facilities, contamination issues, and lack of consumer awareness can hinder recycling efforts and contribute to waste accumulation [78, 79].

3.1.2 End-of-Life Options

1. **Recycling:** Recycling is a key end-of-life option for non-biodegradable sustainable biomaterials, including plastics and composite materials. Mechanical recycling processes involve sorting, shredding, and melting plastic waste to produce recycled pellets or flakes for manufacturing new products. Chemical recycling techniques, such as pyrolysis and depolymerization, break down polymers into their constituent monomers for reuse.
2. **Upcycling:** Upcycling offers a sustainable alternative to traditional recycling by transforming waste materials into higher-value products or materials. Upcycled products retain or enhance the properties of the original materials, reducing the need for virgin resources and minimizing waste generation. Examples include upcycled composite materials made from recycled plastics and fibers for use in construction and manufacturing.
3. **Energy Recovery:** Waste-to-energy technologies, such as incineration and gasification, can be employed to convert non-recyclable biomass and waste materials into heat, electricity, or biofuels. While energy recovery helps reduce landfill volumes and fossil fuel consumption, it may also emit greenhouse gases and pollutants if not properly controlled.
4. **Biodegradation:** Although non-biodegradable sustainable biomaterials are designed to resist degradation in the environment, some materials may undergo biodegradation under specific conditions. Biodegradable additives or enzymes can be incorporated into polymers to facilitate microbial breakdown, leading to the eventual decomposition of the material into harmless by-products.
5. **Landfill Disposal:** Landfilling remains a common disposal method for non-biodegradable biomaterials that cannot be recycled or recovered. However, landfilling poses environmental risks, including leachate contamination, methane emissions, and land degradation. Landfill management practices, such as liner systems and methane capture, can mitigate these impacts to some extent [80, 81].

3.2 APPLICATIONS OF NON-BIODEGRADABLE BIOMATERIALS

1. Recyclable Plastic Alternatives:

   Recyclable plastic alternatives offer sustainable solutions to mitigate the environmental impact of conventional plastics. By incorporating recyclable polymers and biodegradable materials, manufacturers can develop eco-friendly packaging, consumer goods, and construction materials. Case studies highlight successful applications and innovations in recyclable plastic alternatives:

a. **Biodegradable Plastics:** Biodegradable polymers, such as PLA and polyhydroxyalkanoates (PHA), offer compostable alternatives to conventional plastics. Companies like NatureWorks and Danimer Scientific produce biodegradable packaging films, food containers, and disposable cutlery from renewable resources.

b. **Plant-Based Plastics:** Plant-based plastics, derived from sources like corn, sugarcane, and cassava, provide renewable alternatives to petroleum-based plastics. Coca-Cola's PlantBottle™ technology utilizes bio-based polyethylene terephthalate (PET) for beverage bottles, reducing reliance on fossil fuels and lowering carbon emissions.

c. **Recycled Plastics:** Recycling initiatives aim to close the loop on plastic waste by collecting, sorting, and processing post-consumer plastics into new products. Companies like Loop Industries and TerraCycle specialize in advanced recycling technologies to produce high-quality recycled plastics for packaging, textiles, and automotive components [82–84].

2. Upcycled Composite Materials:

Upcycled composite materials combine recycled polymers, fibers, and fillers to create durable and sustainable alternatives to conventional materials. These materials find applications in construction, automotive, and consumer goods industries, reducing waste and resource consumption. Case studies demonstrate the versatility and benefits of upcycled composite materials:

a. **Upcycled Wood–Plastic Composites:** Wood–plastic composites (WPCs) combine recycled wood fibers or flour with thermoplastic polymers to create durable and eco-friendly building materials. Companies like Trex Company and Fiberon produce WPC decking, fencing, and furniture, diverting wood waste from landfills, reducing reliance on virgin timber.

b. **Upcycled Metal–Plastic Composites:** Metal–plastic composites incorporate recycled metals, such as aluminum or steel, with thermoplastic polymers to create lightweight and corrosion-resistant materials. Applications include automotive components, electronic enclosures, and structural reinforcements, offering sustainable alternatives to traditional metal alloys [85].

4   SUSTAINABLE BIOMATERIALS IN BIOPROCESSING 4.0

In the era of Bioprocessing 4.0, the convergence of advanced bioprocessing technologies and sustainable biomaterials is poised to revolutionize industrial processes, driving environmental sustainability, resource efficiency, and technological innovation. This comprehensive exploration delves into the integration of sustainable biomaterials in advanced bioprocessing, the pivotal role of bioprocessing technologies in enhancing sustainability, and the future trends and challenges shaping this dynamic field.

4.1 Integration of Sustainable Biomaterials in Advanced Bioprocessing

The integration of sustainable biomaterials in advanced bioprocessing represents a paradigm shift toward more environmentally friendly and resource-efficient manufacturing practices. This section delves into the definition, scope, advantages, challenges, and considerations associated with the integration of sustainable biomaterials in bioprocessing [86].

4.1.1 Definition and Scope of Sustainable Biomaterials

Sustainable biomaterials encompass a diverse array of materials derived from renewable resources, such as biomass, agricultural residues, and waste streams. These materials exhibit properties such as biodegradability, recyclability, and low environmental impact, making them ideal candidates for integration into advanced bioprocessing systems. Examples include bio-based polymers, biodegradable plastics, biofuels, and biocompatible materials used in pharmaceuticals and medical devices.

Environmental Sustainability: Sustainable biomaterials offer a renewable alternative to traditional fossil-based materials, reducing reliance on finite resources and mitigating environmental impact. By harnessing renewable resources such as plant biomass, algae, and organic waste, bioprocessing can contribute to carbon sequestration, biodiversity conservation, and ecosystem restoration.

Economic Viability: The abundance and availability of biomass-derived biomaterials make them economically attractive for industrial applications, contributing to cost savings and resource efficiency. By leveraging sustainable feedstocks and bioprocessing techniques, companies can enhance their competitiveness, reduce production costs, and create new revenue streams in emerging bio-based markets.

Technological Innovation: The unique properties of sustainable biomaterials, such as biocompatibility, tunable functionality, and renewable origin, enable the development of novel bioprocessing technologies and applications. From bio-based chemicals and materials to renewable energy and healthcare products, sustainable biomaterials drive innovation across diverse industries, unlocking new opportunities for sustainable development and growth [87–89].

4.1.2 Challenges and Considerations

Supply Chain Management: Ensuring a consistent and reliable supply of biomass feedstocks for bioprocessing operations is essential for scaling up sustainable biomaterial production. Challenges include biomass availability, seasonality, geographic distribution, and logistics, as well as competition with other sectors, such as food, feed, and bioenergy.

Process Optimization: Developing efficient and scalable bioprocessing techniques for converting biomass into value-added products requires optimization of process parameters, reactor design, and downstream processing steps. Challenges include substrate variability, product yield, purity, and quality, as well as the integration of multi-step biorefinery processes and co-product utilization.

Regulatory Compliance: Meeting regulatory requirements and standards for sustainable biomaterials, including safety, quality, and environmental regulations, poses challenges for industry stakeholders. Compliance issues include product certification, labeling, traceability, and risk assessment, as well as international harmonization and market access in a rapidly evolving regulatory landscape [90, 91].

4.2 ROLE OF BIOPROCESSING TECHNOLOGIES IN ENHANCING SUSTAINABILITY

Bioprocessing technologies play a pivotal role in enhancing sustainability by enabling the efficient conversion of sustainable biomaterials into value-added products and materials. This section explores the diverse array of bioprocessing techniques, tools, and platforms driving sustainability across various industries.

1. Biorefinery Approaches:

    Biorefinery concepts encompass integrated processes for the conversion of biomass into multiple products, including fuels, chemicals, materials, and energy. Sustainable biomaterials serve as feedstocks for biorefinery platforms, enabling the efficient utilization of biomass resources and the production of high-value bioproducts. Examples include lignocellulosic bioethanol, biodiesel, bio-based polymers, biochemicals, and biogas produced from agricultural residues, forestry waste, and municipal solid waste.

2. Biocatalysis and Enzyme Engineering:

    Enzymatic biocatalysis offers a sustainable and efficient approach to chemical synthesis and bioprocessing, enabling the production of bio-based chemicals, pharmaceuticals, and materials from renewable feedstocks. Enzymes derived from microbial, plant, and animal sources catalyze specific reactions with high selectivity, efficiency, and sustainability. Examples include enzyme-mediated synthesis of biofuels, biopolymers, fine chemicals, and pharmaceutical intermediates, as well as enzyme immobilization, engineering, and optimization for enhanced stability, activity, and specificity.

3. Fermentation and Bioproduction:

    Microbial fermentation processes play a key role in the production of bio-based chemicals, polymers, pharmaceuticals, and nutraceuticals from sustainable biomaterials. Microorganisms such as bacteria, yeast, and fungi convert sugars, carbohydrates, and other organic substrates into value-added products through fermentation, anaerobic digestion, and metabolic engineering. Examples include bioethanol, lactic acid, succinic acid, polyhydroxyalkanoates (PHA), and recombinant proteins produced from agricultural crops, industrial residues, and lignocellulosic biomass.

4. Bioprocess Monitoring and Control:

    Advanced bioprocessing technologies, such as in-situ sensors, data analytics, and process automation, enhance the efficiency and sustainability of biomanufacturing processes. Real-time monitoring and control of key process parameters optimize resource utilization, minimize waste generation, and ensure product quality and consistency. Examples include online sensors for monitoring biomass growth, substrate consumption, and product formation, as well as control strategies for regulating temperature, pH, dissolved oxygen, and agitation in bioreactors [91, 92, 93].

4.3   Future Trends and Challenges

The future of sustainable biomaterials in Bioprocessing 4.0 is shaped by emerging trends, innovations, and challenges that drive continuous improvement and transformation in the bio-based economy. This section explores key trends, challenges, and opportunities shaping the future of sustainable biomaterials and bioprocessing.

1. Emerging Biomaterials and Technologies:

    Continued research and innovation in biomaterials science and bioprocessing technologies are driving the development of novel sustainable biomaterials and bioproduction methods. Emerging trends include the use of synthetic biology, metabolic engineering, and nanotechnology to design and produce advanced biomaterials with tailored properties and functionalities. Examples include bio-based polymers, biodegradable plastics, bioactive compounds, and functional materials for applications in healthcare, agriculture, food, and packaging.

2. Circular Economy and Waste Valorization:

    The transition toward a circular economy model presents opportunities for waste valorization and resource recovery through bioprocessing. Biorefinery platforms and waste-to-value technologies enable the conversion of organic waste streams into bio-based products, energy, and materials, closing the loop on resource utilization and waste generation. Examples include bioconversion of agricultural residues, food waste, and municipal solid waste into bioenergy, biofertilizers, biopolymers, and biochemicals through anaerobic digestion, fermentation, and enzymatic hydrolysis.

3. Sustainability Metrics and Assessment:

    The development of standardized metrics and methodologies for assessing the environmental, social, and economic sustainability of bioprocessing technologies and biomaterials is essential for informed decision-making and policy development. Life cycle assessment, carbon footprint analysis, and sustainability certification schemes provide valuable tools for evaluating the sustainability performance of bioprocesses and products, identifying hotspots, and optimizing resource allocation. Challenges include data availability, methodological complexity, and stakeholder engagement, as well as uncertainty and variability in environmental impact assessments.

4. Policy and Regulatory Considerations:

    Policy frameworks and regulations play a crucial role in shaping the adoption and deployment of sustainable biomaterials and bioprocessing technologies. Governments, industry stakeholders, and research institutions need to collaborate to establish supportive policies, incentives, and standards for promoting sustainable bioeconomy initiatives and fostering innovation in bioprocessing. Examples include renewable energy mandates, bio-based product procurement policies, tax incentives, and research funding programs aimed at accelerating the development and commercialization of sustainable biomaterials and bioproducts. Challenges include policy fragmentation, regulatory uncertainty, and market access barriers, as well as conflicting interests and priorities among stakeholders [88, 94, 95].

REFERENCES

[1] Rodrigo-Navarro A, Sankaran S, Dalby MJ, del Campo A, Salmeron-Sanchez M. Engineered Living Biomaterials. *Nature Reviews Materials.* 2021 Dec;6(12):1175–1190.

[2] Khan A, Alamry KA, Asiri AM. Multifunctional Biopolymers-Based Composite Materials for Biomedical Applications: A Systematic Review. *ChemistrySelect.* 2021 Jan 14;6(2):154–176.

[3] Nehra P, Chauhan RP. Eco-friendly Nanocellulose and Its Biomedical Applications: Current Status and Future Prospect. *Journal of Biomaterials Science*, Polymer Edition. 2021 Jan 2;32(1):112–149.

[4] Chen L, Zhang Y, Chen Z, Dong Y, Jiang Y, Hua J, Liu Y, Osman AI, Farghali M, Huang L, Rooney DW. Biomaterials Technology and Policies in the Building Sector: A Review. *Environmental Chemistry Letters.* 2024 Jan 29:1–36.

[5] Karak N. Fundamentals of Sustainable Nanostructural Materials at Bio-nano Interface. In *Dynamics of Advanced Sustainable Nanomaterials and Their Related Nanocomposites at the Bio-nano Interface*, Karak, Niranjan, Ed . 2019 Jan 1 (pp. 1–24). Elsevier.

[6] Tiza TM, Kpur G, Ogunleye E, Sharma S, Singh SK, Likassa DM. The Potency of Functionalized Nanomaterials for Industrial Applications. *Materials Today: Proceedings.* 2023 Mar 21.

[7] Stephen M, Nawaz A, Lee SY, Sonar P, Leong WL. Biodegradable Materials for Transient Organic Transistors. *Advanced Functional Materials.* 2023 Feb;33(6):2208521.

[8] Biswal T, BadJena SK, Pradhan D. Sustainable Biomaterials and their Applications: A Short Review. *Materials Today: Proceedings.* 2020 Jan 1;30:274–282.

[9] Festas AJ, Ramos A, Davim JP. Medical Devices Biomaterials–A Review. *Proceedings of the Institution of Mechanical Engineers, Part L: Journal of Materials: Design and Applications.* 2020 Jan;234(1):218–228.

[10] Kargozar S, Ramakrishna S, Mozafari M. Chemistry of Biomaterials: Future Prospects. *Current Opinion in Biomedical Engineering.* 2019 Jun 1;10:181–190.

[11] Kulkarni SV, Nemade AC, Sonawwanay PD. An Overview on Metallic and Ceramic Biomaterials. *Recent Advances in Manufacturing Processes and Systems: Select Proceedings of RAM 2021.* 2022 Mar 3:149–165.

[12] Ødegaard KS, Torgersen J, Elverum CW. Structural and Biomedical Properties of Common Additively Manufactured Biomaterials: A Concise Review. *Metals.* 2020 Dec 15;10(12):1677.

[13] Zafar MS. Prosthodontic Applications of Polymethyl Methacrylate (PMMA): An Update. *Polymers.* 2020 Oct 8;12(10):2299.

[14] Soleymani Eil Bakhtiari S, Bakhsheshi-Rad HR, Karbasi S, Tavakoli M, Hassanzadeh Tabrizi SA, Ismail AF, Seifalian A, RamaKrishna S, Berto F. Poly (methyl methacrylate) Bone Cement, Its Rise, Growth, Downfall and Future. *Polymer International.* 2021 Sep;70(9):1182–1201.

[15] Saxena P, Shukla P. A Comparative Analysis of the Basic Properties and Applications of Poly (vinylidene fluoride)(PVDF) and Poly (methyl methacrylate)(PMMA). *Polymer Bulletin.* 2022 Aug;79(8):5635–5665.

[16] Ratner BD, Hoffman AS, Schoen FJ, Lemons JE, Wagner WR, Sakiyama-Elbert SE, Zhang G, Yaszemski MJ. Introduction—Biomaterials Science: An Evolving, Multidisciplinary Endeavor. In *Biomaterials Science: An Introduction to Materials in Medicine*; Wagner, W., Sakiyama-Elbert, S., Zhang, G., Yaszemski, M., Eds. 2020 Jan 1 (pp. 3–19). Elsevier.

[17] Kapusetti G, More N, Choppadandi M. Introduction to Ideal Characteristics and Advanced Biomedical Applications of Biomaterials. *Biomedical Engineering and Its Applications in Healthcare.* 2019:171–204.

[18] Fakhri E, Eslami H, Maroufi P, Pakdel F, Taghizadeh S, Ganbarov K, Yousefi M, Tanomand A, Yousefi B, Mahmoudi S, Kafil HS. Chitosan Biomaterials Application in Dentistry. *International Journal of Biological Macromolecules.* 2020 Nov 1;162:956–974.

[19] Abraham AM, Venkatesan S. A Review on Application of Biomaterials for Medical and Dental Implants. *Proceedings of the Institution of Mechanical Engineers, Part L: Journal of Materials: Design and Applications.* 2023 Feb;237(2):249–273.

[20] Salthouse D, Novakovic K, Hilkens CM, Ferreira AM. Interplay between Biomaterials and the Immune System: Challenges and Opportunities in Regenerative Medicine. *Acta Biomaterialia.* 2023 Jan 1;155:1–8.

[21] Lv B, Zhang X, Yuan J, Chen Y, Ding H, Cao X, Huang A. Biomaterial-supported MSC Transplantation Enhances Cell–Cell Communication for Spinal Cord Injury. *Stem Cell Research & Therapy.* 2021 Dec;12:1–6.

[22] Cheekuramelli NS, Late D, Kiran S, Garnaik B. Biodegradable and Biocompatible Polymer Composite: Biomedical Applications and Bioimplants. In *Lightweight Polymer Composite Structures*, Sanjay Mavinkere Rangappa, Jyotishkumar Parameswaranpillai, Suchart Siengchin, Lothar Kroll, Eds. 2020 Sep 1 (pp. 67–88). CRC Press.

[23] Cometa S, Bonifacio MA, Mattioli-Belmonte M, Sabbatini L, De Giglio E. Electrochemical Strategies for Titanium Implant Polymeric Coatings: The Why and How. *Coatings.* 2019 Apr 20;9(4):268.

[24] Ansari MA, Golebiowska AA, Dash M, Kumar P, Jain PK, Nukavarapu SP, Ramakrishna S, Nanda HS. Engineering Biomaterials to 3D-Print Scaffolds for Bone Regeneration: Practical and Theoretical Consideration. *Biomaterials Science.* 2022;10(11):2789–2816.

[25] Chaudhary K, Kandasubramanian B. Self-healing Nanofibers for Engineering Applications. *Industrial & Engineering Chemistry Research.* 2022 Mar 14;61(11): 3789–3816.

[26] Choudhury M, Bindra HS, Singh K, Singh AK, Nayak R. Antimicrobial Polymeric Composites in Consumer Goods and Healthcare Sector: A Healthier Way to Prevent Infection. *Polymers for Advanced Technologies.* 2022 Jul;33(7):1997–2024.

[27] Ahmad S, Ahmad M, Manzoor K, Purwar R, Ikram S. A Review on Latest Innovations in Natural Gums based Hydrogels: Preparations & Applications. *International Journal of Biological Macromolecules.* 2019 Sep 1;136:870–890.

[28] Tejero-Martin D, Rezvani Rad M, McDonald A, Hussain T. Beyond Traditional Coatings: A Review on Thermal-sprayed Functional and Smart Coatings. *Journal of Thermal Spray Technology.* 2019 Apr 17;28:598–644.

[29] Biobaku-Mutingwende, B. Introduction to Biomaterials. In *Applications of 3D Printing in Biomedical Engineering,* Sharma, Neeta Raj, Karupppasamy Subburaj, Kamalpreet Sandhu, and Vivek Sharma, Eds. 2021 (pp. 1–9). Springer.

[30] Davis R, Singh A, Jackson MJ, Coelho RT, Prakash D, Charalambous CP, Ahmed W, da Silva LR, Lawrence AA. A Comprehensive Review on Metallic Implant Biomaterials and their Subtractive Manufacturing. *The International Journal of Advanced Manufacturing Technology.* 2022 May;120(3):1473–1530.

[31] Goncalves AD, Balestri W, Reinwald Y. Biomedical Implants for Regenerative Therapies. *Biomaterials.* 2020 Mar 19:1–36.

[32] Richter-Bisson ZW, Doktor A, Hedberg YS. Serum Albumin Aggregation Facilitated by Cobalt and Chromium Metal Ions. *ACS Applied Bio Materials*. 2023 Aug 23;6(9):3832–3841.

[33] Contuzzi N, Casalino G, Boccaccio A, Ballini A, Charitos IA, Bottalico L, Santacroce L. Metals Biotribology and Oral Microbiota Biocorrosion Mechanisms. *Journal of Functional Biomaterials*. 2022 Dec 23;14(1):14.

[34] Ratner BD, Hoffman AS, Schoen FJ, Lemons JE, Wagner WR, Sakiyama-Elbert SE, Zhang G, Yaszemski MJ. Introduction—Biomaterials Science: An Evolving, Multidisciplinary Endeavor. *Biomaterials Science: An Introduction to Materials in Medicine*; Wagner, W., Sakiyama-Elbert, S., Zhang, G., Yaszemski, M., Eds. 2020 Jan 1:3–19.

[35] Baranwal J, Barse B, Fais A, Delogu GL, Kumar A. Biopolymer: A Sustainable Material for Food and Medical Applications. *Polymers*. 2022 Feb 28;14(5):983.

[36] Biswas MC, Jony B, Nandy PK, Chowdhury RA, Halder S, Kumar D, Ramakrishna S, Hassan M, Ahsan MA, Hoque ME, Imam MA. Recent Advancement of Biopolymers and their Potential Biomedical Applications. *Journal of Polymers and the Environment*. 2022 Jan 1:1–24.

[37] Veeman D, Sai MS, Sureshkumar P, Jagadeesha T, Natrayan L, Ravichandran M, Mammo WD. Additive Manufacturing of Biopolymers for Tissue Engineering and Regenerative Medicine: An Overview, Potential Applications, Advancements, and Trends. *International Journal of Polymer Science*. 2021 Sep 8;2021:1–20.

[38] Wróblewska-Krepsztul J, Rydzkowski T, Michalska-Pożoga I, Thakur VK. Biopolymers for Biomedical and Pharmaceutical Applications: Recent Advances and Overview of Alginate Electrospinning. *Nanomaterials*. 2019 Mar 10;9(3):404.

[39] Aslam Khan MU, Abd Razak SI, Al Arjan WS, Nazir S, Sahaya Anand TJ, Mehboob H, Amin R. Recent Advances in Biopolymeric Composite Materials for Tissue Engineering and Regenerative Medicines: A Review. *Molecules*. 2021 Jan 25;26(3):619.

[40] Georgeanu VA, Gingu O, Antoniac IV, Manolea HO. Current Options and Future Perspectives on Bone Graft and Biomaterials Substitutes for Bone Repair, from Clinical Needs to Advanced Biomaterials Research. *Applied Sciences*. 2023 Jul 22;13(14):8471.

[41] Ślósarczyk A, Czechowska J, Cichoń E, Zima A. New Hybrid Bioactive Composites for Bone Substitution. *Processes*. 2020 Mar 12;8(3):335.

[42] Tavoni M, Dapporto M, Tampieri A, Sprio S. Bioactive Calcium Phosphate-based Composites for Bone Regeneration. *Journal of Composites Science*. 2021 Aug 27;5(9):227.

[43] Punj S, Singh J, Singh KJ. Ceramic Biomaterials: Properties, State of the Art and Future prospectives. *Ceramics International*. 2021 Oct 15;47(20):28059–28074.

[44] Karadjian M, Essers C, Tsitlakidis S, Reible B, Moghaddam A, Boccaccini AR, Westhauser F. Biological Properties of Calcium Phosphate Bioactive Glass Composite Bone Substitutes: Current Experimental Evidence. *International Journal of Molecular Sciences*. 2019 Jan 14;20(2):305.

[45] Kaur G, Kumar V, Baino F, Mauro JC, Pickrell G, Evans I, Bretcanu O. Mechanical Properties of Bioactive Glasses, Ceramics, Glass-Ceramics and Composites: State-of-the-art Review and Future Challenges. *Materials Science and Engineering: C*. 2019 Nov 1;104:109895.

[46] Ajmal S, Hashmi FA, Imran I. Recent Progress in Development and Applications of Biomaterials. *Materials Today: Proceedings*. 2022 Jan 1;62:385–391.

[47] Jindal S, Manzoor F, Haslam N, Mancuso E. 3D Printed Composite Materials for Craniofacial Implants: Current Concepts, Challenges and Future Directions. *The International Journal of Advanced Manufacturing Technology*. 2021 Jan;112(3):635–653.

[48] Suwardi A, Wang F, Xue K, Han MY, Teo P, Wang P, Wang S, Liu Y, Ye E, Li Z, Loh XJ. Machine Learning-driven Biomaterials Evolution. *Advanced Materials.* 2022 Jan;34(1):2102703.

[49] Altıparmak SC, Yardley VA, Shi Z, Lin J. Extrusion-based Additive Manufacturing Technologies: State of the Art and Future Perspectives. *Journal of Manufacturing Processes.* 2022 Nov 1;83:607–636.

[50] King FL, Baruch J. Review of Properties of Additive Manufactured Materials and Composites. In *Mechanical Properties and Characterization of Additively Manufactured Materials*, Kumar, Ravi K., S. C. Vettivel, and R. Subramanian, Eds. 2023 Sep 13 (pp. 173–210). CRC Press.

[51] Khalid MY, Al Rashid A, Arif ZU, Ahmed W, Arshad H. Recent Advances in Nanocellulose-Based Different Biomaterials: Types, Properties, and Emerging Applications. *Journal of Materials Research and Technology.* 2021 Sep 1;14: 2601–2623.

[52] Thakur V, Guleria A, Kumar S, Sharma S, Singh K. Recent Advances in Nanocellulose Processing, Functionalization and Applications: A Review. *Materials Advances.* 2021;2(6):1872–1895.

[53] Ullah MW, Rojas OJ, McCarthy RR, Yang G. Nanocellulose: A Multipurpose Advanced Functional Material. *Frontiers in Bioengineering and Biotechnology.* 2021 Jul 22;9:738779.

[54] Trache D, Tarchoun AF, Derradji M, Hamidon TS, Masruchin N, Brosse N, Hussin MH. Nanocellulose: From Fundamentals to Advanced Applications. *Frontiers in Chemistry.* 2020 May 6;8:392.

[55] Bacakova L, Pajorova J, Tomkova M, Matejka R, Broz A, Stepanovska J, Prazak S, Skogberg A, Siljander S, Kallio P. Applications of Nanocellulose/Nanocarbon Composites: Focus on Biotechnology and Medicine. *Nanomaterials.* 2020 Jan 23;10(2):196.

[56] Norrrahim MN, Nurazzi NM, Jenol MA, Farid MA, Janudin N, Ujang FA, Yasim-Anuar TA, Najmuddin SU, Ilyas RA. Emerging Development of Nanocellulose as an Antimicrobial Material: An Overview. *Materials Advances.* 2021;2(11):3538–3551.

[57] Ullah MW, Manan S, Ul-Islam M, Revin VV, Thomas S, Yang G. Introduction to Nanocellulose. In *Nanocellulose: Synthesis, Structure, Properties and Applications,* Yang, Guang, Muhammad Wajid Ullah, and Zhijun Shi, Eds. 2021 (pp. 1–50). World Scientific publishing Europe Ltd.

[58] Dufresne A. Nanocellulose Processing Properties and Potential Applications. *Current Forestry Reports.* 2019 Jun 15;5:76–89.

[59] Kabir E, Kaur R, Lee J, Kim KH, Kwon EE. Prospects of Biopolymer Technology as an Alternative Option for Non-degradable Plastics and Sustainable Management of Plastic Wastes. *Journal of Cleaner Production.* 2020 Jun 10;258:120536.

[60] Muniyasamy S, Mohanrasu K, Gada A, Mokhena TC, Mtibe A, Boobalan T, Paul V, Arun A. Biobased Biodegradable Polymers for Ecological Applications: A Move towards Manufacturing Sustainable Biodegradable Plastic Products. *Integrating Green Chemistry and Sustainable Engineering.* 2019 Mar 26;8:215–253.

[61] Geevarghese R, Sajjadi SS, Hudecki A, Sajjadi S, Jalal NR, Madrakian T, Ahmadi M, Włodarczyk-Biegun MK, Ghavami S, Likus W, Siemianowicz K. Biodegradable and Non-biodegradable Biomaterials and their Effect on Cell Differentiation. *International Journal of Molecular Sciences.* 2022 Dec 19;23(24):16185.

[62] Morais FP, Curto JM. Design and Engineering of Natural Cellulose Fiber-Based Biomaterials with Eucalyptus Essential Oil Retention to Replace Non-Biodegradable Delivery Systems. *Polymers.* 2022 Sep 1;14(17):3621.

[63] Andreeßen C, Steinbüchel A. Recent Developments in Non-biodegradable Biopolymers: Precursors, Production Processes, and Future Perspectives. *Applied Microbiology and Biotechnology.* 2019 Jan;103(1):143–157.

[64] Bisht B, Lohani UC, Kumar V, Gururani P, Sinhmar R. Edible Hydrocolloids as Sustainable Substitute for Non-biodegradable Materials. *Critical Reviews in Food Science and Nutrition*. 2022 Jan 25;62(3):693–725.

[65] Ju S, Shin G, Lee M, Koo JM, Jeon H, Ok YS, Hwang DS, Hwang SY, Oh DX, Park J. Biodegradable Chito-beads Replacing Non-biodegradable Microplastics for Cosmetics. *Green Chemistry*. 2021;23(18):6953–6965.

[66] Zeng J, Ma Y, Li P, Zhang X, Gao W, Wang B, Xu J, Chen K. Development of High-Barrier Composite Films for Sustainable Reduction of Non-biodegradable Materials in Food Packaging Application. *Carbohydrate Polymers*. 2024 Apr 15;330:121824.

[67] Zehra T, Kaseem M. Recent Advances in Surface Modification of Plasma Electrolytic Oxidation Coatings Treated by Non-biodegradable Polymers. *Journal of Molecular Liquids*. 2022 Nov 1;365:120091.

[68] Jothimani B, Venkatachalapathy B, Karthikeyan NS, Ravichandran C. A Review on Versatile Applications of Degradable Polymers. *Green Biopolymers and their Nanocomposites*. 2019:403–422.

[69] Jiang DH, Satoh T, Tung SH, Kuo CC. Sustainable Alternatives to Nondegradable Medical Plastics. *ACS Sustainable Chemistry & Engineering*. 2022 Apr 4;10(15):4792–4806.

[70] Singh N, Demirsöz R. Recycling of Traditional Plastics: PP, PS, PVC, PET, HDPE, and LDPE, and their Blends and Composites. In *Nanomaterials in Manufacturing Processes*, Sud, Dhiraj, Anil Kumar Singla, and Munish Kumar Gupta, Eds. 2022 Aug 2 (pp. 235–258). CRC Press.

[71] Chappell B, Pramanik A, Basak AK, Sarker PK, Prakash C, Debnath S, Shankar S. Processing Household Plastics for Recycling–A Review. *Cleaner Materials*. 2022 Dec 1;6:100158.

[72] Okan M, Aydin HM, Barsbay M. Current Approaches to Waste Polymer Utilization and Minimization: A Review. *Journal of Chemical Technology & Biotechnology*. 2019 Jan;94(1):8–21.

[73] Thiounn T, Smith RC. Advances and Approaches for Chemical Recycling of Plastic Waste. *Journal of Polymer Science*. 2020 May 15;58(10):1347–1364.

[74] Sah MK, Mukherjee S, Flora B, Malek N, Rath SN. Advancement in "Garbage in Biomaterials Out (GIBO)" Concept to Develop Biomaterials from Agricultural Waste for Tissue Engineering and Biomedical Applications. *Journal of Environmental Health Science and Engineering*. 2022 Dec;20(2):1015–1033.

[75] Lizundia E, Luzi F, Puglia D. Organic Waste Valorisation towards Circular and Sustainable Biocomposites. *Green Chemistry*. 2022;24(14):5429–5459.

[76] Álvarez-Castillo E, Felix M, Bengoechea C, Guerrero A. Proteins from Agri-food Industrial Biowastes or Co-products and their Applications as Green Materials. *Foods*. 2021 Apr 29;10(5):981.

[77] Abu-Thabit NY, Pérez-Rivero C, Uwaezuoke OJ, Ngwuluka NC. From Waste to Wealth: Upcycling of Plastic and Lignocellulosic Wastes to PHAs. *Journal of Chemical Technology & Biotechnology*. 2022 Dec;97(12):3217–3240.

[78] Molina-Besch K. Use Phase and End-of-life Modeling of Biobased Biodegradable Plastics in Life Cycle Assessment: A Review. *Clean Technologies and Environmental Policy*. 2022 Dec;24(10):3253–3272.

[79] Briassoulis D, Pikasi A, Hiskakis M. End-of-Waste Life: Inventory of Alternative End-of-Use Recirculation Routes of Bio-Based Plastics in the European Union Context. *Critical Reviews in Environmental Science and Technology*. 2019 Oct 18;49(20):1835–1892.

[80] Arif ZU, Khalid MY, Sheikh MF, Zolfagharian A, Bodaghi M. Biopolymeric Sustainable Materials and their Emerging Applications. *Journal of Environmental Chemical Engineering*. 2022 Aug 1;10(4):108159.

[81] Ali SS, Elsamahy T, Abdelkarim EA, Al-Tohamy R, Kornaros M, Ruiz HA, Zhao T, Li F, Sun J. Biowastes for Biodegradable Bioplastics Production and End-of-Life Scenarios in Circular Bioeconomy and Biorefinery Concept. *Bioresource Technology.* 2022 Nov 1;363:127869.

[82] Evode N, Qamar SA, Bilal M, Barceló D, Iqbal HM. Plastic Waste and Its Management Strategies for Environmental Sustainability. *Case Studies in Chemical and Environmental Engineering.* 2021 Dec 1;4:100142.

[83] Jadhav HS, Fulke AB, Giripunje MD. Recent Global Insight into Mitigation of Plastic Pollutants, Sustainable Biodegradable Alternatives, and Recycling Strategies. *International Journal of Environmental Science and Technology.* 2023 Jul;20(7): 8175–8198.

[84] Prata JC, Silva AL, Da Costa JP, Mouneyrac C, Walker TR, Duarte AC, Rocha-Santos T. Solutions and Integrated Strategies for the Control and Mitigation of Plastic and Microplastic Pollution. *International Journal of Environmental Research and Public Health.* 2019 Jul;16(13):2411.

[85] Singh MK, Mohanty AK, Misra M. Upcycling of Waste Polyolefins in Natural Fiber and Sustainable Filler-based Biocomposites: A Study on Recent Developments and Future Perspectives. *Composites Part B: Engineering.* 2023 Jun 17;263:110852.

[86] Silva LP. Current Trends and Challenges in Biofabrication Using Biomaterials and Nanomaterials: Future Perspectives for 3D/4D Bioprinting. *3D and 4D Printing in Biomedical Applications: Process Engineering and Additive Manufacturing.* 2019 Feb 19:373–421.

[87] Miehe R, Full J, Scholz P, Demmer A, Bauernhansl T, Sauer A, Schuh G. The Biological Transformation of Industrial Manufacturing-Future Fields of Action in Bioinspired and Bio-Based Production Technologies and Organization. *Procedia Manufacturing.* 2019 Jan 1;39:737–744.

[88] Gatto F, Re I. Circular Bioeconomy Business Models to Overcome the Valley of Death. A Systematic Statistical Analysis of Studies and Projects in Emerging Bio-Based Technologies and Trends Linked to the SME Instrument Support. *Sustainability.* 2021 Feb 10;13(4):1899.

[89] Miehe R, Bauernhansl T, Beckett M, Brecher C, Demmer A, Drossel WG, Elfert P, Full J, Hellmich A, Hinxlage J, Horbelt J. The Biological Transformation of Industrial Manufacturing–Technologies, Status and Scenarios for a Sustainable Future of the German Manufacturing Industry. *Journal of Manufacturing Systems.* 2020 Jan 1;54:50–61.

[90] Hiloidhari M, Bhuyan N, Gogoi N, Seth D, Garg A, Singh A, Prasad S, Kataki R. Agroindustry Wastes: Biofuels and Biomaterials Feedstocks for Sustainable Rural Development. In *Refining Biomass Residues for Sustainable Energy and Bioproducts,* Kumar, R. Praveen, Edgard Gnansounou, Jegannathan Kenthorai Raman, and Baskar Gurunathan, Eds. 2020 Jan 1 (pp. 357–388). Academic Press.

[91] Usmani Z, Sharma M, Awasthi AK, Sivakumar N, Lukk T, Pecoraro L, Thakur VK, Roberts D, Newbold J, Gupta VK. Bioprocessing of Waste Biomass for Sustainable Product Development and Minimizing Environmental Impact. *Bioresource Technology.* 2021 Feb 1;322:124548.

[92] Boodhoo KV, Flickinger MC, Woodley JM, Emanuelsson EA. Bioprocess Intensification: A Route to Efficient and Sustainable Biocatalytic Transformations for the Future. *Chemical Engineering and Processing-Process Intensification.* 2022 Feb 1;172:108793.

[93] Olguin-Maciel E, Singh A, Chable-Villacis R, Tapia-Tussell R, Ruiz HA. Consolidated Bioprocessing, an Innovative Strategy towards Sustainability for Biofuels Production from Crop Residues: An Overview. *Agronomy.* 2020 Nov 22;10(11):1834.

[94] Aguilar A, Twardowski T, Wohlgemuth R. Bioeconomy for Sustainable Development. *Biotechnology Journal.* 2019 Aug;14(8):1800638.

[95] Dragone G, Kerssemakers AA, Driessen JL, Yamakawa CK, Brumano LP, Mussatto SI. Innovation and Strategic Orientations for the Development of Advanced Biorefineries. *Bioresource Technology.* 2020 Apr 1;302:122847.

7 Smart Sustainable Biomaterials and Their Recent Advancement for Targeting the Nervous System

Sonali S. Shinde and Aniket P. Sarkate
Dr. Babasaheb Ambedkar Marathwada University

Parveen Kumar
Department of Mechanical Engineering, Rawal Institute of
Engineering and Technology, Faridabad, Haryana, India

Samiksha P. Surwade
Marathwada Mitra Mandal's College of Pharmacy Thergaon

Tejal A. Salunkhe
Progressive Education Society's Modern
College of Pharmacy (for Ladies) Moshi

Sanket S. Rathod and Prafulla B. Choudhari
Bharati Vidyapeeth College of Pharmacy

Kalusing S. Padvi
Dr. Babasaheb Ambedkar Marathwada University

LIST OF ABBREVIATIONS

Abbreviations	Full description
0D	Zero dimensional
1D	One dimensional
2D	Two dimensional
3D	Three dimensional
AD	Alzheimer's Disease
APP	Amyloid precursor protein
Arg	Arginine

DOI: 10.1201/9781003434313-7

Abbreviations	Full description
Asp	Aspartate
BACE-1	Beta secretase
BDNF	Brain-derived neurotropic factor
BMP	Bone morphogenetic protein
c-AMP	Cyclic adenosine monophosphate
CD	Carbon dots
Cdc	Central division control protein
CNS	Central Nervous System
CNT	Carbon nano tubes
DRG	Dorsal root ganglion
ECM	Extracellular matrix
EMT	Epithelial–mesenchymal transition
EPS	Exopolysaccharide
FAK	Focal adhesion kinase
FGF-2	Fibroblast growth factor-2
GAG	Glycosaminoglycans
GDNF	Glial cell-derived neurotrophic factor
Gly	Glycine
GTP	Guanosine-5-triophosphate
HA	Hyaluronan
HSPG	Heparin sulfate proteoglycan
IKVAV	Isoleucyl-lysyl-valyl-alanyl-valine
i-NSC	Induced neural stem cell
JNK	C-JUN Kinase
LAB	Lactic acid bacterial
M@CD	Metal-doped carbon dots
MWCNT	Multi-wall carbon nano tubes
NGC	Nerve guidance conduits
NMJ	Neuromuscular Junction
PAN-MA	Poly(acrylontrile-co-methylacrylate, random copolymer, 4 mole percent methylacrylate)
PC 12	Pheochromocytoma cells-12
PCL	Polycaprolactone
PDGFR	Platelet-derived growth factor receptor
PEDOT	Poly(3, 4-ethylenedioxythiophene)
PGA	Poly glycolic acid
PLGA	Poly lactic co-glycolic acid
PLLA	Poly-L-lactic Acid
PND	Peripheral nerve damage
PNN	Perineuronal nets
PNS	Peripheral nervous system

Abbreviations	**Full description**
PPE	Polyphosphoesters
PU	Polyurethane
RGD	Arginine–Glycine–Aspartic
SAPNS	Self-assembling peptide nanofiber scaffolds
SC	Schwan cell
SF	Silk fibroin
SP	Senile plaque
SS	Stainless steel
SVZ	Subventricular zone
SWCNT	Single-wall carbon nano tubes
UHMWPE	Ultra-high molecular weight polyethylene
VG	Vascular grafts
WBDPU	Waterborne biodegradable polyurethane
YIGSR	ile-lys-val-ala-val

1   INTRODUCTION

A biomaterial is a substance developed to interact with biological systems for diagnostic or therapeutic reasons (treating, enhancing, repairing, or replacing a tissue function of the body). The field of biomaterials science incorporates elements of tissue engineering, biology, chemistry, and materials science. Keep in mind that a biomaterial differs from biological materials produced by biological systems, such as bone. Biomaterials can be created in the lab or obtained naturally by a number of chemical processes that use metallic elements, polymers, ceramics, or composite materials. They frequently serve and/or are modified for medical purposes, and as a result, they might be all or a component of an alive system or biomedical device that enhances, replaces, or fulfills a biological work. They could be bioactive with a more interactive capability, like those seen in hydroxyapatite-coated hip implants, or rather passive, like those in a heart valve. Biomaterials are used often in medical, dental, and pharmaceutical operations [1]. Neuronal injuries are the most time-consuming to fully recover from all of the many types of injuries. Communication between the brain and muscles and organs might be hampered by an injury of nerve. Consequently, it is crucial to take care of them. Surgery has been utilized historically up until now to treat nervous system damage brought on by trauma or any other illnesses. Such operations secure the apposition of the edges of wounded tissues with sutures and stitches. These sutures keep the tissues attached while allowing the tissues to heal normally and regenerate. Such operations could end up being useless or taking a very lengthy time to fix the damage. Sutures are made from a variety of biomaterials, including silk, linen, nylon, polyester, catgut, chromic catgut, and others. The healing process is not accelerated or improved by the surgical sutures; they merely hold the tissues together. Additionally, it does not offer a nourishing environment to promote cellular regeneration, consequently, to enhance the treatment of neurodegenerative illnesses and nerve damage. Researchers looked into a variety of materials that could be used as intelligent biomaterials for neuro-regeneration. To increase its effectiveness, several researchers have created biomaterials in systems with different dimensions. The main challenges in formulating biomaterial system are we require a material which is biocompatible; it should be stable inside the body, should be nutritive for neurons and should be biodegradable at the same time so that once work is done, it can be easily removed out from the body without any surgical procedure. The biomaterials should be stable in the external environment during storage. Also the material should be mechanically strong to hold the tissues/cells together at the same time they should have optimum porosity to provide desired space for neuro-regeneration. No any optimum biomaterial or such system is not available yet which fulfills all these criteria are the main research gap. Because of all these, it is important to know various properties of biomaterial to mold them into a therapeutically effective systems. The proposed write up will prove significant to provide information about various types of smart sustainable biomaterials along with their properties. This chapter also describes system of various dimensions in which the biomaterials can be molded to achieve the desired characteristics for encouraging neuronal regrowth. Also, the recent advancements in this field along with applications of biomaterials and real world relevance is discussed.

2 CLASSIFICATION OF SMART BIOMATERIALS

2.1 BIOACTIVE BIOMATERIALS

The phrase "bioactive materials" generally refers to biomaterials that, upon engaging with a biological system, have the capacity to induce and conduct a response. They produce bioactive chemicals that the body is able to use actively and effectively heal and restore the organs' damaged functionality. Additionally, they bring on cell differentiation and propagation as well as gene and tissue regeneration, among other bioactivities or functions [2].

2.1.1 Natural Biomaterials

The biomaterials made from natural resources include plant-based components like cellulose, which is primarily made from wood pulp and cotton, as well as polysaccharides like starch, which are found in grains of maize, rice, or wheat and lignin, which is made from the pulp mill extraction process of wood (softwood, hardwood, or grass). They produce bioactive chemicals that the body is able to actively and effectively heal and restore the organs' damaged functionality, implants, wound dressing, hydrogel, and scaffold formulation. Collagen is a type of extracellular protein that is often derived from the skin, tendons, cartilage, and bones of many animal species. The main components of the fractones were first identified in the extracellular matrix niche of the mouse brain's SVZa or subventricular zone of the lateral ventricle. These molecules are laminin and heparan sulfate proteoglycan (HSPG). A ubiquitous protein called perlecan is present in the stroma of bone, adipose, lymphoreticular, vascular, cartilaginous, and neural tissues. Using different animal tissues, such as the human birth cord, rooster comb, and bovine synovial fluid, hyaluronic acid was extracted from bovine vitreous humor. Gelatin, a collagen derivative is employed as a biomaterial for various intentions, including the fabrication of scaffolds, tissue engineering, nanoparticles, sponges, and gels. Other animal-based biomaterials, such as keratin, silk, and fibrin, are also used in a variety of biological applications. A fibrillar biopolymer called fibrin is created naturally during blood coagulation. Plasma fractionators, blood banks located in hospitals, and single plasma donations can all be used to create fibrin sealant on a large scale. The fibrin matrices have been used in preclinical and clinical settings for tissue engineering utilization including skin, cardiovascular, musculoskeletal, or nerve tissue keratins, a family of proteins normally observed in the epithelial cells of a variety of creatures, are the primary component of wool, fur, feathers, quills, spines, horns, hooves, and reptile scales. Keratin hydrogels are water- and keratin-based three-dimensional gels [3]. Medical uses for keratin films and biopolymers include adjustable drug delivery, regenerative medicine, burn treatment, wound healing, and tissue engineering. A natural fiber called silk is produced by spiders (*Nephila clavipes*) and silkworms (*Bombyx mori*). Together with sericin, silk fibroin, a fibrous protein containing 17 amino acids, forms the silk fibroin. Agarose and alginate, which are obtained from sources like algae, are also included in the category of natural biomaterials. Agar or marine algae (sea kelp), both of which contain agar, are employed to separate and decontaminate the polymer known as agarose. The chemical composition of this polymer which exists in living world, varies between 3,6-anhydro-L-galactose and -D-galactose agarobiose units. Delivering medicines, cancer therapy, tissue engineering and regenerative medicine and disease diagnosis, control, and treatment are just a few of the many applications that agarose has acquired. Ascophyllum, Lessonia, Eclonia, and Durvillea are only a few of the species from which alginate is commercially collected. Alginates are primarily the scaffolding polysaccharides that are derived from brown seaweeds. Alginate is frequently utilized as a hydrogel in biomedicine for purposes such as tissue engineering, drug delivery, wound healing, corneal regeneration, and cardiac applications. Dextran is another biomaterial that is made from bacteria. Exopolysaccharide (EPS) dextran is produced by lactic acid bacteria (LAB) or their enzymes when sucrose is present. Dextran-based hydrogels have several biomedical uses, including as antifungal agents, scaffolds for tissue engineering, and drug delivery systems. Chitosan is a biomaterial that can be found in many different places. Chitin is the source of chitosan, a copolymer of glucosamine and *N*-acetylglucosamine. Cell walls of crustaceans, fungus, insects, as well as some types of algae, bacteria, and some invertebrate creatures contain chitin. Chitosan bio-adhesives have been extensively used in several biomedical applications, including sutureless surgery, hemostasis, and medication delivery [4]. Some clays that can be found naturally include the clay mineral group known as bentonite. Devitrification of volcanic ash that had fallen into the water caused it to develop. A particular kind of clay mineral called montmorillonite is present in bentonite. These clays are used to create hydrogels, scaffolds for 3D printing, and tissue engineering because of their good water absorption capability [5–7] (Figure 7.1).

2.1.2 Synthetic Biomaterials

The synthetic approaches for biomaterials are of four types of ceramics, synthetic polymers, metals, and clays. The ceramics include bioactive glasses, calcium

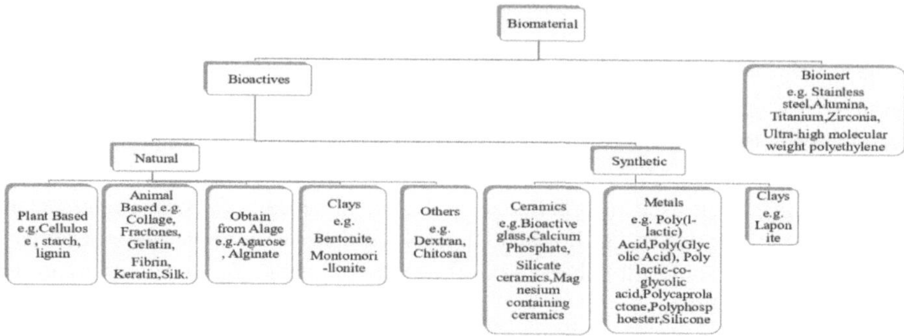

FIGURE 7.1 Classification of biomaterials. The above image demonstrates the classification of biomaterials done on the basis of sources from which they can be obtained.

phosphates, silicate ceramics, and magnesium-containing ceramics. These ceramics are utilized for bone tissue engineering, dentistry, replacement of bone, bone regeneration, etc. The metal used as biomaterial is cobalt–chromium alloy which is utilized for making orthopedic and dental implants. The synthetic polymers are containing poly (L-lactic acid), poly (glycolic acid), poly lactic co-glycolic acid, polycaprolactone, polyphosphoester, silicon are prepared for utilization in tissue engineering and regenerative medicine applications as surgical implants and medication delivery systems, neuro-regeneration, scaffolds hydrogels, and nanocarriers. The laponite is a nanocomposite synthetic clay that can be utilized as drug carrier, hydrogel, scaffolds, biomedical 3D printing, biosensors, and for tissue engineering [8].

2.2 Bioinert Biomaterials

Bioinert biomaterials are a different class of biomaterials. When exposed to biological tissue, bioinert materials—also known as biologically inert or bioinert—do not cause a reaction or interact with it. In other words, the host won't respond when the substance is ingested. Stainless steel (SS), alumina, titanium, zirconia, and ultra-high molecular weight polyethylene (UHMWPE) are a few examples of biomaterials. Due to their high mechanical strength, these materials can be utilized to create bone spacers, joint replacements, dental implants, and other medical devices. Because of its affordable pricing, good mechanical attributes, and biocompatibility, SS can be used to create cardiovascular stents. SS with higher bacterial infection resistance (long-term devices) and blood compatibility (blood-contacting implants). UHMWPE exhibits high extent of biological compatibility, toughness, and wear resistance. Hence, it is usually employed in joint arthroplasty as a bearing material with ceramic or metallic counter surfaces. It is beyond dispute that UHMWPE is important for generating exceptional results in whole joint arthroplasties [9].

3 SMART BIOMATERIALS FOR TARGETING THE NERVOUS SYSTEM

The ability of the brain to be in contact with muscles and organs can be compromised by a nerve injury. Disorders including carpel tunnel syndrome, diabetes, and

Guillain-Barre syndrome. Numerous ailments, including brain injuries, accidents, and self-immune diseases, including lupus, rheumatoid arthritis, and Sjogren's syndrome, can result in nerve damage. Although nerve regeneration is feasible in older persons, it may take longer due to neurological changes brought on by aging. Furthermore, Alzheimer's disease, Parkinson's disease, Huntington's disease, and Lewy body dementia all develop in humans as they get older. These illnesses cause the degeneration and death of brain cells [10].

Therefore, we must enhance the growth and regeneration of neuronal cells in order to cure such disorders. Various natural and artificial biomaterial, including collagen, fractone, lamnin, perlcan, hyaluronan, alginate, and cellulose, are employed for this purpose. Gelatin, chitosan, lignin, and synthetic biomaterials like poly (L-lactic acid), poly (lactic-co-glycolic acid), and poly glycolic acid, as well as polyphospho-ester and polyurethane, among others.

3.1 NATURAL NEURO-REGENERATIVE BIOMATERIALS

3.1.1 Collagen

One of the proteins found in the extracellular matrix (ECM) of peripheral nerves is collagen. Collagen is one of the proteins that occur in peripheral nerves' extracellular matrix (ECM). Cellular activities like migration, growth, differentiation, and survival are managed by the interaction of ECM proteins with cellular membrane receptors. The connective tissues that cover the central (CNS) and peripheral (PNS) nervous systems, the basement membranes (BM) separating the nervous system from other tissues (muscular, endothelial), and the sensory end organs are the three areas of the developed nervous system where collagens are present. The execution of Schwann cell function involves collagen in a significant way. In the PNS, Schwan cells are essential for maintaining and regenerating the axons of the neurons. Out of the 29 different forms of collagen, the mature collagen molecule retains the type-V collagen trimer's *N*-propeptide for Schwann cell mediation adherence to collagen type-V and influence Schwann cell activity. Collagen type-V is located in the basal lamina surrounding the myelinating Schwann cells, where it co-localizes with collagen types I and III. Type-IV collagens normally make up the BM collagens. Underlying epithelial cells in the PNS are BMs that contain type-IV collagen, which surround Schwann cells and their connected axons [11].

3.1.2 Fractone

The dentate gyrus present specialized stem cell niches known as fractones that move toward the olfactory bulb, which serves as a storage of neuroprogenitor cells in the mature brain. Hyaluronan, perlecan, and laminin-5 are elements of the extracellular matrix connected to the stem cell microenvironment of fractone that control stem cells' ability to divide and differentiate into pluripotent migratory progenitor cells, which are capable of taking part in neural tissue repair processes. This method has been reported to significantly improve spinal cord damage and brain trauma repair [12].

3.1.3 Hyaluronan (HA)

The space-filling and hydrating capabilities of HA, a significant supporting element of the nervous system ECM, play a crucial role in the compartmentalization of distinct brain areas and the preservation of confined ionic mini climate that are crucial for healthy cellular pursuit. Additionally, HA offers an exceedingly moist model that promotes cell attachment and migration during the growth of CNS/PNS neuronal networks. HA is required for the development of brain synaptic plasticity and the neuroprotective properties of PNNs, both of which aid in cognitive learning. The regulation of neural stem cell differentiation as well as the procreation, clinging, and displacement of neural cell colonies are among the physiological and pathological processes that HA governs. Additionally, it modifies the surroundings of tissues' cells. HA may adjust its capacity to fill spaces with viscoelastic material in response to factors in its environment, including pH, temperature, and salt levels. Additionally, it preserves the ionic gradients and niche microenvironments essential for the best cellular characteristics required for tissue function [13]. The HA adjusts its hydrogen-bonding and hydrophobic interactions in response to changes in its environment. The CNS/PNS extracellular matrix provides neural cells and neuroregulatory proteins with educational cues for neural development and repair. The ECM of the CNS/PNS is dominated by GAGs, especially HA, in a distinctive way. Early phases of the recovering of vascularized damage need the formation of fibrin clots, which are produced when fibrinogen specifically binds to HA. HA and perlecan in this niche climate also encourage neuron development and differentiation via FGF-2 cell signaling. Through BMP signaling, fibrinogen, a particular ligand for the SVZ niche HA, induces astrogliogenesis after cortical brain harm. Therefore, following TBI, fibrinogen regulates astrogenesis in the SVZ niche.

3.1.4 Perlecan

In the subventricular and sub-granular dentate gyrus existing in the hippocampus, perlecan is found in specialized neuroprogenitor microenvironment of stem cells known as fractones, where it stimulates neural cell endurance and expansion through hidden FGF-2. A modular multiple role PG with five separate domains is called perlecan. Domain V of perlecan has been recommended to treat Alzheimer's disease (AD), patients of stroke, vascular dementia, and brain trauma due to its neuroprotective and anti-inflammatory effects. Bio scaffolds that contain perlecan and VEGF encourage angiogenesis and tissue healing. The damaged blood-brain barrier that results from an ischemic stroke is encouraged to be repaired by perlecan domain V. Perlecan domain V interacts with glycoproteins having pro-angiogenic properties including progranulin and matrix protein 1 (ECM1) the cells promote angiogenesis and tissue healing. The neuromuscular junction (NMJ) contains complexes of perlecan with dystroglycan and acetylcholinesterase, which are important for the development and operation of this structure. In addition to promoting pericyte movement and enhancing PDGF-BB-persuade activation of platelet-derived growth factor receptor (PDGFR), Src homology region 2 domain-containing phosphatase-2 (SHP-2), and focal adhesion kinase (FAK), interactions between perlecan domain V 51 integrins are important. This promotes the preservation of the blood-brain barrier in its normal state and the procedures by which this structure is repaired after an ischemic stroke. Perlecan, an omnipresent molecule occurring in various distinct forms is engaged in the control of several somatic activities in tissues of various forms and functions, is a component of the basement membrane of blood vessels, the blood-brain barrier, and nerve basal structures in moto-neuron synapses of the NMJ [3].

3.1.5 Laminin

One of the foundational elements of the basement membrane is laminin. In vitro studies have demonstrated that SC migration is aided by laminin-1, laminin-2, and fibronectin. Axonal growth, myelination, and nerve regeneration—all necessary for the efficient operation of the nervous system—are facilitated by laminin, in addition to its function in conveyed relocation. Large and flexible laminin-1 is made up of three polypeptide chains with the numbers 1, 1, and 1. The laminin sequence's segments RGD, YIGSR, IKVAV, RNIAEIIKDI, and PDSGR have all been discovered and tested for bioactivity. For instance, RGD, which has been fastened to synthetic and natural proteins is known to be a flexible cell clinging site in fibronectin and numerous other proteins. Integrins 11, 31, 41, 61, and Cdc42 GTPase are activated when laminin-111 chain connect to cell surface receptors. The phosphorylation of Jun and the stimulation of c-Jun kinases are the results of the activated GTPase activating Cdc42.When c-Jun kinases are activated, there is an increase in c-Jun expression and neurite outgrowth. It is unknown where in the route the synthesis of nitric oxides takes place. According to Weston et al. (2000), the production of nitric oxide might occur before Cdc42 is activated. However, it is demonstrated that nitric oxide production plays a crucial role in laminin-mediated neurite propagation [14].

3.1.6 Gelatin

A molecular byproduct of type I collagen is gelatin. Collagen's triple helical structure is typically irreversibly hydrolyzed by processes like heat and enzymatic denaturation, resulting in random coil domains. Gelatin is consequently less structured than collagen but shares many of its molecular characteristics. Gelatin can thus take the role of collagen and carry out similar biomaterial tasks for cellular growth in vitro. The following justifications explain why gelatin has become more popular than pure, undamaged ECM proteins: gelatin has several advantages over ECM proteins, including: (i) being more widely available and cheaper; (ii) being extra soluble than other ECM proteins, making it simpler to use for biomedical applications; (iii) only a small portion of the entire ECM protein order being necessary for cell connection and bringing cellular reciprocation; and (iii) gelatin having a structure that is very similar to that of ECM proteins. Gelatin-based materials have weak mechanical qualities, little temperature-based stability, and a rather quick rate of disintegration, which is their main shortcoming. Materials made of gelatin might not endure long when employed in research that call for a prolonged time period, such as controlled drug release, cell differentiation, and palliate injury. Additionally, gelatin is far more unsafe to various proteases than collagen is, which could result in gelatin degrading more quickly [15]. To readily overcome these drawbacks, gelatin can be modified and made into composites to boost its mechanical stability, biocompatibility, and bioactivity. These demerits have become not much significant in comparison with the innumerable advantages of employing gelation for biomedical applications due to improvements in fabricating technology and our understanding of material chemistry. Gelatin and collagen almost exactly match each other in terms of their amino acid content. Meanwhile gelatin is a linear protein having molecular weights ranging from 15 to 250 kDa that is degraded collagen. Gelatin and collagen are similar to one another yet have diverse functions in culturing of cells and tissues. For example, the linear tripeptides Arginine (Arg), Glycine (Gly), and Aspartate (Asp) or the Arg–Gly–Asp identification series found in gelatin facilitate cell adhesion, migration, and survival by binding to numerous integrin proteins. The absence or very less content of the aromatic amino acids tryptophan, tyrosine, and phenylalanine is another significant commonality between collagen and gelatin. One of the main causes of the truncated antigenicity and lethalness of both gelatin and collagen is the absence of these amino acids [16].

3.1.7 Silk Fibroin

Particularly SF from silkworms and orb-weaving spiders possesses excellent mechanical characteristics to aid in the formation of functional tissues. Its use as a biomaterial is predicated on excellent biocompatibility. To aid in the deposition of extracellular matrix and tissue regeneration, the majority of biomaterials should degrade with a speed that corresponds to the growth of the new tissue. Additionally, silk fibroin can be altered through the chemistry of the amino acid side chains to change its surface characteristics and affect cell proliferation. In contrast to other fibrous proteins, SF provides a variety of sterilization choices, which is remarkable. It can keep its original shape and structure in high-temperature environments [17].

3.1.8 Cellulose

For the transport of growth factors into the nervous system's tissues as well as for the cultivation of nerve cells, cellulosic materials have been employed as scaffolds. It has been demonstrated that cellulose-based biomaterials encourage the reformation of neurons following spinal cord damage. Neuronal stem cells are differentiated in vitro using NFC scaffolds in research. The NFC scaffolds' adjustable porosity can enable the best possible release of growth factors into the damaged spinal cord region in vivo. For the transportation and release of growth factor, scaffolds have been used. This is helpful for mending injured nerve tissues and for the heterogeneous neuronal differentiation of large populations of stem cells.

3.1.9 Chitosan

A polysaccharide, chitosan is primarily made of D-glucosamine and, at low extent, *N*-acetyl-D-glucosamine units that are randomly—(1–4)—linked. It can be made by deacetylating chitin, which is known to be nature's second-most prevalent polysaccharide after cellulose. Depending on the intended use, it is vital to take into account the chitosan source and obtention method. These elements specify the properties of the finished product. Its molecular weight (Mw), crystallinity, deacetylation degree, and lack of impurity, are crucial for medicinal applications. The mechanical and biological properties of chitosan are strongly correlated with these variables.

Chitosan and its derivatives have demonstrated in tissue engineering and regenerative medicine to support axonal reconstruction, reduce inflammation, and provide neurotrophic factors and cells with a corresponding functional recuperation. The transfer of stem cells onto chitosan-based vehicles offers a promising future for neuro-regeneration. Senile plaques SPs are made up of—amyloid (A) peptides, which proteolysis enzymes and break from amyloid precursor proteins (APPs). Since it starts the synthesis of A during APP proteolysis—secretase, sometimes referred to as—amyloid cleavage enzyme (BACE-1), appears to be a crucial enzyme. As a result, BACE-1 is a potential biomarker for AD and a potential therapeutic target. Studies on the BACE-1 inhibition and the formation inhibition properties of chitosan and its derivatives have been conducted over a decade have found that chitosan compounds exhibit BACE-1 inhibitory activity [18].

3.1.10   Alginate

In one study, it was discovered that free hydratation and dissolving conditions may be the most crucial for achieving and maintaining alginate's neuro-regeneration-stimulating effects. Schwann cells and axons are growing simultaneously. It might imply that the development of nerve tissue took place concurrently with a significant amount of alginate that was freely dissolved.

3.1.11   Agarose

Agarose is a naturally occurring polysaccharide polymer with remarkable characteristics that make it a good option for tissue engineering implementation. Agarose has unique properties that are significant for its application as a biomaterial for growth of cells and/or controlled/localized transport of medicine, for instance its superb compatibility with biologicals, gelation behavior based on temperature, and physical and chemical features. Due to its similarities to the extracellular matrix, this natural carbohydrate polymer has enticing features that generate significant interest in its application in the field [19]. Agarose can be used to create a gel that can replicate the morphological characteristics of tissues' ECM. Cell attraction to agarose is increased by blending, combining with other proteins and polymers such as collagen, chitosan, and bacterial cellulose. One of agarose's primary problems is its inability to be broken down because the body lacks the necessary enzymes. However, adding an exogenous enzyme like agarase before use can help break down agarose. Three different kinds of peripheral nerve conduits were created by Abidian et al. utilizing agarose and poly(3, 4-ethylenedioxythiophene, or PEDOT). Because of conductivity and sufficient nutritional penetration, the agarose coated partially with PEDOT demonstrated superior reformation than pure agarose and agarose fully covered with PEDOT. In order to maximize the dorsal root ganglion (DRG) neurite extension, the stiffness of the agarose gel should be optimized. Structure of agarose was altered using photolithography and micro molding approach to immobilize neural cells, which paved the way for the development of neurons. Agarose conduit containing MSC that secrete the brain-derived neurotrophic factor (BDNF) promotes the regeneration of peripheral nerves. Additionally, the effect was exacerbated by BNDF loaded singly in agarose [20].

3.1.12   Lignin

When used in peripheral nerve applications, lignin has appealing characteristics. For the regeneration of new nerves, lignin exhibits an antioxidant property and, thus, acts as an oxygen-free radical scavenger. This property stabilizes oxygen radical-generated chemical processes. Therefore, it is highly appreciated to fabricate a suitable substrate that contains lignin. In one study, aligned electrospun VG fibers with varying amounts of lignin were created for neural duties. Based on the scaffolds' mechanical characteristics, degradation assessment, and water uptake behavior, it was concluded that nerve tissue building could be assured with the embodiment of 5% lignin. The electrospinning technique was used to create aligned polycaprolactone (PCL) fibers with different lignin nanoparticle content percentages. The application of PCL with 15% lignin nanoparticles for nerve regeneration has a lot of promise.

3.1.13 Keratin

As a natural biomaterial, keratin can control the biological exertion of many cell types, secure the microenvironment of the injured region, and speed up axon permutation. As a result, keratin has good encouraging effects on both cell culture and tissue regeneration, but further research is needed to understand its potential regulating mechanism. SCs are crucial in the restoration of neuronal structure because they act as "repair cells" in the aftermath of injury. Keratin was found to have neuro-inducible activity in in vitro studies, and SCs produced in the keratin club underwent a number of morphological and proliferation-related alterations. These alterations resembled the activation of the SCs' epithelial–mesenchymal transition (EMT) program, the primary physiological reaction of cells involved in repair. Additionally, SCs can release neurotrophic factors through a variety of routes to support axons, prevent axon apoptosis, and promote cell procreation. In one investigation, it was found that keratin can encourage neuronal axon elongation in a cultured environment. DRG neurons seeded on the keratin club outperformed those of the control group in terms of both abundance and length. This adjustment might be explained by keratin-activating intracellular c-AMP levels and so enhancing neurons' intrinsic growth potential [21]. The regeneration environment contains essential cells besides SCs. Macrophages are drawn to the site of a nerve injury after it occurs, where they carefully control the local inflammatory retort to prevent excessive tissue devastation and develop an ideal microenvironment for nerve reformation.

3.2 SYNTHETIC NEURO-REGENERATIVE BIOMATERIALS

Natural materials have been used in neural TE applications for spinal cord injury and peripheral nerve damage (PND). However, natural materials often have a limited mechanical strength and degrade quickly in living things. Poly-L-lactic acid (PLLA), polycaprolactone (PCL), polyethylene glycol, and many other substances are examples of synthetic materials. They are easier to modify than natural materials in terms of their qualities. Porousness, rigidness, and debasement rate, for example, can be changed to accommodate the needs of various tissues. However, because artificial materials typically lack integrin-binding molecules, surface functionalization may be necessary [22].

3.2.1 Poly(L-Lactic Acid)

To maximize its surface bioactivity, PLLA can be simply combined with additional substances or bioactive polymers. In addition, PLLA-based materials could be used to create two-dimensional or three-dimensional (3D) scaffolds with secure nanofibrous and spongy arrangement that offer the bio-resembling structures for cells of nerve adherence and growth while promoting the generation of neural tissue. Additionally, PLLA-based scaffolds have the capability to be merge with a range of healing modalities, including stem cell therapy, electrical titillation, magnetic titillation, and drug delivery, which have all been investigated in the context of nerve restoration and reformation. The augmentation and production of glial cell-derived neurotrophic factor (GDNF), which is necessary for the existence of injured neurons and the recreation of axons in PNI, might be stimulated by the electrospun PLLA/gelatin scaffold, according to a report [23]. Yao et al. investigated the impact of the microstrips' wideness and arc radius on directed cell migration. In contrast, broad or curvy microstrips with a lower radius induced numerous-row cell arrangement, while tapering micro strips hindered cell adhesion, which relatively reduced migration, according to their findings. According to Xuan et al. (2021) aligned electrospun PLLA fibers might boost the degree of alignment of PC12 cells' neurite outgrowth and neurite length. In comparison with random nanofibrous structures, aligned electrospun PLLA fibers having coating of decellularized tangential nerve matrix gel promoted additional SC adherence and movement and accelerated extension of neurite.

3.2.2 Poly (Glycolic Acid)

A synthetic, braided polymer is called PGA. With suitable biocompatibility and excellent tensile strength, PGA is a biodegradable aliphatic polyester and has the capability to be used in a diverse tissue engineering exertion. The practicality of PGA-based absorbable suture revealed that soft tissues might be securely filled with polymers containing PGA. When adopted in different tissue engineering techniques, like vascular tissue engineering, PGA is also beneficial. According to Ichihara et al. (2015), polyglycolic acid possesses good mechanical qualities for the healing of a lengthy nerve lesion. The extracellular and natural matrix polymers, including collagen, could be combined with PGA polymer to enhance this polymer's physical properties for use in both animal models and people. According to one study, the PGA nerve conduit exhibits nerve proliferation and regeneration. The study also demonstrates that using collagen and PGA together yields even greater results [24].

3.2.3 Poly Lactic co-Glycolic Acid

After bridging nerve defects using autogenous nerve grafts and poly-(lactic-co-glycolic acid) conduit transplantation for sciatic nerve injury, tensile stress and strain impact directly on the quality of nerve regeneration. Li et al. confirmed that, after treating with alcohol, the glass transition temperature and temperature of thermolysis of poly (lactic-co-glycolic acid) tubular scaffolds having diameter of 1,660,218 nm and 80.6% porosity were raised; increased temperature stability was improved; and shattering power, bursting strength, and suture strength were also augmented. According to Zhao et al.'s research, polylactic acid plays a significant role in boosting the tensile strength of poly(lactic-co-glycolic acid) scaffolds as they rose in tensile strength as the amount of polylactic acid was raised. Poly(lactic-co-glycolic acid) has strong biodegradability, biocompatibility, and mechanical qualities. It is nontoxic and nonantigenic, and the amount of its constituents can affect how quickly it degrades. Because of this, scaffolds of poly(lactic-co-glycolic acid) have good reproducibility, mechanical caliber, and they are widely employed in tissue-engineered materials [25].

3.2.4 Polycaprolactone

The use of synthetic PCL has been deemed suitable for numerous biomedical applications, including implants, sutures, medicine release systems, and tissue engineering scaffolds. Additionally, the usage of this polymer is appealing for neural applications due to its good mechanical properties and straightforward processing. Finally, the slow dilution method has been used to study the effects of PCL on nerves. Due to the PCL-amnion membrane's nanofibrous structure, which contains distinctive fibers and large holes, the end-to-end epineural suture significantly increased Schwann cell propagation and axonal reproduction and decreased the formation of scar at the nerve recreation area, according to study results. The M2 phenotype of macrophages was important in driving macrophage polarization into the PCL-amnion nanofibrous membrane, which supported reformation of nerve and functional recovery. Our research revealed a unique therapeutic strategy to encourage nerve reformation and reduce marks at the site of nerve restore, which should help in the development of next generation nerve enfolds. The absence of studies for the immunological protection, cell compatibility, and cell toxicity of PCL amniotic nanofiber membrane is a limitation of the current study [26]. They examine their PCL amniotic nanofiber membrane for immunological safety, compatibility with cells, and toxicity in cells in our ongoing research. We'll also be looking into how the PCL-amnion nanofibrous membrane controls and activates macrophages, how M1 macrophages effect Wallerian deterioration, and how M2 macrophages affect Schwann cell differentiation and axonal recreation. The slow rate of degradation of PCL is another drawback. After epineurium recovery and nerve fibers have passed through the restoration area, the optimum wrapping material starts to degrade. In order to increase the rate of biodegradation, PCL and other natural polymer materials can be mixed to create nerve enfolds.

3.2.5 Polyphosphoesters

The class of biodegradable polymers known as polyphosphoesters (PPE) exhibits good biocompatibility with neurons and can breakdown in somatic conditions through hydrolysis or enzymatic breaking of the PPE linkages. A hydrophilic derivative of PPE called poly(ethyl ethylene phosphate, or PEEP), can copolymerize with PCL to form an amphiphilic block copolymer called poly(ethyl ethylene phosphate)-block-poly(ethyl ethylene phosphate, or PCL-PEEP. Paclitaxel is loaded into a hybrid ransferrin-conjugated polyphosphoester micelle for delivery to the brain, and researchers have come to the conclusion that the hybrid micelle polyphosphoester could one day be a good method of glioma chemotherapy drug delivery that specifically targets the brain [27].

3.2.6 Polyurethane

Polyurethane (PU) is produced as a result of the tough section (diisocyanate and chain extender) and smooth section (long-chain diol or oligodiols') microphase separated structure. High tensile strength, flexibility, tear power, and wear refusal are just a few of its exceptional mechanical qualities. In particular, the plasticity, compatibility with biologicals, and toughness of PU may be optimized for usage in biomedical gadgets, such artificial hearts and different kinds of conduits by carefully adjusting the ratios of hard and soft segment. Biodegradable oligodiols can be used to create waterborne biodegradable polyurethane (WBDPU) without the need for crosslinking. We previously created a family of biodegradable PU through environmentally friendly aqueous techniques. In comparison with polylactic acid (PLA) sponges, freeze-dried sponges of PU have shown a stronger capacity to encourage cartilage matrix synthesis. Because they offer good mechanical qualities and appropriate biocompatibility, biologically degradable polyurethanes have been employed as nerve transplant materials. Peripheral nerve regeneration has been observed to be supported by the polyurethane nerve conduit made of poly(E-caprolactone) and poly(ethylene glycol). Waterborne polyurethane scaffolds have also been created for the reformation of rat brain tissue [28]. Glial cells have been cultured on electrospun nonwovens made of polyurethane and polylactide. These tests confirmed that polyurethanes are compatible with glial cells. WBDPU has a customizable debasement speed and is easily made using freeze-drying. For infiltration in cell and mass interchange to support peripheral nerve reproduction, the highly pervious structure and permeable web are crucial.

4 BIOMATERIAL SYSTEMS BASED ON NANOSCALE DIMENSIONS

To improve the mechanical and biological performance of biomaterials, they have to be formulated into a particular nanoscale systems that can be called as nano-biomaterials. In simple language, nano-biomaterials are the nanostructures made up of biomaterials. Based on dimensions, nano-biomaterials are classified into zero dimensional, one dimensional, two dimensional, and three dimensional. Formulating biomaterials into nanostructure has various benefits in neuronal tissue engineering like it improves the interaction of biomaterial with cells and hence facilitating the transfer of nutrients becomes easier, also nanostructures like nanofibers and nanotubes can act as a guidance cue and provide necessary topographical guidance for rejoining and reproduction of neuronal cells and axons. 3D nano-biomaterials can mimic the biological system. Such approaches enhance the treatment of damaged nerves. The classification of nano-biomaterial along with its structure is further explained in detail under each type of biomaterial.

4.1 ZERO DIMENSIONAL

Zero-dimensional biomaterials are nanoparticles with all three dimensions stringently contained to the nanoscale. Carbon dots are a part of this. Carbon dots (CD) are a potentially useful biomaterial. Recent studies on CD with fluorescent metal doping demonstrated good bioavailability and a favorable impact on neuronal differentiation and growth. In this investigation, metals including Ga, Sn, Zn, Ag, and Au were used to dope the CD (2–10 nm). Ga@CD demonstrated antibacterial activity as well as neuronal development when compared with all other CDs that have been treated with various metals (M@CD). The adrenal phaeochromocytoma (PC12) (PC12) had good viability with CDs, whereas the M@CDs demonstrated 72 hours of greater than 100% cell viability. This means that we can employ metal-doped CDs for neural tissue engineering applications because they are not harmful to PC12 cells. After conducting pharmacodynamics investigations, a novel type of CD, designated as CrCi-Cn's, were discovered and isolated from the charcoal medicine Crinis Cabonisatus, which showed neuroprotective consequences on brain ischemia–reperfusion injury. This may be linked with controlling neurotransmitters and lowering neurotoxic excitatory stimulation [29]. There are still numerous questions that need to be researched or clarified about the implementation of CDs in the clinical therapy of AD. These include establishing ways for long-term monitoring of delivery, demonstrating improved efficacy and fewer side effects in comparison with existing treatments, and optimizing the size, surface charge, lethalness, biocompatibility, and route of administration of CDs. Although they will necessitate a significant amount of additional work, we are certain that these objectives can be met.

4.2 ONE DIMENSIONAL

One-dimensional biomaterials are nanomaterials that have just two dimensions and are only present at the nanoscale (less than 100 nm). These materials can be further separated into nanowires, nanotubes, etc. One-dimensional biomaterials feature a very high degree of anisotropy and a distinctive morphology (for example, a high length-to-diameter ratio and nano-topography), which leads to a variety of specific characteristics. A lot of one-dimensional biomaterials have also been used as the foundation for the creation of higher-dimensional biomaterials. Nanofibers are helpful for tissue reformation and also for the prolonged release of encapsulated drugs or growth factors because they imitate the porosity topology of natural extracellular matrix. In order to create or build nanofibrous scaffolds, synthetic polymers are used, including poly(L-lactic acid), poly lactic co-glycolic acid, polycaprolactone, and polyurethane. The natural biomaterials include collagen, hyaluronan, lamnin, gelatin, silk fibroin, cellulose, alginate, lignin, and keratin [30]. For greater mechanical strength and therapeutic efficacy, two or more biomaterials are typically utilized in combination.

As they are fine uninterrupted fibers having a high surface up on volume ratio, these fibers match with the natural ECM. The topographical guidance necessary to rejoin axons reproducing from the nearest section of the damaged neuron to its primary objective can be provided by electrospun nano- and micro-scale fibers. These one-dimensional nanofibers can be used to create three-dimensional scaffolds, two-dimensional nanofiber meshes, and nerve conduits. They can also be loaded into hydrogels. Nanotubes are another one-dimensional system helpful for neuro-regeneration. The cylindrical shape of carbon nanotubes (CNTs) is similar to that of distal neuronal dendrites which are tiny cellular structures vital to neurons' capacity to display sophisticated processing abilities. This resemblance, together with the topographic features, physical traits, such as conductivity, and surface-to-volume ratio of carbon nanotubes, pave the way for the employment of carbon nanotubes as interfaces for neuronal physiology. To date, CBN has

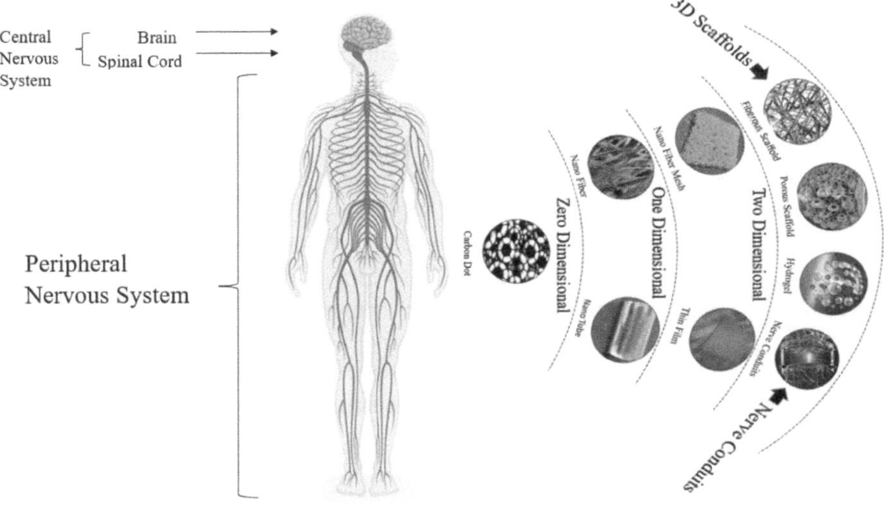

FIGURE 7.2 Biomaterial systems based on dimensions. The above image demonstrates the complete nervous system along with various dimensional systems in which the biomaterials can be formulated. As shown in the diagram, the first circle containing a black dot-like structure represents CD (zero dimensional), the another division having a circle containing a pink thread-like structure represents nanofibers while the hollow tube-like structure represents nanotubes. The diagrams in the next section representing the sheet with a net-like structure is a nanofibrous mesh and the plane sheet represents thin films. (These are 2D systems.) The next portion with a complex web structure is fibrous scaffolds; the structure with holes is porous scaffold; the interlinked structure with water molecules is hydrogel; and the cylindrical structure is nerve conduits. (These are 3D systems.)

employed carbon nanotubes (CNTs) most frequently for nerve regeneration. In contrast to other CBNs, CNTs have chemical (surface functionalization and high electrochemical surface area) characteristics and tunable physical (length, diameter, single-walled SWCNTs vs. multiwalled MWCNTs, chirality). Graphene layers wrapped up into nanometer-sized cylinders are what CNTs look like. CNTs have a long conjugated sp2 carbon network that results in an electron system that extends over the nanostructure and creates either strongly conducting CNTs or semiconducting ones. They also have changeable band gaps that are suitable for neurological pursuit. The application of CNTs specifically for brain stimulation and regeneration is covered in more detail in the section after. Fabbro et al. examined hippocampal neurons grown on multiwalled carbon nano tubes (MWCNTs). Paxillin, a membrane protein implicated in focal adhesions-mediated intracellular signaling pathways, is expressed more frequently, indicating that electrophysiological inputs from CNTs can be converted into particular neural messages. To create more CBN uses, however, more in-depth mechanistic research at the interface between neural tissue and CNTs are still needed (Figure 7.2).

4.3 TWO DIMENSIONAL

As just one of its dimensions is inside the nanoscale range (100 nm), a two-dimensional (2D) biomaterial is defined as such. According to the definition, 2D biomaterials have remarkable qualities such as a high capacity for absorption due to their high diameter-to-thickness ratio, for instance, thin films and nanofiber mesh. In addition to assisting with direct nerve tissue reconnection, 2D meshes of nanofiber can be utilized to direct neurite direction in the CNS, allowing for high throughput screening as well as the study of decomposition of the CNS neuron circuitry. Neurites (axons and dendrites) are highly responsive to topographic signals. Thin films are another 2D technology helpful for neuro-regeneration. Fundamentally, a thin film is a less-dimensional material created by arranging atomic, molecular, and ionic species, typically with a final width in the nm range. These fine films can be oriented along 3D structures like nerve guidance channels for neuro-regeneration, in addition to being employed alone [31].

4.4 THREE DIMENSIONAL

Here each of the x, y, and z dimensions is larger than 100 nm. Three-dimensional biomaterials are defined as biomaterials having all dimensions bigger than the nanoscale, and the majority of implants utilized in clinical settings fall into this category. Three-dimensional (3D) biomaterials with customizable spatial structure and biochemical characteristics may function as extracellular matrices that simulate extracellular ones and control cell behavior. It is important to note that in order to integrate their remarkable biological effects, 0D, 1D, and 2D biomaterials systems are typically merged into 3D systems. Examples include hydrogels, scaffolds, and nerve conduit.

By bridging the gap left by a severed nerve, tubular structures called nerve guidance conduits (NGCs) serve as a guide, a protective microenvironment for the regenerating axons, and a barrier against the ingrowth of tissue that forms scars. The issues with direct nerve suturing and nerve grafting can be efficiently resolved by the use of neural conduits. In the nerve conduit bridging strategy, proximal and distal nerve stumps are placed into the two ends of a nerve conduit, and regenerated axons from the proximal stump pass through the conduit and specifically develop into the distal stump's original pathways. Both stumps are supported tropically by the conduit, which also prevents neighboring tissues from penetrating the space between them. Additionally, nerve conduits create a microenvironment and enrich the neurotrophic factors in the chamber, which improves axonal regeneration following injury. To create NGCs, different zero-dimensional nanofiber biomaterials made from natural and manmade materials can be used. Thin films in two dimensions are also applicable. Films constituting of aligned poly(acrylontrile-co-methylacrylate, random copolymer, 4 mole percent methylacrylate) (PAN-MA) fibers were formed by using an electrospinning technique. Koch Membrane Systems' polysulfone nerve guiding channels were altered to make upgraded nerve guides with line up thin films built into their lumens. For the repair of peripheral nerves, thin-film enhanced nerve guide channels have been created. In this study, regeneration was shown to occur in guide channels with one or three thin films, with the one-film channels displaying considerably greater regeneration. Reformed nerve chains through the one-film channels had cellular distribution and alignment patterns that were more akin to normal nerve. These outcomes spotlight the prospective impact of internal scaffolding on the progress of endogenous healing [32].

In order to restore the defective tissue, the scaffold should be designed and built to closely resemble the anatomical makeup, biological properties, and biomechanics of the original tissue.

Scaffolds with certain mechanical firmness or bio-decomposition, sufficient size, surface roughness, and porosity are required for successful tissue regeneration in order to offer an ideal milieu for the required cell–cell contact, cell movement, increase, and differentiation. The scaffold pore diameters, which are crucial for waste elimination and food and oxygen diffusion, have a major impact on how well transplanted cells continue to operate. In addition, depending on the varied goals of tissue reformation, pore diameters have a significant impact on cell fixing, cell–cell contact, and cell transmigration over the membrane [33]. Porous, fibrous, and hydrogel scaffolds are examples of 3D scaffold types with good potential for neuro-regeneration based on their geometry. Sponge or foam are examples of the porous scaffolds.

Porous scaffolds frequently have an oriented or random interconnected pore structure that is very helpful for neuro-regeneration. The following biomaterials can be utilized to create neuro-regenerative sponges: collagen, hyaluronic, gelatin, fibroin, chitosan, alginate, poly(L-lactic acid-co-ε-caprolactone)/silk fibroin, poly-lactic co-glycolic acid, polyurethane. The scaffolds made with nanofibers are what make up the fibrous scaffolds. Three-dimensional (3D) scaffolds can help CNS nerve regeneration by promoting neurite guidance. As a result, adding a nanofiber mesh to 3D nanofibrous scaffolds has the potential to help close the gap left by injury and restore normal CNS function. Using nanofibers, Ellis-Behnke et al. created a model of the mammalian visual system. Self-assembling peptide nanofiber scaffolds (SAPNS) were injected to a viaduct with regard to a brain tissue injury and to stimulate axonal renewal; it did not induce inflammation and helped restore vision. Complete tissue reconnection was seen along with biodegradable and nontoxic products. Liu et al. created scaffolds for researching rat spinal cord damage using poly(lactic-co-glycolic acid) (PLGA) and PLGA adapted with polyethylene glycol (PEG) [34]. A gelatin sponge was combined with electrospun PLGA or PLGA-PEG nanofibers to create 3D scaffolds that were used for an in vivo transplantation to heal a totally transected defect. Induced neural stem cells (iNSCs) that were restored from mouse embryonic fibroblasts were sown into scaffolds. When compared with PLGA scaffolds, PLGA-PEG scaffolds produced greater iNSC adhesion and proliferation. Additionally, they promoted the development of iNSC into neurons and glial cells. It was confirmed that iNSC-derived neurons produce action potentials using the patch clamp technique. Practical resumption was noticed in both PLGA-PEG and PLGA transplants starting two weeks after surgery. This work shows how electrospun nanofibers can assist in vivo cell adhesion and proliferation to speed up CNS healing. To mimic natural ECM, Yang et al. created a 3D construct using hydrogel and transportable electrospun nanofiber meshes [35]. The hydrogel base was covered with several layers of nanofibers, with filter sheets acting as spacers between the layers. The constructed nanofibers were seeded with cells (and hydrogel), and the frames carrying extra nanofibers were cut off. A few hours later, the cells in various layers arranged independently and at various angles with respect to one another along nanofibers. With the help of this novel method, we can control the orientation of a single cell and keep it stable within 3D structures.

Hydrogel scaffolds, another type of 3D scaffold, are commonly used for neuro-regeneration. Substantial volumes of water can be soaked up by three-dimensional web structures called hydrogels. Hydrogels frequently do not decompose because of entanglement in physical, chemical, or chain reactions. Biomaterials like collagen, hyaluronic acid, gelatin, silk fibroin, cellulose, chitosan, alginate, and keratin can be produced as a hydrogel, and low-dimensional systems like nanofibers and nanotubes can be added. The nerve conduit is frequently filled with hydrogel. An interlinked, water-soaking polymeric web known as a hydrogel is prepared by the easy reaction of one or further monomers [36]. Due to their high water content, they are remarkably similar to genuine tissues in terms of flexibility, and their 3D shape resembles the original extracellular matrix (ECM) in the CNS. A hydrogel is regarded as the ideal carrier system for a variety of biological implementation, including nerve restore. Comprehensive studies have revealed that hydrogel carriers are effective at repairing stroke-induced brain damage [37].

5   RECENT ADVANCEMENT AND APPLICATIONS OF BIOMATERIALS

In the field of neuro-regeneration, application of biomaterials in humans is not practiced yet but it's effects in animals have been studied in recent years. Some of such studies are discussed here. Yen CM et al. (2019) implanted the PCL+ type I collagen nanofiber conduits in rats. This study showed that, this nerve conduit created the natural microenvironment for neurons, stimulated the nerve regeneration, axon growth was accelerated, got higher number of myelin sheaths without any inflammatory immune response. Yang et al. (2023) transplanted a hyaluronic acid granular hydrogel nerve guidance conduit to repair a 10-mm long sciatic nerve gap. Its results recommend that tissue-engineered nerve conduits having hyaluronic acid granular hydrogels effectively promote the morphological and functional retrieval of the damaged sciatic nerve. In a study of aligned polycaprolactone (PCL) fibers with various percentages of lignin nanoparticles were made using the electrospinning method on in vivo studies showed that cell viability and augmentation and extension of length of neurite were extended by increasing lignin concentration. Peter et al. (2008) studied a Mouse Model—Nerves that regenerated through the keratin hydrogel had lower conduction delays, greater amplitudes, more myelinated axons, and larger axons than nerves that regenerated through empty conduits. Other advances in biomaterials are also briefly mentioned under each biomaterial. Other applications of biomaterial include intraocular lenses, and joint replacements all involve biomaterials: bone marrow, artificial tendons and ligaments, dental implants for the fixing of teeth, heart valves, artificial tissue skin healing devices, blood vessel prosthesis, and breast implants, contact lenses, and cochlear implants delivery systems.

6 CONCLUSION

The environment and nutrients needed for cell growth and division can be provided by natural biomaterials. Natural biomaterials have integrin-binding molecules. It is simple to change the porosity, mechanical strength, and rate of degradation of synthetic biomaterials to meet the needs of various tissues. In recent researches, it is proved that combining natural and synthetic biomaterials can lead to the formation of stable biomaterials having good physical, mechanical, and biological properties. Low-dimensional nanoscale biomaterial systems are used as the building blocks for three-dimensional systems. Low-dimensional nano-biomaterials can occasionally be fed into three-dimensional systems. Formulating biomaterials into nanostructures of various dimensions can structurally mimic the biological micro-environment; also these systems can direct the growth of neurons and axons. After examining all factors, it is clear that choosing biomaterials based on their nutritional qualities, biocompatibility, biodegradability, mechanical strength and molding them into nanostructures of various dimensions is the best way to design an efficient neuro-regenerative system.

REFERENCES

[1] Krishnaveni, R., and K. Senthilkannan. *Preamble to Biomaterials and Its Applications in Science and Technology.* Lulu. com, 2019.

[2] Ajay, Parveen, Sharif Ahmad, Jyotsna Sharma, and Victor Gambhir. *Handbook of Sustainable Materials: Modelling, Characterization, and Optimization.* CRC Press, 2023. https://doi.org/10.1201/9781003297772

[3] Hayes, Anthony J., Brooke L. Farrugia, Ifechukwude J. Biose, Gregory J. Bix, and James Melrose. "Perlecan, a Multi-Functional, Cell-Instructive, Matrix-Stabilizing Proteoglycan with Roles in Tissue Development Has Relevance to Connective Tissue Repair and Regeneration." *Frontiers in Cell and Developmental Biology* 10 (2022): 856261. https://doi.org/10.3389/fcell.2022.856261

[4] Hamedi, Hamid, Sara Moradi, Samuel M. Hudson, Alan E. Tonelli, and Martin W. King. "Chitosan Based Bioadhesives for Biomedical Applications: A Review." *Carbohydrate Polymers* 282 (2022): 119100. https://doi.org/10.1016/j.carbpol.2022.119100

[5] Boopathi, Sampath, and Parveen Kumar. "5 Advanced Bioprinting Processes Using Additive Manufacturing Technologies: Revolutionizing Tissue Engineering." *3D Printing Technologies: Digital Manufacturing, Artificial Intelligence, Industry 4.0* (2024): 95. https://doi.org/10.1515/9783111215112-005

[6] Yadav, Priyanka, Parveen Kumar, Mayank Saxena, Yashi Dwivedi, and Abhishek Dubey. "15 Application of Three-Dimensional Printing in Medical, Agriculture, Engineering, and Other Sectors." *3D Printing Technologies: Digital Manufacturing, Artificial Intelligence, Industry 4.0* (2024): 311. https://doi.org/10.1515/9783111215112-015

[7] Srivastava, Ashish Kumar, Ajay Kumar, Parveen Kumar, Preeti Gautam, and Namrata Dogra. "Research Progress in Metal Additive Manufacturing: Challenges and Opportunities." *International Journal on Interactive Design and Manufacturing (IJIDeM)* (2023): 1–17. https://doi.org/10.1007/s12008-023-01661-6

[8] Kumar, Ajay, Parveen Kumar, Ravi Kant Mittal, and Victor Gambhir. "Materials Processed by Additive Manufacturing Techniques." In *Advances in Additive Manufacturing*, pp. 217–233. Elsevier, 2023. https://doi.org/10.1016/B978-0-323-91834-3.00014-4

[9] Hussain, Muzamil, Rizwan Ali Naqvi, Naseem Abbas, Shahzad Masood Khan, Saad Nawaz, Arif Hussain, Nida Zahra, and Muhammad Waqas Khalid. "Ultra-High-Molecular-Weight-Polyethylene (UHMWPE) as a Promising Polymer Material for Biomedical Applications: A Concise Review." *Polymers* 12, no. 2 (2020): 323. https://doi.org/10.3390/polym12020323

[10] Berman, Taryn, and Armin Bayati. "What Are Neurodegenerative Diseases and How Do They Affect the Brain?." *Frontiers for Young Minds* 6 (2018). https://doi.org/10.3389/frym.2018.00070

[11] Koopmans, Guido, Birgit Hasse, and Nektarios Sinis. "The Role of Collagen in Peripheral Nerve Repair." *International Review of Neurobiology* 87 (2009): 363–379. https://doi.org/10.1016/S0074-7742(09)87019-0

[12] Melrose, James. "Fractone Stem Cell Niche Components Provide Intuitive Clues in the Design of New Therapeutic Procedures/Biomatrices for Neural Repair." *International Journal of Molecular Sciences* 23, no. 9 (2022): 5148. https://doi.org/10.3390/ijms23095148

[13] Melrose, James, Anthony J. Hayes, and Gregory Bix. "The CNS/PNS Extracellular Matrix Provides Instructive Guidance Cues to Neural Cells and Neuroregulatory Proteins in Neural Development and Repair." *International Journal of Molecular Sciences* 22, no. 11 (2021): 5583. https://doi.org/10.3390/ijms22115583

[14] Rialas, Christos M., Motoyoshi Nomizu, Miquelle Patterson, Hynda K. Kleinman, Christi A. Weston, and Benjamin S. Weeks. "Nitric Oxide Mediates Laminin-Induced Neurite Outgrowth in PC12 Cells." *Experimental Cell Research* 260, no. 2 (2000): 268–276. https://doi.org/10.1006/excr.2000.5017

[15] Gorgieva, Selestina, and Vanja Kokol. "Collagen-vs. Gelatine-Based Biomaterials and their Biocompatibility: Review and Perspectives." *Biomaterials Applications for Nanomedicine* 2 (2011): 17–52. https://doi.org/10.5772/24118

[16] Gudimetla, Avinash, Parveen Kumar, S. Sambhu Prasad, Satish Geeri, and V. V. N. Sarath. "Towards Smart Materials: Enhancing the Efficiency of the Materials." In *Modeling, Characterization, and Processing of Smart Materials*, pp. 1–30. IGI Global, 2023. https://doi.org/10.4018/978-1-6684-9224-6.ch001

[17] Meinel, Lorenz, Sandra Hofmann, Vassilis Karageorgiou, Ludwig Zichner, Robert Langer, David Kaplan, and Gordana Vunjak-Novakovic. "Engineering Cartilage-like Tissue Using Human Mesenchymal Stem Cells and Silk Protein Scaffolds." *Biotechnology and Bioengineering* 88, no. 3 (2004): 379–391. https://doi.org/10.1002/bit.20252

[18] Pangestuti, Ratih, and Se-Kwon Kim. "Neuroprotective Properties of Chitosan and Its Derivatives." *Marine Drugs* 8, no. 7 (2010): 2117–2128. https://doi.org/10.3390/md8072117

[19] Zarrintaj, Payam, Saeed Manouchehri, Zahed Ahmadi, Mohammad Reza Saeb, Aleksandra M. Urbanska, David L. Kaplan, and Masoud Mozafari. "Agarose-based Biomaterials for Tissue Engineering." *Carbohydrate Polymers* 187 (2018): 66–84. https://doi.org/10.1016/j.carbpol.2018.01.060

[20] Kumar, Ajay, Parveen Kumar, Ashish Kumar Srivastava, and Vikas Goyat. *Modeling, Characterization, and Processing of Smart Materials.* IGI Global, 2023. https://doi.org/10.4018/978-1-6684-9224-6

[21] Soto, Jennifer, and Paula V. Monje. "Axon Contact-Driven Schwann Cell Dedifferentiation." *Glia* 65, no. 6 (2017): 864–882. https://doi.org/10.1002/glia.23131

[22] Ajay, Parveen, Ashwini Kumar, Ravi Kant Mittal, and Rajesh Goel. *Waste Recovery and Management: An Approach Toward Sustainable Development Goals.* CRC Press, 2023. https://doi.org/10.1201/9781003359784

[23] Niu, Yuqing, Florian J. Stadler, and Ming Fu. "Biomimetic Electrospun Tubular PLLA/ Gelatin Nanofiber Scaffold Promoting Regeneration of Sciatic Nerve Transection in SD Rat." *Materials Science and Engineering: C* 121 (2021): 111858. https://doi. org/10.1016/j.msec.2020.111858

[24] Dehnavi, Navid, Kazem Parivar, Vahabodin Goodarzi, Ali Salimi, Kourosh Mansoori, and Mohammad Reza Nourani. "Regeneration of Sciatic Nerve Injury by Polyglycolic Acid/Collagen/Bioglass Conduit." *Journal of Applied Biotechnology Reports* 8, no. 3 (2021): 283–292. https://doi.org/10.30491/JABR.2020.227237.1213

[25] Tiwari, Himanshu Kumar, Ashish Kumar Srivastava, Parveen Kumar, Manish Kumar Singh, Hritik Kumar, and Akshit Bhadauria. "Materials and Technology for Implant Manufacturing: Challenges and Opportunity." In *Modeling, Characterization, and Processing of Smart Materials*, pp. 83–106. IGI Global, 2023. https://doi.org/ 10.4018/978-1-6684-9224-6.ch004

[26] Sant, Shilpa, Dharini Iyer, Akhilesh K. Gaharwar, Alpesh Patel, and Ali Khademhosseini. "Effect of Biodegradation and De Novo Matrix Synthesis on the Mechanical Properties of Valvular Interstitial Cell-Seeded Polyglycerol Sebacate– Polycaprolactone Scaffolds." *Acta biomaterialia* 9, no. 4 (2013): 5963–5973. https://doi. org/10.1016/j.actbio.2012.11.014

[27] Zhang, Pengcheng, Luojuan Hu, Yucai Wang, Jun Wang, Linyin Feng, and Yaping Li. "Poly (ε-caprolactone)-block-poly (ethyl ethylene phosphate) Micelles for Brain-Targeting Drug Delivery: In Vitro and In Vivo Valuation." *Pharmaceutical Research* 27 (2010): 2657–2669.

[28] Kumar, Ajay, Parveen Kumar, Namrata Dogra, and Archana Jaglan. "Application of Incremental Sheet Forming (ISF) toward Biomedical and Medical Implants." In *Handbook of Flexible and Smart Sheet Forming Techniques: Industry 4.0 Approaches*, pp. 247–263. 2023. https://doi.org/10.1002/9781119986454.ch13

[29] Kumar, Vijay Bhooshan, Raj Kumar, Aharon Gedanken, and Orit Shefi. "Fluorescent Metal-Doped Carbon Dots for Neuronal Manipulations." *Ultrasonics Sonochemistry* 52 (2019): 205–213. https://doi.org/10.1016/j.ultsonch.2018.11.017

[30] Evans, Gregory RD. "Peripheral Nerve Injury: A Review and Approach to Tissue Engineered Constructs." *The Anatomical Record: An Official Publication of the American Association of Anatomists* 263, no. 4 (2001): 396–404. https://doi.org/10.1002/ ar.1120

[31] Clements, Isaac P., Young-tae Kim, Arthur W. English, Xi Lu, Andy Chung, and Ravi V. Bellamkonda. "Thin-film Enhanced Nerve Guidance Channels for Peripheral Nerve Repair." *Biomaterials* 30, no. 23–24 (2009): 3834–3846. https://doi.org/10.1016/j. biomaterials.2009.04.022

[32] Muheremu, Aikeremujiang, and Qiang Ao. "Past, Present, and Future of Nerve Conduits in the Treatment of Peripheral Nerve Injury." *BioMed Research International* 2015 (2015). https://doi.org/10.1155/2015/237507

[33] Loh, Qiu Li, and Cleo Choong. "Three-dimensional Scaffolds for Tissue Engineering Applications: Role of Porosity and Pore Size." (2013). https://doi.org/10.1089/ten. teb.2012.0437

[34] Liu, Chang, Yong Huang, Mao Pang, Yang, Shangfu Li, Linshan Liu, Tao Shu et al. "Tissue-Engineered Regeneration of Completely Transected Spinal Cord Using Induced Neural Stem Cells and Gelatin-Electrospun Poly (Lactide-co-Glycolide)/Polyethylene Glycol Scaffolds." *PloS One* 10, no. 3 (2015): e0117709. https://doi.org/10.1371/journal. pone.0117709

[35] Yang, Ying, Ian Wimpenny, and Mark Ahearne. "Portable Nanofiber Meshes Dictate Cell Orientation throughout Three-Dimensional Hydrogels." *Nanomedicine: Nanotechnology, Biology and Medicine* 7, no. 2 (2011): 131–136. https://doi.org/10.1016/ j.nano.2010.12.011

[36] Kumar, Ajay, Parveen Kumar, Naveen Sharma, and Ashish Kumar Srivastava, eds. *3D Printing Technologies: Digital Manufacturing, Artificial Intelligence, Industry 4.0.* Walter de Gruyter GmbH & Co KG, 2024. https://doi.org/10.1515/9783111215112

[37] Gopalakrishnan, Aswathi, Sahadev A. Shankarappa, and G. K. Rajanikant. "Hydrogel Scaffolds: Towards Restitution of Ischemic Stroke-Injured Brain." *Translational Stroke Research* 10 (2019): 1–18.

8 Processing of Sustainable Biomaterials by Additive Manufacturing Methods

Aditya Singh and Purba Mandal
Integral University

Shubhrat Maheshwari
Rama University and Sam Higginbottom University
of Agriculture, Technology and Sciences

Juber Akhtar
Integral University

Bhupendra G. Prajapati
Ganpat University

Sudarshan Singh
Chiang Mai University

1 INTRODUCTION

The market has come to recognize the requirement for rapid, dependable working components that can be supplied to clients due to the significant industrial and technological developments that have been made public over the last many decades (1). This technique is used in many other industries, and it was named rapid prototyping (RP). Technologies utilizing digital data to create physical goods are called "RP" technologies (2). Developments in new manufacturing processes, especially three-dimensional (3D) printing, often known as additive manufacturing (AM), have laid the foundation for numerous engineering and biomedical uses because of these processes' efficiency, accuracy, and precision (3). Tissue engineering may be accomplished by 3D printing, also referred to as additive manufacturing (AM), which involves precisely layer-by-layer applying different "inks" to biomaterials and cells and producing three-dimensional things using computer-aided design and prototypes. AM, 3D bioprinting is creating functional 3D objects with predetermined porosity architectures by layer-by-layer extrusion of hydrogel fluids containing living cells, often known as bioinks, by a nozzle. The ability to precisely manipulate the printed structure's microarchitecture by maximizing the inks' physical, biological, and mechanical characteristics as well as the distribution of their cells is a key benefit

DOI: 10.1201/9781003434313-8

of additive manufacturing (AM) (4). AM is a versatile and complex manufacturing technique that is extensively employed in producing customized biopolymer-based goods and sophisticated healthcare system architecture. These sustainable materials are printed in 3D for practical therapeutic contexts, such as implants, tissue engineering, drug delivery devices, and wound dressing (5). To create intricate 3D construct models, AM utilizes computer-assisted manufacturing (CAM) and computer-aided design (CAD). Following that, they are transformed into Standard Triangulation Language (STL) format, which allows for the slicing and binding of 2D layers. A 3D printer can then be used to manufacture the sliced components layer by layer by automatically depositing ink onto a substrate. Typical 3D printing techniques include those that utilize stereolithography, inkjet, extrusion, and laser assistance (6). Without the use of molds or machining, AM technology creates 3D personalized items like patient-specific implants using imaging methods or software for computer-aided design (CAD) (5). While products manufactured using 3D technology closely approximate better physiological traits, biomedical equipment created with this method of manufacturing are not designed for dynamic situations and are static by nature (7). The original tissues' micro-environment within a living body regulates their biological processes and aids in their development (8). Bioactive materials must exhibit exceptional biodegradability, biocompatibility, and adaptability to ensure optimal performance in a changing environment (9).

To address this issue, four-dimensional (4D) printing technology has been created. The idea of 4D printing was first put forth by Professors Jerry and Tibbits in 2013 (10). In this concept, sophisticated materials change their structure over time in a simulated environment. As a more sophisticated kind of 3D printing, 4D printing is regarded as a cutting-edge method of production (11). When utilizing a range of materials, 4D printing may construct any complicated product with greater precision, performance, quality, and accuracy capabilities than 3D printing techniques (9). Advanced 4D printing materials, often known as active origami systems or shape-morphing elements, deposit materials, such as polymer composites, layer by layer using 3D printing processes to create 3D items (12). The sorts of materials and their characteristics should be carefully considered while selecting bioprinters (13). 4D printing extends these advanced materials (AMs) into new dimensions when a variety of environmental stimuli, like biomolecule, cell traction force, electric field, enzymes, light, magnetic field, moisture, pH, and temperature, are present (9). The 3D bioprinting technique produces intricate, three-dimensional, cell-filled tissue designs that observe natural tissues by using bio-ink functional ingredients. Using this technique, various synthetic skin, cartilage, and bone can be produced (14). Bioprinting makes use of three main approaches: inkjet bioprinting based on extrusion and laser technology (15). On the other hand, the idea of 4D bioprinting suggests that 3D bio-printed structures can deform in response to external stimuli (9). Additionally, the 3D-printed components go through predefined shape adjustments during post-printing to get the desired outcome. It permits regulated as well as accurate tissue replication (16). Table 8.1 and Figure 8.1 show the main differentiation between 4D and 3D printing. In addition, it facilitates the achievement of a partially dynamic contact between native cells (17). One benefit of AM is its ability to fabricate items from digital designs with nearly any material and intricate shapes. Metals, composites, meta-materials, ceramics, cement and concrete, food, polymers, clays, and organ tissue are examples of materials. Electronics, homes, turbines, organs, urban furniture, art, robots, prosthetics, architecture electromagnetic shielding, and engines, are just a few of the incredibly varied items that have been printed thus far. AM is transforming manufacturing globally in several ways, bringing the production and design processes to a level everyone can use. These days, technology's effect on the environment is one of the key factors determining its viability and sustainability. Analysis has been done on AM's energy use, recycling, metrology, and circular economy (18, 19).

TABLE 8.1

A meta-analysis of 3D and 4D printing technologies

Category	3D printing technology	4D printing technology	References
Printing technique	Printing makes a 2D structure repeat, laid out layer by layer, bottom to top	Printing is a derivative of 3D printing	(20)
Type of printer	3D printer	4D printer with smart or multi-material capabilities	(20)
Materials resources	Biomaterials, ceramics, food, metals, nanomaterials, paper, polymers, and thermoplastics	Smart materials include self-actuating, self-assembling, and self-sensing materials, as well as magneto strictive materials, shape memory polymers, and advanced materials	(21)
Concept of design	3D digital item (scan or drawing)	3D digital item with the stretching feature	(22)
Flexibility	No	Yes, afterward printing in different shapes, colors, functions, and conditions	(22)
Status of the product	Static structure	Smart, dynamic structure	(22)
Cost of equipment	Low	High	(22)
Estimates of the market	Medium	Medium–high	(22)
Applications	Aerospace, consumer items, defense, education, engineering and design, fashion, industrial goods, medical, military, robotics, and more	Aerospace, aviation, biomedical devices, construction, furniture, medical, soft robotics, transportation, and more	(23)

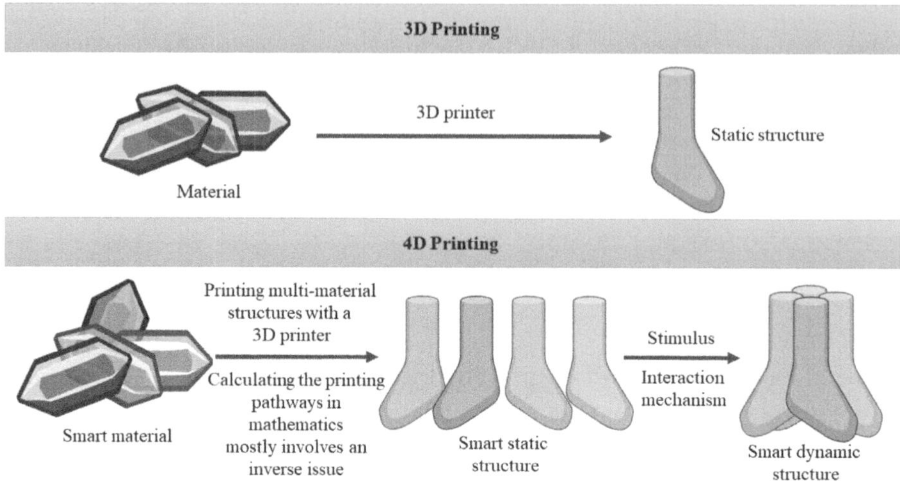

FIGURE 8.1   Basic differences between 3D and 4D printing.

2   ADDITIVE MANUFACTURING TECHNOLOGIES

The technology for 3D printing has advanced recently. Since this industry is still in its early stages, many technological developments and discoveries are continually being made. It's possible that technology won't be able to significantly revolutionize the production sector just yet. 3D printers construct the layers that eventually comprise the final things using other technologies and additive manufacturing processes 3D printing technology variations serve at different purposes (24). The AM technique is a manufacturing process that involves layering materials at a time to create products (25). CAD is utilized by researchers, pharmacists, and surgeons to create the instructions that control the printing trajectory of the nozzle. Following a pre-designed 3D model, the printer nozzle uses this command to stack the ink, containing the active pharmaceutical ingredients (APIs) with the binder, one layer at each stage, producing a 3D printing product or dosage form (26). They are divided into seven basic categories as shown in Figure 8.2.

2.1   MATERIAL EXTRUSION

The cost-effectiveness of this technique makes it extensively usable. An extrusion-based material manufacturing process can be applied for printing a variety of substances, colorful plastics, as well as living cells (27). Material extrusion engineering techniques include fused deposition modeling (FDM) or fused filament fabrication (FFF). The most basic AM technique for RP and 3D printing is FDM. Because of its high accuracy, low cost, and short turnaround time, the FFF principle is generally used by companies and researchers in products (8). Filament-type thermoplastic is the material of input used in FDM. It is introduced through the heated nozzle, wherever it melts and is squeezed out in the form of layers. The printing software includes layer thickness and design pattern, and the bed automatically descends when the first layer is finished, allowing for the printing of the next layer until the finished

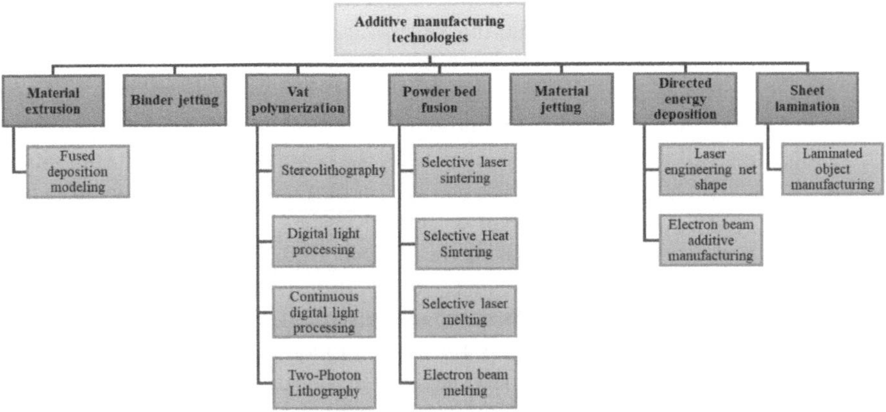

FIGURE 8.2 Categories of additive manufacturing technologies.

product is created. FDM based on extrusion is crucial for producing functionally graded goods (28).

When performing an investigation, process variables are essential; thus it's critical to fine-tune them to produce parts of enough quality. The qualities of the finishing output are greatly influenced through a wide range of process factors used in FDM. Layer thickness is the most often used process variable. It is the height among the extruded layers measured vertically along their Z-axis. The raster position or angle and the difference in the deposition bed relative to the X-axis represented the platform. The deposition layer gap between the two corresponding rasters is known as the air gap. When two neighboring layers overlap, there is also a negative air gap (29). Raster width is the term used to describe the breadth of the deposited beds on the machine platform. It is dependent on the extrusion nozzle's diameter. Positioning a product on a machine platform for many axes is known as "build orientation." Density of infill: solid layers surround the component. Infill, or internal structure layers, are the solid layers that encircle the unseen internal portion. The quantity of infill and the amount of filament used to create the interior structure of the component is known as the infill density. To provide a solid internal structure and bonding, infill patterns are typically used in building components. The printing speed is determined by how far the nozzle extruder travels in a given amount of time during extrusion. Only print speed affects printing time (30). FDM technology enters the picture due to its affordable price and good printing speed. Simple prototyping and proof-of-concept models work well with this technology. FDM has a variety of drawbacks, such as limited usage of different thermoplastic materials, excessive surface roughness, and poor mechanical properties (31). Resolution in FDM ranges from 50 to 200 μm (18).

2.2 BINDER JETTING (BJ)

BJ is a multistep AM technique that was first created by the Massachusetts Institute of Technology at the beginning of the 1990s. Despite being developed in the 1990s, it took a long time for it to become commercially available in 2010 (32). BJ involves a RP technique. It also entails making several components that complement the binding agent being used. Metals and alloys (such as those based on aluminum, iron, copper, cobalt, and nickel), as well as ceramics (such as graphite, glass, sand, etc.), can be processed with this technology. All materials that can be printed in color and are available as powder, according to reports, should be compatible. A binder material is used to attach the metal or ceramic powder component to both sides of the layers, whereas the metal or ceramic-based substance to be manufactured is the primary material used in the BJ process. Typically, a liquid binder is used, whereas a solid powder of metal or ceramic is used. To manufacture an AM part, the method of printing is comparable with all other printing processes. Depending on the needs identified by the CAD model, a layer of binder is put on top of the ceramic or powdered metal layer. The ceramic or metal is dispersed. This method is used repeatedly to construct the full section (33). Liquid binding substances are used to bind together particles of powder. The binding agent is applied, layer by layer, to the top of the metallic powder material, using the 3D model as a guide (34). Ceramics, metals, sand, as well as granular polymers, are among the materials that are frequently used (35). Although a variety of materials are available, they are still only helpful for prototyping because the printed components do not perform to the requirements. Because of the low packaging factor of the powder material, the printed pieces have a low density and a high volume of pores (36). Usually, post-processing methods like sintering and infiltration are needed for these parts (37). Sintering is one of the most widely used post-processing procedures out of the two. However, this method is partially responsible for porosity, creep development, and dimensional inaccuracies (38). The printed part is submerged in a multiphase fluid solution throughout the infiltration process, allowing the infiltrants to enter the component through capillary action. Ceramics are typically infused with one of the substances utilized in BJ to reduce the percentage of the porosity in the painted portion (39). The primary issue with this approach is that the infiltrant is not sufficiently deposited into the holes in the binder-jetted printing part size, which is smaller than the standard size of the particles of the solid loadings, resulting in an incomplete reduction of porosity (40). The raw ingredients preparation, fabrication, and post-processing phases of the process begin at this point. After the roller was initially utilized for each coating of powder from the powdered stock to the developing platform, the printer head determines the layers with its use of the binding agents as specified by the 3D design or model file. After the first layer is finished, its height is decreased, and the process is repeated for the subsequent layer in the same way. Once the full geometry is produced, the process is repeated; this section called the "green part," is then isolated from the powder to prepare it for post-processing. Additional post-processing steps include curing, sintering, de-binding, and densification. After being heated to 200°C, the green component undergoes polymerization and becomes tougher. Following the conversion of the green portion into the brown portion, the binder is burned and allowed to sinter. This is the "finished product," where the density and mechanical attributes have both been raised (41).

2.3 VAT PHOTOPOLYMERIZATION

Photo-polymerization of materials is used in this 3D printing method to create solid pieces. In a vat, materials like liquid as well as photopolymers are gathered. Following that, the materials are exposed to light sources in successive layers to create 2D layered patterns (42). The process of selectively curing a photopolymer liquid in a vat through light-activated polymerization is known as vat photopolymerization (43). There are four main techniques for producing it: 2-photon polymerization (2PP), continuous liquid interface production, stereolithography (SLA), and digital light processing (DLP) (44).

2.3.1 Stereolithography (SLA)

In a 3D printer that uses SLA technology, light-curable resin (henceforth commonly referred to as resin) is placed into a clear tank. The laser light beam tracks the structure area and cures parts using the cut STL template when the platform is immersed in resin. After a layer forms, the platform is generally elevated or lowered in the direction of the Z by a set amount equal to the height of the layer, based on how the machine performs the top-down as well as bottom-up technique. Layer by layer, the curing process is continued until a 3D model is completed. A layer can vary in height from 12 to 150 μm (45). SLA offers significant advantages, one of which is its large spatial resolution. 100 μm is the most often employed layer height. 10–20 millimeters may be printed each hour by standard SLA printers. The accuracy of SLA production is influenced by the spot size or the diameter of the beam of a laser at the curing point. For example, 140 μm is the ideal spot size for the Formlabs Form 2-SLA (46).

2.3.2 Digital Light Processing

UV light from the DLP is projected using the DMD (digital micro-mirror device), which then projects light waves onto the vat resin's top or bottom surface. By utilizing the DMD, the UV light can retain its excellent dimensional precision while curing greater parts of resin per unit of time compared with standard SLA (47). DLP is a three-dimensional, autonomously standing item created by carefully connecting light to the photosensitive resin through a progressive layering procedure. DLP exposes each resin stratum to light, just like SLA does. DLP shows all the layers at once, in contrast to SLA. Consequently, the DLP process is less susceptible to oxygen inhibition and faster than the SLA. The resolution of DLP printers, typically 10–50 micrometers, is determined by the total amount of pixels or reflections within the DMD and the optics used for projecting the design patterns through the building platform.

In conclusion, DLP is a quick and precise 3D printing method that works well for producing finely detailed, high-quality things (48).

2.3.3 Continuous Digital Light Processing (CDLP) or Continuous Liquid Interface Production (CLIP)

A more advanced and modern form of DLP technology is called CDLP/CLIP technology. It employs LEDs for digital projection in place of conventional glass windows, and, in contrast to SLA and DLP technology, it makes use of an oxygen-permeable window (49). The building platform of CDLP/CLIP machines is also in continuous motion, enabling the printing of prototypes at velocities of numerous 100 mm/h without interval. There are a few significant distinctions between DLP and CDLP/CLIP technologies. For example, DLP employs a laser for digital displays, but CDLP/CLIP uses LEDs; moreover, DLP lacks an oxygen-permeable window, whereas CDLP/CLIP does. Additionally, DLP lacks the continual mobility of the build platform that CDLP/CLIP offers. Because of these modifications, 3D printing prototypes with CDLP/CLIP technology is quicker and more effective. Furthermore, compared with DLP technology, it provides more versatility, allowing for the printing of more materials (50). Similar to DLP, CDLP/CLIP uses a DMD to project digital light through an O_2 window into the polymer vat. This prevents the solidified resin from adhering to the window by preventing the cross-linking in the resin layer nearest to the window, or "dead zone" (51).

2.3.4   Two-Photon Lithography (2PL)

A laser beam is used in 2PL, a 3D printing method, to harden polymer resin in a vat. Using this technique, complex 3D microstructures can be produced with a finer resolution than the limit of diffraction, which is not possible with traditional 3D printing. In contrast to the sequential layering technique used in conventional procedures, 2PL allows polymer curing to occur anywhere in the resin vat. Because of this, it can create complicated 3D structures with overhangs and detailed features that are challenging to produce using conventional techniques. The print resolution in this technology is determined by the smallest polymer unit, known as a voxel, which acts as an ellipsoidal 3D point. The range of 2PL includes producing various items, including prototypes, medical implants, and microscale gadgets. While 2PL is slower than conventional techniques, it produces incredibly high-resolution structures. Despite its high initial cost, the cost decreases as more people use the device (52).

2.4   Powder Bed Fusion (PBF)

PBF is the method utilizing which parts of a powdered bed are selectively fused by heat energy (43). Products are produced perfectly with this process. The method uses a laser and electron beams to melt and fuse the powdered material. This helps in the production of many different geometrically complex goods. Ceramics, metals, composites, hybrid, and polymer materials are used here (53). Several promising technologies, including electron beam melting (EBM), selective heat sintering (SHS), selective laser melting (SLM), and selective laser sintering (SLS) are presented by powder bed fusion. Lasers were used for the sintering process in the 3D printing process known as SLS (54). To create solid structures, the powdered coalesce components are sintered one layer at a time. Pressurized air and brushes are used to remove the finished items, and loose powder helps to envelop them. SLS is essentially a 3D printing method that adjusts to the surface finishing while operating at incredibly fast rates and improved precision. SHS is a 3D printing method that doesn't require a high-intensity laser (55). Additionally, thermal print heads are used to melt the material required to make 3D objects. SHS 3D printers employ thermoplastic materials.

Rollers are used to apply a layer of thermoplastic powder on these heated surfaces. To sinter higher layers, thermal print heads trace things under cross-sectional sections over powder. This process is repeated until a three-dimensional item is produced. Technology principles are used in SLM, a direct sintering method on metal, to manufacture metallic components (56). SLM is responsible for the full melting of powder obtained for metallic items with a single component, such as aluminum. Usually, alloys are the only materials that can be sintered. The occurrence and distortion of excessive residual tension will be reduced when additional support is provided by applying various solutions. A further 3D printing method that adds a source of energy to the material that is heated is called EBM (56).

2.5 Material Jetting (MJ)

MJ works similarly to inkjet printers to create objects. There are two methods for dropping material onto a construction platform: drop on demand and continuous drop. Liquid photopolymer material is deposited across the platform as the printer head rotates horizontally over it onto it for every model layer. After that, the material is either allowed to cool down and harden or is cured by UV light. Depending on the complexity and methods employed, different machines control material deposition differently. The following procedure is to cure or become stiff the material's layers using UV radiation. After that, the process is performed once again until the component is completed. The print head is made to deposit the component material along with any additional material that will later need to be taken out. The fact that the materials must be placed in limits the amount of materials that may be used. Waxes and polymers are commonly used materials due to their viscosity and ability to form droplets (57, 58).

2.6 Directed Energy Deposition (DED)

DED technologies create varied parts by melting materials layer-by-layer, moving nozzles in multiple directions, and using excellent grain structure control. This sophisticated process yields high-quality goods with excellent grain structure control (59). DED relates to different technologies that differentiate from one another based on the substance that became fused (27). It makes material printing possible by just melting the powder. This process is referred to as deposition of metal technology since it has been used extensively on metal powder but has also been utilized on ceramics and polymers. A DED technique called (LENS) Laser Engineering Net Shaping depositing uses controlled computers lasers to construct objects (60). LENS technology is used to achieve 3D AM DED by applying layers of powdered materials like alloys, metal, and ceramics, while Electron Beam Additive Manufacturing (EBAM) uses electron beam guns for other 3D printing methods (61). As a result, layers of metallic materials containing wired feedstock are deposited. This continues until the desired net shape is reached. It is possible to produce massive metallic structures with EBAM. The primary problem related to DED is the residual strains brought on by the uneven expansion and contraction of heat, which lead to the distortion-induced creation of cracks. To solve the previously described problem, researchers created a laser scanning and stereo-based planning path system (62). The other problem is with the surface polishing of the pieces made by DED. Lower geometrical tolerance along with surface finish is displayed by the DED-printed parts (63, 64).

2.7 SHEET LAMINATION

Different materials can be used to fusion this 3D printing procedure. Over an existing layer, the material is adhered to in place using adhesives. A laser-guided by digital data is then used to cut out the appropriate shape. Essentially, this method creates colorful things with excellent resolution. The laminated object manufacturing (LOM) process is adopted by other technologies. Because lasers are used to bind stacks of metal sheets together, LOM has the potential to construct three-dimensional objects utilizing localized energy sources. This process involves feeding a lot of sheet metal rolls into the construction sections, where heat is supplied via a hot roller to the other layers, causing them to be sliced into preset shapes by a laser. Even complex geometrical pieces can be produced using LOM at low cost and with shorter operating times. The two most often used production processes for sheet lamination operations are ultrasonic condensation and ultrasonic additive manufacturing (UAM) (64, 65).

3 APPLICATION OF ADDITIVE MANUFACTURING TECHNOLOGIES

Biopolymeric materials are commonly used in clinical and biological settings. Sustainable substances possess inherent features, such as renewability, non-toxicity, flexibility, biodegradability, and biocompatibility, making them suitable for environmental and public health investigations. Biopolymers, both synthetic and natural, are being researched for 3D and 4D anatomical modeling, artificial implants, scaffold design, surgical equipment, and tissue engineering, with applications in hard and soft tissue. Hard tissue 3D-printed biodegradable polymers, both natural and synthetic, have great biocompatibility and cytotoxicity, making them ideal for bone regeneration. Furthermore, these 3D printing technologies may manufacture complex structures using biopolymers while retaining acceptable biological, physical, and mechanical properties. Furthermore, 3D-printed biopolymeric composites can imitate real soft tissues. The human body's soft tissues, such as cartilage, gut, ligament, liver, nerve, skin, tendon, urethra, and vascular system, continue to function. Soft tissues differ from hard tissues in compatible modulus of elasticity, flexibility, and low mechanical characteristics. As a result, semicrystalline bio-polymeric substances are not appropriate for soft tissue purposes (5). Table 8.2 summarizes 3D-printed biopolymer composites used for soft and hard regeneration of tissues. This section highlights significant uses for 3D/4D-printed biopolymeric composite materials.

4 ADVANCED BONE REGENERATION TECHNIQUES

3D/4D bioprinting is an extremely interesting approach for creating scaffolds for bone grafting since the polymeric materials it produces have varying functionalities

TABLE 8.2

3D-printed biopolymer composites are utilized to grow soft and hard tissues

Target tissue	Method of printing	Biopolymeric components	*In vitro* study	Structures	References
Hard Tissue					
Bone	FDM	PVA/BC	Human osteoblast cells	–	**(66)**
		PLA	hBMSCs		**(67)**
	Extrusion-based 3D printing	Gel/PVA	MG63 cells		**(68)**
		PEG/Silk/PCL	BMSCs	Crypt formations	
Soft Tissue					
Cartilage	FDM	PCL/PLA/PEG	hBMSCs	Layer-by-layer based honey-comb formations	**(69)**
	Extrusion-based 3D printing	SF/Gel	hMSCs	Layer-based 3D formations	**(70)**
		SF/PEG	Chondrocytes	Diskormeniscus-shaped scaffold	**(71)**
Nasal cartilage	Extrusion-based 3D printing	Collagen	Human chondrocytes	Microporous formations	**(72)**
Skin	DLP	PEG/SF	NIH/3T3	3D lattice development with a thin coating of keratin	**(73)**
	Extrusion-based 3D printing	Keratin/glycol chitosan methacrylate	hASCs	3D model based on "NTU"	**(74)**
Cornea	Extrusion-based 3D printing	GelMA	Human keratocytes	Complex porous	**(75)**
Nerve	Electrohydro-dynamic jet-based 3D printing	PCL	PC12	Tubular multilayered complex	**(76)**
Lung	Extrusion-based 3D printing	SF/CNF	Lung epithelial stem cells	Two crossing layers	**(77)**
Liver	Extrusion-based 3D printing	SF/Gel	Huh7, hepatocytes	Six-layered-based scaffolds	**(78)**

over time. Black phosphorus nanosheets (BPNSs) and osteogenic peptides are added to photothermally sensitive β-tricalcium phosphate (β-TCP) and poly lactic acid-co-trimethylene carbonate (PLA-TMC)-based scaffolds. The outcomes showed that the inclusion of light-sensitive BPNSs caused the β-TCP and PLA-TMC-based scaffolds' form to be reconfigured by the application using near-infrared radiation.

Furthermore, at physiological temperature, the scaffolds' mechanical properties were equivalent to those of the native trabecular bone. Furthermore, using these scaffolds to treat rat cranial lesions increased bone repair, according to in-vivo studies (79). Swelling polymeric materials is another way to tackle the problem of bone grafting. This can help to expand the pore size, which in response allows for the transfer of nutrients as well as oxygen to the scaffolds' interior regions using plant-inspired hydrogel composite bio-ink with cellulose fibrils and acrylamide matrices (80–83), creating a dynamic, biomimetic, 4D-printed structure. Additionally, this printing technique oriented the cellulose fibrils in a single orientation, resulting in longitudinal swelling deformation. Similar to this, scaffold tissues are packed with living cells utilizing polymers, such as polyethylene glycol diacrylate (PEGDA), methacrylate gelatin (GelMA), and methacrylate alginate (MA), as well as live cell, UV absorber, and photo-initiator (PI) (9). To meet the world's demands, 4D bioprinting's in-vivo reaction is crucial. The literature has several in-vivo research. One such study uses hyaluronic acid/poly(ε-caprolactone) [HA/PCL]-based shape memory polymer (SMP) to create porous scaffolds that are implanted in the mandibular bones to repair bone tissue engineering. According to the findings, scaffolds based on SMP temporarily distorted their pore shapes before recovering to become porous scaffolds at room temperature. Furthermore, there is great potential to implant these intelligent 4D-printed scaffolds in dynamic in vivo situations (84).

5 CONSTRUCTING CARDIAC TISSUE

A viable approach to addressing the difficulties associated with vascularization tissues is 4D bioprinting. The adaptability of this method has been demonstrated by the creation of vascularized models in both small and large sizes, utilizing the self-folding mechanisms that convert 2D planar shape into 3D micro-scaled hollow tubes (MHTs) in the response for external stimuli (85, 86). Using bio-ink material, which modifies physical properties over time, causes localized remodeling in these models. 4D bioprinting, in contrast to 3D bioprinting, aids in achieving homogeneous cell distribution in MHTs (87). Today, 4D bioprinting is the main method used to create self-folding architectures using, mesenchymal stem cells, photo cross-linkable hydrogels, and SMPs (88). Furthermore, photo cross-linkable hydrogels have the ability to self-fold in response to water stimulation to create microvascular scaffolds (89).

6 ADVANCED NEURAL TISSUE ENGINEERING (NTE)

To promote cellular proliferation, adhesion, and differentiation, NTE scaffolds can imitate composition as well as structure of (ECM) extracellular matrix (90). Moreover, as nerve function depends in the reconnecting of the fibers of nerves, 4D-printed micro-grooves aid in the growth of nerves in particular directions (91). The NTE method uses 4D bioprinting and bio-ink comprising acrylated epoxidized soybean oil, human mesenchymal stem cells (hMSCs), and graphene to create nerve guidance conduits (NGCs) (92). Without using a crosslinker, thermosensitive polyurethane-based hydrogel shows improved stiffness qualities. Before adding gelatin, neural stem cells are distributed across polyurethane (PU)-based hydrogel.

According to the findings, at physiological temperature, printed bioproducts demonstrated remarkable differentiation and proliferation. Biodegradable hydrogel's in-vivo reaction in a zebra-fish model has shown that 4D-printed goods can effectively protect zebrafish from nerve damage. Similarly, used poly(glycerol sebacate)/methacrylated alginate-hyaluronic acid [(PGS)-PCL/MA-HA]-based bilayer mats to develop NGCs. Analysis showed that the print softer NGCs have outstanding degradation resistance and biocompatibility. Moreover, PC-12 neuron cells grown on synthetic nerve grafts demonstrated superior proliferation, differentiation, and cellular adhesion. In a similar vein, a hydrogel that heals itself via 4D bioprinting, which may find use in nerve regeneration was also created (93).

7 ADVANCED TISSUE ENGINEERING FOR CARTILAGE

To restore tissues cartilage in the sites of damage, 4D bio-printing technologies creates the proper scaffolds for cartilage tissue engineering (CTE) (94). Curved trachea implants have also been developed for cartilage repair using 4D bioprinting technology (95). For example, when acrylamide matrices containing uni-aligned nano-cellulose fibrils swell, the resulting deformation is longitudinal. After the swelling process, intricate curves are filled (96) with 4D-printed trachea implant using a hydrogel based on photo-crosslinked silk fibroin. The authors created a hydrogel structure loaded with cells using DLP technology, which was then implanted into rabbits to assess the structure's in-vivo reaction. The rabbit trachea showed good cartilage repair, according to the results. Through real-time matrix remodeling during printing, the orientation of collagen fibers was regulated by taking CTE applications into account. They subsequently produced artificial fibro-cartilage using this method. The study used several bio-ink layers that were loaded with magnetic chondrocytes to produce complex cartilage scaffolds. Certain structures have diverse properties and a heterogeneous makeup, such as tendon-bone connection and fish scale (97, 98). Functionally graded polymeric materials are printed in four dimensions with mechanical properties that can be customized and these materials can be used to print the soft muscles around the bones (98, 99).

8 DRUG DELIVERY SYSTEMS (DDS)

4D bioprinting technology has a varied range of DDS, such as multi-layered DDS, tablets, transdermal system, dermal skin patches, orodispersible film, nanosuspensions, vaginal delivery systems and rectal, which may be produced with extremely high accuracy (100). In addition to customizing the mechanical strength or shape, 4D bioprinting aids in the delivery of biomolecules by reacting to outside cues (94). Because drug delivery patterns can now be better controlled, there has been a constant increase in interest in stimuli-responsive DDS (101). Smart automated bandages with pH-sensitive sensors have recently been created; these bandages can release medication in response to specific signals, abnormal disease conditions, and external stimuli (99) by incorporating a stimulus-dependent response to 4D bio-printed DDS that regulates drug release (102). For example poly(N-isopropyl acrylamide-co-acrylic acid)/poly(propylene fumarate) or PNIPAM-AAc/PPF that is a thermoresponsive-based

DDS that treats gastrointestinal disorders by releasing medications in a regulated manner. By adding MNPs to these photocurable polymers, 4D-printed helical micro-swimmers for targeted DDS are also produced (103). In a different investigation, CS-based coiled micro-swimmer was found to be activated using UV light. The created micro-swimmer demonstrated outstanding drug-releasing and biodegradability, according to the results. In a similar vein, 4D printed capsules featuring hydrogel core-shell structure may release drug in pre-determined times and amounts. The central portion of the capsule held biomolecules, poly(vinyl alcohol), and ethylene glycol; on the other hand, the shell held poly(lacticco-glycolic acid) (PLGA) and gold nanorods (AuNRs). Under laser stimulation, these photo-responsive AuNRs burst and discharged medication (104, 105).

9   MEDICAL DEVICES

Stents, orthopedic implants, and complicated medical equipment tailored to individual patients have all been made possible by 4D bioprinting (106). Drug release has been accurately controlled in the medical field with the use of SMPs-based drug-eluting stents. Vascularization is another main use for these stents (107). In addition to opening blood arteries, these stents facilitate the release of antiplatelet and antiproliferative medications. Stents that are 4D printed can also reduce surgical invasiveness. Helical goods are typically made from thermoresponsive and chemo-triggered polymeric substances because of their exceptional biomechanical resistance (108). In addition, other elements like adherence with platelets, release performance, and biocompatibility are also considered significant (109). Because of their exceptional biomechanical resistance, helical goods are typically made from thermoresponsive and chemo-triggered polymeric substances. Shape programmable 3D format microfluidic systems, especially for a variety of geometries, including open-mesh topologies, which utilize a bilayer-containing channel embedded SMPs and polydimethylsiloxane (PDMS). Numerous three-dimensional structures for microfluidic instruments were created. Under temperature stimulation, reverse-form morphing behavior and programming were seen (110). Both in damaged and recovered geometries, fluid flow was well maintained in the microfluidic channels. Moreover, the addition of magnetic particles to the PDMS layer was shown to cause the quick action of 3D microfluidic structures, which were then managed by a portable magnet. As a result, open-mesh 3D microfluidic strop restructures with fully shape-programmable behavior were created, which are especially useful for the engineering of tissues-based biomedical uses and drug delivery (111, 112). Functional biorobots used live cardiomyocytes in a different study, with adjustable acoustic fields. These biorobots could be controlled by an external stimulus that combines chemical substance and electrical elements, such as natural contraction–relaxation cycles and the controlled actuation due to soft skeleton or microparticle pumping. For the creation of various products based on reprocessable and self-healing polythiourethane (PTU), dynamic thiocarbamate linkages are included in the photocurable methacrylate. Through printing a hand model, the shape memorized PTU showed outstanding shape recovery, shape fixity, surface wettability, and great biocompatibility. 4D-printed PTU had a variety of applications, particularly in the bio-implant industry (100, 113).

10 OTHER APPLICATIONS IN BIOMEDICINE

The technology of 4D bio-printing is highly promising for the creation of further 4D-printed items (114, 115). Additional biomedical uses comprise the creation of bionic elbow guards, hands, and splints. Another ground-breaking study, using radioscopic characterization showed a 4D-printed dense polyurethane (PU)-based SMPs. The combination of tungsten and sodium chloride salt created radiopacity and porosity, respectively. The result was a 4D-printed bio-product with opaque pores and the endovascular coil was cleaned with distilled water. Expensive experts are trying to produce and customize gland cells using 4D bio-printing (116). The creation of in-vitro modeling for the tumor micro-environment is the current oncological problem (117). SUM159 cell seeding for a bi-layered hydrogel based on polyethylene glycol diacrylate (PEGDA)/methacrylated gelatin (GelMA). These biocompatible, curved, tubular, and folded structures are thought to be essential for assessing ductal cancer in vitro models. Cellular applications of 4D printing technology have surfaced. The oil droplet method prints a tissue-like structure using water. Researchers successfully replicated the functional transmission of neurons by printing cell-free expressions in an oily atmosphere and applying external light stimulation (117).

11 CONCLUSIONS

Excellent prospects for the economical production of a variety of novel and cutting-edge products for biomedical applications, as well as for the simplification of manufacturing processes, have been made possible by AM technology for the creation of medical devices and biomaterials. Except for related materials like metals, polymers, and ceramics, the benefits, drawbacks, strengths, and limitations of each technique, as well as aspects like achievable resolution, anisotropy levels, cell embedding in feedstock substances, part sizes, post-processing, printing costs, printing speed, and support structures, are critical for the AM fabrication of biomaterials. The micro-architecture structure of biomaterials, which is a crucial part of their creation, and the choice of appropriate AM techniques, in addition to process parameter optimization within the available AM machines, are vital to the success of 3D printing techniques. For biological purposes, porous and lattice structures are frequently required. To ease the transfer of oxygen and nutrients to cells, the method entails making sure that biomaterial holes have certain morphologies and sizes, as well as being fully open and linked. Significant breakthroughs for the biomedical engineering sector are possible, thanks to AM technologies, which also improve material qualities and present new prospects for product development and biomedical device manufacture. Customizing design and characteristics is the main emphasis of AM research and development; nevertheless, existing biomaterials are not suitable for artificial organs, and tissue engineering; therefore, further research is required in the future. More study is required in this field since it is still frequently challenging to keep biomaterials within the body for extended periods without risking immune rejection. It is expected that improved materials will be able to meet the wear-reduction and durability requirements of artificial joints. The process of tissue engineering involves the cellular reaction to biomaterials that are both biocompatible and compatible with the host tissue. This means that bioactive ingredients must be included to stimulate some processes or block others, which might lead to new developments in the field of biomaterial development for AM. The goal of Advanced Manufacturing (AM) research is to overcome the constraints of traditional technologies such as welding and soldering by investigating multi-material 3D printing, different metal-joining techniques, and nano- and micro-printing with multi-materials, especially bimetals.

Abbreviations: bacterial cellulose (BC); cellulose nanofibrils (CNF); digital light processing (DLP); fused deposition modeling (FDM); gelatin (Gel); human adipose-derived stem cells (hASCs); human bone marrow stromal cells (hBMSCs); human hepatoma-derived cell line (Huh7); human mesenchymal stem cells (hMSCs); methacrylated gelatin (GelMA); National Institutes of Health/3-day transfer, inoculum 3×10^5 cells (NIH/3T3); pheochromocytoma cell line (PC12); polycaprolactone (PCL); polyethylene glycol (PEG); polylactide (PLA); polyvinyl alcohol (PVA); silk fibroin (SF); AM; computer-assisted manufacture (CAM); computer-aided design (CAD); Standard Triangulation Language (STL); fused filament fabrication (FFF); stereolithography (SLA).

REFERENCES

1. Rejeski D, Zhao F, Huang Y. Research needs and recommendations on environmental implications of additive manufacturing. *Additive Manufacturing*. 2018;19:21–28.
2. Matta A, Raju DR, Suman K. The integration of CAD/CAM and rapid prototyping in product development: A review. *Materials Today: Proceedings*. 2015;2(4–5):3438–3445.
3. Eyers DR, Potter AT. Industrial additive manufacturing: A manufacturing systems perspective. *Computers in Industry*. 2017;92:208–218.
4. Zhu Y, Joralmon D, Shan W, Chen Y, Rong J, Zhao H, et al. 3D printing biomimetic materials and structures for biomedical applications. *Bio-Design and Manufacturing*. 2021;4:405–428.
5. Arif ZU, Khalid MY, Noroozi R, Hossain M, Shi HH, Tariq A, et al. Additive manufacturing of sustainable biomaterials for biomedical applications. *Asian Journal of Pharmaceutical Sciences*. 2023:100812.
6. Poukens J, Laeven P, Beerens M, Nijenhuis G, Sloten JV, Stoelinga P, et al. A classification of cranial implants based on the degree of difficulty in computer design and manufacture. *The International Journal of Medical Robotics and Computer Assisted Surgery*. 2008;4(1):46–50.
7. Santoni S, Gugliandolo SG, Sponchioni M, Moscatelli D, Colosimo BM. 3D bioprinting: Current status and trends—A guide to the literature and industrial practice. *Bio-Design and Manufacturing*. 2022;5(1):14–42.
8. Li H, Fan W, Zhu X. Three-dimensional printing: The potential technology widely used in medical fields. *Journal of Biomedical Materials Research Part A*. 2020;108(11):2217–2229.
9. Arif ZU, Khalid MY, Zolfagharian A, Bodaghi M. 4D bioprinting of smart polymers for biomedical applications: Recent progress, challenges, and future perspectives. *Reactive and Functional Polymers*. 2022:105374.
10. Gupta S, Bissoyi A, Bit A. A review on 3D printable techniques for tissue engineering. *BioNanoScience*. 2018;8:868–883.
11. Mahmood A, Akram T, Shenggui C, Chen H. Revolutionizing manufacturing: A review of 4D printing materials, stimuli, and cutting-edge applications. *Composites Part B: Engineering*. 2023:110952.

12. Yang E, Miao S, Zhong J, Zhang Z, Mills DK, Zhang LG. Bio-based polymers for 3D printing of bioscaffolds. *Polymer Reviews.* 2018;58(4):668–687.

13. Salerno A, Netti PA. Review on computer-aided design and manufacturing of drug delivery scaffolds for cell guidance and tissue regeneration. *Frontiers in Bioengineering and Biotechnology.* 2021;9:682133.

14. Jiang D, Ning F, Wang Y. Additive manufacturing of biodegradable iron-based particle reinforced polylactic acid composite scaffolds for tissue engineering. *Journal of Materials Processing Technology.* 2021;289:116952.

15. Morouço P, Azimi B, Milazzo M, Mokhtari F, Fernandes C, Reis D, et al. Four-dimensional (bio-) printing: a review on stimuli-responsive mechanisms and their biomedical suitability. *Applied Sciences.* 2020;10(24):9143.

16. Mostafaei A, Elliott AM, Barnes JE, Li F, Tan W, Cramer CL, et al. Binder jet 3D printing—Process parameters, materials, properties, modeling, and challenges. *Progress in Materials Science.* 2021;119:100707.

17. Mehrpouya M, Dehghanghadikolaei A, Fotovvati B, Vosooghnia A, Emamian SS, Gisario A. The potential of additive manufacturing in the smart factory industrial 4.0: A review. *Applied Sciences.* 2019;9(18):3865.

18. Ngo TD, Kashani A, Imbalzano G, Nguyen KT, Hui D. Additive manufacturing (3D printing): A review of materials, methods, applications and challenges. *Composites Part B: Engineering.* 2018;143:172–196.

19. Quanjin M, Rejab M, Idris M, Kumar NM, Abdullah M, Reddy GR. Recent 3D and 4D intelligent printing technologies: A comparative review and future perspective. *Procedia Computer Science.* 2020;167:1210–1219.

20. Cai J, Wang S, Su Z, Li T, Zhang X, Bai G. Meta-analysis of QTL for Fusarium head blight resistance in Chinese wheat landraces. *The Crop Journal.* 2019;7(6):784–798.

21. Wang Y, Qian L. Three-or four-dimensional hysterosalpingo contrast sonography for diagnosing tubal patency in infertile females: A systematic review with meta-analysis. *The British Journal of Radiology.* 2016;89(1063):20151013.

22. Valenti C, Federici MI, Masciotti F, Marinucci L, Xhimitiku I, Cianetti S, et al. Mechanical properties of 3D-printed prosthetic materials compared with milled and conventional processing: A systematic review and meta-analysis of in vitro studies. *The Journal of Prosthetic Dentistry.* 2022.

23. Johnstone S, Dainty A, Wilkinson A. Integrating products and services through life: an aerospace experience. *International Journal of Operations & Production Management.* 2009;29(5):520–538.

24. Kharat VJ, Singh P, Raju GS, Yadav DK, Gupta MS, Arun V, et al. Additive manufacturing (3D printing): A review of materials, methods, applications and challenges. *Materials Today: Proceedings.* 2023. https://doi.org/10.1016/j.matpr.2023.11.033

25. Hajare DM, Gajbhiye TS. Additive manufacturing (3D printing): Recent progress on advancement of materials and challenges. *Materials Today: Proceedings.* 2022;58:736–743.

26. Buchanan C, Gardner L. Metal 3D printing in construction: A review of methods, research, applications, opportunities and challenges. *Engineering Structures.* 2019;180:332–348.

27. Rouf S, Malik A, Singh N, Raina A, Naveed N, Siddiqui MIH, et al. Additive manufacturing technologies: Industrial and medical applications. *Sustainable Operations and Computers.* 2022;3:258–274.

28. Goh GD, Agarwala S, Goh G, Dikshit V, Sing SL, Yeong WY. Additive manufacturing in unmanned aerial vehicles (UAVs): Challenges and potential. *Aerospace Science and Technology.* 2017;63:140–151.

29. Syrlybayev D, Zharylkassyn B, Seisekulova A, Akhmetov M, Perveen A, Talamona D. Optimisation of strength properties of FDM printed parts-A critical review. *Polymers (Basel).* 2021;13(10):1587.

30. Bouzaglou O, Golan O, Lachman N. Process design and parameters interaction in material extrusion 3D printing: A review. *Polymers (Basel).* 2023;15(10):2280.

31. Luo X, Cheng H, Wu X. Nanomaterials reinforced polymer filament for fused deposition modeling: A state-of-the-art review. *Polymers (Basel).* 2023;15(14):2980.

32. Abdulhameed O, Al-Ahmari A, Ameen W, Mian SH. Additive manufacturing: Challenges, trends, and applications. *Advances in Mechanical Engineering.* 2019;11(2): 1687814018822880.

33. Babu SS, Love L, Dehoff R, Peter W, Watkins TR, Pannala S. Additive manufacturing of materials: Opportunities and challenges. *MRS Bulletin.* 2015;40(12):1154–1161.

34. Prabhakar MM, Saravanan A, Lenin AH, Mayandi K, Ramalingam PS. A short review on 3D printing methods, process parameters and materials. *Materials Today: Proceedings.* 2021;45:6108–6114.

35. Saxena KK, Awasthi A. Novel additive manufacturing processes and techniques in industry 4.0. *Handbook of research on integrating industry 40 in business and manufacturing.* IGI Global. 2020:439–455.

36. Uysal E. *Development of Biocompatible and Flexible Polymers for DLP Processing and improvement of Mechanical Properties.* Marmara Universitesi (Turkey); 2021.

37. Tang Y, Zhou Y, Hoff T, Garon M, Zhao Y. Elastic modulus of 316 stainless steel lattice structure fabricated via binder jetting process. *Materials Science and Technology.* 2016;32(7):648–656.

38. Crane NB, Wilkes J, Sachs E, Allen SM. Improving accuracy of powder-based SFF processes by metal deposition from a nanoparticle dispersion. *Rapid Prototyping Journal.* 2006;12(5):266–274.

39. Farzadi A, Solati-Hashjin M, Asadi-Eydivand M, Abu Osman NA. Effect of layer thickness and printing orientation on mechanical properties and dimensional accuracy of 3D printed porous samples for bone tissue engineering. *PloS One.* 2014;9(9): e108252.

40. Maleksaeedi S, Eng H, Wiria F, Ha T, He Z. Property enhancement of 3D-printed alumina ceramics using vacuum infiltration. *Journal of Materials Processing Technology.* 2014;214(7):1301–1306.

41. Du W, Ren X, Ma C, Pei Z, editors. Binder jetting additive manufacturing of ceramics: A literature review. ASME International Mechanical Engineering Congress and Exposition; 2017: American Society of Mechanical Engineers.

42. Bagheri A, Jin J. Photopolymerization in 3D printing. *ACS Applied Polymer Materials.* 2019;1(4):593–611.

43. Mancilla-De-la-Cruz J, Rodriguez-Salvador M, An J, Chua CK. Three-dimensional printing technologies for drug delivery applications: Processes, materials, and effects. *International Journal of Bioprinting.* 2022;8(4).

44. Li Y, Zhang X, Zhang X, Zhang Y, Hou D. Recent progress of the vat photopolymerization technique in tissue engineering: A brief review of mechanisms, methods, materials, and applications. *Polymers.* 2023;15(19):3940.

45. Jamróz W, Szafraniec J, Kurek M, Jachowicz R. 3D printing in pharmaceutical and medical applications–recent achievements and challenges. *Pharmaceutical Research.* 2018;35:1–22.

46. Li W, Wang M, Ma H, Chapa-Villarreal F, Lobo A, Zhang Y. Stereolithography apparatus and digital light processing-based 3D bioprinting for tissue fabrication. *iScience.* 2023;26(2):106039.

47. Emami MM, Rosen DW. Modeling of light field effect in deep vat polymerization for grayscale lithography application. *Additive Manufacturing.* 2020;36:101595.

48. Deng W, Xie D, Liu F, Zhao J, Shen L, Tian Z. DLP-based 3D printing for automated precision manufacturing. *Mobile Information Systems.* 2022;2022:2272699.

49. Sarabia-Vallejos MA, Rodríguez-Umanzor FE, González-Henríquez CM, Rodríguez-Hernández J. Innovation in additive manufacturing using polymers: A survey on the technological and material developments. *Polymers*. 2022;14(7):1351.
50. Pagac M, Hajnys J, Ma Q-P, Jancar L, Jansa J, Stefek P, et al. A review of vat photopolymerization technology: Materials, applications, challenges, and future trends of 3d printing. *Polymers*. 2021;13(4):598.
51. Sharma A, Anand S, Bharti PS. 13 3D printing insight: Techniques. *3D Printing Technologies: Digital Manufacturing, Artificial Intelligence, Industry 40*. 2024:259. De Gruyter.
52. Hossain MI, Khan MS, Khan IK, Hossain KR, He Y, Wang X. Technology of additive manufacturing: A comprehensive review. *Kufa Journal of Engineering*. 2024;15(1): 108–146.
53. Shahrubudin N, Lee TC, Ramlan R. An overview on 3D printing technology: Technological, materials, and applications. *Procedia Manufacturing*. 2019;35:1286–1296.
54. Singh DD, Mahender T, Reddy AR. Powder bed fusion process: A brief review. *Materials Today: Proceedings*. 2021;46:350–355.
55. Tian X, Peng G, Yan M, He S, Yao R. Process prediction of selective laser sintering based on heat transfer analysis for polyamide composite powders. *International Journal of Heat and Mass Transfer*. 2018;120:379–386.
56. Zhang L, Xiang Y, Zhang H, Cheng L, Mao X, An N, et al. A biomimetic 3D-self-forming approach for microvascular scaffolds. *Advanced Science*. 2020;7(9):1903553.
57. Mendoza-Duarte ME, Estrada-Moreno IA, López-Martínez EI, Vega-Rios A. Effect of the addition of different natural waxes on the mechanical and rheological behavior of PLA—A comparative study. *Polymers*. 2023;15(2):305.
58. Chohan JS, Singh R, Boparai KS, Penna R, Fraternali F. Dimensional accuracy analysis of coupled fused deposition modeling and vapour smoothing operations for biomedical applications. *Composites Part B: Engineering*. 2017;117:138–149.
59. Galati M, Iuliano L. A literature review of powder-based electron beam melting focusing on numerical simulations. *Additive Manufacturing*. 2018;19:1–20.
60. Hu Y, Ning F, Cong W, Li Y, Wang X, Wang H. Ultrasonic vibration-assisted laser engineering net shaping of ZrO2-Al2O3 bulk parts: Effects on crack suppression, microstructure, and mechanical properties. *Ceramics International*. 2018;44(3):2752–2760.
61. Raplee J, Plotkowski A, Kirka MM, Dinwiddie R, Okello A, Dehoff RR, et al. Thermographic microstructure monitoring in electron beam additive manufacturing. *Scientific Reports*. 2017;7(1):43554.
62. Wang Z, Liu R, Sparks T, Liu H, Liou F. Stereo vision based hybrid manufacturing process for precision metal parts. *Precision Engineering*. 2015;42:1–5.
63. Oyelola O, Crawforth P, M'Saoubi R, Clare AT. On the machinability of directed energy deposited Ti6Al4V. *Additive Manufacturing*. 2018;19:39–50.
64. Oyelola O, Crawforth P, M'Saoubi R, Clare AT. Machining of additively manufactured parts: Implications for surface integrity. *Procedia Cirp*. 2016;45:119–122.
65. Miao S, Nowicki M, Cui H, Lee S-J, Zhou X, Mills DK, et al. 4D anisotropic skeletal muscle tissue constructs fabricated by staircase effect strategy. *Biofabrication*. 2019;11(3):035030.
66. Aki D, Ulag S, Unal S, Sengor M, Ekren N, Lin C-C, et al. 3D printing of PVA/hexagonal boron nitride/bacterial cellulose composite scaffolds for bone tissue engineering. *Materials & Design*. 2020;196:109094.
67. Grémare A, Guduric V, Bareille R, Heroguez V, Latour S, L'heureux N, et al. Characterization of printed PLA scaffolds for bone tissue engineering. *Journal of Biomedical Materials Research Part A*. 2018;106(4):887–894.

68. Kim H, Yang GH, Choi CH, Cho YS, Kim G. Gelatin/PVA scaffolds fabricated using a 3D-printing process employed with a low-temperature plate for hard tissue regeneration: Fabrication and characterizations. *International Journal of Biological Macromolecules.* 2018;120:119–127.

69. Heichel DL, Tumbic JA, Boch ME, Ma AW, Burke KA. Silk fibroin reactive inks for 3D printing crypt-like structures. *Biomedical Materials.* 2020;15(5):055037.

70. Sharma A, Desando G, Petretta M, Chawla S, Bartolotti I, Manferdini C, et al. Investigating the role of sustained calcium release in silk-gelatin-based three-dimensional bioprinted constructs for enhancing the osteogenic differentiation of human bone marrow derived mesenchymal stromal cells. *ACS Biomaterials Science & Engineering.* 2019;5(3):1518–1533.

71. Li Z, Zhang X, Yuan T, Zhang Y, Luo C, Zhang J, et al. Addition of platelet-rich plasma to silk fibroin hydrogel bioprinting for cartilage regeneration. *Tissue Engineering Part A.* 2020;26(15–16):886–895.

72. Lan X, Liang Y, Erkut EJ, Kunze M, Mulet-Sierra A, Gong T, et al. Bioprinting of human nasoseptal chondrocytes-laden collagen hydrogel for cartilage tissue engineering. *The FASEB Journal.* 2021;35(3):e21191.

73. Kwak H, Shin S, Lee H, Hyun J. Formation of a keratin layer with silk fibroin-polyethylene glycol composite hydrogel fabricated by digital light processing 3D printing. *Journal of Industrial and Engineering Chemistry.* 2019;72:232–240.

74. Yu K-F, Lu T-Y, Li Y-CE, Teng K-C, Chen Y-C, Wei Y, et al. Design and synthesis of stem cell-laden keratin/glycol chitosan methacrylate bioinks for 3D bioprinting. *Biomacromolecules.* 2022;23(7):2814–2826.

75. Bektas CK, Hasirci V. Cell loaded 3D bioprinted GelMA hydrogels for corneal stroma engineering. *Biomaterials Science.* 2020;8(1):438–449.

76. Vijayavenkataraman S, Zhang S, Thaharah S, Sriram G, Lu WF, Fuh JYH. Electrohydrodynamic jet 3D printed nerve guide conduits (NGCs) for peripheral nerve injury repair. *Polymers.* 2018;10(7):753.

77. Huang L, Yuan W, Hong Y, Fan S, Yao X, Ren T, et al. 3D printed hydrogels with oxidized cellulose nanofibers and silk fibroin for the proliferation of lung epithelial stem cells. *Cellulose.* 2021;28:241–257.

78. Sharma A, Rawal P, Tripathi DM, Alodiya D, Sarin SK, Kaur S, et al. Upgrading hepatic differentiation and functions on 3d printed silk–decellularized liver hybrid scaffolds. *ACS Biomaterials Science & Engineering.* 2021;7(8):3861–3873.

79. Wang C, Yue H, Liu J, Zhao Q, He Z, Li K, et al. Advanced reconfigurable scaffolds fabricated by 4D printing for treating critical-size bone defects of irregular shapes. *Biofabrication.* 2020;12(4):045025.

80. Sydney Gladman, A., Elisabetta A. Matsumoto, Ralph G. Nuzzo, Lakshminarayanan Mahadevan, and Jennifer A. Lewis. Biomimetic 4D printing. *Nature Materials.*2016; 15(4):413–418.

81. Ding A, Lee SJ, Ayyagari S, Tang R, Huynh CT, Alsberg E. 4D biofabrication via instantly generated graded hydrogel scaffolds. *Bioactive Materials.* 2022;7:324–332.

82. Liu X, Zhao K, Gong T, Song J, Bao C, Luo E, et al. Delivery of growth factors using a smart porous nanocomposite scaffold to repair a mandibular bone defect. *Biomacromolecules.* 2014;15(3):1019–1030.

83. Gil-Castell O, Ontoria-Oviedo I, Badia J, Amaro-Prellezo E, Sepúlveda P, Ribes-Greus A. Conductive polycaprolactone/gelatin/polyaniline nanofibres as functional scaffolds for cardiac tissue regeneration. *Reactive and Functional Polymers.* 2022;170: 105064.

84. Liu E, He W, Yan C. 'White revolution' to 'white pollution'—Agricultural plastic film mulch in China. *Environmental Research Letters.* 2014;9(9):091001.

85. Zolfagharian A, Kaynak A, Bodaghi M, Kouzani AZ, Gharaie S, Nahavandi S. Control-based 4D printing: Adaptive 4D-printed systems. *Applied Sciences.* 2020; 10(9):3020.

86. Spiegel CA, Hippler M, Münchinger A, Bastmeyer M, Barner-Kowollik C, Wegener M, et al. 4D printing at the microscale. *Advanced Functional Materials.* 2020;30(26):1907615.

87. Yang GH, Yeo M, Koo YW, Kim GH. 4D bioprinting: Technological advances in bio-fabrication. *Macromolecular Bioscience.* 2019;19(5):1800441.

88. Khalaj R, Tabriz AG, Okereke MI, Douroumis D. 3D printing advances in the development of stents. *International Journal of Pharmaceutics.* 2021;609:121153.

89. Cui C, An L, Zhang Z, Ji M, Chen K, Yang Y, et al. Reconfigurable 4D printing of reprocessable and mechanically strong polythiourethane covalent adaptable networks. *Advanced Functional Materials.* 2022;32(29):2203720.

90. Yang L, Lou J, Yuan J, Deng J. A review of shape memory polymers based on the intrinsic structures of their responsive switches. *RSC Advances.* 2021;11(46):28838–28850.

91. Papadimitriou L, Manganas P, Ranella A, Stratakis E. Biofabrication for neural tissue engineering applications. *Materials Today Bio.* 2020;6:100043.

92. Miao S, Cui H, Nowicki M, Xia L, Zhou X, Lee SJ, et al. Stereolithographic 4D bio-printing of multiresponsive architectures for neural engineering. *Advanced Biosystems.* 2018;2(9):1800101.

93. Bodaghi M, Damanpack A, Liao W. Triple shape memory polymers by 4D printing. *Smart Materials and Structures.* 2018;27(6):065010.

94. Agarwal T, Chiesa I, Presutti D, Irawan V, Vajanthri KY, Costantini M, et al. Recent advances in bioprinting technologies for engineering different cartilage-based tissues. *Materials Science and Engineering: C.* 2021;123:112005.

95. Rehmani SS, Al-Ayoubi AM, Ayub A, Barsky M, Lewis E, Flores R, et al. Three-dimensional-printed bioengineered tracheal grafts: Preclinical results and potential for human use. *The Annals of Thoracic Surgery.* 2017;104(3):998–1004.

96. Lin N, Dufresne A. Nanocellulose in biomedicine: Current status and future prospect. *European Polymer Journal.* 2014;59:302–325.

97. Kumari G, Abhishek K, Singh S, Hussain A, Altamimi MA, Madhyastha H, et al. A voyage from 3D to 4D printing in nanomedicine and healthcare: Part I. *Nanomedicine.* 2022;17(4):237–253.

98. Yarali E, Baniasadi M, Zolfagharian A, Chavoshi M, Arefi F, Hossain M, et al. Magneto-/electro-responsive polymers toward manufacturing, characterization, and biomedical/soft robotic applications. *Applied Materials Today.* 2022;26:101306.

99. Falahati M, Ahmadvand P, Safaee S, Chang Y-C, Lyu Z, Chen R, et al. Smart polymers and nanocomposites for 3D and 4D printing. *Materials Today.* 2020;40:215–245.

100. Wang J, Soto F, Ma P, Ahmed R, Yang H, Chen S, et al. Acoustic Fabrication of Living Cardiomyocyte-based Hybrid Biorobots. *ACS Nano.* 2022;16(7):10219–10230.

101. Qasim, Muhammad, Dong Sik Chae, and Nae Yoon Lee. Advancements and frontiers in nano-based 3D and 4D scaffolds for bone and cartilage tissue engineering. *International Journal of Nanomedicine.* 2019:4333–4351. 102. Noroozi, Reza, Zia Ullah Arif, Hadi Taghvaei, Muhammad Yasir Khalid, Hossein Sahbafar, Amin Hadi, Ali Sadeghianmaryan, and Xiongbiao Chen. 3D and 4D bioprinting technologies: a game changer for the biomedical sector? *Annals of Biomedical Engineering.* 2023;51(8):1683–1712.

103. Mohammed A, Elshaer A, Sareh P, Elsayed M, Hassanin H. Additive manufacturing technologies for drug delivery applications. *International Journal of Pharmaceutics.* 2020;580:119245.

104. Wang, Yu, Huaiyuan Zhang, Huifen Qiang, Meigui Li, Yili Cai, Xuan Zhou, Yanlong Xu et al. Innovative biomaterials for bone tumor treatment and regeneration: tackling postoperative challenges and charting the path forward." *Advanced Healthcare Materials.* 2024: 2304060. 105. Francis W. Stimuli-controlled manipulation of synthetic discrete micrometre sized "vehicles." PhD thesis, Dublin City University; 2017.

106. Naniz MA, Askari M, Zolfagharian A, Bodaghi M. 4D bioprinting: Fabrication approaches and biomedical applications. *Smart Materials in Additive Manufacturing.* Elsevier; 2022. pp. 193–229.

107. Aldawood Faisal Khaled. A comprehensive review of 4D printing: State of the arts, opportunities, and challenges. *Actuators,* 2023;12(3):101, MDPI.

108. Rajasekar R, Moganapriya C, Kumar PS. *Additive manufacturing with novel materials: Process, properties and applications.* John Wiley & Sons; 2024.

109. Cao Q, Chen W, Zhong Y, Ma X, Wang B. Biomedical applications of deformable hydrogel microrobots. *Micromachines.* 2023;14(10):1824.

110. Hashemi A, Ezati M, Nasr MP, Zumberg I, Provaznik V. *Extracellular vesicles and hydrogels: An innovative approach to tissue regeneration.* ACS Omega; 2024.

111. Wang Z, Jiang H, Wu G, Li Y, Zhang T, Zhang Y, et al. Shape-programmable three-dimensional microfluidic structures. *ACS Applied Materials & Interfaces.* 2022;14(13):15599–15607.

112. Daly R, Harrington TS, Martin GD, Hutchings IM. Inkjet printing for pharmaceutics– A review of research and manufacturing. *International Journal of Pharmaceutics.* 2015;494(2):554–567.

113. Rajabi N, Rezaei A, Kharaziha M, Bakhsheshi-Rad HR, Luo H, RamaKrishna S, et al. Recent advances on bioprinted gelatin methacrylate-based hydrogels for tissue repair. *Tissue Engineering Part A.* 2021;27(11–12):679–702.

114. Yu C, Schimelman J, Wang P, Miller KL, Ma X, You S, et al. Photopolymerizable biomaterials and light-based 3D printing strategies for biomedical applications. *Chemical Reviews.* 2020;120(19):10695–10743.

115. Shirazi, Seyed Farid Seyed, Samira Gharehkhani, Mehdi Mehrali, Hooman Yarmand, Hendrik Simon Cornelis Metselaar, Nahrizul Adib Kadri, and Noor Azuan Abu Osman. A review on powder-based additive manufacturing for tissue engineering: selective laser sintering and inkjet 3D printing. *Science and Technology of Advanced Materials.* 2015;16(3):033502.

116. Li H, Fan W, Zhu X. Three-dimensional printing: The potential technology widely used in medical fields. *Journal of Biomedical Materials and Research A.* 2020;108(11):2217–2229.

117. Mohamed OA, Masood SH, Bhowmik JL. Optimization of fused deposition modeling process parameters: A review of current research and future prospects. *Advances in Manufacturing.* 2015;3:42–53.

9 Comprehensive Characterization of Biochar from Coconut Shell Pyrolysis

Effect of Temperature

V. Karuppasamy Vikraman
Amrita School of Agricultural Sciences

D. Praveen Kumar
Bannari Amman Institute of Technology

G. Boopathi
Amrita School of Agricultural Sciences

1 INTRODUCTION

Biochar is a solid carbonaceous material produced by the thermal decomposition of lignocellulosic biomass in the absence of oxygen [1–3]. Biochar can be produced through slow pyrolysis, fast pyrolysis, flash pyrolysis, gasification, and hydrothermal carbonization [4–6]. Among these technologies, slow pyrolysis is widely favored for biochar production due to higher yield and lower production costs [7]. Slow pyrolysis thermally degrades the components of lignocellulosic biomass, leading to devolatilization and resulting in biochar as the energy-dense solid product [8].

Biochar has diverse applications in the energy, environmental, and agricultural sectors due to its renewable and versatile nature [9]. It can enhance crop productivity when applied to soil due to improvements in soil texture, increased pH, nutrient retention, and water holding capacity [10, 11]. Biochar with an aromatic structure is resistant to biological and chemical decomposition, indicating its potential for carbon sequestration when applied to soil [12]. The porous structure of biochar facilitates the adsorption of pollutants in both land and water [13–15] and its energy value makes it a suitable fuel substitute for low-rank coals. Sivamani et al. [16] reported a maximum copper removal efficiency and adsorption capacity of 99.61% and 116.28 mg g^{-1}, respectively for biochar derived from orange zest. Nallasivam et al. [17] achieved a 98.65% reduction of Cr(VI) within 180 minutes of reaction time using COVID-19 waste char. Kumar et al. [18] found the yield and

DOI: 10.1201/9781003434313-9

fixed carbon content of biochar from coconut palm residues to range from 26.4% to 29.66% and from 65.4% to 78.2%, respectively. Additionally, biochar can serve as a precursor for activated carbon, support for catalysts, and electrodes in energy storage devices [8]. Rawat et al. [19] fabricated supercapacitor electrodes using litchi seed biochar, reporting a specific capacitance of 49 F g^{-1} with 92% capacitance retention after 10,000 cycles. The specific capacitance and energy density of a supercapacitor made from jamun seed biochar were reported as 42.3 F g^{-1} and 19.8 Wh kg^{-1}, respectively [20]. These studies highlight the versatility of biochar in fulfilling various applications.

The versatility of biochar depends on its physicochemical characteristics, particularly its proximate and ultimate composition. The physicochemical characteristics of biochar are influenced by the pyrolysis process conditions such as pyrolysis temperature, residence time, heating rate, and the nature of biomass [21]. The heating rate of pyrolysis controls the yield of biochar and bio-oil. A slower pace of heating in pyrolysis limits biomass degradation, leading to a higher biochar yield, while faster heating rates can generate significant quantities of bio-oil. Residence time also has a similar effect on pyrolysis, in addition to controlling the porous nature of biochar [22]. Temperature was reported to have a significant effect on the yield and characteristics of biochar [23]. Higher aromaticity and recalcitrance potential were achieved for biochar derived at higher pyrolysis temperatures, while low-temperature biochar had less condensed carbon structures [24]. Therefore, it is necessary to investigate the influence of temperature on the attributes of biochar to guide its application in different sectors.

Hence, the major objective of the present study was to explore the influence of temperature on the characteristics of biochar derived from coconut shell pyrolysis. The characteristics of biochar namely proximate composition, elemental composition, atomic ratios (H/C and O/C), bulk density, heating value, pH, electrical conductivity (EC), stability, aromaticity factor, CO_2 reduction potential, and recalcitrance potential were evaluated. The study aims to address the existing research gap by systematically exploring the impact of pyrolysis temperature on the critical physicochemical characteristics of biochar produced from coconut shells. Special attention has been given to determining the suitable pyrolysis temperature for maximizing the utility of biochar in energy, environmental, and agricultural applications.

2 MATERIALS AND METHODS

The methodology adopted for the characterization of coconut shells, production of biochar at different temperatures, and biochar characterization are discussed in this section. Each experiment was repeated three times, and the values are reported as the mean ± standard deviation.

2.1 COCONUT SHELL COLLECTION AND CHARACTERIZATION

Coconut shells gathered from local restaurants in Coimbatore were powdered and sieved to a particle size of <0.2 mm before analysis. The proximate composition was analyzed using ASTM methods: moisture (ASTM E871–82) [25], volatile matter (ASTM E872–82) [26], and ash (ASTM E1755-01) [27]. The ultimate composition was estimated using a CHNO analyzer (Flash 2000, Thermo Scientific, United States). Fixed carbon and elemental oxygen content were calculated by difference. The biochemical composition analysis was performed following the Van Soest method [28]. The heating value of the coconut shell was determined using a digital bomb calorimeter (C200, IKA, United States). The bulk density of the coconut shell was estimated using the weight-to-volume ratio method [29]. The thermal degradation profile of coconut shells was determined using a thermogravimetric analyzer (Q50, TA Instruments, United States). The parameters studied included a temperature range of 30°C to 900°C, a heating rate of 10°C min^{-1}, and a N_2 flow rate of 40 mL min^{-1}.

2.2 PYROLYSIS EXPERIMENTS

The pyrolysis of coconut shells was performed in a fixed-bed pyrolyzer with a biomass capacity of 1 kg. The fixed-bed pyrolyzer was constructed from stainless steel (SS 310) and had internal dimensions of 12 cm in diameter and 34 cm in height. Electric heating coils were used to heat the pyrolyzer, with a thermocouple inside for temperature control. The thermocouple was connected to a proportional controller to maintain the pyrolyzer temperature with an accuracy of ± 1°C. The temperature range for the pyrolysis of coconut shells was determined from the thermal degradation profiles obtained from the thermogravimetric analyzer. The heating rate, N_2 flow rate, and residence time maintained were 10°C min^{-1}, 300 mL min^{-1}, and 1 h, respectively [30]. Nitrogen gas (N_2) with a purity of 99.9% was supplied from a gas cylinder, and its flow rate was regulated using a rotameter. The heating rate of 10°C min^{-1} was chosen based on previous literature, as this specific heating rate has been shown to maximize biochar yield by facilitating volatilization [31]. The biochar yield (%) was estimated using the Eq. (9.1):

$$\text{Biochar yield } (\%) = \left(\frac{W_{BC}}{W_{CS}} \right) \times 100 \tag{9.1}$$

where, W_{BC} and W_{CS} represent the weight of biochar remaining after pyrolysis (kg) and the input weight of the coconut shell (kg), respectively.

2.3 BIOCHAR CHARACTERIZATION

The proximate composition of biochar, namely volatile matter, ash, and fixed carbon was determined using ASTM D1762–84 [32]. The ultimate composition, heating value, and bulk density of biochar were estimated using the same procedure as that followed for coconut shells (Section 2.1).

The pH of the biochar was analyzed by mixing biochar with distilled water (1:20 w/v) for 10 min on a magnetic stirrer and analyzed using a pH meter (2001, Digisun Electronics Systems). For the determination of EC, a mixture of biochar and distilled water (1:10 w/v) was allowed to stand for 24 hours and was then measured using an EC meter (LT-16, Labtronics) [33].

The stable C mass fraction (C_S), aromaticity factor (f_a), and recalcitrance potential (R_{50}) were determined using the following equations (Eqs. 9.2–9.4) [34]:

$$C_s = 1.00 - 2.24 \times \frac{O}{C} \tag{9.2}$$

$$f_a = 0.967 \times \frac{FC}{C} \tag{9.3}$$

$$R_{50} = \frac{0.170 \times (0.474 \times VM + 0.963 \times FC + 0.067 \times A)}{(100 - A) + 0.00479} \tag{9.4}$$

The mean residence time (MRT) and the percentage of carbon in biochar after 100 years (BC_{+100}) were calculated using Eqs. (9.5) and (9.6) [33]:

$$MRT = 4501 \times e^{-3.2 \times \left(\frac{H}{C}\right)} \tag{9.5}$$

$$BC_{+100} = 1.05 - 0.616 \times \frac{H}{C} \tag{9.6}$$

The CO_2 reduction potential (CO_2 eq kg of biochar^{-1}) was estimated using Eq. (9.7) [35]:

$$CO_2 \text{ reduction potential} = g \text{ of C in biochar} \times \frac{80}{100} \times \frac{44}{12} \tag{9.7}$$

3 RESULTS AND DISCUSSION

3.1 Characteristics of Coconut Shell

The proximate, ultimate, biochemical composition, and heating value of the coconut shell are presented in Table 9.1. The moisture content of the coconut shell, which is 5.48%, falls within the acceptable limit (<10%) for all thermochemical conversion processes [36]. The volatile matter (75.52%) and fixed carbon (18.12%) of the coconut shell indicate their suitability for bio-oil and biochar production, respectively. The lower ash content (0.88%) could result in less clinker formation during the thermochemical conversion of the coconut shell. The elemental carbon and hydrogen content were 48.58% and 6.12%, respectively. The lower nitrogen content (0.59%) facilitates the lesser formation of oxides of nitrogen. The lower heating value of the coconut shell was 18.63 MJ kg^{-1}, which could be a result of a higher quantity of carbon and hydrogen in the feedstock. The cellulose (39.38%) and hemicellulose (25.97%) show the viability of coconut shells for higher bio-oil generation through pyrolysis. The higher lignin content (29.72%) could result in higher biochar yield [37]. The characteristics of the coconut shell such as higher elemental carbon, fixed carbon, and lignin corroborate its feasibility for biochar production.

TABLE 9.1

Characteristics of coconut shell

Analysis	Value
Proximate composition (wt%)	
Moisture	5.48 ± 0.77
Volatile matter	75.52 ± 0.91
Ash	0.88 ± 0.08
Fixed carbon[a]	18.12 ± 0.74
Ultimate composition (wt%)	
Carbon	48.58 ± 1.05
Hydrogen	6.12 ± 0.14
Nitrogen	0.59 ± 0.03
Oxygen[b]	44.71 ± 0.93
H/C ratio	1.48 ± 0.01
O/C ratio	0.69 ± 0.02
Biochemical composition (wt%)	
Cellulose	39.38 ± 1.21
Hemicellulose	25.97 ± 1.15
Lignin	29.72 ± 0.84
Lower heating value (MJ kg^{-1})	18.63 ± 0.55
Bulk density (kg m^{-3})	510.3 ± 7

[a] Fixed carbon (%) = 100 – [Moisture (%) + Volatile matter (%) + Ash (%)].

[b] Oxygen (%) = 100 – [Carbon (%) + Hydrogen (%) + Nitrogen (%)].

3.2 THERMAL DEGRADATION BEHAVIOR

Figure 9.1 shows the thermal degradation behavior of the coconut shell in the temperature from 30°C to 900°C. The DTG curve shows the presence of three weight loss regions in the pyrolysis of the coconut shell. The broad hump observed below 175°C indicates the release of adsorbed moisture and lower-volatility components from the biomass. Following the moisture release, the thermal degradation process accelerated, resulting in two sharp peaks occurring between 200°C and 400°C. This temperature range can be considered the active pyrolysis zone of the coconut shell. The first peak in the temperature range of 200°C to 315°C can be attributed to the pyrolysis of hemicellulose in coconut shells, while the cellulose decomposition occurred in the temperature range of 315°C to 400°C [38]. The overall weight loss in the active pyrolysis zone was 51.35%. The peak pyrolysis temperature and maximum derivative weight loss were 356°C and 7.12% min^{-1}, respectively. The stable weight loss zone after active pyrolysis suggests the decomposition of lignin.

The slow and steady decomposition of lignin was observed until 900°C with a derivative weight loss of 0.08% min^{-1}. The weight (%) of residual char after the completion of the pyrolysis process was 26.77%. Lignin is the primary constituent of biomass that leads to higher biochar yield [39]. Though the active pyrolysis of coconut shells occurred between 200°C and 400°C, the authors decided to extend the pyrolysis temperature to 700°C. This decision was made to explore the deviations in the properties

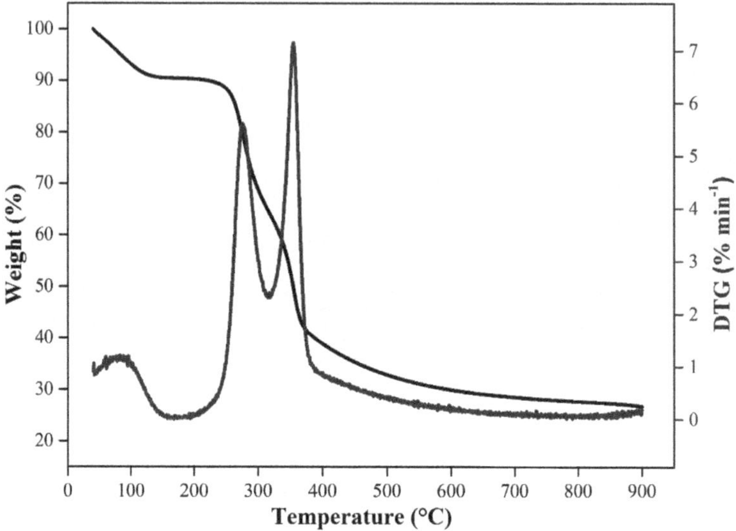

FIGURE 9.1   Thermal degradation behavior of coconut shell.

of biochar due to lignin decomposition at higher pyrolysis temperatures. The coconut shell was exposed to pyrolysis at temperatures varying between 300°C and 700°C in 100°C increments. The resulting characteristics are discussed in the following sections.

3.3 BIOCHAR YIELD

The pyrolysis temperature had a notable impact on the yield of biochar from coconut shells (Table 9.2).

The biochar yield reduced with an increase in temperature. The biochar yield was 86.13% at 300°C, while it was reduced to 27.74% at 700°C. The high biochar yield at 300°C suggests that at this temperature, primarily drying and minimal devolatilization occurred. The biochar yield reduced steeply from 86.13% to 38.05% when the temperature was increased from 300°C to 400°C, indicating the release of volatiles and the formation of biochar. Concurrently, the volatile matter of biochar reduced from 76.30% at 300°C to 53.58% at 400°C. With further increase in temperature, the biochar yield reduced gradually to 33.42% at 500°C and 29.81% at 600°C. This yield reduction with the temperature rise can be attributed to the complete degradation of hemicellulose and cellulose with the partial decomposition of lignin [33]. The volatile matter of biochar at 700°C was 9.18%, indicating the near-complete removal of volatile vapors. Pyrolysis of coconut shells at higher temperatures led to a diminishment in biochar yield and the formation of stable carbon structures.

3.4 PHYSICOCHEMICAL CHARACTERISTICS

Table 9.2 displays the physicochemical characteristics of biochar at various pyrolysis temperatures. The volatile matter decreased at elevated pyrolysis temperatures while an

TABLE 9.2
Physicochemical characteristics of biochar at different pyrolysis temperatures

	Temperature (°C)				
Parameters	300	400	500	600	700
Biochar yield (wt.%)	86.13 ± 0.82	38.05 ± 0.25	33.42 ± 0.95	29.81 ± 0.65	27.74 ± 0.63
Proximate composition (wt.%)					
Volatile matter	76.30 ± 1.79	53.58 ± 2.09	19.70 ± 0.59	10.40 ± 0.57	9.18 ± 0.27
Ash	1.85 ± 0.10	3.93 ± 0.17	5.67 ± 0.25	8.65 ± 0.31	9.16 ± 0.21
Fixed carbon	21.85 ± 1.89	42.48 ± 1.92	74.64 ± 0.80	80.95 ± 0.84	81.66 ± 0.08
Ultimate composition (wt.%)					
Carbon	57.33 ± 0.96	66.57 ± 0.87	81.59 ± 0.48	83.47 ± 0.53	83.60 ± 0.07
Hydrogen	5.73 ± 0.09	4.41 ± 0.13	2.34 ± 0.03	1.73 ± 0.03	1.65 ± 0.02
Oxygen	35.87 ± 0.81	25.28 ± 0.97	9.59 ± 0.25	5.09 ± 0.24	4.49 ± 0.14
Nitrogen	1.07 ± 0.06	3.73 ± 0.22	6.48 ± 0.23	9.71 ± 0.29	10.25 ± 0.22
Atomic ratios					
H/C	1.20	0.80	0.34	0.25	0.24
O/C	0.47	0.28	0.09	0.05	0.04
Lower heating value (MJ kg^{-1})	19.37 ± 0.84	25.84 ± 0.50	31.52 ± 0.18	30.87 ± 0.17	30.79 ± 0.07
Bulk density (kg m^{-3})	450.7 ± 13	346 ± 8	329.7 ± 4.5	308.3 ± 8.5	288.3 ± 3.8

opposite trend was observed for ash and fixed carbon. The volatile matter content at 300°C was 76.3%, and at 400°C, it decreased to 53.58%. This suggests that organic substances in the coconut shell exhibited resistance to volatilization at these temperatures. When the pyrolysis temperature exceeded 500°C, a sharp reduction in volatile matter was observed. The minimum volatile matter of 9.18% was observed at 700°C. Maximum fixed carbon content of 81.66% and ash content of 9.16% was attained at 700°C. The increase in ash and fixed carbon of biochar can be related to the reduction in volatile matter. A similar relative increase in fixed carbon and ash was also reported for other biomass [34]. A slight increment in the fixed carbon content as the temperature escalated from 600°C (80.95%) to 700°C (81.66%), indicated the heightened thermal stability of the biochar. The rise in fixed carbon was directly reflected in the increase in the elemental carbon. The reduction in elemental oxygen and hydrogen content could be due to the dehydration and devolatilization reactions at higher pyrolysis temperatures [40].

The H/C ratio reduced from 1.20 to 0.24, and the O/C ratio reduced from 0.47 to 0.04 with the rise in temperature from 300°C to 700°C. This could be a result of deoxygenation and decarboxylation reactions. The reduction in H/C and O/C shows the loss of oxygen functionalities from the biochar and its reducing hydrophilicity. These are the significant aspects that show the chemical and biological stability of biochar. Due to reduced hydrophilicity, there will be less transfer of moisture from the atmosphere to the biochar, making it a suitable solid biofuel. Biochars with H/C

< 0.6 and O/C < 0.4 are recommended for C-sequestration [41]. Hence, the biochars obtained from the pyrolysis of coconut shells ≥500°C can be feasible for long-term sequestration of carbon. The increase in N content with temperature indicates the presence of recalcitrant heterocyclic N compounds in the biochar [42]. The heating value of the biochar increased from 19.37 MJ kg^{-1} at 300°C to 31.52 MJ kg^{-1} at 500°C, but it decreased gradually with a further increment in the pyrolysis temperature. The pyrolysis of coconut shell up to 500°C released the lighter low-energy compounds leading to increased energy content in the biochar. However, when the temperature rises above 500°C, the energy content decreases due to the breakdown of the carbon structure. From this phenomenon, it can be inferred that the pyrolysis of coconut shells can be limited to 500°C if the biochar is intended to be used as solid biofuel. The bulk density of the coconut shell biomass was 510.3 kg m^{-3} while the bulk density of the coconut shell biochars decreased with an increase in temperature. The bulk density of the biochars at 300°C, 400°C, 500°C, 600°C, and 700°C were 450.7, 346, 329.7, 308.3, and 288.3 kg m^{-3}, respectively. The reduction in bulk density could be a result of organic matter decomposition transforming the biomass matrix into a lighter and more porous form.

3.5 pH AND ELECTRICAL CONDUCTIVITY

The pH and EC of the biochars derived at various pyrolysis temperatures are portrayed in Figure 9.2.

The pH of the biochar increased with temperature, while the EC of the biochar showed no clear trend. The pH of the biochar at 300°C, 400°C, 500°C, 600°C, and 700°C were 3.13, 4.61, 7.77, 8.14, and 8.21, respectively. The biochar was acidic until

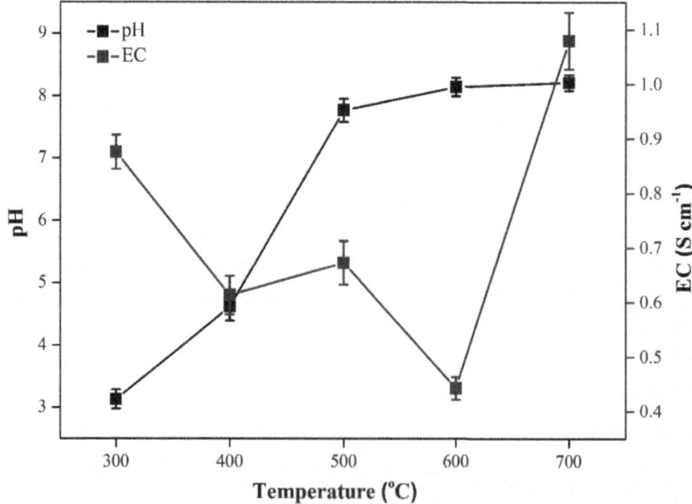

FIGURE 9.2   pH and EC of biochar at different pyrolysis temperatures.

400°C and alkaline at ≥500°C. The rise in biochar pH can be ascribed to the concurrent increase in inorganic ash content as the temperature rises. Similar results were reported elsewhere [33]. The alkaline biochar has potential application as a soil conditioner in the acidic soils by improving the cation exchange capacity and lowering greenhouse gas emissions. Contrary to pH, the EC values of the biochar were low (0.46–1.06 S cm^{-1}) and showed no clear trend with the temperature. Ortiz et al. [34] also reported low EC values for almond and nut shells.

3.6 BIOCHAR STABILITY

The biochar stability parameters, including stable C mass fraction (C_S), aromaticity factor (f_a), MRT, recalcitrance potential (R_{50}), and the percentage of carbon in biochar after 100 years (BC_{+100}) were obtained at various pyrolysis temperatures and are presented in Table 9.3.

From a carbon sequestration and agronomic perspective, biochar should be resistant to biotic and abiotic degradation to maintain its stability and functionality. The recalcitrance nature of biochar depends mainly on its aromaticity and availability of stable carbon [43]. From Table 9.3, it can be observed that the C_S, f_a, and R_{50} showed an increasing trend with pyrolysis temperature. The C_S was negative (−0.052) at 300°C, indicating incomplete pyrolysis and an elevated quantity of volatile carbon fractions. A sharp increase in C_S, f_a, and R_{50} was observed with the elevation of temperature from 300°C to 400°C. The C_s (0.910), f_a (0.945), and R_{50} (0.156) were maximum at 700°C, indicating near-complete pyrolysis and recalcitrant biochar with an aromatic carbon structure. The biochar stability studies suggest that higher pyrolysis temperatures produce a smaller amount of recalcitrant biochar, while lower pyrolysis temperatures generate a larger quantity of biochar with poor stability. The MRT of the biochar ranged between 97.62 and 2106.37 years while the BC_{+100} of the biochar ranged between 31.15% and 90.38%. Hence, the coconut shell biochars produced at 600°C or 700°C can be applied in carbon-poor soils, offering the additional benefit of extended carbon sequestration lasting up to 2,000 years.

TABLE 9.3

Stability parameters of the biochar at different pyrolysis temperatures

Temperature (°C)	Stable C mass fraction (C_S)	Aromaticity factor (f_a)	Recalcitrance potential (R_{50})	Mean residence time in years (MRT)	% of C in biochar after 100 years (BC_{+100})
300	−0.052 ± 0.041	0.368 ± 0.026	0.099 ± 0.002	97.62 ± 12.29	31.15 ± 2.39
400	0.362 ± 0.032	0.617 ± 0.020	0.118 ± 0.001	354.80 ±3 7.67	56.02 ± 2.04
500	0.802 ± 0.006	0.885 ± 0.004	0.147 ± 0.001	1496.75 ± 32.94	83.80 ± 0.42
600	0.898 ± 0.005	0.938 ± 0.004	0.155 ± 0.001	2026.95 ±3 9.74	89.64 ± 0.37
700	0.910 ± 0.002	0.945 ± 0.001	0.156 ± 0.002	2106.37 ± 17.42	90.38 ± 0.16

3.7 CO$_2$ REDUCTION POTENTIAL

Figure 9.3 shows the CO$_2$ reduction potential (CO$_2$ eq kg of biochar^{-1}) of the biochars obtained at multiple pyrolysis temperatures.

An increase in pyrolysis temperature exhibited a favorable impact on the CO$_2$ reduction potential of the biochars. The CO$_2$ reduction potential increased from 1681.71 CO$_2$ eq kg of biochar^{-1} at 300°C to 2452.40 CO$_2$ eq kg of biochar^{-1} at 700°C. This indicates the inherent potential of the coconut shell biochar for CO$_2$ storage in soil that otherwise would have increased the atmospheric CO$_2$ through decomposition or burning. The coconut shell biochars pyrolyzed at higher temperatures (≥600°C) can be considered as a potential option to store carbon in soil for climate change mitigation. Based on the yield of biochar and its fixed carbon, the total potential carbon (g of C kg of feedstock^{-1}) in coconut shell biochar were 188.20 at 300°C, 161.64 at 400°C, 249.41 at 500°C, 241.34 at 600°C, and 226.50 at 700°C. These values agree well with the total potential carbon reported for rice husk at 450°C (180 g of C kg of feedstock^{-1}) and maize straw at 550°C (194 g of C kg of feedstock^{-1}) [35].

3.8 FUTURE DIRECTIONS AND SCOPE

This study has provided valuable insights into the pyrolysis temperature-dependent attributes of coconut shell biochar, offering guidance for its application in energy, environmental, and agricultural sectors. However, there are several avenues for future research in this field. Examining alternative pyrolysis factors, including residence time and heating rate, can enhance our understanding of biochar properties.

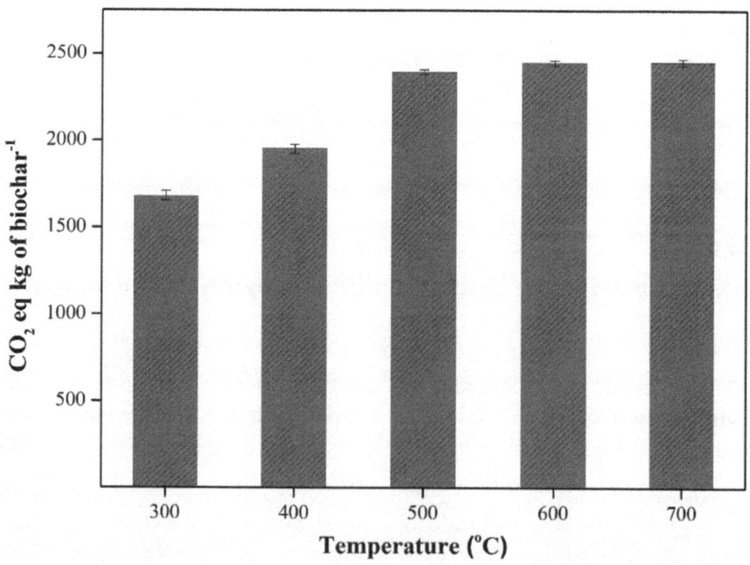

FIGURE 9.3 CO$_2$ reduction potential of biochar from various pyrolysis temperatures.

Subsequent investigations should emphasize the fine-tuning of biochar characteristics through post-production techniques to enhance its suitability for specific applications. The long-term effects on soil health and greenhouse gas emissions should be explored for the sustainable adoption of biochar on a larger scale. The scalability and economic feasibility of biochar production at different temperatures will be essential for practical applications and commercial-scale implementation.

4   CONCLUSION

The coconut shell biomass was pyrolyzed in the temperature range between 300°C and 700°C and the characteristics of biochar were discussed. Thermogravimetry was utilized to estimate the optimal range of temperature for pyrolysis of coconut shells. The influence of pyrolysis temperature was positive and linear for fixed carbon, elemental carbon, stability, and CO_2 reduction potential. The optimal pyrolysis temperature for utilizing coconut shell biochar as solid fuel was found to be 500°C. This temperature yields the maximum heating value (31.52 MJ kg^{-1}), making it an excellent choice for clean and efficient energy production. From an environmental standpoint, the biochar produced at temperature ≥600°C exhibits superior stability (>2,000 years) and has the potential to sequester 2,400 CO_2 eq kg of biochar^{-1}, contributing significantly to carbon mitigation efforts. In the agricultural context, high-temperature biochars exhibit enhanced aromaticity and porous nature. Thus, biochars produced at temperatures ≥600°C can be recommended as soil amendments, thereby enhancing crop productivity and promoting sustainable agricultural practices.

REFERENCES

1. Ajay, Parveen, Sharif Ahmad, Jyotsna Sharma, and Victor Gambhir, eds. *Handbook of Sustainable Materials: Modelling, Characterization, and Optimization* (1st ed.). CRC Press, 2023. https://doi.org/10.1201/9781003297772
2. Kumar, Ajay, Parveen Kumar, Ashish Kumar Srivastava, and Vikas Goyat, eds. *Modeling, Characterization, and Processing of Smart Materials*. IGI Global, 2023.
3. Kumar, Ajay, Ravi Kant Mittal, and Abid Haleem, eds. *Advances in Additive Manufacturing: Artificial Intelligence, Nature-Inspired, and Biomanufacturing*. Elsevier, 2022.
4. Gudimetla, Avinash, Parveen Kumar, S. Sambhu Prasad, Satish Geeri, and V. V. N. Sarath. "Towards Smart Materials: Enhancing the Efficiency of the Materials." In *Modeling, Characterization, and Processing of Smart Materials*, pp. 1–30. IGI Global, 2023.
5. Tiwari, Himanshu Kumar, Ashish Kumar Srivastava, Parveen Kumar, Manish Kumar Singh, Hritik Kumar, and Akshit Bhadauria. "Materials and Technology for Implant Manufacturing: Challenges and Opportunity." In *Modeling, Characterization, and Processing of Smart Materials*, pp. 83–106. IGI Global, 2023.
6. Kumar, Ajay, Parveen Kumar, Ravi Kant Mittal, and Victor Gambhir. "Materials Processed by Additive Manufacturing Techniques." In *Advances in Additive Manufacturing Artificial Intelligence, Nature-Inspired, and Biomanufacturing*, pp. 217–233. Elsevier, 2023. https://doi.org/10.1016/B978-0-323-91834-3.00014-4

7. Wang, Liang, Maria N. P. Olsen, Christophe Moni, Alba Dieguez-Alonso, José María de la Rosa, Marianne Stenrød, Xingang Liu, and Liangang Mao. "Comparison of Properties of Biochar Produced from Different Types of Lignocellulosic Biomass by Slow Pyrolysis at 600°C." *Applications in Energy and Combustion Science* 12 (2022): 100090. https://doi.org/10.1016/j.jaecs.2022.100090

8. Makepa, Denzel C., Chido H. Chihobo, and Downmore Musademba. "Advances in Sustainable Biofuel Production from Fast Pyrolysis of Lignocellulosic Biomass." *Biofuels* 14, no. 5 (2023): 529–550.

9. Arshad, Usman, Muhammad Raheel, Waqas Ashraf, Atta Ur Rehman, Muhammad Salman Zahid, Mahmoud Moustafa, and Muhammad Amjad Ali. "Influence of Biochar Application on morpho-physiological attributes of tomato (Lycopersicon esculentum Mill) and soil Properties." *Communications in Soil Science and Plant Analysis* 54, no. 4 (2023): 515–525. https://doi.org/10.1080/00103624.2022.2118295

10. Kapoor, Aanchal, Rakesh Sharma, Anil Kumar, and Swapana Sepehya. "Biochar as a Means to Improve Soil Fertility and Crop Productivity: A Review." *Journal of Plant Nutrition* 45, no. 15 (2022): 2380–2388. https://doi.org/10.1080/01904167.2022.2027980

11. Kumar, Ajay, Parveen Kumar, Namrata Dogra, and Archana Jaglan. "Application of Incremental Sheet Forming (ISF) Toward Biomedical and Medical Implants." In *Handbook of Flexible and Smart Sheet Forming Techniques: Industry 4.0 Approaches.* Wiley, 2023, p. 247.

12. Bolan, Nanthi, Son A. Hoang, Jingzi Beiyuan, Souradeep Gupta, Deyi Hou, Ajay Karakoti, Stephen Joseph et al. "Multifunctional Applications of Biochar beyond Carbon Storage." *International Materials Reviews* 67, no. 2 (2022): 150–200. https://doi.org/10.1080/09506608.2021.1922047

13. Rushimisha, Iranzi Emile, Xiaojing Li, Ting Han, Xiaodong Chen, Abdoul Salam Issiaka Abdoul Magid, Yan Sun, and Yongtao Li. "Application of Biochar on Soil Bioelectrochemical Remediation: Behind Roles, Progress, and Potential." *Critical Reviews in Biotechnology* 44 (2022): 1–19. https://doi.org/10.1080/07388551.2022.2119547

14. Aziz, Sadia, Shabana Bibi, Mohammad Mehedi Hasan, Partha Biswas, Muhammad Ishtiaq Ali, Muhammad Bilal, Hitesh Chopra, Nobendu Mukerjee, and Swastika Maitra. "A Review on Influence of Biochar Amendment on Soil Processes and Environmental Remediation." *Biotechnology and Genetic Engineering Reviews* (2023): 1–35. https://doi.org/10.1080/02648725.2022.2122288

15. Kumar, Ajay, Hari Singh, Parveen Kumar, and Bandar AlMangour, eds. *Handbook of Smart Manufacturing: Forecasting the Future of Industry 4.0.* CRC Press, 2023. https://doi.org/10.1201/9781003333760

16. Sivamani, S., B. S. Naveen Prasad, K. Nithya, N. Sivarajasekar, and A. Hosseini-Bandegharaei. 2022. "Back-Propagation Neural Network: Box–Behnken Design Modelling for Optimization of Copper Adsorption on Orange Zest Biochar." *International Journal of Environmental Science and Technology* 19 (5): 4321–4336. https://doi.org/10.1007/s13762-021-03411-1

17. Nallaselvam, Tamilarasan, Sakthivel Rajamohan, Balaji Kalaiarasu, and Anh Tuan Hoang. "High Efficient COVID-19 Waste Co-pyrolysis char/TiO_2 Nanocomposite for Photocatalytic Reduction of Cr (VI) under Visible Light." *Environmental Science and Pollution Research* (2023): 1–17. https://doi.org/10.1007/s11356-023-29281-3

18. Praveen Kumar, D., D. Ramesh, V. Karuppasamy Vikraman, and P. Subramanian. "Comprehensive Characterization of Cocos Nucifera Palm Residues for Biochar Production." *Journal of Agricultural Engineering* 59, no. 4 (2022): 356–368. http://doi.org/10.52151/jae2022594.1788

19. Rawat, Shivam, T. Boobalan, M. Sathish, Srinivas Hotha, and Bhaskar Thallada. "Utilization of CO_2 Activated Litchi Seed Biochar for the Fabrication of Supercapacitor Electrodes." *Biomass and Bioenergy* 171 (2023): 106747. https://doi.org/10.1016/j.biombioe.2023.106747

20. Rawat, Shivam, Luo Jinlin, Anuradha A. Ambalkar, Srinivas Hotha, Akinori Muto, and Thallada Bhaskar. "Syzygium Cumini Seed Biochar for Fabrication of Supercapacitor: Role of Inorganic Content/Ash." *Journal of Energy Storage* 60 (2023): 106598. https://doi.org/10.1016/j.est.2022.106598

21. Tang, Jia Yong, Boaz Yi Heng Chung, Jia Chun Ang, Jia Wen Chong, Raymond R. Tan, Kathleen B. Aviso, Nishanth G. Chemmangattuvalappil, and Suchithra Thangalazhy-Gopakumar. "Prediction Model for Biochar Energy Potential Based on Biomass Properties and Pyrolysis Conditions Derived from Rough Set Machine Learning." *Environmental Technology* 45 (2023): 1–15. https://doi.org/10.1080/09593330.2023.2192877

22. Amalina, Farah, Abdul Syukor Abd Razak, Santhana Krishnan, Haspina Sulaiman, A. W. Zularisam, and Mohd Nasrullah. "Advanced Techniques in the Production of Biochar from Lignocellulosic Biomass and Environmental Applications." *Cleaner Materials* 6 (2022): 100137. https://doi.org/10.1016/j.clema.2022.100137

23. Labanya, Rini, Prakash Chandra Srivastava, Satya Pratap Pachauri, Arvind Kumar Shukla, Manoj Shrivastava, and Prashant Srivastava. "Effect of Three Plant Biomasses and Two Pyrolysis Temperatures on Structural Characteristics of Biochar." *Chemistry and Ecology* 38, no. 5 (2022): 430–450. https://doi.org/10.1080/02757540.2022.2075858

24. Wu, Liang, Jinzhi Ni, Huiying Zhang, Shuhan Yu, Ran Wei, Wei Qian, Weifeng Chen, and Zhichong Qi. "The Composition, Energy, and Carbon Stability Characteristics of Biochars Derived from Thermo-Conversion of Biomass in Air-Limitation, CO_2, and N_2 at Different Temperatures." *Waste Management* 141 (2022): 136–146. https://doi.org/10.1016/j.wasman.2022.01.038

25. ASTM E871–82. "Standard Test Method for Moisture Analysis of Particulate Wood Fuels." Annual Book of ASTM Standard, American Society for Testing and Materials, 2004.

26. ASTM E872–82. "Standard Test Method for Volatile Matter in the Analysis of Particulate Wood Fuels." Annual Book of ASTM Standard, American Society for Testing and Materials, 2013.

27. ASTM E1755-01. "Standard Test Method for Ash in Biomass." Annual Book of ASTM Standard, American Society for Testing and Materials, 2015.

28. van Soest, P. J., James B. Robertson, and Betty A. Lewis. "Methods for Dietary Fiber, Neutral Detergent Fiber, and Nonstarch Polysaccharides in Relation to Animal Nutrition." *Journal of Dairy Science* 74, no. 10 (1991): 3583–3597.

29. Shrivastava, Pranshu, Phonthip Khongphakdi, Arkom Palamanit, Anil Kumar, and Perapong Tekasakul. "Investigation of Physicochemical Properties of Oil Palm Biomass for Evaluating Potential of Biofuels Production via Pyrolysis Processes." *Biomass Conversion and Biorefinery* 11 (2021): 1987–2001. https://doi.org/10.1007/s13399-019-00596-x

30. Praveen Kumar, D., D. Ramesh, P. Subramanian, S. Karthikeyan, and A. Surendrakumar. "Activated Carbon Production from Coconut Leaflets through Chemical Activation: Process Optimization Using Taguchi Approach." *Bioresource Technology Reports* 19 (2022): 101155.

31. Vikraman, V. Karuppasamy, D. Praveen Kumar, G. Boopathi, and P. Subramanian. "Kinetic and Thermodynamic Study of Finger Millet Straw Pyrolysis through Thermogravimetric Analysis." *Bioresource Technology* 342 (2021): 125992.

32. ASTM D1762–84. "Chemical Analysis of Wood Charcoal." Annual Book of ASTM Standard, American Society for Testing and Materials, 2007.

33. Venkatesh, Govindarajan, Kodigal A. Gopinath, Kotha Sammi Reddy, Baddigam Sanjeeva Reddy, Mathyam Prabhakar, Cherukumalli Srinivasarao, Venugopalan Visha Kumari, and Vinod Kumar Singh. "Characterization of Biochar Derived from Crop Residues for Soil Amendment, Carbon Sequestration and Energy Use." *Sustainability* 14, no. 4 (2022): 2295. https://doi.org/10.3390/su14042295

34. Ortiz, Leandro Rodriguez, Erick Torres, Daniela Zalazar, Huili Zhang, Rosa Rodriguez, and Germán Mazza. "Influence of Pyrolysis Temperature and Bio-Waste Composition on Biochar Characteristics." *Renewable Energy* 155 (2020): 837–847.

35. Tesfamichael, B., N. Gesesse, and S. A. Jabasingh. "Application of Rice Husk and Maize Straw Biochar for Carbon Sequestration and Nitrous Oxide Emission Impedement." *Journal of Scientific and Industrial Research* 77 (2018): 587–591.

36. Vikraman, V. Karuppasamy, P. Subramanian, D. Praveen Kumar, S. Sriramajayam, R. Mahendiran, and S. Ganapathy. "Air Flowrate and Particle Size Effect on Gasification of Arecanut Husk with Preheated Air through Waste Heat Recovery from Syngas." *Bioresource Technology Reports* 17 (2022): 100977.

37. Chen, Dengyu, Kehui Cen, Xiaozhuang Zhuang, Ziyu Gan, Jianbin Zhou, Yimeng Zhang, and Hong Zhang. "Insight into Biomass Pyrolysis Mechanism Based on Cellulose, Hemicellulose, and Lignin: Evolution of Volatiles and Kinetics, Elucidation of Reaction Pathways, and Characterization of Gas, Biochar and Bio-Oil." *Combustion and Flame* 242 (2022): 112142.

38. Boussetta, Abdelghani, Anass AIT Benhamou, Francisco J. Barba, Mohammed EL Idrissi, Nabil Grimi, and Amine Moubarik. "Valorization of Solanum Elaeagnifolium Cavanilles Weeds as a New Lignocellulosic Source for the Formulation of Lignin-Urea-Formaldehyde Wood Adhesive." *The Journal of Adhesion* 99, no. 1 (2023): 34–57. https://doi.org/10.1080/00218464.2021.1999232

39. Adeniyi, Adewale George, Victor Temitope Amusa, Ebuka Chizitere Emenike, and Kingsley O. Iwuozor. "Co-carbonization of Waste Biomass with Expanded Polystyrene for Enhanced Biochar Production." *Biofuels* 14, no. 6 (2023): 635–643. https://doi.org/1 0.1080/17597269.2022.2161133

40. Singh, Arshdeep, Sonil Nanda, Jesus Fabricio Guayaquil-Sosa, and Franco Berruti. "Pyrolysis of Miscanthus and Characterization of Value-Added Bio-oil and Biochar Products." *The Canadian Journal of Chemical Engineering* 99 (2021): S55–S68. https://doi.org/10.1002/cjce.23978

41. de Jager, Megan, Frank Schröter, Michael Wark, and Luise Giani. "The Stability of Carbon from a Maize-Derived Hydrochar as a Function of Fractionation and Hydrothermal Carbonization Temperature in a Podzol." *Biochar* 4, no. 1 (2022): 52. https://doi.org/10.1007/s42773-022-00175-w

42. Liu, Zhuangzhuang, Zhiwei Yan, Fen Liu, and Jun Fang. "Speciation and Transformation of Nitrogen for Swine Manure Thermochemical Liquefaction." *Scientific Reports* 12, no. 1 (2022): 12056. https://doi.org/10.1038/s41598-022-16101-w

43. Ali, Liaqat, Arkom Palamanit, Kuaanan Techato, Asad Ullah, Md Shahariar Chowdhury, and Khamphe Phoungthong. "Characteristics of Biochars Derived from the Pyrolysis and Co-pyrolysis of Rubberwood Sawdust and Sewage Sludge for Further Applications." *Sustainability* 14, no. 7 (2022): 3829. https://doi.org/10.3390/su14073829

10 Applications of Sustainable Biomaterials in the Pharmaceutical and Medical Fields

Mahalakshmi Suresha Biradar
Rajiv Gandhi University of Health Sciences

Shankar Thapa
Universal College of Medical Sciences

Sonali S. Shinde
Dr. Babasaheb Ambedkar Marathwada University

Parveen Kumar
Department of Mechanical Engineering, Rawal Institute of Engineering and Technology, Faridabad, Haryana, India

Aniket P. Sarkate
Dr. Babasaheb Ambedkar Marathwada University

LIST OF ABBREVIATIONS

Abbreviation	Full Form
ABDS	Affinity-based delivery systems
ASTM	American Society for Testing and Materials
BNC	Bacterial nanocellulose
BrdU	Bromodeoxyuridine
CEN	European Committee for Standardization
CMC	Carboxymethylcellulose
CMV	Cytomegalovirus
DIN	German Institute for Standardization
ECM	Extracellular matrix
ELISA	Enzyme-linked immunosorbent assay
EtO	Ethylene oxide
EVA	Ethylene-vinyl acetate

DOI: 10.1201/9781003434313-10

Abbreviation	**Full Form**
HA	Hydroxyapatite
ISO	International Organization for Standardization
LCA	Life cycle assessment
MDR	Medical Device Regulation
MFC	Microbial fuel cells
MRI	Magnetic resonance imaging
MTT	3-(4,5-Dimethylthiazol-2-yl)-2,5-diphenyltetrazolium bromide
ODT	On-demand therapeutics
PEG	Polyethylene glycol
PGA	Poly (lactic acid)
PLGA	Poly (lactic-co-glycolic acid)
PVA	Polyvinyl alcohol
RT-PCR	Reverse transcription-polymerase chain reaction
SPR	Surface plasmon resonance
TCP	Tricalcium phosphate
USFDA	U.S. Food and Drug Administration
VOC	Volatile organic compounds

1   INTRODUCTION

The term "any material used as an implant" refers to biomaterials. A pharmacologically inert material that is systematically constructed for integration or implantation into a live system is called a biomaterial. Since ancient times, right up to the dawn of modern medicine, biomaterials have been utilized. Ancient civilizations made use of natural resources like metals, wood, and bone for medicinal purposes. Dental implants, prostheses, and splints were all made with these materials. There have been finds of metal and wood prosthetic fingers and toes from ancient Egypt. Dental implants made of metal and ivory were utilized in ancient Rome. The use of materials like gutta-percha, a natural latex, for dental fillings in the 19th century marked the beginning of the creation of increasingly sophisticated biomaterials. During this time, the first attempts were made to create prosthetic limbs out of materials like rubber and ivory. Early to mid-20th century, the interest in biomaterials grew as a result of the development of antibiotics and better surgical methods. Artificial organs and prosthetic devices are products of the invention of synthetic polymers such as nylon and polyethylene. The Second World War accelerated the development of materials for tissue healing and wound treatment. In the late 20th century, the subject of biomaterials advanced significantly as a result of the introduction of biocompatible materials like silicone rubber and other metals (titanium, stainless steel). The creation of biodegradable materials for medication delivery systems and sutures increased in frequency. As interest in tissue engineering and regenerative medicine grew, scientists investigated the application of biomaterial scaffolds to promote tissue development. In the 21st century, biocompatible polymers, ceramics, and composites are being used more often, which has resulted in advancements in tissue engineering, medication delivery, and medical devices.

The demand for socially and ecologically conscious materials in a variety of industries, such as biomedical engineering, pharmaceuticals, and healthcare, gave rise to the idea of sustainable biomaterials. The goal of sustainable biomaterials is to have a minimal ecological imprint throughout the course of their whole useful life. They are made from renewable resources, such as bacteria, plants, and animals. Increasing awareness of the detrimental effects of conventional materials on the environment is the driving force behind the push for sustainable biomaterials [1]. Conventional materials, like plastics and synthetic polymers, are typically derived from fossil fuels and have significant ecological implications, including resource depletion, pollution, and long-lasting waste. Additionally, their use in biomedical applications can lead to adverse effects on environment and the human health. The field of biomaterials, which focuses on developing materials that interact with biological systems, has embraced the concept of sustainability to address these concerns. Biomass, such as plants, algae, and animal byproducts, serves as the basis for sustainable materials. These sources can be sustainably cultivated, harvested, or obtained as waste products from other industries, reducing the reliance on non-renewable resources. Furthermore, the production processes for sustainable biomaterials aim to minimize energy consumption, greenhouse gas emissions, and the use of toxic chemicals. Biocompatibility and biodegradability are crucial characteristics of sustainable biomaterials. Biocompatible materials reduce the risk of immune response, inflammation, and other complications, promoting better patient outcomes whereas biodegradable biomaterials can naturally break down over time through enzymatic or microbial action, eliminating the need for surgical removal and reducing waste. These properties are required in specific applications.

In recent years, advancements in material science, bioengineering, and sustainability have contributed to the development and utilization of sustainable biomaterials. Researchers and industry professionals are actively exploring new biomaterial sources, improving their processing methods, and optimizing their functional properties to meet the demands of the pharmaceutical and medical fields while ensuring environmental responsibility. Overall, the field of sustainable biomaterials aims to provide innovative and eco-friendly solutions to address the environmental challenges associated with traditional materials, promote human health and well-being, and contribute to a more sustainable future.

While sustainable biomaterials offer promising solutions for addressing environmental concerns and advancing healthcare, there are still several unknowns and challenges associated with their development and implementation. Some of the key areas that remain uncertain or problematic include performance, durability, scalability, cost, public perception and acceptance, regulation, and long-term environmental impact. By addressing these knowledge gaps and obstacles, sustainable biomaterials can realize their full potential as eco-friendly alternatives in the pharmaceutical and medical fields.

There are several examples of sustainable biomaterials that have gained attention and are being used in various applications. One of the sustainable biomaterials is chitosan; it is derived from chitin, a polymer found naturally in the exoskeletons of crustaceans such as prawns and crabs. Because of its biocompatible, biodegradable, and antibacterial properties, it has potential uses in wound healing, medication delivery, and tissue engineering scaffolding [2].

Sustainable biomaterials are especially significant in the pharmaceutical and medical industries for overcoming environmental, social, and economic challenges. Because of the positive effects on ecology, waste reduction, creativity, and scientific progress.

2   CLASSIFICATION OF SUSTAINABLE BIOMATERIALS

The classification of biomaterials is not mutually exclusive, as there can be an overlap between the categories. The biomaterial used is determined by the application's unique needs, desirable qualities, and compatibility with the biological environment. Researchers continue to explore novel biomaterial compositions and combinations to advance the field of biomedical engineering and improve patient outcomes.

2.1   NATURALLY EXTRACTED BIOMATERIALS

Nature-derived biomaterials are critical in the creation of medications. These biomaterials are derived from plant, animal, and microbial sources. They are used for producing pharmaceutical products with therapeutic effects in various forms, including extracts, isolates, and derivatives. The derived compounds have bioactive constituents which are helpful in treating specific diseases.

2.1.1   Protein-Based Biomaterials

Proteins are big, complex molecules made up of amino acids that play important functions in a variety of biological processes. Since they can replicate the intricate activities of natural tissues and are inherently biocompatible, protein-based biomaterials have attracted a lot of interest in the field of biomedical engineering. These biomaterials are made of proteins, which are vital components of all living things. Biomaterials based on proteins provide several benefits. First of all, proteins may be obtained from a wide range of natural sources, including plants, animals, and recombinant protein expression systems. This diversity of sources enables the creation of materials with various characteristics and functions. It is possible to create and modify proteins to display particular qualities, including mechanical strength, biodegradability, or cell adhesion.

The capacity of protein-based biomaterials to facilitate cellular connections is one of their main advantages. Proteins are useful for tissue engineering and regenerative medicine applications because they may stimulate cell adhesion, differentiation, and proliferation. Biomaterials based on proteins frequently have strong biodegradability. It is possible to gradually redesign the biomaterial and replace it with fresh tissue growth through the controlled enzymatic degradation of proteins. This characteristic is especially useful in applications like scaffolds or temporary implants, where the patient's tissue eventually replaces the material over time.

The potential for biological uses of biomaterials formed from proteins is considerable. Their suitability for tissue engineering, drug administration, and regenerative medicine stems from their biocompatibility, natural tissue-mimicking capabilities, and tunability. Protein-based biomaterials are expected to produce novel solutions to a variety of biological problems and improve patient care with continued research and development. Some of the examples of protein-based biomaterials are fibrin, collagen, and gelatin. Collagen is often used in tissue engineering and wound-healing applications due to its biocompatibility and ability to promote cell adhesion and tissue regeneration.

2.1.2 Polysaccharide-Based Biomaterials

Polysaccharide-based biomaterials have garnered significant attention in the field of biomedical engineering due to their versatile nature and distinct characteristics. Long chains of sugar molecules called polysaccharides are common in nature and provide various benefits for the creation of biomaterials. One of the main benefits of polysaccharide-based biomaterials is their biocompatibility. Numerous naturally occurring polysaccharides, such hyaluronic acid, alginate, and chitosan, are intrinsically biocompatible. This indicates that live tissues may tolerate them well, reducing negative effects and immunological responses. Because of their hydrophilic characteristics, polysaccharides may absorb and hold onto water. Because of this property, they may be used to create hydrogels, which are three-dimensional networks that have a large water-retaining capacity. The moist and supportive environment that polysaccharide hydrogels provide cells encourages cell adhesion, proliferation, and tissue regeneration. It is simple to functionalize and alter polysaccharides to modify their characteristics for certain uses. Chemical alterations may add bioactive motifs, regulate the pace of disintegration, or increase their mechanical strength. These modifications enable polysaccharide-based biomaterials to be tailored to specific needs for a variety of biomedical applications, including as scaffolds for tissue engineering, wound dressings, and drug delivery systems.

Additionally, polysaccharides have outstanding biodegradability. Since the body can break down many polysaccharides enzymatically, the biomaterial may gradually dissolve and be eliminated over time. Because the patient's own tissue may progressively replace temporary implants or scaffolds as the healing process progresses, this characteristic is very helpful in these situations. Furthermore, certain binding sites on polysaccharides allow them to interact with biological substances and cells. This enables the addition of bioactive compounds to polysaccharide-based biomaterials, such as peptides or growth factors. Improved treatment results may be achieved by actively modulating cellular behavior and tissue regeneration processes via the use of these bioactive polysaccharide materials.

2.1.3 Marine, Microbial, and Plant-Origin Gum-Based Biomaterials

Marine, microbial, and plant-based gum-based biomaterials have garnered significant attention in the biomaterials area because of their special qualities and potential uses in a range of biomedical and pharmaceutical industries. Gums of marine origin are used to make natural biomaterials, such as seaweeds and algae, from marine sources. These biomaterials have exceptional gel-forming, biodegradability, and biocompatibility qualities. Examples of these include alginate and carrageenan. Gum-based biomaterials derived from marine sources have many benefits, including affordability, ease of use, and the presence of naturally occurring bioactive substances that may improve cellular connections and tissue regeneration processes.

Certain bacteria use fermentation processes to make biomaterials of a microbial origin, such as gums. Examples include of dextran, gellan gum, and xanthan gum. These biomaterials have a high viscosity, distinct rheological characteristics, and gelling ability. Their benefits include simple processing, high stability, and controllable viscosity and gelation behavior.

Natural gums that are taken from plants are the source of plant-origin gum-based biomaterials. Guar gum, gum arabic, and pectin are a few examples. These biomaterials have qualities that make them mucoadhesive, biocompatible, and biodegradable. They have many benefits, including affordability, accessibility, and the capacity to change their physicochemical characteristics chemically.

Natural, biocompatible, and ecologically benign gum-based biomaterials derived from microorganisms, plants, and oceans make them appealing substitutes for synthetic polymers. Their many qualities, including gelation, controlled release, adhesion, and structural support, qualify them for a range of pharmacological and biological uses [3].

2.2 SYNTHETICALLY DERIVED BIOMATERIALS

Synthetically derived biomaterials are biomaterials that are chemically synthesized or engineered in the laboratory rather than being obtained from natural sources. These biomaterials offer unique advantages, including precise control over their properties, tailored functionality, and enhanced reproducibility. They are designed to mimic or surpass the properties of natural materials, making them valuable for various biomedical applications. Through careful design and engineering, the synthesized biomaterials provide tailored solutions to meet the complex requirements of biomedical applications [4]. The synthetically derived biomaterials are classified based on their material type. The most abundant materials used in medical applications are polymer, ceramic, and peptide-based biomaterials.

2.2.1 Polymer-Based Biomaterials

Polymer-based biomaterials, including those used in prostheses and medical implants, have played a major role in pushing the boundaries of biomedical engineering. The purpose of these biomaterials is to engage with biological systems in a way that enhances compatibility and offers tailored capabilities to fulfill the specific needs of every application. For better integration and a reduced immune response, these materials may be engineered to exhibit a broad range of physical and chemical properties that are comparable to, or even identical to, those of genuine tissues. Their potential applications may also be expanded by shaping them into a variety of different forms, including films, fibers, polymers, and scaffolds. Regenerative medicine and tissue engineering are two areas where biodegradable polymers are very helpful. Over time, these components may break down, allowing new tissue to progressively develop in place of the scaffold. Common biodegradable polymers utilized in the creation of scaffolds for tissue engineering include poly (lactic-co-glycolic acid) (PLGA), poly(lactic acid) (PGA), and their copolymer poly(lactic acid) (PLGA). Research and development on polymer-based biomaterials is still ongoing because of its great potential to improve healthcare and solve a variety of biological issues. They are widely used in biomedical engineering because of their flexibility, variety, and interaction with biological systems [5].

2.2.2 Ceramic-Based Biomaterials

Ceramic-based biomaterials have emerged as crucial components in biomedical engineering owing to their unique mix of characteristics, making them amenable to a broad range of applications. These biomaterials have shown remarkable mechanical strength, chemical stability, and biocompatibility. They are mostly generated from inorganic chemicals. The bio inertness of ceramic biomaterials—their low reactivity when in contact with biological tissues—is one of their most remarkable features. This characteristic makes them perfect for long-term implanted devices like bone scaffolds, prosthetic joints, and dental implants since they lower the risk of immunological responses, inflammation, and unpleasant reactions.

Conversely, bioactive ceramics can actively engage with the biological world around them. The composition and structure of materials such as tricalcium phosphate (TCP) and hydroxyapatite (HA) are similar to those of natural bone minerals. These bioactive ceramics may help with osseointegration, which is the process by which new bone grows and fuses with the implant to improve stability and function, when utilized in bone tissue engineering. Excellent mechanical qualities, such as strong compressive strength and stiffness, are also shown by ceramic biomaterials. This qualifies them for load-bearing uses where they can tolerate the stresses involved in everyday activities, including hip or knee replacements. Ceramics also wear down slowly, which lowers the possibility of particle production and the ensuing inflammatory reactions. Biomaterials based on ceramics can last lengthy periods of time under physiological conditions because of their exceptional resistance to deterioration, corrosion, and erosion. When thermal or electrical conductivity is undesirable, such as in dental or orthopedic implants, their thermal and electrical insulating qualities make them appropriate for use [6]. From orthopedics and dentistry to tissue engineering, medication delivery, and diagnostic devices, ceramic-based biomaterials have a wide variety of uses in the medical and pharmaceutical industries.

2.2.3 Peptide-Based Biomaterials

Peptide-based biomaterials have garnered significant attention lately because of their special qualities and possible uses in a wide range of biological fields. Short sequences of amino acids called peptides provide a flexible framework for precisely controlling the structure, function, and bioactivity of biomaterials. Peptide-based biomaterials' intrinsic biocompatibility is one of their main benefits. The natural proteins that make up the human body are the source of peptides, which may be modified to resemble certain amino acid sequences found in extracellular matrix proteins or other bioactive patterns. As a result, peptides may interact with biological systems in an advantageous way, encouraging tissue regeneration, cell adhesion, and proliferation. Peptides may self-assemble through particular peptide–peptide interactions or non-covalent interactions to form well-defined nanostructures like nanofibers, nanotubes, or nanoparticles. Because of its capacity for self-assembly, hierarchical structures with architectures that mirror real tissues may be built. Since they provide both structural support and a conducive environment for cell growth and differentiation, peptide-based nanofibrous scaffolds have been thoroughly studied for use in tissue engineering applications.

It is simple to alter or functionalize peptide sequences to add desirable characteristics, such as improved stability, regulated rates of breakdown, or greater mechanical strength. Because of their adaptability, biomaterials may be designed with precise properties to satisfy the needs of certain applications. Because they provide special possibilities for creating biomaterials with specialized qualities and functions, peptide-based biomaterials remain a burgeoning field of study. Their flexibility, biocompatibility, and interaction with biological systems may be advantageous for drug delivery, tissue engineering, and other biomedical purposes [7].

2.3 METAL BIOMATERIALS

Distinguished by their high electrical and thermal conductivity, brightness, malleability, and ductility, metals are a class of chemical elements. People have been using them for thousands of years because of their unique qualities and wide range of uses. In the realm of biomaterials, metals have been essential to many biomedical applications, including implants, medical devices, and diagnostics.

The metal composition of biomaterials varies according to their intended use. Due to its exceptional strength-to-weight ratio, resistance to corrosion, and biocompatibility, orthopedic and dental implants often employ titanium and titanium alloys. The alloy of iron, chromium, and nickel known as stainless steel is often used to make surgical instruments and other medical equipment because it provides corrosion resistance, biocompatibility, and strong mechanical qualities. Dental prostheses and orthopedic implants often employ cobalt–chromium alloys. Their superior strength and resistance to wear make them appropriate for use in load-bearing scenarios. Drug delivery methods, medical equipment, and diagnostics have all made use of gold and silver. Because of their antibacterial qualities, they may be used in antimicrobial coatings, catheters, and wound dressings. Metals are valuable biomaterials with a wide range of applications in the biomedical field. Their unique combination of mechanical, electrical, and chemical properties, along with their biocompatibility, makes them indispensable in areas such as implants, medical devices, diagnostics, and drug delivery systems [8].

2.4 Biomimetic Materials

The purpose of biomimetic materials is to mimic the properties and functions of living things to improve their efficiency, usefulness, and compatibility with biological systems. Numerous sectors, including tissue engineering, regenerative medicine, drug delivery, and bioinspired engineering, have shown a great deal of interest in biomimetic materials. Bioactive compounds, including growth factors, enzymes, and signaling molecules, are often included with biomimetic materials to augment biological responses and promote tissue regeneration. These molecules, which may be attached to the material's surface or enclosed inside it, provide signals that replicate the natural cellular milieu and encourage certain cellular processes including angiogenesis, differentiation, and proliferation.

Inspired by biological processes, biomimetic materials may demonstrate self-assembly and self-healing capabilities. The spontaneous arrangement of molecules or other constituents into distinct shapes or patterns is known as self-assembly. When exposed to mechanical stress or environmental circumstances, self-healing materials may repair damage or restore their integrity, mimicking the regeneration powers of live creatures [9].

3 BASIC CONSIDERATION TO DESIGN SUSTAINABLE BIOMATERIALS

When designing sustainable biomaterials, several key considerations should be considered.

- Selecting biomaterials made from renewable resources, including waste materials, quickly growing plants, or agricultural byproducts.
- Aiming for biodegradable biomaterials, meaning they can be broken down naturally by microorganisms and return to the environment without leaving harmful residues.
- A life cycle assessment (LCA) is used to evaluate the environmental impact of the biomaterial across its whole life cycle, which includes the extraction of raw materials, processing, production, consumption, and disposal.
- Making certain that the biomaterial is not hazardous to persons or the environment. Use products that do not include toxic compounds, heavy metals, or harmful additions.
- Improving the manufacturing process to reduce energy usage and greenhouse gas emissions.
- Developing biomaterials with sufficient performance and durability attributes for the uses for which they are scheduled [10].

While these factors are considered while designing and developing biomaterials, they may help create ecologically friendly and sustainable alternatives that are less harmful to the environment and support a circular economy. A few essential elements are essential for taking better action.

3.1 BIOCOMPATIBILITY

"Biocompatibility" means that a substance may serve its purpose without harming any living things or the environment. Biomaterials must be suitable for the tissues or organs they are going to work with. Making sure the material's chemical and physical characteristics align with those of the surrounding tissues is part of this. For instance, the material used to create an artificial hip implant should have mechanical qualities similar to those of bone and be physiologically inert. Degradability of biodegradable materials is a crucial factor in sustainable biomaterials. Implanted gadgets and materials that the body can naturally absorb and break down over time may be removed with fewer follow-up operations. Biodegradable biomaterials include polymers such as polyglycolic acid (PGA) and polylactic acid (PLA).

The biomaterial should not cause the body to get inflamed or mount an immunological reaction. This is especially important for long-term implants since problems might arise from persistent inflammation. It is possible to lessen immunogenicity by applying coatings and surface changes. When sustainable biomaterials come into touch with bodily fluids or tissues, they should not release any hazardous compounds or byproducts. Thorough testing and material selection are required to guarantee that the body is not exposed to any hazardous materials. When required, biomaterials need to encourage cell adhesion and proliferation. Scaffolds for tissue engineering applications should provide an environment that promotes tissue regeneration and cell proliferation [11]. This approach promotes the development of biomaterials that have minimal impact on human health and contribute to a sustainable future.

3.2 MECHANICAL PROPERTIES

The mechanical properties of a biomaterial include strength, stiffness, elasticity, ductility, and toughness. A material's strength is its capacity to bear an applied load without cracking or buckling. Biomaterials must possess the strength to endure the pressures and strains that they will experience within the body. For example, materials used to replace missing bone must be strong enough to sustain the weight of the body. Young's modulus, another name for stiffness, is a property that quantifies a material's resistance to deformation in the presence of an external force. It shows how much a material will compress or stretch when subjected to force. A biomaterial's stiffness needs to be adjusted to correspond to the surrounding tissues' or organs' stiffness. The optimal biomaterial ought to resemble the natural tissue it is supplanting or enhancing in terms of rigidity. Because silicone elastomers mirror the softness of real tissue and have a low Young's modulus, they are employed in soft tissue replacements (e.g., breast implants). The capacity of a substance to regain its previous shape after deformation is referred to as elasticity. A biomaterial needs to be elastic enough to withstand deformation without suffering long-term harm. Heart implants, for instance, must be sufficiently elastic to allow for the heart's pumping action. A material is said to be ductile if it can flex without breaking, and tough if it can absorb energy before breaking. These characteristics are especially crucial for materials that are used in applications that sustain weight. Materials with excellent ductility and toughness include metals and certain polymers. Shape-memory alloy nitinol is ductile and utilized in blood artery stents because it may be crimped during delivery and then reverts to its original form after it is deployed [12].

3.3 STERILIZATION

Effective sterilization ensures that biomaterials are free from viable microorganisms, reducing the risk of infection and contamination. Steam sterilization (autoclaving), ethylene oxide (EtO) sterilization, gamma irradiation, hydrogen peroxide gas plasma sterilization, and electron beam sterilization are all frequently used sterilization methods. Each sterilizing technique has its benefits and drawbacks, and the ideal biomaterial would be one that would not be significantly affected by the chosen technique. Some sterilization methods, such as high temperatures or harsh chemicals, can lead to changes in material properties, including degradation, loss of mechanical strength, alteration of surface characteristics, or changes in bioactivity [13]. It is important to assess and validate the impact of sterilization on the specific biomaterial to maintain its integrity and functionality.

3.4 REGULATORY CONSIDERATION

In order to guarantee that sustainable biomaterials for medical applications fulfill safety and effectiveness requirements, several regulatory factors must be taken into account. Before these materials may be employed in clinical practice, regulatory permission is essential. Testing for biocompatibility evaluates the biomaterial's relationship to biological systems. Tests are conducted both in vitro and in vivo to assess the material's safety and suitability for use with live tissues. An international set of standards called ISO 10993 specifies the conditions for biocompatibility testing. For example, the broad guidelines and processes for assessing biocompatibility are outlined in ISO 10993-1. Biomaterials used for medical purposes in the United States need to be approved or cleared by the US Food and Drug Administration (FDA). A CE mark is needed in the EU to verify compliance with health and safety regulations. When developing a sustainable biomaterial for a novel medical device, a company must apply for premarket clearance (PMA) from the FDA or follow European CE marking guidelines. Numerous biomaterials-related standards have been produced by the International Organization for Standardization (ISO), such as ISO 14971 for medical device risk management and ISO 13485 for quality management. The ISO 13485 certification guarantees that a firm has successfully established a quality management system that is appropriate for the design and manufacturing of medical devices. The laws established by Good Manufacturing Practices (GMP) aim to ensure the consistent manufacturing of biomaterials, according to rigorous standards of quality and safety at every stage of the production process. Manufacturers are required to comply with GMP when developing sustainable biomaterials intended for use in medical devices, to ensure that these materials fulfill predetermined specifications and quality requirements. Evaluation of biomaterials' possible toxicity and environmental effects during production, use, and disposal may be mandated by regulatory bodies [14].

4   EVALUATION OF SUSTAINABLE BIOMATERIAL BEHAVIOR

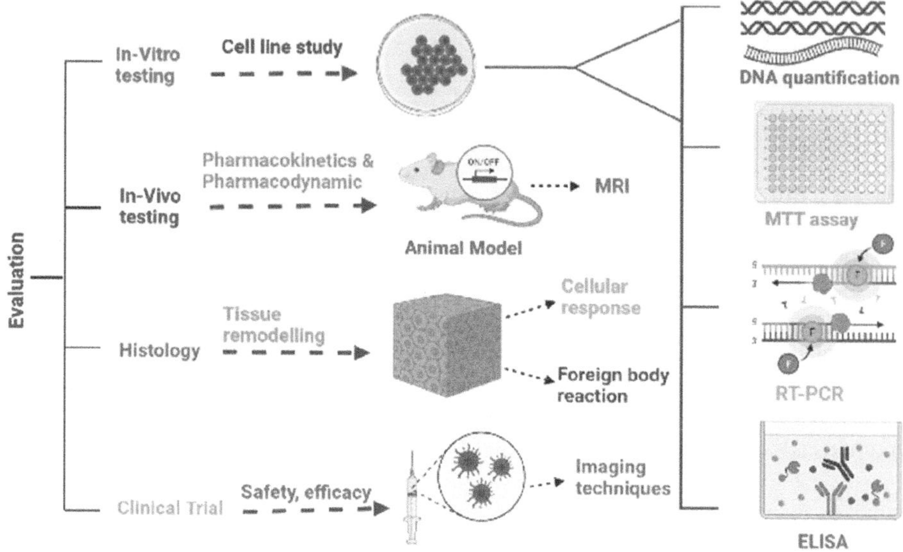

FIGURE 10.1   Evaluation of sustainable biomaterials behaviors. (The evaluation of biomaterials is a four-step process. As part of the in vitro testing, we examine cell lines using DNA quantification, MTT Assay, RT-PCR, and ELISA to determine how the cell lines interact with biomaterials. Animals are subjected to in vivo testing to examine their behavior and responses to biomaterials. To examine cellular responses and body reaction, histology evaluation is performed. After the biomaterials have passed all tests, a clinical trial will be conducted to determine their safety and efficacy.)

4.1   IN VITRO TESTING

In vitro testing plays a crucial role in evaluating the behavior and performance of sustainable biomaterials. It involves conducting experiments and assessments in controlled laboratory settings using cell cultures, tissue models, or simulated body fluids. Commonly used in vitro assay include cell adhesion assay, proliferation assay, and cytotoxicity testing. Assays for cell adhesion evaluate a cell's capacity to stick to the surface of a biomaterial. Cells are grown on the substance, and cell attachment and spreading are used to measure the degree of adhesion. Since cell adhesion is a crucial component of tissue integration, these tests are essential for determining tissue engineering products' suitability for biocompatibility (Figure 10.1).

Assays for cell proliferation assess how cells grow and multiply over time on surfaces made of biomaterials. They evaluate the material's ability to promote cell division. The biomaterial's suitability for tissue regeneration and other applications where cell growth is crucial is determined in part by these experiments. Cytotoxicity tests determine whether biomaterials discharge harmful compounds that damage cells. The assurance of material safety depends on them. Because biomaterials should not damage neighboring cells or tissues, cytotoxicity testing is crucial to the assessment of biocompatibility.

Assessing protein adsorption onto the biomaterial surface provides insights into its interactions with the surrounding biological environment. Techniques like enzyme-linked immunosorbent assay (ELISA) or surface plasmon resonance (SPR) can be used to quantify protein adsorption. Additionally, immunostaining or fluorescence microscopy can visualize the interaction between cells and the biomaterial surface. In vitro gene expression analysis, such as reverse transcription-polymerase chain reaction (RT-PCR), allows the evaluation of the biomaterial's effect on specific cellular processes. It provides insights into how the biomaterial influences the expression of genes related to tissue regeneration, inflammation, extracellular matrix production, or cell-signaling pathways. In vitro, mechanical testing involves subjecting the biomaterial–cell constructs to mechanical forces to assess their mechanical behavior. Techniques such as tensile testing, compression testing, or shear stress testing can be performed to evaluate the biomaterial's mechanical properties in the presence of cells. In vitro testing provides valuable insights into the interaction between biomaterials and biological systems. It allows for controlled experimentation and preliminary assessments of the material's biocompatibility, cell response, and functionality. However, it is important to note that in vitro testing has limitations, and results should be validated in relevant in vivo models to ensure the biomaterial's performance and safety [15].

4.2 In Vivo Testing

In vivo testing means looking at the biomaterial's response and interactions within living organisms or animal models. In vivo testing provides valuable insights into the biomaterial's biocompatibility, safety, efficacy, and long-term behavior. In vivo implantation studies involve surgically inserting the biomaterial into animal models to assess its biocompatibility, tissue integration, and immune response. The biomaterial's interaction with surrounding tissues, inflammation response, wound healing, and potential adverse reactions are evaluated over a specific duration.

In vivo studies can be conducted to evaluate the biomaterial's degradation behavior within living organisms. These studies monitor the degradation rate, changes in mechanical properties, and breakdown products of the biomaterial over time. Imaging techniques, such as radiography or magnetic resonance imaging (MRI), can be used to visualize the degradation process and changes in the implant site. In vivo studies will assess the host response, biocompatibility, functional performance, pharmacokinetics, drugs delivery, and long-term performance to biomaterials [16]. In vivo testing helps bridge the gap between in vitro evaluations and real-world applications by providing a more comprehensive understanding of how biomaterials behave within living systems. In vivo studies provide valuable data to optimize design, validate safety, and assess the potential clinical efficacy of sustainable biomaterials.

4.3 Histology

Histology evaluation is an important method for assessing the behavior and performance of sustainable biomaterials within living tissues. Histological analysis involves the microscopic examination of tissue sections to study the interactions between the biomaterial and the host tissue, cellular responses, and tissue remodeling.

Histology evaluation allows the assessment of how well the biomaterial integrates with the surrounding tissue, inflammatory response, foreign-body reaction, tissue remodeling processes, insight into the vascularization of biomaterials. Tissue sections can be stained and examined to observe the presence of a well-defined interface between the biomaterial and the host tissue, as well as the extent of tissue ingrowth or cellular infiltration into the biomaterial. Histology evaluation provides visual and quantitative information about the interactions between sustainable biomaterials and host tissues. Histology, when combined with other characterization techniques, contributes to a comprehensive understanding of sustainable biomaterial behavior in vivo [17].

4.4 CLINICAL TRIALS

Clinical trials assess the safety of the biomaterial by monitoring adverse events, potential side effects, and complications associated with its use. Patient reports, physical examinations, laboratory tests, and imaging studies are conducted to detect any adverse reactions or safety concerns.

Clinical trials evaluate the biocompatibility of the biomaterial by monitoring the host response to the material. This includes assessing tissue integration, inflammation response, and the absence of adverse tissue reactions through clinical evaluations, histological analysis, and imaging techniques. The biomaterial's functional performance is evaluated in clinical trials to determine its ability to fulfill its intended purpose. For example, if the biomaterial is used for bone repair, clinical assessments may include radiographic evaluations of bone healing, functional outcome measures, or assessments of range of motion. Clinical trials provide insights into the long-term stability and behavior of the biomaterial. Follow-up evaluations over an extended period are conducted to monitor the biomaterial's durability, degradation, tissue integration, and any potential complications that may arise over time.

Clinical trial evaluation provides real-world data on the behavior, safety, and efficacy of sustainable biomaterials in human subjects. It helps assess the biomaterial's performance in diverse patient populations, contributes to evidence-based medicine, and guides the decision-making process for clinical use and regulatory approval [18].

5 PROPERTIES OF SUSTAINABLE BIOMATERIALS ASSESSED THROUGH IN VIVO EXPERIMENTS

5.1 BIOCOMPATIBILITY

The first criterion for every biomedical material is biocompatibility. A biomaterial when meets the human body should not show any adverse effects. A biomaterial must be biocompatible, which may be defined as having appropriate material functioning without causing any unwanted effect on the immune system or tissue reactions. If a substance meets all of these criteria, it is said to be biocompatible [19] (Table 10.1).

It is necessary to remember that the biocompatibility requirements for biomaterials vary based on the types and uses of the materials. The characteristics of

TABLE 10.1

Categorization of biomaterials or biocompatibility assessment

Biomaterial category	Device classification	Definition	Biocompatibility requirement
Class I	• Surface devices	• These substances come into contact with the tissue's surface. Examples include low-burn dressings, catheters, sutures, and so on.	• Low
Class II	• Externally communicating devices	• These include things like dialysis machines, ventilators, and the like, which encounter the body's tissues on occasion. (This class includes many medical tools.)	• Medium
Class III	• Implanted devices	• These include things like bone scaffolds and hip implants, which are designed to be in continuous interaction with the body.	• High

TABLE 10.2

Categories or biocompatibility assessment

Biocompatibility assessment	Short definition
Cytotoxicity	• Ability to trigger cell or tissue demise
Carcinogenicity	• Capability to induce cancer development
Mutagenicity	• Ability to harm genomes
Pyrogenicity	• Capacity to elicit a negative immune response, such as ever
Allergenicity	• Possibility to trigger sensitization and allergic reactions
Thrombogenicity	• Ability to cause blood coagulation

biomaterials based on metal or ceramic are often different from those of polymers and composites. Overall, a biomaterial's clinical usage determines the level of biocompatibility that is necessary. In general, the biocompatibility of biomaterials is defined by the classes associated with them, as shown in Table 10.2.

According to Ramakrishna et al., class III biomaterials have the most stringent biocompatibility requirements because they come into direct contact with living tissues and are anticipated to initiate bio-integration processes. However, as indicated in Table 10.3, any material must meet the acceptable performance standards for each category in order to be considered biocompatible. According to these evaluations, satisfactory biocompatibility is the state in which the biomaterial has no detectable negative effects. Biocompatibility is basically the absence of injury to the host body as a result of the biomaterial.

For example, Aramwit et al. assessed the inflammatory response of a sericin-based cream that was administered to rat wounds that were intentionally created.

After using the cream for seven days, interleukin 1 beta (IL1) and TNF levels were quantified. The findings showed that the release of these cytokines had significantly decreased. The effects of sericin on the amounts of cytokines generated by macrophages and monocytes were also demonstrated in the studies done by Dash et al. and Mandal et al. Thus, it is feasible to verify that sericin is a substance that is biocompatible and does not elicit a strong immune response [20].

5.2 DEGRADATION

Biomaterials must deteriorate at a regulated pace and without exhibiting harmful effects in order to reach their intended state. One essential aspect of the biomaterial is its rate of deterioration. The rapid breakdown of biomaterials may pose a hazard to their mechanical qualities. It is crucial to treat stiffness and degeneration separately as a consequence. It is easy to adjust the copolymerization ratio and molecular weight to maximize the rate of degradation. According to Kong et al., modifying alginates via various cross-linking methods may preserve stiffness while promoting breakdown, which in turn may improve bone production by mesenchymal stem cells derived from bone marrow [21].

5.3 MECHANICAL PROPERTIES

The biomaterial's response to external forces is determined by the material's mechanical characteristics, such as stresses and strains, which are often defined in terms of displacements in defects. Insufficiently stiff biomaterials may impede bone tissue regeneration due to the stress shielding effect, whereas biomaterials with insufficient mechanical properties cannot provide sufficient mechanical support for bone tissue regeneration, especially in load-bearing applications. For the regeneration of bone tissue, it is necessary to optimize the mechanical properties of biomaterials. To achieve ideal tissue regeneration, mechanical properties of substituting biomaterials need to regulate effectively along with static mechanical properties. Mechanical properties of materials are determined by subjecting specimens to standardized mechanical tests in which they are loaded in predefined ways. In order to provide better comparisons between mechanical tests performed in various laboratories, specimens of the material being evaluated with standard dimensions are collected, taking into consideration 'International Standards' (e.g., ISO, CEN, ASTM, DIN). To better understand the actual behavior of the material, certain working conditions (such as accelerated aging, wear, and fatigue) can be recreated in some circumstances. Based on their mechanical properties' biomaterials are mainly classified into composite, metals, polymers, and ceramics thereof. Ceramics, metals, polymers, and their composites are the primary classes of clinical biomaterials in terms of their mechanical behavior. Mechanical characteristics of biomaterials can be affected by a wide variety of factors, including chemical composition, element type, physical combination, chemical cross-linking or phase transition, processing, transformation techniques, and more. The mechanical behavior of a material is more affected by the type of tension and how it is applied.

The mechanical characteristics of metals cause more impact during the designing of orthopedic implants and load-bearing dental implants. However,

Co, Cr, and Mo alloys are employed to achieve the desired results when the implant needs with non-biodegradable sutures and lessen significant wear resistance, such as in artificial joints. Metals, in contrast to ceramics and polymeric materials, allow engineers to create implants that can tolerate substantial mechanical pressures because of their high tensile strength and fatigue limit. Metals have higher tensile strength, elastic modulus, and lower stresses at failure than polymers. After being implanted in the body, metal-based biomaterials may undergo surface changes and deterioration that cause the production of byproducts. Because of this releasing mechanism, interactions take place between the metallic implant surface and cells or tissues. In order to create biocompatible materials, this feature has encouraged modern researchers to place a high priority on comprehending the surface characteristics of metallic products [22, 39, 40].

5.4 Surface Interaction

The biomaterial surface is the first site to come into contact with the physiological environment; the first biological reaction depends on interfacial contact and the enhancement of surface characteristics through alteration of the biomaterial's surface. According to Liu et al. (2014), multiscale surfaces and surface chemistry work in concert to provide exceptional surface wettability. When a biomaterial comes in touch with water for even a nanosecond or microsecond, it forms a water layer or shell on its surface. Adsorption of biomolecules can be significantly influenced by the characteristics and thickness of this aqueous shell. Adsorption of lipids, carbohydrates, and proteins to a surface occurs rapidly, on the order of milliseconds to seconds. This protein layer regulates tissue cell communication and blood vessel compatibility. The ideal properties of an implant material would be for it to be both hemo- and biocompatible. The ability to tell the difference between nonspecific adsorption (hydrophobic effect, weak van der Waals forces, and electrostatic double-layer forces produce physisorption) and selective adsorption (binding to highly well-defined chemistry and steric pockets) is vital. Most investigations of protein adsorption are conducted over time scales that allow for the assumption of equilibrium, which simplifies the calculation of the free energy of adsorption. However, in practice, protein adsorption may not always take place in a vacuum and may not reach equilibrium at a biomaterial surface prior to the occurrence of significant biological processes. Within this crucial window, it is necessary to take these facts into consideration and conduct more study.

Cell adhesion, migration, and differentiation may take place if the implant is applied over a period of several days and hours. This process is affected by several factors, including the material's microstructure (porosity), the biophysical environment, and the material's altering surface qualities (chemistry, microtopographies, and nanoparticles). Biological molecules, such as proteins found in the ECM, the cytoskeleton, and cell membranes, are also involved. Although atomic terraces and kinks exist and can create active spots where interactions are encouraged, one can mistakenly believe that a material's surface is smooth and level. According to research on mitochondria, such action can provide nanoparticles with substantial biological activity.

The surface porosity, surface topology, fiber network structure, and fiber density of tissue engineering scaffolds are all important characteristics that influence cell behavior as a result of their interactions with biomaterials. Immortalized human vein endothelial cells showed varied behavior when cultivated on two distinct BNC surfaces, as discovered by Berti et al. Both BNC surfaces were successful in maintaining the viability of endothelial cells; however, after 20 days in culture, the cells were more robust on the porous bacterial nanocellulose (BNC) surface. This difference may be due to the arrangement or density of the fiber network. Carboxy methyl cellulose, pectin, and methyl cellulose-based cellulose scaffolds with extraordinarily high porosity were also produced by lyophilization [23].

5.5 BIODEGRADATION

Materials that could be broken down by nature through hydrolytic mechanisms without the use of enzymes or enzymatic mechanisms are generally referred to as "biodegrading." In the literature, other words like "absorbable," "erodible," and "resorbable" have also been used to denote biodegradation. Although it is normal practice to use non-biodegradable materials, because of the negative impacts of their component (usually metal), biodegradable materials are preferred. Polymers are being employed because they can degrade themselves over time within the body without causing any negative consequences. The development of biodegradable biomaterials is now the focus of study in the period of new developments in material science. Over the past three decades, biodegradable materials have shown to be extremely important in medical applications. In the 1960s, the first biodegradable sutures were approved. Biomaterials have two important advantages over non-biodegradable. First off, because they progressively integrate into the body and do not leave behind residue at the sites of implantation, they do not cause persistent, long-lasting foreign-body responses. Second, through the interaction of their biodegradation with immune cells like macrophages, some of them have recently been shown to be capable of tissue regeneration, or so-called "tissue engineering." Therefore, temporary scaffolds for tissue regeneration might be created using surgical implants comprising biodegradable biomaterials. One of the most exciting disciplines of the 21st century is this method of reconstructing damaged, ill, or old tissues.

Natural biomaterials are made of plant and animal sources and are biodegradable and biocompatible. Natural biomaterials are mostly composed of proteins (such as collagen, elastin, and silk, fibrin) and/or polysaccharides (such as cellulose, glycosaminoglycans, chitosan, and dextran) as well as biomorphic carbon material produced from plants. The ECM functions as an essential element of the stem cell environment. Besides providing mechanical support for cell adhesion, the ECM is composed of bioactive chemicals that regulate cell development and differentiation by direct contact with specific integrins on the cell surface or non-canonical presentation of growth factors. Natural biomaterials could have intrinsic biological signals that replicate the ECM composition and create favorable conditions for tissue engineering applications. In addition to biochemical features, rigidity, absorbency, and topology of the ECM can affect stem cell development [24].

5.6 Immunogenicity

When the body's immune system perceives a biomaterial as alien, it becomes immunogenic or immunoreactive. Immunoreactive biomaterials, especially wear particles, are recognized by antigenic reactions on cells. After this metabolic chain reaction, T-helper cells migrate toward the biomaterial. The biomaterial may be rejected as a result of this immune response, and the biomaterial and wound site may not come together. This would be a significant barrier to the efficient use of bio-integrative products. Biomaterials that are no longer non-immunogenic due to oxidation or deterioration may also have immunogenic properties. Beyond simple cell binding, tailored biomaterials may be able to affect the intracellular localization of immunotherapy medications, vary the release kinetics of the treatments, and/or concentrate them in specific tissue locations. Adaptive immune response modulation at tissue transplant sites by biomaterials is a current area of researchTreg cell production in vitro and Treg cell penetration into allogeneic pancreatic-cell transplants for the treatment of an in vivo diabetes animal were both stimulated by biodegradable polymer microspheres releasing TGF. In a rat model of vascularized hindlimb allogeneic tissue transplantation, local injection of microspheres releasing TGF, IL-2, and rapamycin boosted generation and proliferation of Treg cells at the implant site, leading to long-term tissue engraftment [25].

6 PHARMACEUTICAL AND MEDICAL APPLICATIONS OF SUSTAINABLE BIOMATERIALS

6.1 Orthopedic Biomaterials

Orthopedic biomaterials are intended to be inserted into the body as parts of devices that are intended to carry out specific biological tasks by replacing or mending different tissues like bone, cartilage, ligaments, and tendons, or even by providing guidance for bone repair when necessary. Orthopedic implants can be used for a variety of procedures, including reconstruction, fracture management, rehabilitation, arthroscopy, electrical stimulation, and casting. Common applications include artificial joint replacement, dynamic stabilization, and better fracture repair.

Scaffolds that replicate the extracellular matrix and aid in the regeneration of missing or injured bone tissue are made from sustainable biomaterials (Figure 10.2). A biodegradable polymer called polycaprolactone (PCL) is used as a scaffold in bone tissue engineering. Long-term support for bone regeneration is provided by its gradual rate of deterioration. In bone tissue engineering, sustainable biomaterials minimize environmental impact and eliminate the requirement for several procedures that are necessary with standard materials. It may be difficult to regulate the pace of deterioration and get mechanical characteristics that are close to those of natural bone.

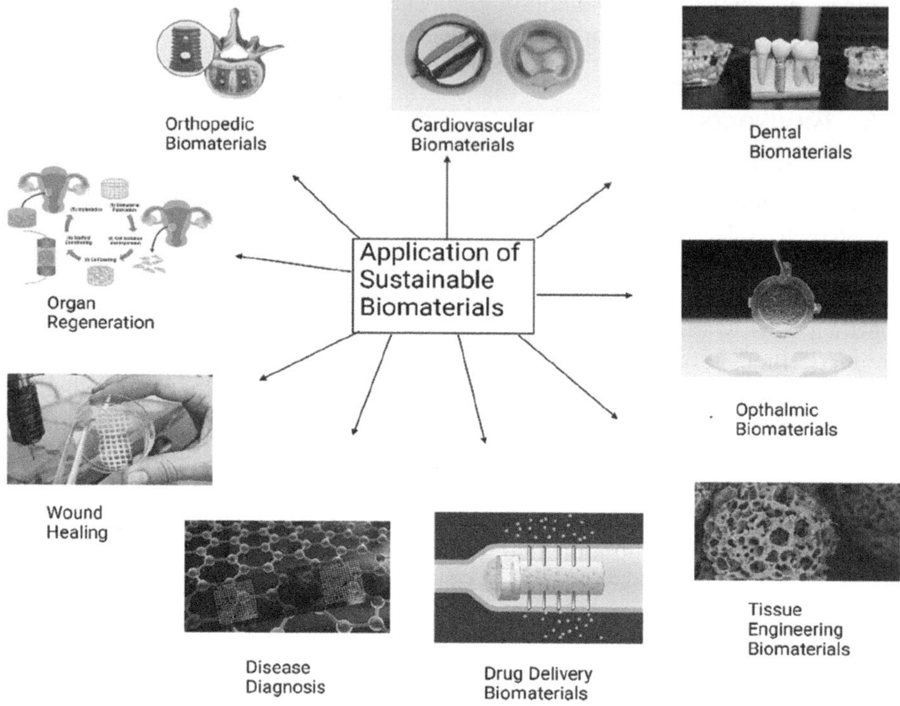

FIGURE 10.2 Various applications of sustainable biomaterials. (Applications for sustainable biomaterials are extensively utilized in numerous medical and pharmaceutical fields, for instance, orthopedic, dental, and drug delivery.)

Biomaterials that are sustainable find use in orthopedic implants, including spinal implants, fixation devices, and joint replacements. Because of its superior mechanical qualities and resistance to corrosion, titanium alloys are often used in joint replacements. Although these implants are not biodegradable by nature, their lifespan lessens the need for replacement surgery. By extending the life of implants, sustainable biomaterials may lessen the environmental impact of both production and disposal. For sustainable orthopedic implant materials, finding a balance between biodegradability and mechanical strength continues to be a problem.

Ligament repair and orthopedic sutures both use sustainable biomaterials. One biodegradable polymer that is often used for sutures is polydioxanone (PDO). It gives strength as a wound heals and deteriorates progressively over time. Sustainable sutures minimize environmental waste associated with non-biodegradable sutures and lessen the necessity for suture removal operations. It is crucial to strike the ideal balance between suture strength and deterioration rate [26].

6.2 Cardiovascular Biomaterials

Coronary artery disease is treated with biodegradable stents. These stents lessen the possibility of long-term problems connected to permanent metal stents by progressively deteriorating over time. Biodegradable polymers called PGA and PLA are utilized to make stents. Biodegradable stents minimize the environmental effect of permanent metal stents, lower the risk of restenosis, and remove the requirement for long-term dual anti-platelet medication. The difficulties lie in regulating the pace of biodegradation to correspond to vessel healing and preserving mechanical integrity throughout the degradation process.

Heart valve replacements that may be seeded with the patient's own cells to encourage tissue regeneration are designed using sustainable biomaterials. Animal tissues' decellularized extracellular matrix (ECM) scaffolds serve as the foundation for tissue-engineered heart valves. Sustainable tissue-engineered heart valves are superior to mechanical or conventional bioprosthetic valves in terms of biocompatibility and thrombosis risk. In heart valve tissue engineering, sourcing, processing, and immunogenicity of ECM scaffolds are problems.

Biodegradable vascular grafts are used for the purpose of substituting or circumventing impaired blood arteries. The grafts undergo a progressive degradation process, facilitating the healing of the patient's blood vessels. Poly(ε-caprolactone) (PCL) and poly (lactic-co-glycolic acid) (PLGA) are often used biodegradable polymers in the fabrication of vascular grafts. The use of biodegradable grafts has been shown to effectively mitigate the potential for long-term problems and the need for further surgical interventions aimed at the removal or replacement of permanent grafts. The task of ensuring suitable mechanical qualities, degradation rates, and compatibility with native tissue presents significant challenges [27].

6.3 DENTAL BIOMATERIALS

Materials used in dentistry fall under the umbrella term "dental biomaterials," which includes a wide variety of different materials. A whole tooth, together with dental pulp-like tissues, may be manufactured. Dental restorations, such as fillings and crowns, often use sustainable biomaterials as viable substitutes for conventional materials such as amalgam or metal alloys. Dental fillings often use bioactive glass-based restorative materials, namely glass ionomer cements. These substances facilitate the release of fluoride ions and facilitate the process of remineralization. Sustainable restorative materials provide the ability to replicate the inherent characteristics of the natural tooth structure, exhibit strong adhesion to dental tissue, and enable the controlled release of advantageous ions to enhance oral well-being. The task of attaining mechanical strength and durability that is on par with conventional materials, while also assuring biocompatibility, is a formidable challenge.

Orthodontic equipment, including as braces and aligners, include sustainable biomaterials to provide patients a more environmentally conscious alternative. Thermoplastic materials, such as polycaprolactone (PCL), are used in the production of clear aligners as a substitute for conventional acrylic materials. Sustainable orthodontic materials provide patients with enhanced visual appeal, heightened comfort, and increased user-friendliness. Additionally, these materials possess the characteristic of being biodegradable or recyclable. The task of ensuring the necessary mechanical qualities and long-term durability of orthodontic materials poses a significant challenge [28].

6.4 OPHTHALMIC BIOMATERIALS

The eye has several anatomical and physiological obstacles, which makes ocular medication administration rather difficult. When treating disorders of the eye's anterior segments, such as keratitis or anterior uveitis, topical eye treatments are most successful; however, medications sometimes fail to achieve therapeutic concentration in the posterior segments. Diseases of the posterior segment have a direct effect on the patient's eyesight. According to estimates, there will be at least 7 million more visually handicapped or blind individuals worldwide by the year 285 million. Because of this, treating these illnesses has become a health issue. Neovascular age-related illness, diabetic retinopathy, and macular degeneration are among the disorders for which intravitreal injections are now used as a form of treatment. However, the treatment's effectiveness is diminished by the high risk of major adverse reactions from repeated injections. These include vitreous hemorrhage, cataract formation, retinal detachment, endophthalmitis, and a rise in intraocular pressure caused using corticosteroids. Systems for sustained administration might eliminate these negative effects. Implants and inserts have been investigated as delivery systems to lessen the need for repeated intraocular injections, boost medication concentration in certain ocular tissues, and boost patient compliance. Noriyuki Kunou et al. found that Ganciclovir-loaded PLGA implants appeared to be effective in the long-term administration of Ganciclovir for the treatment of CMV retinitis. In order to treat proliferative vitreoretinopathy intraocularly, several medicines were put into PLGA implants. These implants contained human recombinant tissue plasminogen activator, triamcinolone, and 5-fluorouridine. Vitrasert is made up of a polyvinyl alcohol (PVA) permeability layer and an ethylene-vinyl acetate (EVA) impermeability layer. Vitrasert is the first non-biodegradable scleral implant to get FDA approval. It was discovered to be helpful for treating CMV retinitis in AIDS patients for five to eight months.

Retisert is the first intravitreal implant/insert used to treat chronic non-infectious uveitis. It is made of silicon laminate and PVA. For almost three years, the system might emit 0.3–0.4 mcg of fluocinolone acetonide per day. Iluvien™ is a non-biodegradable implant that, once placed in the vitreous, releases fluocinolone acetonide over a poly (vinyl alcohol) membrane for up to 36 months to treat persistent diabetic macular edema. On-demand therapeutics (ODT) has created an implant that responds to light in preparation for vitreous administration. In stimulus–responsive systems, changes in environmental conditions comprise light, pH, temperature, magnetic field, ultrasound image or electric current, or control drug release. The production of contact lenses uses sustainable biomaterials to lessen their negative environmental effects while yet offering secure and pleasant vision correction. One common eco-friendly material for contact lenses is silicone hydrogel. They have a high oxygen permeability, which lowers the possibility of eye disorders linked to hypoxia. Wearers of sustainable contact lenses benefit from increased comfort, greater hydration, and increased oxygen transfer. It is difficult to strike an ideal balance between oxygen permeability, mechanical characteristics, and comfort of the lens.

In order to produce intraocular lenses (IOLs), which are utilized in refractive lens exchange and cataract surgery, sustainable biomaterials are being investigated. For sustainable IOLs, bioinspired and biocompatible materials like hydrophilic acrylics are considered. Better biocompatibility and optical clarity are provided by sustainable IOLs, which enhance patient outcomes. Ensuring long-term stability and good optical quality for sustainable IOLs might present difficulties [29].

6.5 Tissue Engineering Biomaterials

Tissue engineering aims to restore and repair damaged tissues in various ways. Any object, structure, or surface that interacts with biological processes is referred to as a biomaterial. The tissue engineering biomaterials are used in bone, cartilage, skin, organ, and vascular region. In bone tissue engineering, scaffolds made of sustainable biomaterials are used to encourage the regeneration of missing or injured bone tissue. Scaffold materials made of biodegradable polymers, such as poly (lactic-co-glycolic acid) (PLGA), are often used. With time, these polymers break down, enabling fresh bone to grow in place of the scaffold. Sustainable materials for bone tissue engineering minimize environmental impact and eliminate the requirement for several procedures that are necessary with standard materials. It may be difficult to regulate the pace of deterioration and get mechanical characteristics that are close to those of natural bone. Scaffolds that facilitate the regeneration of articular cartilage in joints are being developed by cartilage tissue engineering using sustainable biomaterials. Scaffolds for cartilage tissue engineering are made of natural polymers harvested from sustainable sources, such as agarose and chitosan. Better biocompatibility and the potential to encourage spontaneous tissue regeneration are provided by sustainable cartilage tissue engineering materials. Achieving long-term durability and preserving the mechanical qualities necessary for load-bearing joints are difficult tasks.

Pitavastatin calcium and tedizolid were added to dual zein in situ forming implants by Eldeeb et al. for bone regeneration in order to give the implants osteogenic and antibacterial qualities, respectively. A titanium-doped bioactive glass was added to the implants in order to increase their bone-proliferative properties. As a porogenic agent, sodium hyaluronate was also added to provide the porosity required for cell penetration and proliferation. The fabricated implant displayed a steady release of both medications throughout the course of 28 days. Studies conducted in vivo on Sprague Dawley rats showed a sizable bone-regenerating effect. Zein was successfully introduced as a potential implant matrix in this study; however, further research is needed to improve the designed implant so that it may be used to treat large-sized bone abnormalities [30].

6.6 Drug Delivery Biomaterials

Antibodies, peptides, vaccines, medicines, and enzymes are just a few of the pharmacological substances whose distribution and effectiveness have been enhanced by biomaterials. To maximize therapeutic advantages, it is possible to carefully adapt the physicochemical characteristics of biomaterials and the desired route of administration. Biomaterials have enhanced the two most common routes of medication administration: orally and intravenously. Additionally, novel drug delivery methods have been developed, including those through the lungs, skin, eyes, and nose. To reduce the need for frequent dosing, drug delivery systems that release therapeutic chemicals over a prolonged period of time are made using sustainable biomaterials.

One biodegradable polymer that is often used in sustained-release medication delivery systems is poly(lactic-co-glycolic acid) (PLGA). By lowering the frequency of medication administration, sustainable drug delivery systems minimize environmental waste, increase patient compliance, and decrease adverse effects. It may be difficult to maintain material stability throughout the release period and to precisely manage the kinetics of drug release.

Transdermal drug delivery patches, which enable medication absorption via the skin and provide a non-invasive administration approach, are made of sustainable biomaterials. Transdermal medication delivery methods may make use of natural hydrogels or sustainable polymers like chitosan. When compared with oral administration, transdermal patches provide regulated and sustained medication release, improved patient comfort, and less environmental effect. Transdermal medication administration presents issues in maintaining adhesion and achieving optimum drug penetration through the skin [31].

6.7 DISEASE DIAGNOSIS

The development of biosensor-based diagnostics, which can offer a useful method for quick, accurate diagnosis, can use biomaterial-based techniques. Compared with traditional analytical methods, optical biosensors have clear benefits in the real-time, label-free detection of biological and chemical components in a manner that is extremely sensitive, focused, and economical. Volatile organic compounds (VOCs) from human breath are of tremendous value in non-invasive medical diagnostics for prompt identification of numerous illnesses, such as cancer. Gas sensors played a significant role in this process.

Banerjee et al. created functional magnetic nanosensors and coupled them to ZIKV and DENV proteins for the precise detection of Zika and Dengue antibodies with cross-reactivity. Nanosensors have excellent specificity and sensitivity for the detection of ZIKA and DENV in both basic and complex media. Biosensors and assay systems that identify certain biomarkers or infections linked to illnesses use sustainable biomaterials. Sustainable nanoparticles are used as labels in a variety of biosensing applications. One example of this is gold nanoparticles functionalized with ligands obtained from green chemistry. High sensitivity, specificity, and simplicity of functionalization are provided by sustainable biosensors, which also employ less harmful components. It may be difficult to get consistent and repeatable outcomes using sustainable nanoparticles because of differences in surface chemistry, size, and shape. Microfluidic devices, such as blood and urine testing, that are used for sample analysis and illness diagnosis. To create microfluidic chips, sustainable polymers such as poly(methyl methacrylate) (PMMA) are used. Microfluidic devices that are sustainable minimize waste production, boost diagnostic precision, and automate assays. It is difficult to fabricate accurate, sustainable microfluidic devices while preserving their biocompatibility [32].

6.8 REGENERATIVE MEDICINES

For the advancement of regenerative medicine, novel biomaterials that are bioactive in vivo and can influence biological processes in the target and adjacent tissues will

TABLE 10.3

A few biomaterials that are being developed for use in regenerative medicine

Organs	Type of cells	Hydrogels type	Applications
• Eye	• –	• Hyaluronic acid	• Used for regeneration of bone, drug delivery
• Skin	• Fibroblast	• Hyaluronic acid, collagen, fibrin	• Reconstruction of the throat, grafting, ear, nose
• Heart	• Cardiomyocytes, embryonic stem cells, BMC	• Superabsorbent polymer, alginate, hyaluronic acid, fibrin	• Used for tissue engineering of heart cells
• Blood vessels	• Endothelial cells stem cells	• Alginate, PEG, hyaluronic acid	• Vessels grafting

be indispensable. Traumatic injuries can stop muscles from contracting by harming the peripheral nerves and skeletal muscles. Scar tissue may develop during natural healing. Thus, the use of three-dimensional scaffolds will result in the elongation, orientation, fusion, and striation of muscle cells. Muscle repair has employed electro spun chitosan microfibers as new biomaterials. Traditional porous scaffolds might not be sufficient since they do not replicate the usual cardiac environment. Given the importance of topography in this process, one strategy is to replicate the natural tissue's microenvironment. In this regard, bioartificial blends based on alginate, gelatin, and a new poly (*N*-isopropyl acrylamide) based copolymer were used to build microfabricated scaffolds employing soft lithography.

The mechanical characteristics of this scaffold were anisotropic, similar to those of natural tissue. Interesting aspects of hydrogels include their high level of compatibility with the live cells found within the human body. Because it is a soft biomaterial, its rigidity is comparable to that of human tissues, making it ideal for use in tissue engineering and regeneration [33].

6.9 WOUND HEALING

Gelatin and collagen, two natural biomaterials, have demonstrated good skin wound-healing outcomes. Given that they mimic the extracellular matrix, collagen, and gelatin are extremely biocompatible with human tissues. While gelatin is a hydrolyzed version of collagen, collagen is naturally present in the human body. During wound healing, researchers discovered that native collagen provided a three-dimensional environment that promoted cell proliferation and facilitated the migration of keratinocytes and fibroblasts to physiologic areas. They discovered via their in vitro research that fibroblasts and keratinocytes affixed to Col with great affinity.

Burns to the skin can also be treated with the use of biomaterials. Bioengineered skin replacements for severe burn injuries are now under development, thanks to advances in skin tissue engineering. Skin replacements are made up of diverse groupings of materials (which may be cellular or acellular) with characteristics like those of skin. They cover wounds either temporarily or permanently and encourage autologous regenerative repair with a low inflammatory response. The synthetic or natural polymers utilized to create bioengineered skin substitutes include collagen, gelatin, chitosan, quercetin, polylactic-co-glycolic acid (PLGA), and polyethylene glycol (PEG), and hyaluronic acid. There are three main types of skin replacements: 'epidermal cover, dermo-epidermal replacement, and dermal replacement.' Tissue-engineered skins help the wound microenvironment by acting as a barrier, supplying blood and nutrients, being elastic and supportive, and firmly attaching to the dermis, all of which are necessary for wound healing and skin regeneration [34].

6.10 Organ Regeneration

Organ regeneration following disease detection and treatment has been proposed as a viable option for the replacement of vital structures, including the heart, trachea, and lungs. A cardiac patch, constructed from stem cells and biomaterials, was designed, and evaluated for implantation. It has been shown that cardiomyocytes implanted in hydrogel may contract in response to mechanical stimulation. Decellularized scaffolds from rat lungs were seeded with pulmonary epithelium and vascular endothelium to regenerate the lungs ex vivo. Total knee replacement operations can be made safer and more successful with the use of two different kinds of ceramic-based functional biomaterials designed to reinforce the bone prosthesis and prevent aseptic loosening. Alumina and zirconia biomaterials with functionally graded surfaces fared exceptionally well [35].

A wide variety of biomaterials have been utilized to construct tracheal epithelial grafts, including hyaluronan poly (ethylene glycol), chitosan-collagen, fibrin glue, collagen vitrigel membrane, gelatin, and silk fibroin. This was a huge help for people with breathing problems. Biomaterials that can function as substitutes for the liver, kidney, airways, trachea, and larynx may one day be created because of advances in organ engineering [36].

7 FUTURE DIRECTIONS IN BIOMATERIALS

In addition to computational and automated applications, to better predict how biomaterials might interact with physiological factors such serum proteins, cell membranes, and cellular receptors, it is possible that learning technologies could have an impact on biomaterial design. Computational methods and artificial intelligence (AI) have the potential to transform biomaterial design and development by speeding up the discovery of novel materials with specific properties. Machine learning algorithms can predict the performance of biomaterials by examining enormous datasets and swiftly selecting potential candidates. AI-driven simulations have the potential to model complex molecular interactions and improve material properties for medicinal applications such as drug delivery or tissue engineering. This reduces the need for costly experimental iterations and speeds up the research process. Additionally, AI can help us understand biological processes better, which will facilitate the development of materials that imitate biological processes. Computational methods and AI generally have the potential of promoting efficiency and creativity in the research of biomaterials, which will benefit the biotechnology and healthcare industries. AI and computational techniques have the potential to have a big influence on biomaterial research and design in several ways.

The development of nanoparticles for the administration of drugs in cancer treatment: in the past, scientists would painstakingly create many nanoparticle formulations via trial-and-error procedures, often taking years to discover an ideal design. AI and computational techniques together may significantly speed up this process. AI can analyze large databases of materials, their characteristics, and biological reactions. It can determine the relationships between the properties of nanoparticles (such as size, shape, and surface charge) and how well they carry drugs. Predictive models that quantify the effects of various nanoparticle characteristics on drug release, circulation time, and targeting efficiency may be created using machine learning algorithms. This enables researchers to concentrate on potential options and reduce the size of the design space. Materials Studio is a complete software package designed for modeling and simulating materials at the atomic and molecular levels. It was created by BIOVIA (previously Accelrys). It is useful for the creation of biomaterials as it contains several modules for molecular simulations, material design, and property prediction [37].

The molecular interactions between nanoparticles and biological systems may be simulated using computational methods. In order to create materials that maximize medication delivery, researchers can better understand how nanoparticles interact with biological fluids, tissues, and cells, thanks to this. An open-source software platform called CompuCell3D is used to study and simulate the behavior of tissues and cells. The study of tissue engineering and the interplay between biomaterials and biological systems makes extensive use of it. AI-powered optimization algorithms may repeatedly improve nanoparticle designs to find the ideal blend of characteristics for a certain therapeutic use. Virtual prototypes may be rapidly created and tested by researchers without requiring a lot of lab labor. Materia is a materials informatics platform that speeds up the discovery of new materials by using AI and machine learning. By evaluating and forecasting material qualities based on available data, it may help in the design and optimization of biomaterials.

AI can also help in biomimetic design, which is the process of creating nanoparticles that resemble natural processes and structures. AI, for instance, may be used to create virus-like nanoparticles that improve medicine delivery effectiveness and cellular absorption. Utilizing AI and computational techniques, scientists can significantly cut down on the time and costs needed to create medication delivery nanoparticles that work. They can create materials with more accuracy and therapeutic results, which will eventually result in more focused and effective cancer therapies. This strategy not only expedites the development of biomaterials but also provides access to personalized medicine, in which patient profiles are used to customize therapies for improved health outcomes.

Industries are changing as a result of new developments in bioinspired designs and smart materials. Shape-memory alloys and self-healing structures are examples of advances made possible by responsive materials' ability to adapt to changing environments. Biomimetic designs are influenced by nature and produce materials with amazing qualities. Electronics and medication distribution are impacted by nanotechnology's ability to precisely govern at the nanoscale. Artificial organs and complex biomimetic constructions are made possible by 3D printing. Biohybrid systems provide new advances in healthcare by fusing synthetics with biological green materials that are driven by sustainability, which lessens the effect on the environment. IoT gadgets are powered by materials that capture energy. AI speeds up the optimization and finding of materials. Implantable and wearable technology enhances medical treatment. Prosthetics and healthcare are advanced by biodevices and soft robotics. These developments portend the development of materials with a wide range of uses that are intelligent, sustainable, and biologically inspired [38].

8   CONCLUSIONS

Biomaterials may be the most interdisciplinary subject area. They have been utilized constructively in the health and pharmaceutical industries since their invention. One goal of smart materials research is the development of biocompatible materials that may respond to external signals or the environment and find applications in areas, such as medical devices, medicine delivery, immunological engineering, tissue engineering, and medical equipment. Smart materials utilized in tissue engineering should be able to interact with tissues and cells, whether by stimulating adherence or dissolving when commanded. Smart materials have the potential to be used in medication delivery to deliver drugs selectively to different tissues and in response to different stimuli over extended periods of time before being eliminated from the body. Smart materials are needed in the field of immune engineering to ensure the accurate distribution of immune cells and the prevention of unwanted immunological reactions and inflammation. Constructing smart materials with multiple characteristics at once can be challenging. It may be challenging, for instance, to design a smart material with optimal mechanical, chemical, and biological capabilities without compromising any of those features. Biomedical applications may benefit from the development of bioinspired materials that mimic biological structures and function by drawing inspiration from nature. Improved host tissue integration helps reduce rejection and inflammation, and smart materials can make 3D printing of organs even more stable and easy. The human body makes extensive use of numerous biomaterial varieties. Organ regeneration, tissue science, wound healing, medicine delivery, illness diagnostics, and tissue and organ modeling are just few of the many applications for implants manufactured from biomaterials. Biomaterials have recently been used in urology and in the field of cell regeneration. Drug delivery scaffolds like nano-carriers are becoming increasingly important. Biomaterials are gaining prominence in the healthcare and pharmaceutical sectors.

A revolutionary change in the pharmaceutical and medical industries is being sparked by sustainable biomaterials. Pharmaceutical items packaged sustainably, implantable devices, tissue engineering structures, and environmentally responsible drug delivery systems are just a few of the uses for them. These biomaterials support a greener, more conscientious healthcare sector by mitigating the environmental impact of pharmaceutical and medical practices. Additionally, by providing biocompatibility, better medication release profiles, and creative tissue healing options, they improve patient outcomes. In the realm of cell regeneration and urology, biomaterials have lately been used. Nano-carriers and other drug delivery scaffolds are becoming more and more significant. In the pharmaceutical and healthcare industries, biomaterials are becoming more and more popular. The overall effect of sustainable biomaterials is about more than simply supporting ecological sustainability; it's also about improving healthcare services and creating a more wholesome future for Earth and its people.

REFERENCES

1. Marin, Elia, Francesco Boschetto, and Giuseppe Pezzotti. 2020. "Biomaterials and Biocompatibility: An Historical Overview." *Journal of Biomedical Materials Research Part A* 108 (8): 1617–1633. https://doi.org/10.1002/jbm.a.36930.
2. Baharlouei, Parnian, and Azizur Rahman. 2022. "Chitin and Chitosan: Prospective Biomedical Applications in Drug Delivery, Cancer Treatment, and Wound Healing." *Marine Drugs* 20 (7): 460. https://doi.org/10.3390/md20070460.
3. Troy, Eoin, Maura A. Tilbury, Anne Marie Power, and J. Gerard Wall. 2021. "Nature-Based Biomaterials and Their Application in Biomedicine." *Polymers* 13 (19): 3321. https://doi.org/10.3390/polym13193321.
4. Deng, Zhengyu, Qiangqiang Shi, Jiajia Tan, Jinming Hu, and Shiyong Liu. 2021. "Sequence-Defined Synthetic Polymers for New-Generation Functional Biomaterials." *ACS Materials Letters* 3 (9): 1339–1356. https://doi.org/10.1021/acsmaterialslett.1c00358.
5. Banoriya, Deepen, Rajesh Purohit, and R. K. Dwivedi. 2017. "Advanced Application of Polymer Based Biomaterials." *Materials Today: Proceedings* 4 (2): 3534–3541. https://doi.org/10.1016/j.matpr.2017.02.244.
6. Piconi, Corrado, and Simone Sprio. 2021. "Oxide Bioceramic Composites in Orthopedics and Dentistry." *Journal of Composites Science* 5 (8): 206. https://doi.org/10.3390/jcs5080206.
7. Punj, Shivani, Jashandeep Singh, and K. Singh. 2021. "Ceramic Biomaterials: Properties, State of the Art and Future Prospectives." *Ceramics International* 47 (20): 28059–28074. https://doi.org/10.1016/j.ceramint.2021.06.238.
8. Park, Joon B., and Young Kon Kim. "Metallic biomaterials." In *Biomaterials*, pp. 1-1. CRC press, 2007.
9. Abdelhamid, Mohamed A. A., and Seung Pil Pack. 2021. "Biomimetic and Bioinspired Silicifications: Recent Advances for Biomaterial Design and Applications." *Acta Biomaterialia* 120: 38–56. https://doi.org/10.1016/j.actbio.2020.05.017.
10. Reeve, Lesley, and Paul Baldrick. 2017. "Biocompatibility Assessments for Medical Devices–Evolving Regulatory Considerations." *Expert Review of Medical Devices* 14 (2): 161–167. https://doi.org/10.1080/17434440.2017.1280392.
11. Liu, Shuai, Jiang-Ming Yu, Yan-Chang Gan, Xiao-Zhong Qiu, Zhe-Chen Gao, Huan Wang, Shi-Xuan Chen, Yuan Xiong, Guo-Hui Liu, and Si-En Lin. 2023. "Biomimetic Natural Biomaterials for Tissue Engineering and Regenerative Medicine: New Biosynthesis Methods, Recent Advances, and Emerging Applications." *Military Medical Research* 10 (1): 16. https://doi.org/10.1186/s40779-023-00448-w.
12. Zhai, Wenzheng, Lichun Bai, Runhua Zhou, Xueling Fan, Guozheng Kang, Yong Liu, and Kun Zhou. 2021. "Recent Progress on Wear-Resistant Materials: Designs, Properties, and Applications." *Advanced Science* 8 (11): 2003739. https://doi.org/10.1002/advs.202003739.
13. Dilmani, Sara Asghari, Sena Koc, Demet Çakır, and Menemşe Gümüşderelioğlu. 2023. "Organomodified Nanoclay with Boron Compounds Is Improving Structural and Antibacterial Properties of Nanofibrous Matrices." *European Journal of Pharmaceutics and Biopharmaceutics* 184: 125–138. https://doi.org/10.1016/j.ejpb.2023.01.015.

14. Huzum, Bogdan, Bogdan Puha, Riana Maria Necoara, Stefan Gheorghevici, Gabriela Puha, Alexandru Filip, Paul Dan Sirbu, and Ovidiu Alexa. 2021. "Biocompatibility Assessment of Biomaterials Used in Orthopedic Devices: An Overview." *Experimental and Therapeutic Medicine* 22 (5): 1–9.

15. Wu, Tianyi, Fuli Yin, Nan Wang, Xin Ma, Chaolai Jiang, Lihui Zhou, Yang Zong, et al. 2021. "Involvement of Mechanosensitive Ion Channels in the Effects of Mechanical Stretch Induces Osteogenic Differentiation in Mouse Bone Marrow Mesenchymal Stem Cells." *Journal of Cellular Physiology* 236 (1): 284–293. https://doi.org/10.1002/jcp.29841.

16. Kutner, Neta, Konda Reddy Kunduru, Luna Rizik, and Shady Farah. 2021. "Recent Advances for Improving Functionality, Biocompatibility, and Longevity of Implantable Medical Devices and Deliverable Drug Delivery Systems." *Advanced Functional Materials* 31 (44): 2010929. https://doi.org/10.1002/adfm.202010929.

17. Karami, Peyman, Theofanis Stampoultzis, Yanheng Guo, and Dominique P. Pioletti. 2023. "A guide to preclinical evaluation of hydrogel-based devices for treatment of cartilage lesions." *Acta Biomaterialia* 158: 12–31. 18.Tian, Jiaxin, Xu Song, Yongqing Wang, Maobo Cheng, Shuang Lu, Wei Xu, Guobiao Gao, Lei Sun, Zhonglan Tang, and Minghui Wang. 2022. "Regulatory Perspectives of Combination Products." *Bioactive Materials* 10: 492–503. https://doi.org/10.1016/j.bioactmat.2021.09.002.

19. Kumar, Ajay, Parveen Kumar, Ashish Kumar Srivastava, and Vikas Goyat. 2023. *Modeling, Characterization, and Processing of Smart Materials.* IGI Global. https://doi.org/10.4018/978-1-6684-9224-6

20. Mandal, Biman B., Anjana S. Priya, and S. C. Kundu. 2009. "Novel Silk Sericin/Gelatin 3-D Scaffolds and 2-D Films: Fabrication and Characterization for Potential Tissue Engineering Applications." *Acta Biomaterialia* 5 (8): 3007–3020. https://doi.org/10.1016/j.actbio.2009.03.026.

21. Kong, Hyun Joon, Darnell Kaigler, Kibum Kim, and David J. Mooney. 2004. "Controlling Rigidity and Degradation of Alginate Hydrogels via Molecular Weight Distribution." *Biomacromolecules* 5 (5): 1720–1727. https://doi.org/10.1021/bm049879r.

22. Joyce, Kieran, Georgina Targa Fabra, Yagmur Bozkurt, and Abhay Pandit. 2021. "Bioactive Potential of Natural Biomaterials: Identification, Retention and Assessment of Biological Properties." *Signal Transduction and Targeted Therapy* 6 (1): 122. https://doi.org/10.1038/s41392-021-00512-8.

23. Gudimetla, Avinash, Parveen Kumar, S. Sambhu Prasad, Satish Geeri, and V. V. N. Sarath. 2023. "Towards Smart Materials: Enhancing the Efficiency of the Materials." In *Modeling, Characterization, and Processing of Smart Materials*, pp. 1–30. IGI Global. https://doi.org/10.4018/978-1-6684-9224-6.ch001.

24. Geevarghese, Rency, Seyedeh Sara Sajjadi, Andrzej Hudecki, Samad Sajjadi, Nahid Rezvani Jalal, Tayyebeh Madrakian, Mazaher Ahmadi, et al. 2022. "Biodegradable and Non-Biodegradable Biomaterials and Their Effect on Cell Differentiation." *International Journal of Molecular Sciences* 23 (24): 16185. https://doi.org/10.3390/ijms232416185.

25. Fisher, James D., Stephen C. Balmert, Wensheng Zhang, Riccardo Schweizer, Jonas T. Schnider, Chiaki Komatsu, Liwei Dong, et al. 2019. "Treg-Inducing Microparticles Promote Donor-Specific Tolerance in Experimental Vascularized Composite Allotransplantation." *Proceedings of the National Academy of Sciences of the United States of America* 116 (51): 25784–25789. https://doi.org/10.1073/pnas.1910701116.

26. Chahal, Sugandha, Anuj Kumar, and Fathima Shahitha Jahir Hussian. 2019. "Development of Biomimetic Electrospun Polymeric Biomaterials for Bone Tissue Engineering. A Review." *Journal of Biomaterials Science, Polymer Edition* 30 (14): 1308–1355. https://doi.org/10.1080/09205063.2019.1630699.

27. Mohtashami, Zahra, Zahra Esmaili, Molood Alsadat Vakilinezhad, Ehsan Seyedjafari, and Hamid Akbari Javar. 2020. "Pharmaceutical Implants: Classification, Limitations and Therapeutic Applications." *Pharmaceutical Development and Technology* 25 (1): 116–132. https://doi.org/10.1080/10837450.2019.1682607.

28. Tiwari, Himanshu Kumar, Ashish Kumar Srivastava, Parveen Kumar, Manish Kumar Singh, Hritik Kumar, and Akshit Bhadauria. 2023. "Materials and Technology for Implant Manufacturing: Challenges and Opportunity." In *Modeling, Characterization, and Processing of Smart Materials*, pp. 83–106. IGI Global.

29. Mansoor, Saffar, Baruch D. Kuppermann, and M. Cristina Kenney. 2009. "Intraocular Sustained-Release Delivery Systems for Triamcinolone Acetonide." *Pharmaceutical Research* 26: 770–784. https://doi.org/10.1007/s11095-008-9812-z.

30. Majid, Qasim A., Annabelle T.R. Fricker, David A. Gregory, Natalia Davidenko, Olivia Hernandez Cruz, Richard J. Jabbour, Thomas J. Owen, et al. 2020. "Natural Biomaterials for Cardiac Tissue Engineering: A Highly Biocompatible Solution." *Frontiers in Cardiovascular Medicine*, 192. https://doi.org/10.3389/fcvm.2020.554597.

31. Adeosun, Samson O., Margaret O. Ilomuanya, Oluwashina P. Gbenebor, Modupeola O. Dada, and Cletus C. Odili. 2020. "Biomaterials for Drug Delivery: Sources, Classification, Synthesis, Processing, and Applications." *Advanced Functional Materials*, 141–167.

32. Banerjee, Tuhina, Truptiben Patel, Oleksandra Pashchenko, Rebekah Elliott, and Santimukul Santra. 2021. "Rapid Detection and One-Step Differentiation of Cross-Reactivity between Zika and Dengue Virus Using Functional Magnetic Nanosensors." *ACS Applied Bio Materials* 4 (5): 3786–3795. https://doi.org/10.1021/acsabm.0c01264.

33. Eldeeb, Alaa Emad, Salwa Salah, Mostafa Mabrouk, Mohammed S. Amer, and Nermeen A. Elkasabgy. 2022. "Dual-Drug Delivery via Zein In Situ Forming Implants Augmented with Titanium-Doped Bioactive Glass for Bone Regeneration: Preparation, In Vitro Characterization, and In Vivo Evaluation." *Pharmaceutics* 14 (2): 274. https://doi.org/10.3390/pharmaceutics14020274.

34. Elfawy, Loai A., Chiew Yong Ng, Ibrahim N. Amirrah, Zawani Mazlan, Adzim Poh Yuen Wen, Nur Izzah Md Fadilah, Manira Maarof, Yogeswaran Lokanathan, and Mh Busra Fauzi. 2023. "Sustainable Approach of Functional Biomaterials–Tissue Engineering for Skin Burn Treatment: A Comprehensive Review." *Pharmaceuticals* 16 (5): 701. https://doi.org/10.3390/ph16050701.

35. Kumar, Ajay, Parveen Kumar, Namrata Dogra, and Archana Jaglan. 2023. "Application of Incremental Sheet Forming (ISF) Toward Biomedical and Medical Implants." In *Handbook of Flexible and Smart Sheet Forming Techniques: Industry 4.0 Approaches*, edited by Vishal Gulati, Hari Singh, Parveen Kumar, Ajay Kumar, and Pravin Kumar, p. 247.

36. Fishman, Jonathan M., Katherine Wiles, Mark W. Lowdell, Paolo De Coppi, Martin J. Elliott, Anthony Atala, and Martin A. Birchall. 2014. "Airway Tissue Engineering: An Update." *Expert Opinion on Biological Therapy* 14 (10): 1477–1491. https://doi.org/10.1517/14712598.2014.938631.

37. Kumar, Ajay, Ravi Kant Mittal, and Abid Haleem, eds. 2022. *Advances in Additive Manufacturing: Artificial Intelligence, Nature-Inspired, and Biomanufacturing*. Elsevier. https://doi.org/10.1016/C2020-0-03877-6

38. Kumar, Ajay, Parveen Kumar, Ravi Kant Mittal, and Victor Gambhir. 2023. "Materials Processed by Additive Manufacturing Techniques." In *Advances in Additive Manufacturing Artificial Intelligence, Nature-Inspired, and Biomanufacturing*, pp. 217–233. Elsevier. https://doi.org/10.1016/B978-0-323-91834-3.00014-4.

39. Kumar, Ajay, Parveen Kumar, Naveen Sharma, and Ashish Kumar Srivastava, eds. 2024. *3D Printing Technologies: Digital Manufacturing, Artificial Intelligence, Industry 4.0.* Walter de Gruyter GmbH & Co KG. https://doi.org/10.1515/9783111215112.
40. Ajay, Hari Singh, Parveen, and Bandar AlMangour. 2023. *Handbook of Smart Manufacturing: Forecasting the Future of Industry 4.0.* CRC Press. https://doi.org/10.1201/9781003333760.

11 The Utilization of Biomaterials in Diagnostic, Preventive, and Therapeutic Approaches for the Management of COVID-19

Odangowei Inetiminebi Ogidi
Bayelsa Medical University, Yenagoa

LIST OF ABBREVIATIONS

3D	three-dimensional
ACE2	angiotensin-converting enzyme 2
ARDS	acute respiratory disease syndrome
AT2	alveolar type II
AuNP	gold nanoparticles
CCL 2	chemokine (C-C motif) ligand 2
CDC	Center for Disease Control
CT	computer tomography
E	envelope
ECM	extracellular matrix
ECMO	extracorporeal membrane blood oxygen
ELISAs	enzyme-linked immunosorbent assays
FAD	Food and Drug Administration
FET	field-effect transistor
FTO	fluorine-doped tin oxide
HA	hemagglutinin
HCV	hepatitis C virus
HFNO	high flow nasal oxygen
HiPSC	human-induced pluripotent stem cells

DOI: 10.1201/9781003434313-11

ICU	intensive care unit
LNP	lipid nanoparticle
M	membrane
MERs	Middle East respiratory syndrome
MHC	major histocompatibility
MOD	multi-organ dysfunction
MSC	mesenchymal stem cell
N	nucleocapsid
Nab	neutralizing antibody
NIV	noninvasive ventilation
NPs	nanoparticles
NSPs	nonstructural proteins
NW-FET	nano wire field-effect transistor
OoC	organ on a chip
PBAEs	poly (b-amino) esters
PC	polycarbonate
PDMS	polydimethylsiloxane
PEG	polyethylene glycol
PEI	polyethyleneimine
PET	polyethylene terephythalate
PLA	polylactide
PLGA	polyglycolide
PP	polypropylene
PTFE	polytetrafluoro ethylene
PVDF	polyvinylidene fluoride
QD	quantum dot
R0	reproduction number
RBD	receptor-binding domain
RT-PCR	reverse-transcription polymerase chain reaction
S	spike
SARS-CoV-2	severe acute respiratory syndrome corona virus 2
SPR	surface plasmon resonance
T	thamus
WHO	World Health Organization

1 INTRODUCTION

Pandemics may be caused by infections, which have significantly impacted
human life and health throughout history [1]. Important infections include those
brought on by viruses. Through the creation of suitable preventative measures and
immunization programs, many viral diseases are now under control [1]. Others
continue to be very difficult, such as the newly discovered SARS-CoV-2 coronavirus, which has so far resulted in significant fatality rates around the globe [2].
Just a few months after it first appeared, the World Health Organization (WHO)
proclaimed a pandemic [1]. The COVID-19 pandemic was caused by the coronavirus SARS-CoV-2, which had a substantially greater infection rate than the

SARS-CoV of 2003 with 1.7 million fatalities and over 80 million sick persons worldwide [3].

There is currently no effective treatment for COVID-19, and it is difficult for people who recover from the illness to heal their damaged tissues and organs. On the preventative side, it is unclear whether long-lasting protection can be attained despite major advancements in vaccine research. Since then, our understanding of this unique virus has increased, although it is still insufficient. Additionally, the ability of the disease to be transmitted by asymptomatic people underscores the need for creating an efficient and user-friendly diagnostic procedure. To curb its spread, there is an urgent need for novel diagnostic and therapeutic treatments [4]. To promptly isolate affected individuals and stop the spread of the virus, development of fast, accurate, and laboratory-free point-of-care home-test equipment should be taken into consideration [5]. Patients must depend on their own immune systems if there are no potent antiviral medications or other therapies that can stop or reduce the viral infections. The situation is serious, and only worldwide coalitions of leaders from many interdisciplinarity disciplines can equip us with the means to defeat this shared foe.

New treatment approaches have been discovered, thanks in large part to the work of the field of biomaterials research [6]. Natural or artificial materials that have been created specifically to interact with biological systems for therapeutic or diagnostic reasons are called biomaterials. They provide a special way to bioengineer intricate microenvironments, for example, to better mimic and research human illnesses. As high-throughput screening platforms, these microenvironments may also be utilized to evaluate the efficacy of novel and developing treatments. Additionally, biomaterials may be created to have antiviral properties and be employed as efficient delivery systems for vaccines and medications [7].

Hydrogels, cryogels, and nanoparticles (NPs) are some of the biomaterial kinds that are particularly pertinent. According to Rogers et al. [8] and Gsib et al. [9], hydrogels are hydrophilic, three-dimensional (3D) cross-linked polymers. Hydrogels have a high water content, which makes them friendly and often employed in biomedical applications including medication delivery and tissue engineering [10]. Cryogels, a subclass of hydrogels made at subzero temperatures, have been used for bioseparation, tissue engineering, and immunotherapy among other bio-related applications. They have special properties like a macroporous network, shape memory, and syringe injectability. Modern medicine has made extensive use of NPs, such as liposomes, which are nanoscale biomaterials. Their therapeutic applications vary from contrast agents to nanocarriers for medication and gene delivery [1].

These biomaterials provide a variety of opportunities for creating disease models, preventative measures, diagnostic tools, therapeutic treatments, and vaccinations. Controlled and targeted drug administration might be used as an alternative to standard antiviral medications due to their poor bioavailability and short circulation times. As a result, this chapter covers the etiology, epidemiology, and morphology of COVID-19, as well as the mechanism of infection and transmission, a general overview of biomaterials, their significance for COVID-19, applications of biomaterials against COVID-19, the benefits and drawbacks of using biomaterials for COVID-19 diagnosis and therapy, current challenges, and future prospects.

2   ETIOLOGY, EPIDEMIOLOGY, AND MORPHOLOGY OF COVID-19

In Wuhan Province, China, in December 2019, a cluster of pneumonia cases led to the discovery of the SARS-CoV-2 virus [11]. Since then, the virus has spread at an astounding pace around the world. Some intriguing insights may be gained from a quick comparison of SARS-CoV-2 to similar viruses SARS coronavirus (SARS-CoV) and Middle East respiratory syndrome (MERS) coronavirus (MERS-CoV). The first SARS epidemic started in Guangdong Province, China, in November 2002, and by July 5, 2003, the WHO had proclaimed the outbreak to be under control. In all, 29 nations reported 774 fatalities (9.2% mortality) and almost 8,000 SARS cases. While there was no overall human-to-human transmission, the bulk of these cases included persons who had frequent close contact with animals and healthcare staff who were involved in aerosol-producing activities [12].

The MERS epidemic started in Saudi Arabia in September 2012, and by June 2019, 27 countries have reported a total of over 2,500 cases and 845 fatalities (34.5% mortality). There is evidence of a SARS-like pattern of transmission with very little transfer from person to person [13]. Although all three are unique coronaviruses that are transmitted through zoonosis, COVID-19 stands out due to how fast and extensively it has spread and how little mortality it has compared with the other two viruses [12,13]. To demonstrate this point, consider that while COVID-19 is currently estimated to be at least twice as contagious as SARS and MERS, SARS and MERS were found to have basic reproduction numbers (R0), a measure of how contagious an infectious disease is, between 2.5 and 3.5 [14].

Not only is this number crucial for estimating the rate of spread, but R0 is also closely tied to the idea of herd immunity, which states that in a given community, a certain percentage of vaccinated people provide immunity against disease transmission to the unimmunized people. The "herd immunity threshold" is the fundamental connection between R0 and herd immunity and is computed as $\dfrac{1-1}{R0}$. According to this equation, illnesses with a R0 of 3 would need the immunization of 67% of the population, but diseases with a R0 of 5.5, such as COVID-19, would require the immunization of 82% of the population [15].

According to taxonomy, SARS-CoV-2 is a member of the beta-coronavirus family and is an enveloped positive-sense RNA virus. Around 30 kb in size, the viral genome is broken into a number of open reading frames that encode both structural and nonstructural proteins [16]. After successfully entering cells, viral RNA is released into the cytoplasm, where host cells transform positive-sense RNA into proteins. Spike (S) glycoprotein, envelope (E) glycoprotein, membrane (M) glycoprotein, and nucleocapsid (N) protein are some of the structural proteins. In a nutshell, the S protein interacts with ACE2 to fuse with host cells, and this interaction is of considerable importance to current studies. In the endoplasmic reticulum, the E protein aids in the assembly and budding of new virions. The primary structural protein is the M protein. In addition to digesting and stabilizing viral RNA, the N protein binds to it [17].

Enzymes that combine to produce replication–transcription complexes important in viral replication are examples of nonstructural proteins (NSPs). It is thought that replication occurs inside a network of altered endoplasmic reticulum membranes. Following assembly, virions are released into the extracellular space through exocytosis, where they continue to infect other host cells until being evacuated or eliminated from the body [18].

3 MECHANISM OF TRANSMISSION AND INFECTION OF COVID-19

Although a zoonotic virus at first, the main means of transmission have changed to close contact, fomites, and droplets [19]. The "asymptomatic carrier" is a patient who sheds the virus for days to weeks before symptoms, if any, appear. This patient is of tremendous concern. Additionally, there is mounting evidence that SARS-CoV-2 is extremely infectious in part because it may move through aerosols, which are generally characterized as tiny respirable droplets with a diameter of between 5 and 10 m. These tiny aerosols, in contrast to big (>20 m) droplets, are able to travel great distances in the air without instantly settling owing to gravity [20].

Pathogenesis starts in the lungs as soon as the particles are ingested because host epithelial cells are directly invaded. This results in a localized inflammatory response that is mediated by several cytokine pathways. It has been shown that the binding of the S protein with ACE2 initiates a signaling pathway that dramatically increases the production of the cytokine CCL2, which attracts T cells, monocytes, and basophils [21]. Another crucial mechanism includes the presentation of intracellular antigenic peptides to cytotoxic T cells by the major histocompatibility complex (MHC). These immune triggers are also in charge of inducing a systemic inflammatory response that can range from a typical immune response to a cytokine storm, or the unchecked production and movement of pro-inflammatory cytokines that can harm other organ systems and result in superinfection, acute respiratory disease syndrome (ARDS), cardiac injury, stroke, and multi-organ dysfunction [22].

The Chinese Center for Disease Control (CDC) released the biggest clinical report on COVID-19 (72,314 patients), which was described in the Journal of the American Medical Association in February 2020. According to clinical presentation, patients were divided into three groups: moderate (81%), severe (14%), and critical infection (5%) [23]. For moderate infections, symptoms included headache, diarrhea, and cough. For those who were gravely infected, symptoms included ventilator-dependent respiratory failure, sepsis, and multi-organ dysfunction (MOD). For light and severe cases, there were no documented deaths, but in critical infections, the case fatality rate was close to 50%. These figures are comparable to those from smaller American research. Treatment for these individuals are focused secondary injury management and supportive respiratory care [24].

Acute respiratory failure that progresses to severe stages requires more invasive therapies. In actual practice, blood oxygen saturation (SpO2), a measurement of the percentage of oxygenated hemoglobin, is assessed using a pulse oximeter to monitor respiratory status with a goal saturation of >90%. The use of a low flow nasal cannula is a component of the most fundamental oxygen therapy. High flow nasal oxygen (HFNO) and noninvasive ventilation (NIV) therapies are the next stage. These more recent treatments have the potential to reduce intensive care unit (ICU) death rates and duration of stay. HFNO uses a specific nasal cannula to supply large amounts of oxygen, and because of the high flow, it also enhances breathing physiology. Without the requirement for intubation, NIV employs an oro-nasal mask to provide mechanical breathing and oxygenation [25].

However, intubation with invasive mechanical ventilation becomes essential if a patient remains hypoxic after receiving these treatments. Extracorporeal membrane blood oxygenation (ECMO) is a last resort rescue treatment if a patient remains hypoxic despite receiving the most ventilatory assistance possible. Nearly all of these critically sick individuals have ARDS, which substantially impairs oxygen transport across their lung alveolar membranes. By employing a machine to drain, exchange gas on, and then recirculate oxygenated blood back to the patient, ECMO bypasses the lungs and gives the patient time to recuperate [26]. All of these oxygen treatments, despite their need, pose a risk to medical personnel, with virus particle aerosolization during intubation being the greatest danger.

4 BIOMATERIAL OVERVIEW

As of right now, biomaterials are those that interact with biological systems to assess, heal, or replace any tissue or bodily function. According to Hussein et al. [27], a biomaterial may stimulate an appropriate response from the host in a given setting, thanks to its biocompatibility. Depending on the performance or function needed, several definitions of biocompatibility might be used. In a study by Chen and Thouas [28], biocompatibility was assessed by its capacity to cause tumor development, genetic damage, immunological response, or blood clot formation in cells or live tissues. The Food and Drug Administration (FDA) stated that a substance must not damage the patient in order to be deemed biocompatible, taking into consideration the variety of problems that have been raised. As a result, a medical device's biocompatibility must take into account both the biological compatibility of the materials employed and the design of the device, such as its geometry, electric control, and mechanical performance [28].

Biomaterials may be categorized as follows, while generalizing and attempting to place them historically [29]:

- 1960–1970: designated as inert biomaterials, these were the first generation, whose objective was to replace damaged tissue and provide structural support with minimal impact on the patient;
- 1980–1990: still used currently in numerous commercial products, the bioactive biomaterials have made it possible, through the use of coatings, a biological reaction of chemical nature in the contact area between the material and the host with an increase in the effectiveness of the medical device;
- 2000–2010: in order to overcome the setbacks caused by bioactive biomaterials, generally associated with infections caused by toxic or immunological processes, biodegradable biomaterials were developed with the ability to degrade and be absorbed by the patient, thereby minimizing or avoiding the disadvantages of permanent implants and avoiding a new surgery or intervention to remove the device with foreseeable consequences for the patient;

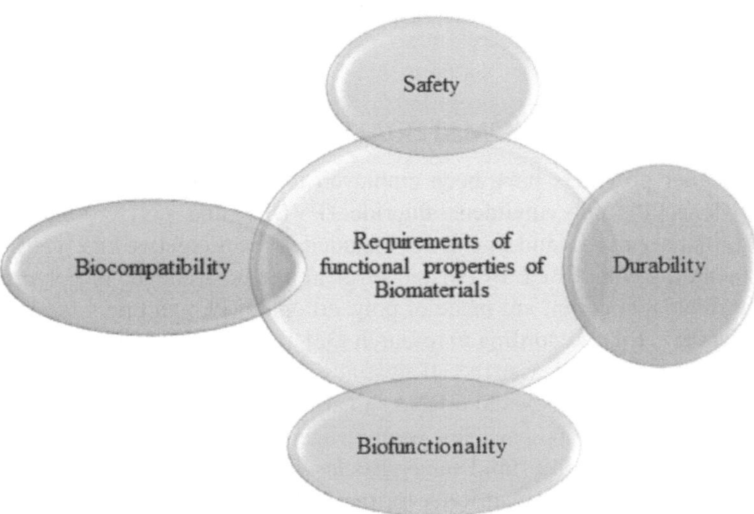

FIGURE 11.1   Requirements of functional properties of biomaterials.

- 2010–now: the fourth generation named as smart biomaterials that have been made possible with the advances in different fields, has the characteristic of being able to emulate the natural structures and mechanisms, repairing and regenerating the damaged tissues by promoting specific reactions of the cells. This property is also known as "biomimetic biomaterials." They may be temporary devices, like plates or screws, or they might develop a permanent presence, like prostheses. Although they are the most recent generation of biomaterials and offer certain benefits over others, their applications are still quite limited.

Despite the progress of biomaterials, first-generation biomaterials are still often and frequently used in modern applications. The functional properties of a biomaterial must be in compliance with certain requirements, some of which were outlined (Figure 11.1) by Reinwald et al. [29]. These include safety, which is regarded as the most crucial aspect of a medical device, durability to reduce the number of surgical interventions, biocompatibility, where the host must tolerate the imposed material regardless of the length of contact, and bio functionality, where the suitability of the intervention must be guaranteed to prevent adverse effects.

Toxicology still allows for the differentiation of biomaterials based on the kind of reaction as toxic when adjacent and surrounding tissues are destroyed, as nontoxic and biologically inert when fibrous tissue of varying thickness develops, as nontoxic and biologically active when bonding forms in the interface zone, and as nontoxic and biodegradable when the involving tissue takes the place of the implant. Metal, ceramics, polymers, and composites are the four primary materials utilized in the manufacture of medical equipment. The choice of material for each application must take into account a variety of factors, including cost, availability, performance during processing and use, and natural properties of the material, such as biocompatibility and corrosion resistance. These factors also include mechanical and metallurgical properties.

5 IMPORTANCE OF BIOMATERIALS FOR COVID-19

Recently, other polymers have been employed to create protective kits, including polypropylene(PP), polyvinylidene fluoride (PVDF), and polytetrafluoroethylene (PTFE) [30]. Face masks and shields are included in the protective kits. The approved masks, including N95, FFP2, FFP3, and surgical masks, are made of polypropylene. The face shields, however, are made of polycarbonate (PC) and poly (ethylene terephthalate) (PET) [31]. According to research [32], the use of nano biomaterials in the drug delivery system increases the effectiveness of antiviral medications.

For the transport of DNA and mRNA, a variety of organic nanoparticles, including lipid and polymer nanoparticles, were utilized. Due to their spherical form and improved biocompatibility, lipid nanoparticles like liposomes are ideal for drug administration [33]. Recently, mRNA for the SARS-CoV-2 spike protein was delivered using negatively charged ionized liposomes. Polyethyleneimine (PEI) with modified fat chains and poly (b-amino) esters (PABEs) were modified among polymeric nanoparticles to carry DNA and mRNA. Another research claimed to have employed chitosan nanoparticles to deliver antiviral medications to the sick individuals. Novochizol, which contains chitosan nanoparticles, is being utilized to treat diseased regions, particularly lung infections brought on by COVID-19 [34]. Due to its effectiveness against bacteria and viruses, graphene has recently been acknowledged as a substitute material for the treatment of COVID-19 infection [35].

6 APPLICATIONS OF BIOMATERIALS AGAINST COVID-19

The use of biomaterials in COVID-19 diagnosis and treatment has been investigated in a variety of forms and sources (natural or synthetic). According to Talebian et al. [36], the majority of research on the utilization of biomaterials focused on the creation of fast, ultra-sensitive field-effect transistor (FET)-based biosensing, antiviral platforms, vaccinations, and nano-based materials. In the past, fungus and bacteria have been the focus of the majority of research into using biomaterials to treat illnesses. More virus-related research has to be done as a result of the MERS, influenza, and new coronaviruses becoming more common [1]. In the most recent COVID-19 pandemic, biosafety issues such as insufficient protective gear, a lack of transportation equipment for infected patients, and difficulties in making a quick viral diagnosis have come to light as significant obstacles. In this context, a broad range of materials with peculiar properties have been explored in order to address the problem of biosafety concerns.

6.1 PROTECTION (BIOSAFETY) MATERIALS

The biosafety material idea has recently been proposed for building materials that contain medication delivery systems, COVID-19 detection, and disinfection [37]. Safety biomaterials are used for protective gear in the COVID-19 pandemic, including

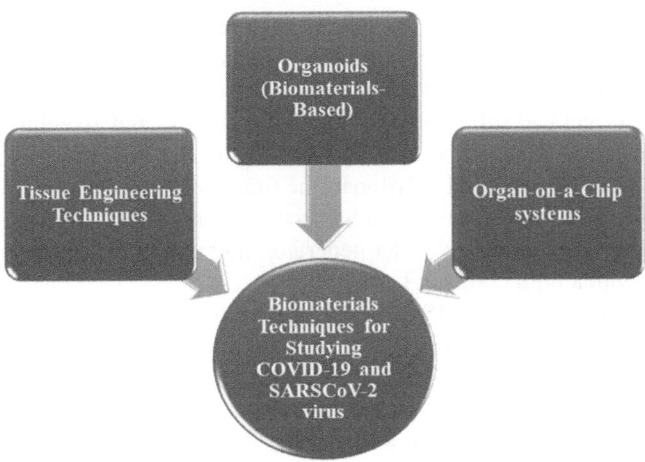

FIGURE 11.2 Biomaterials techniques for studying COVID-19 and SARS-CoV-2 virus.

garments, masks, and goggles. These items are thought to be crucial instruments in the fight against the virus. Shields and face masks have been made from a variety of polymers, including polypropylene, polyvinylidene fluoride (PVDF), and polytetra-fluoroethylene (PTFE). Additionally, polypropylene is employed in the production of surgical, FFP2, FFP3, and other types of masks [38].

6.2 BIOMATERIAL TECHNIQUES FOR STUDYING COVID-19 AND SARS-CoV-2 VIRUS

Figure 11.2 shows the biomaterials techniques for studying COVID-19 and SARS-CoV-2 virus.

6.2.1 Tissue Engineering Techniques

According to Langer and Vacanti [39], tissue engineering applies engineering ideas to biological applications, such as the study of COVID-19. Relevant tissue engineering research for coronavirus therapy may be classified as in vitro models to assess the effectiveness of suggested treatments, drug delivery strategies, and vaccinations. Among various tissue engineering techniques, 3D bioprinting, organoid engineering, and microfluidic organ-on-a-chip (OoC) systems are accepted to be the best approaches to design and develop an effective in vitro tissue models [40].

While many different cell types may be employed for tissue engineering, stem cells are a key source. Because they may develop into several cell types, pluripotent stem cells in particular are very important. Personalized disease models may now be created using induced pluripotent stem cells (iPSCs), which were recently produced. iPSCs can also be used to study the pathogenesis of COVID-19 [41]. For instance, to evaluate the severity and infectivity of the SARS-CoV-2 infection, 3D human lung organoid models that were generated from iPSCs were differentiated into the endodermal lung tissue [42].

Alginate, Matrigel, collagen, gelatin, polydopamine, and hydrogels are among the biomaterials used in the creation of lung models employing pluripotent stem cells: for instance, utilizing human iPSCs (hiPSCs) of lung bud organoids that were then enclosed in Matrigel and cultured. The lung disease model was then infected with a virus after Matrigel had solidified and branching airways and alveolar structure had formed. A customized lung disease model was created by Wilkinson et al. [43] using iPSCs grown on collagen I and polydopamine functionalized alginate beads to investigate the illness mechanism and conduct medication testing. Alginate beads were created using an electrostatic droplet generator, and they were afterward immersed in a solution with a high collagen I content.

Mesenchymal stem cell (MSC)-based approaches have been applied in recent research to treat viral infection [44]. Certain cytokines found in MSCs may be effective in treating COVID-19 patients. According to Harikrishnan and Krishnan [45], MSCs may assist in the regeneration of injured tissues and guard against cytokine storms. MSCs are ACE2-negative at the gene level and, as a result of their immunomodulatory effects, anti-inflammatory properties [46], and regenerative abilities [47]. Additionally, viral infections may cause acute lung damage and inflammation; MSCs might lessen these effects. In order to cure COVID-19, MSCs may thus be a beneficial technique [48].

The presence of antigens as well as certain components that might activate the immune system should be taken into consideration for the creation of a vaccination platform against pathogenic agents in order for it to be stored in memory cells [49]. Since the size, shape, and physico-chemical characteristics of biomaterials can affect the behavior of immune cells [50], researchers may use tissue engineering and biomaterials to stimulate host immune responses to create vaccination strategies. Such technologies are used to target antigen-presenting cells and trigger certain immune responses [49].

To research viral infection, a number of tissue-engineered in vitro human lung models have been developed. But these models need to be improved, and the 3D scaffolds need to be made to resemble real pulmonary architecture. Additionally, extracellular matrix (ECM) must be produced using culture techniques prior to viral injection [49]. To evaluate the efficacy of medications, antiviral medicines, and vaccines, tissue engineering may be employed to better understand COVID-19 disease processes [51]. It is necessary to expose a model tissue or organ to the virus in order to analyze the molecular mechanisms behind COVID-19 pathogenesis and therapeutic approaches.

Due to their natural tissue biomimicking qualities, using 3D bioprinting technique and cell self-organization to make spheroids and organoids [40], 3D-engineered tissues may either be (1) scaffold-based or (2) scaffold-free. As a result, tissue engineering techniques may be effective tools for examining SARS-CoV-2 infection and host–pathogen interactions. They call for a multidisciplinary approach and provide the chance to create fresh concepts and procedures to improve our understanding, find new therapeutic targets and medications.

6.2.2 Organoids (Biomaterial-Based)

Organoids are collections of cells with a particular organ focus, including bladder, stomach, liver, intestine, and liver cells [52]. They have represented organ characteristics and are generated from progenitor or stem cells [52]. In order to study viral infection and potential interactions between the host and pathogenic agents, cell self-organization methods or 3D organoid cultures are useful models. SARS-CoV-2 infection has previously been studied using organoids. According to the findings [53], human recombinant soluble angiotensin-converting enzyme 2 (hrsACE2) may prevent SARS-CoV-2 infections. In order to facilitate organoid development and culture, hydrogel matrixes were also used. It is possible to produce organoids using a variety of synthetic and natural hydrogels. The mechanical strength of synthetic hydrogels is often greater than that of natural hydrogels, but they are deficient in key bioactive molecules that facilitate cell adhesion and proliferation [54]. Laminin and the arg-gly-asp (RGD) sequence are molecules found in natural hydrogels that act as biological signals and cell-anchoring mechanisms for the formation of organoids. The survival of organoids can also be supported by synthetic hydrogels like polyethylene glycol (PEG), modified PEG using RGD and laminin, PEG-based hydrogels, alginate, hyaluronic acid, collagen, gelatin, and fibrin [55].

To investigate and comprehend COVID-19, 3D organoids made from bronchial transient epithelial cells, which are often targeted by SARS-CoV-2, might be beneficial. A full adult stem cell-derived lung organoid model was produced by Tindle et al. [56]. Adult lung organoids, primary airway cells, and alveolar type-II (AT2) pneumocytes produced from human-induced pluripotent stem cells (hiPSCs) were used to create the in vitro model. Organoids were infected with COVID-19 patient-derived respiratory samples while they were fixed on a Matrigel surface. To assess how medications affect lung fibrosis, researchers published a model of pulmonary fibrosis. Collagen hydrogels, lung matrixes made from decellularized fibroblasts, lung spheroids and organoids, and precisely cut lung slices were used to create 3D in vitro tissue models [57].

In a model made from organoids generated from human airway cells, Mykytyn et al. [58] showed that the medication camostat mesylate inhibited serine protease activity, preventing SARS-CoV-2 access to cells. To test the drug's impact on SARS-CoV-2, they implanted human airway cells obtained from organoids onto collagen-coated transwell. To examine human parainfluenza virus infection using an in vitro model, lung organoids were seeded on Matrigel. As a result, lung organoids produced from stem cells may serve as an effective model for researching viral infection and host–pathogen interactions in the lung. Giobbe et al. [59] looked at the stomach's sensitivity to SARS-CoV-2 infection using a gastric organoids model in a different investigation. In order to explore the viral infection of the stomach, gastric organoids were embedded in Matrigel and infected with viral suspension. For the purpose of growing the virus and researching stomach SARS-CoV-2 infection, this gastric organoid system offers a special microenvironment.

6.2.3 OoC Systems

Due to the expression of various receptors in animals compared with humans, it is difficult to evaluate enterovirus infection using animal models. Additionally, the utilization of static cell culture techniques does not replicate the in vivo environment's inherent complexity. In order to better predict clinical outcomes, evaluating new therapeutic molecules and compounds utilizing microfluidic OoC devices might provide useful findings [60]. OoC systems are capable of accurately simulating human organ physiology, disease states, and pharmacological treatment responses [61]. Additionally, OoC and microfluidic-based 3D models provide biomimetic environments with air and fluid fluxes similar to those seen in the lung and may be helpful for researching viral infection. Therefore, it is critically necessary to develop OoC-based technologies to investigate how human tissues and organs react to SARS-CoV-2 [62].

In order to investigate health and disease conditions, gut-on-a-chip systems, for instance, are a desirable tool [63]. They may also be used to research SARS-CoV-2infection. The establishment of a SARS-CoV-2 infection in vitro utilizing a biomimetic human intestine-on-a-chip device for researching COVID-19 was recently described by Guo et al. [62]. The intestine-on-a-chip device had two parallel cell culture microchannels: an upper channel lined with human intestinal epithelial cells and cells that secrete mucin, and a lower channel lined with vascular endothelial cells. These channels were separated by a thin, flexible polydimethylsiloxane (PDMS) membrane that was coated with ECM. As a result, they offered a quick and affordable intestine-on-chip infection model for researching how SARS-CoV-2 infection affects people.

In a different application, Zhang et al. [64] developed a biomimetic lung injury-on-a-chip model of SARS-CoV-2-induced illness in humans and assessed immune responses to viral infection. Collagen type I was applied to the chip device and cells were implanted therein. This chip was used to research the inhibition of viral replication by Remdesivir. The most recent developments in tissue engineering for COVID-19 show that MSC therapy, exosomes, organoids, and on-chip disease models may be very helpful for examining SARS-CoV-2 viral infection, host–pathogen interaction, and evaluating medications for COVID-19 treatment.

6.3 BIOMATERIALS FOR THE DIAGNOSIS OF COVID-19 INFECTION

To assist curb the spread of the COVID-19 illness, researchers have been attempting to create quick diagnostic methods. Real-time reverse-transcription polymerase chain reaction (RT-PCR) has been used extensively as a reliable approach for screening infected individuals for the identification of target sequences linked to SARS-CoV-2. The chest computed tomography (CT) scan is another method for a quick and accurate identification of SARS-CoV-2 infection. Small patchy shadows and interstitial alterations in the lung may be signs of infection. According to reports, positive chest CT results were obtained for 97% of patients with positive PCR results and for only 75% of patients with negative PCR results. While PCR is the first-line method to detect individuals at the early stages of illness [65].

The drawbacks of CT scans include the need for high-quality imaging equipment and qualified medical personnel for precise diagnosis. In order to determine the concentrations of IgG and IgM in blood samples from patients who had SARS-CoV-2 infection, immunoassay is another diagnostic method that has been utilized. ELISAs, or enzyme-linked immunosorbent tests, are often used in its administration. Additionally, plasmas containing antibodies may be given to individuals who are already infected. ELISA is a time-consuming method [1]. Furthermore, it might take up to three weeks for high antibody levels to manifest after the onset of the first symptoms, making this approach unsuitable for use as an early diagnostic tool. To solve

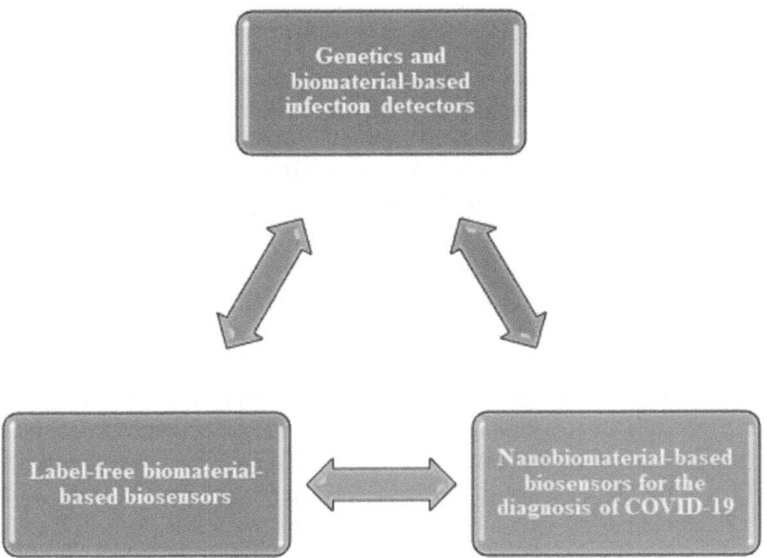

FIGURE 11.3 Biomaterials for the diagnosis of COVID-19 infection.

these problems, scientists have been striving to create a quick, inexpensive, dependable, and portable device [66]. As an alternative, techniques based on biomaterials have shown promise for application in viral detection as shown in Figure 11.3 [66].

6.3.1 Genetics and Biomaterial-Based Infection Detectors

The creation of targeted methodologies for the quick diagnosis of the virus' target RNA/DNA is the goal of genetics-based detection. When exposed, the intended biomaterials and the target sequences interact, resulting in the production of genes and the creation of a signal that can be detected by relatively low-tech equipment. There are already a number of technologies available for this purpose that may be used to determine SARS-CoV-2 infection.

6.3.2 Nanobiomaterial-Based Biosensors for the Diagnosis of COVID-19

The creation of biosensor-based diagnostics, which brings about a useful strategy for an expedient and sensitive diagnosis, may make use of biomaterial-based techniques [1]. In this context, the use of nanoparticles for the magnetic field-assisted separation of RNA or DNA from biological samples may be used as a substitute technique for quick diagnosis. The creation of nucleic acid biosensors, immunoassays for the evaluation of viral antigens, and antibody detection, which allow improved COVID-19 detection in both symptomatic and asymptomatic infected persons, might be considered an additional contribution of the biomaterials sector.

Nanomaterial-based biosensors are based on the utilization of single- or multiple-specific binding for infection diagnosis. The objective while creating a biosensor is to create a fast, inexpensive, portable platform that selectively binds to the target and reports the outcome as a signal. For instance, since they may electrostatically attach to antigens and antibodies and exhibit localized surface plasmon resonance shift, gold nanoparticles (AuNP) are often utilized in diagnostic systems. The Middle East respiratory syndrome coronavirus (MERS-CoV) was identified in one research by Kim et al. [67] utilizing a colorimetric readout bioassay that focused on the optical characteristic of AuNP through salt-induced aggregation [67]. Similar to this, Moitra et al. [68] developed a platform for SARS-CoV-2 naked-eye detection based on the application of AuNP [68].

6.3.3 Label-Free Biomaterial-Based Biosensors

Another platform for the electrical detection of certain DNA or protein sequences is provided by ultra-sensitive, label-free, and quick nanowire field-effect transistor (NW-FET) devices. A platform for the detection of particular oligomers related to the hepatitis C virus (HCV) was presented by Cho et al. [69] before using a sample in this platform, DNA/RNA extraction and amplification should be carried out since the Si NW-FET design has been suggested for identifying target sequences in viruses. Uhm et al. [70] stated that the detection of viral surface protein may be a more sensitive infection diagnostic tool, without the need for extra processes, as a POC biosensor. They used a functionalized SiNW-FET platform to detect the hemagglutinin (HA) surface protein released by the swine flu (H1N1) virus to accomplish this goal [70].

A FET made of coated graphene sheets with immobilized antibodies against the SARS-CoV-2 spike protein was first described by Seo et al. [71]. At concentrations of 100 fg/ml, the suggested device demonstrated a very sensitive platform for selective spike detection. Additionally, it has been shown that biomaterial-based surface biosensors, such as those that use quantum dots (QD) or gold nanoparticles for biosensing, can recognize certain DNA sequences. These techniques may be used to create RNA detection systems based on biomaterials. In this context, RNA toehold detection techniques based on enzymatic reporter and RNA converted to cDNA have been used for the detection of viral RNA in conjunction with ribosome or isothermal amplification. Viral biosensors, such as those based on CRISPR, may provide fast on-site assessment of clinical samples from the COVID-19 viewpoint [72].

The development of a CRISPR-chip and a graphene-based FET recently made it possible to detect nucleic acids quickly and without amplification within 15 minutes. In addition, Mahari et al. [73] created a fluorine-doped tin oxide (FTO) immunosensor based on gold nanoparticles (AuNPs) for the detection of the COVID-19 spike antigen. Within 10–30 seconds, the gadget presented the findings [73]. In applications for antibody detection, surface plasmon resonance (SPR)-based biosensors have also been used. Recombinant SARS-CoV-2 nucleocapsid was used by Djaileb et al. [74] to evaluate anti-SARS-CoV-2 antibodies using an SPR-based biosensor. This biosensor is a label-free, quick test for SARS-CoV-2 antibodies based on human serum; results appear within 15 minutes [74].

6.4 Biomaterials for Developing Prophylaxis and Treatment for COVID-19

Several medications have been shown to shorten patients with COVID-19's hospital and critical care unit stays as well as their requirement for intubation and ventilation.

Different medications may be given, depending on the disease's stage. Patients often get supportive therapy in order to maintain a healthy calorie intake, water consumption, and electrolyte balance. The use of therapeutic techniques including immunotherapy, oxygen therapy, antiviral therapy, and organ support may prevent and manage organ failure, acute respiratory distress syndrome, cytokine storms, and secondary hospital infections.

6.4.1 Therapeutics

Antiviral medications, anti-inflammatory medications, and monoclonal antibodies are the most often prescribed medications for COVID-19 patients. The antiviral medications work in a variety of ways, including by targeting both viral and cellular proteins and by boosting the body's immunological response to the viral infection [75]. Some of the clinically utilized medications include Ribavirin [76], Lopinavir/Ritonavir, Remdesivir, Favipiravir, Sofosbuvir, Dexamethasone, and Chloroquine [77]. According to Cojocaru et al. [78], these medications may be purchased over-the-counter as pills, capsules, oral suspensions, ocular drops, topical creams, intravenous injections, or nasal ointments.

Unfortunately, the majority of antiviral medications have poor bioavailability and short circulation times, making them unsuitable for oral or intravenous delivery. Furthermore, transport of antibodies, cytokines, and other protein-based medicines to target areas is often hampered by their short half-lives. The inadequacies of the present treatment platforms may be addressed in a number of ways with the help of biomaterials. Different biomaterials are used as drug carriers in viral diseases in the form of micro- or nanoparticles, or capsules [79].For instance, advances in biomaterials for drug delivery will enable the development of new approaches. In order to overcome the difficulties associated with administering newly discovered therapies, biomaterial-based drug delivery devices may be helpful [80].

It's highly promising to use medication carriers that can specifically target the lung. Additionally, systemically active medications may be easily administered via inhalation because of the large surface area of the alveoli. Additionally, individual antibodies may be functionalized into lung carriers that contain medicines or vaccinations to target certain cell types. As a result, there is a lower chance of adverse medication reactions. Examples of biomaterial-based drug delivery systems that have been used in the treatment of conditions including lung cancer and influenza may be found in the literature [81]. For the creation of pulmonary therapeutic molecule delivery systems, natural biodegradable polymers, such as cyclodextrin, albumin, alginate, chitosan, gelatin, and collagen as well as synthetic polymers such as polylactide (PLA), polyglycolide (PLGA), polyacrylates, and polyanhydrides are suitable [82].

6.4.2 Vaccines

The development of COVID-19 vaccines via extensive research has been effective, and efforts to find a cure are currently underway. Understanding the etiology of COVID-19 and the impact of the SARS-CoV-2 virus on infected individuals is crucial for achieving further success. Utilizing nanoparticles, many techniques to suppress viral surface receptors have been proposed. For instance, mesoporous silica rods (MSRs) were used in the fabrication of a biomaterial vaccine by Langellotto et al.

[83]. It was determined that the developed MSR vaccine may result in significant protection against SARS-CoV-2 when its effects were assessed in vivo [83]. Another example is the utilization of mRNA vaccines that are encased in lipid nanoparticles (LNPs) [84].

Until now, several techniques have been used in vaccination trials. Utilizing technical advancements has reduced the amount of time needed for vaccine production, preclinical testing, and clinical applications. Recombinant proteins and viral vectors are being used instead of traditional methods, which involve live attenuated viruses or vaccines conjugated with proteins. A design with the CoV spike receptor-binding domain (RBD) was reported by Dai et al. [85] for use in several beta-coronavirus vaccines, including COVID-19. Here, MERS-CoV was used to create the RBD dimer design.

Mice infected with MERS-CoV (RBD) study revealed an anti-infection effect, neutralizing antibody (NAb) titers also showed a considerable rise [85]. Future research should focus on determining how immunopathogenic COVID-19 is, particularly by doing characterization studies to aid in the creation of COVID-19 vaccines. Preclinical and clinical research should continue to be done to create new vaccines. Two businesses have reportedly succeeded in creating an mRNA-based vaccine that codes for essential components of the SARS-CoV-2 protein and stimulates the immune system.

7 DRAWBACKS OF BIOMATERIALS APPLIED FOR COVID-19 DIAGNOSIS

The pandemic led to the introduction of various laboratory-based diagnostic methods. The drawbacks of these diagnostic approaches include longer turnaround times and the need for equipped instruments. The availability of self-testing kits, similar to those for home pregnancy tests, will be intriguing. Such kits have only lately been released [86]. As a result, the illness may be rapidly identified and quarantined. The COVID-19 conflict has also resulted in travel restrictions. As a result, creating a POC platform with precise SARS-CoV-2 detection at the airports would enable individuals to resume their usual lives. Biomedical researchers have so far attempted to process a biological cascade using engineering methods in order to identify viruses. By identifying certain volatile organic chemicals in exhaled air, [87] proposed a multiplexed biosensor with good accuracy for the detection of SARS-CoV-2 infection [87].

8 ADVANTAGES OF BIOMATERIALS USED FOR COVID-19 TREATMENT

Following the COVID-19 crisis, scientists have been studying the virus in terms of genome sequencing and surface proteins, enabling them to plan the creation of effective therapeutics with a wide range of activity. Immunomodulatory medicines, viral protease inhibitors, and antiviral medications, including Ribavirin, Remdesivir, and Chloroquine, have all been studied on infected individuals so far, and some of the findings were encouraging. Due to their brief half-lives, several specialized medicines cannot be administered by traditional oral or intravenous methods [1].

Due to their high hydrophobicity and limited stability, many newly introduced small molecules circulate for brief periods of time. Additionally, certain medicines

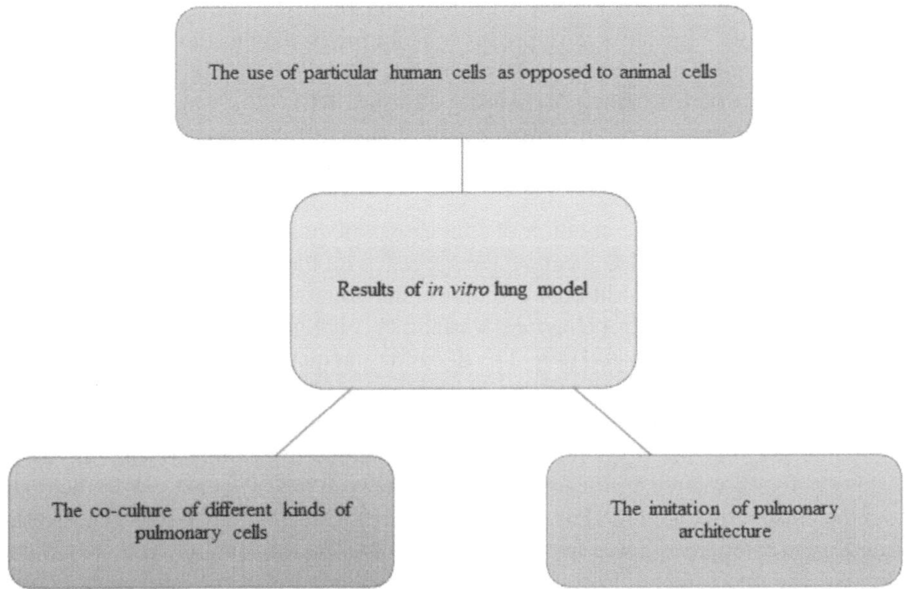

FIGURE 11.4   Results of creation of a trustworthy in vitro lung model for SARS-CoV-2.

have been shown to have a beneficial impact on SARS-CoV-2 infection, but their clinical use has failed as a result of unfavorable off-target effects. In light of this, the development of tailored drug delivery systems based on biomaterials in conjunction with sustained hydrophobic drug delivery and the distribution of efficient agents that take into account viral receptors might be an intriguing strategy [88].

For the purpose of researching SARS-CoV-2, human organoid systems for the small intestine, gut, kidney, and lung were created. The multicellularity of organs like the lung is a formidable barrier for developing a trustworthy and biomimetic in vitro model. The creation of a trustworthy in vitro lung model for SARS-CoV-2 investigations enables the following results as shown in Figure 11.4. All of these strategies depend on the close collaboration of biomedical engineers and medical scientists to be successful [49].

Some biodegradable biomaterials, including chitosan, MSRs, and PLGA nanoparticles, have been used as vaccine platforms and can be used to develop a SARS-CoV-2 vaccine. Notably, the suggested platform's size, shape, and other physicochemical characteristics should be carefully planned to accomplish the desired effects on the immune system. The majority of the newly developed preparations, however, are still in the preclinical testing stage and have not yet entered the clinical setting [49].

9   CURRENT CHALLENGES AND FUTURE PROSPECTS

Biomaterials engineers have been working hard to provide novel techniques for disease identification and treatment since the COVID-19 pandemic. For example, scientists created 3D-printed nose swabs for collecting samples. Some nanoparticle-based formulations still have difficulties with biocompatibility, however. The high viral mutation rate, which results in improper functioning of diagnostic, therapeutic, and vaccination targets, may be the most significant obstacle [89]. For instance, the use of RBD-based vaccines may not be adequate since RBD is a changeable sequence in the CoV genome. In order to reduce the danger of adverse effects, substantial research and studies are also necessary before scaling up biomaterial-based constructions [89].

On the other hand, the creation of innovative medicines and gadgets as well as the design of resilient biological systems need a multidisciplinary approach. The SARS-CoV-2 virus has additional difficulties due to its variable behavior in various hosts. In order to quantify behavioral variability, highly biomimetic organoids and OoC systems must be created [89]. The dynamic interactions between the virus and the host tissue/organ components should be modeled in in vitro experiments.

Alveolospheres, which have recently been developed by researchers, however, indicate that the difficulties in creating reliable organoid models still exist. Additionally, there are significant differences between how organs work in the body and the present organoid models. Organoids and OoC technology have been investigated by researchers as a way to close these gaps and more accurately imitate natural organ shape and function [90]. Different medication delivery methods must be utilized depending on the SARS-CoV-2 virus infection stage.

A thorough knowledge of the pathophysiology of COVID-19 is necessary for the effective implementation of biomaterial-based medication delivery systems since various extrapulmonary issues arise as a result of viral disease. Additionally, this would increase the therapeutic impact and lessen the possibility of negative side effects. The primary targeting of receptors used by the virus to enter cells, such as ACE-2 and TMPRSS2, is a crucial concern since it might interfere with SARS-CoV-2's interaction with receptors [90]. Additionally, the interaction between the S protein and the proteoglycans prevalent in human lung tissue, which include heparin sulfate chains, provides a possible target for treatment with an antithrombotic drug like heparin. Additionally, it is believed that a combined delivery method can help individuals with acute symptoms. Overall, it can be said that a variety of factors are increasing the use of biomaterials as a viable alternative for detecting, preventing, and treating SARS-CoV-2 infection. Integration of several fields is necessary to advance our understanding of COVID-19, forecast viral activity, and create efficient drugs and vaccines [90].

10   CONCLUSION

The field of biomaterials, which is currently undergoing rapid development, holds significant potential for enhancing our capacity to prevent and treat perilous viral diseases such as COVID-19. Prominent advancements in the field of biomaterials research have been observed in various areas, such as the development of biosensors for nucleic acids (RNA and DNA) and proteins (antigens and antibodies), as well as the utilization of biomaterials for vaccine and antiviral delivery. Furthermore, the application of biomaterials in the manufacture of organoids has proven valuable in studying the intricacies of SARS-CoV-2 infection, its associated issues, and its subsequent consequences. The proliferation of innovative and intelligent materials, including CRISPR-based biosensors, FET-based transistors, protein-based vaccinations, metallic and self-assembled nanovaccines, and organoid technologies, still has a lot of potential applications. The creation of extremely high-precision, lab-independent POC biosensors is necessary for diagnostic systems. Although there has been significant advancement in the use of biomaterials for the creation of diagnostic biosensors and vaccines, more thorough research is still required in order to create therapies for the management of COVID-19. It is difficult to design tailored antiviral delivery systems for viruses since they have poor therapeutic targets and undergo repeated alterations. Thus, the creation of an intelligent antiviral system based on biomaterials will have a big effect. Future research on the COVID-19 infection and drug testing are expected to benefit greatly from the development of OoC technologies. However, the development of innovative biomaterial-based techniques for the prevention, treatment, and monitoring of COVID-19 requires the collaboration of several disciplines.

REFERENCES

[1] Chakhalian, D., R.B. Shultz, C.E. Miles, and J. Kohn. 2020. Opportunities for biomaterials to address the challenges of COVID-19. Journal of Biomedical Materials Research A 108(10):1974–1990. https://doi.org/10.1002/jbm.a.37059

[2] Ogidi, O.I. 2022. Omicron (B.1.1.529): Variant of concern - A mini review. Biomedical Journal of Scientific & Technical Research 41(5):33106–33112. https://doi.org/10.26717/BJSTR.2022.41.006672

[3] Ogidi, O.I., W.L. Berefagha, and E. Okara. 2021. Covid-19 vaccination: The pros and cons. World Journal of Biology Pharmacy and Health Sciences 7(1):15–22. https://doi.org/10.30574/wjbphs.2021.7.1.0072

[4] Ogidi, O.I. 2020. A review on the use of herbal remedies and clinical therapeutics for the management of covid-19 pandemic. ASIO Journal of Pharmaceutical & Herbal Medicines Research 6(2):68–77. http://doi-ds.org/doilink/10.2020-85722646/

[5] Vashist, S.K. 2020. In vitro diagnostic assays for COVID-19: Recent advances and emerging trends. Diagnostics (Basel, Switzerland) 10(4):202. https://doi.org/10.3390/diagnostics10040202

[6] Uludag, H., A. Pandit, and L. Kuhn. 2020. Editorial: Enabling biomaterials for new biomedical technologies and clinical therapies. Frontiers in Bioengineering and Biotechnology 8(559):1–4. https://doi.org/10.3389/fbioe.2020.00559

[7] Tang, H., Y. Abouleila, L. Si, et al. 2020. Human organs-on-chips for virology. Trends in Microbiology 28(11):934–946. https://doi.org/10.1016/j.tim.2020.06.005

[8] Rogers, Z.J., M. Zeevi, and S.A. Bencherif. 2020. Electroconductive hydrogels for tissue engineering: Current status and future perspectives. Bioelectricity 2(3):279–292. https://doi.org/10.1089/bioe.2020.0025

[9] Gsib,O., C. Egles, and S.A. Bencherif. 2017. Fibrin: An underrated biopolymer for skin tissue engineering. Journal of Molecular Biology and Biotechnology 2:1–3.

[10] Li, J., and D.J. Mooney. 2016. Designing hydrogels for controlled drug delivery. Nature Reviews Materials 1(12):16071. https://doi.org/10.1038/natrevmats.2016.71

[11] Huang, H., C. Fan, M. Li, et al., 2020. COVID-19: A call for physical scienutists and engineers. ACS Nano 14(4):3747–3754. https://doi.org/10.1021/acsnano.0c02618

[12] WHO. 2003. Consensus document on the epidemiology of severe acute respiratory syndrome (SARS): World Health Organization. Retrieved from https://www.who.int/publications/i/item/consensus-document-on-the-epidemiology-of-severe-acute-respiratory-syndrome-(-sars)

[13] WHO. 2019. WHO MERS-CoV global summary and assessment of risk, July 2019 (WHO/MERS/RA/19.1). Geneva, Switzerland: World Health Organization. Retrieved from https://apps.who.int/iris/bitstream/handle/10665/326126/WHO-MERS-RA-19.1-eng.pdf?ua=1

[14] Sanche, S., Y.T. Lin, C. Xu, E. Romero-Severson, N. Hengartner, and R. Ke. 2020. High contagiousness and rapid spread of severe acute respiratory syndrome coronavirus 2. Emerging Infectious Diseases 26(7):1470–1477. https://doi.org/10.3201/eid2607.200282

[15] Metcalf, C.J.E., M. Ferrari, A.L. Graham, and B.T. Grenfell. 2015. Understanding herd immunity. Trends in Immunology 36(12):753–755. https://doi.org/10.1016/j.it.2015.10.004

[16] Wrapp, D., N. Wang, K.S. Corbett, et al. 2020. Cryo-EM structure of the 2019-nCoV spike in the prefusion conformation. Science 367(6483):1260–1263. https://doi.org/10.1126/science.abb2507

[17] Astuti, I., and S. Ysrafil. 2020. Severe acute respiratory syndrome coronavirus 2 (SARS-CoV-2): An overview of viral structure and host response. Diabetes & Metabolic Syndrome 14(4):407–412. https://doi.org/10.1016/j.dsx.2020.04.020

[18] Zumla, A., J.F. Chan, E.I. Azhar, D.S. Hui, and K.Y. Yuen. 2016. Coronaviruses - drug discovery and therapeutic options. Nature Reviews Drug Discovery 15(5):327–347. https://doi.org/10.1038/nrd.2015.37

[19] Huang, C., Y. Wang, X. Li, et al. 2020. Clinical features of patients infected with 2019 novel coronavirus in Wuhan, China. Lancet 395(10223):497–506. https://doi.org/10.1016/S0140-6736(20)30183-5

[20] Anderson, E.L., P.Turnham, J.R. Griffin, and C.C. Clarke. 2020. Consideration of the aerosol transmission for COVID-19 and public health. Risk Analysis 40(5):902–907. https://doi.org/10.1111/risa.13500

[21] Kang, S., W. Peng, Y. Zhu, et al. 2020. Recent progress in understanding 2019 novel coronavirus (SARS-CoV-2) associated with humanrespiratory disease: Detection, mechanisms and treatment. International Journal of Antimicrobial Agents 55(5):105950. https://doi.org/10.1016/j.ijantimicag.2020.105950

[22] Markus, H.S., and M. Brainin. 2020. COVID-19 and stroke-A global World Stroke Organization perspective. International Journal of Stroke 15(4):361–364. https://doi.org/10.1177/1747493020923472

[23] Wu, Z., and J.M. McGoogan. 2020. Characteristics of and important lessons from the coronavirus disease2019 (COVID-19) outbreak in China: Summary of a report of 72314 cases from the Chinese center for disease control and prevention. JAMA 323(13):1239–1242. https://doi.org/10.1001/jama.2020.2648

[24] Aggarwal, S., N. Garcia-Telles, G. Aggarwal, C. Lavie, G. Lippi, and B.M. Henry. 2020. Clinical features, laboratorycharacteristics, and outcomes of patients hospitalized with coronavirus disease 2019 (COVID-19): Early report from the United States. Diagnosis 7(2):91–96. https://doi.org/10.1515/dx-2020-0046

[25] Price, S., S. Singh, S. Ledot, et al. 2020. Respiratory management insevere acute respiratory syndrome coronavirus 2 infection. European Heart Journal Acute Cardiovascular Care 9(3):229–238. https://doi.org/10.1177/2048872620924613

[26] Ramanathan, K., D. Antognini, A. Combes, et al. 2020. Planning and provision of ECMO services for severe ARDS during the COVID-19 pandemic and other outbreaks of emerging infectious diseases. The Lancet Respiratory Medicine 8(5):518–526. https://doi.org/10.1016/S2213-2600(20)30121-1

[27] Hussein, A.H., M.A.H. Gepreel, M.K. Gouda, et al. 2016. Biocompatibility of new Ti-Nb-Ta base alloys. Materials Science and Engineering: C 61:574–578. https://doi.org/10.1016/j.msec.2015.12.071

[28] Chen, Q., and G.A. Thouas. 2015. Metallic implant biomaterials. Materials Science and Engineering: R: Reports 87:1–57. https://doi.org/10.1016/j.mser.2014.10.001

[29] Reinwald, Y., K. Shakesheff, and S. Howdle. 2016. Biomedical devices. Epub ahead of print. https://doi.org/10.1002/9781119267034

[30] Correa, H., and D.G. Correa. 2020. Polymer applications for medical care in the COVID-19 pandemic crisis: Will we still speak Ill of these materials? Frontiers in Materials 7:283. https://doi.org/10.3389/fmats.2020.00283

[31] Piedmont Plastics. 2020. Surgical Face Shields - Polycarbonate and polyester sheets. Retrieved from https://www.piedmontplastics.com/applications/surgical-face-shield

[32] Kumar, N., D. Sood, P.J. van der Spek, H.S. Sharma, and R. Chandra. 2020a. Molecular binding mechanism and pharmacology comparative analysis of noscapine for repurposing against SARS-CoV-2 protease. Journal of Proteome Research 19(11):4678–4689. https://doi.org/10.1021/acs.jproteome.0c00367

[33] Islam, M.A., E.K. Reesor, Y. Xu, H.R. Zope, B.R. Zetter, and J. Shi. 2015. Biomaterials for mRNA delivery. Biomaterials Science 3:1519–1533. https://doi.org/10.1039/c5bm00198f

[34] Zhou, F., T. Yu, R. Du, et al. 2020a. Clinical course and risk factors for mortality of adult inpatients with COVID-19 in Wuhan, China: A retrospective cohort study. Lancet 395(10229):1054–1062. https://doi.org/10.1016/S0140-6736(20)30566-3

[35] Palmieri, V., and M. Papi. 2020. Can graphene take part in the fight against COVID-19? Nano Today 2020(33):100883. https://doi.org/10.1016/j.nantod.2020.100883

[36] Talebian, S., G.G. Wallace, A. Schroeder, F. Stellacci, and J. Conde. 2020. Nanotechnology-based disinfectants and sensors for SARS-CoV-2. Nature Nanotechnology 15(8):618–621. https://doi.org/10.1038/s41565-020-0751-0

[37] Maghdid, H.S., K.Z. Ghafoor, A.S. Sadiq, K. Curran, and K. Rabie 2020. A novel AI-enabled framework to diagnose coronavirus Covid 19 using smartphone embedded sensors: Design study. arXiv preprint arXiv 2003: 07434. https://doi.org/10.48550/arXiv.2003.07434

[38] Cao, P., J. Shi, J. Zhang, et al. 2020. Piezoelectric PVDF membranes for use in anaerobic membrane bioreactor (AnMBR) and their antifouling performance. Journal of Membrane Science 603:118037. https://doi.org/10.1016/j.memsci.2020.118037

[39] Langer, R., and J.P. Vacanti. 1993. Tissue engineering. Science 260(5110):920–926. https://doi.org/10.1126/science.8493529

[40] Shpichka, A., P. Bikmulina, M. Peshkova, et al. 2020. Engineering a model to study viral infections: Bioprinting, microfluidics, and organoids to defeat coronavirus disease 2019 (COVID-19). International Journal of Bioprinting 6(4):302. https://doi.org/10.18063/ijb.v6i4.302

[41] Chrzanowski, W., S.Y. Kim, and L. McClements. 2020. Can Stem Cells Beat COVID-19: Advancing stem cells and extracellular vesicles toward mainstream medicine for lung injuries associated with SARS-CoV-2 infections. Frontiers in Bioengineering and Biotechnology 8:554. https://doi.org/10.3389/fbioe.2020.00554

[42] Bose, B. 2020. Induced Pluripotent Stem Cells (iPSCs) derived 3D human lung organoids from different ethnicities to understand the SARS-CoV2 severity/infectivity percentage. Stem Cell Reviews and Reports 17(1):293–295. https://doi.org/10.1007/s12015-020-09989-2

[43] Wilkinson, D.C., J.A. Alva-Ornelas, J.M. Sucre, et al. 2017. Development of a three-dimensional bioengineering technology to generate lung tissue for personalized disease modeling. Stem Cells Translational Medicine 6(2):622–633. https://doi.org/10.5966/sctm.2016-0192

[44] Leng, Z., R. Zhu, W. Hou, et al. 2020. Transplantation of ACE2-mesenchymal stem cells improves the outcome of patients with COVID-19 pneumonia. Aging and Disease 11(2):216–228. https://doi.org/10.14336/AD.2020.0228

[45] Harikrishnan, P., and A. Krishnan. 2020. Tissue engineering strategies in Covid-19 research. Trends in Biomaterials & Artificial Organs 34:6–17.

[46] Lin, Z., Q. Gao, F. Qian, et al. 2020b. The nucleocapsid protein of SARSCoV-2 abolished pluripotency in human induced pluripotent stem cells. http://doi.org/10.2139/ssrn.3561932

[47] Parekh, K.R., J. Nawroth, A. Pai, S.M. Busch, C.N. Senger, and A.L. Ryan. 2020. Stem cells and lung regeneration. American Journal of Physiology Cell Physiology 319(4):C675–C693. https://doi.org/10.1152/ajpcell.00036.2020

[48] Zayed, M., and K. Iohara. 2020. Immunomodulation and regeneration properties of dental pulp stem cells: A potential therapy to treat coronavirus disease 2019. Cell Transplant 29:0963689720952089. https://doi.org/10.1177/0963689720952089

[49] Tatara, A.M. 2020. Role of tissue engineering in COVID-19 and future viral outbreaks. Tissue Engineering Part A 26(9–10):468–474. https://doi.org/10.1089/ten.TEA.2020.0094

[50] Chen, X., Y. Yan, M. Müllner, et al. 2016. Shape-dependent activation of cytokine secretion by polymer capsules in human monocyte-derived macrophages. Biomacromolecules 17(3):1205–1212. https://doi.org/10.1021/acs.biomac.6b00027

[51] Park, S.J., K.M. Yu, Y.I. Kim, et al. 2020. Antiviral efficacies of FDA-approved drugs against SARS-CoV-2 infection in Ferrets. mbio 11(3):e011142–e011120. https://doi.org/10.1128/mBio.01114-20

[52] He, J., X. Zhang, X. Xia, et al. 2020. Organoid technology for tissue engineering. Journal of Molecular Cell Biology 12(8):569–579. https://doi.org/10.1093/jmcb/mjaa012

[53] Monteil, V., H. Kwon, P. Prado, et al. 2020. Inhibition of SARS-CoV-2 infections in engineered human tissues using clinical-grade soluble human ACE2. Cell 181(4):905–913. https://doi.org/10.1016/j.cell.2020.04.004

[54] Jabbari, E. 2019. Challenges for natural hydrogels in tissue engineering. Gels 5(2):30. https://doi.org/10.3390/gels5020030

[55] Capeling, M.M., M. Czerwinski, and S. Huang. 2019. Nonadhesive alginate hydrogels support growth of pluripotent stem cell-derived intestinal organoids. Stem Cell Reports 12(2):381–394. https://doi.org/10.1016/j.stemcr.2018.12.001

[56] Tindle, C., M. Fuller, A. Fonseca, et al. 2020. Adult stem cell-derived complete lung organoid models emulate lung disease in COVID-19. Medicine Stem Cells and Regenerative Medicine. https://doi.org/10.7554/eLife.66417

[57] Sundarakrishnan, A., Y. Chen, L.D. Black, B.B. Aldridge, and D.L. Kaplan. 2018. Engineered cell and tissue models of pulmonary fibrosis. Advanced Drug Delivery Reviews 129:78–94. https://doi.org/10.1016/j.addr.2017.12.013

[58] Mykytyn, A.Z., T.I. Breugem, S. Riesebosch, et al. 2021. SARS-CoV-2 entry into human airways organoids is serine protease-mediated and facilitates by the multibasic cleavage site. Elife 10:e64508. https://doi.org/10.7554/eLife.64508

[59] Giobbe, G.G., F. Bonfante, E. Zambaiti, et al. 2020. SARSCoV-2 infection and replication in human fetal and pediatric gastric organoids. Nature Communications 12(1):6610. https://doi.org/10.1038/s41467-021-26762-2

[60] Bein, A., W. Shin, S. Jalili-Firoozinezhad, et al. 2018. Microfluidic organ-on-a-chip models of human intestine. Cellular and Molecular Gastroenterology and Hepatology 5(4):659–668. https://doi.org/10.1016/j.jcmgh.2017.12.010

[61] Ashammakhi, N., K. Wesseling-Perry, A. Hasan, E. Elkhammas, and Y.S. Zhang. 2018a. Kidney-on-a-chip: Untapped opportunities. Kidney Interntional 94(6):1073–1086. https://doi.org/10.1016/j.kint.2018.06.034

[62] Guo, Y., R. Luo, Y. Wang, et al. 2020. Modeling SARS-CoV-2 infection in vitro with a human intestine-on-chip device. bioRxiv 9(1):1–22. https://doi.org/10.1101/2020.09.01.277780

[63] Ashammakhi, N., R. Nasiri, N.R. Barros, et al. 2020. Gut-on-achip: Current progress and future opportunities. Biomaterials 255:120196. https://doi.org/10.1016/j.biomaterials.2020.120196

[64] Zhang, F., O.O. Abudayyeh, J.S. Gootenberg, C. Sciences, and L. Mathers. 2020a. A protocol for detection of COVID-19 using CRISPR diagnostics. Bioarchive 20200321:1–8.

[65] Poortahmasebi, V., M. Zandi, S. Soltani, and S.M. Jazayeri. 2020. Clinical performance of RT-PCR and chest CT scan for Covid-19 diagnosis; A systematic review. Advanced Journal of Emergency Medicine 4(2, S):1–7.

[66] Sharifi, M., A. Hasan, S. Haghighat, et al. 2021. Rapid diagnostics of coronavirus disease 2019 in early stages using nanobiosensors: Challenges and opportunities. Talanta 223(P1):121704–121704. https://doi.org/10.1016/j.talanta.2020.121704

[67] Kim, H., M. Park, J. Hwang, J.H. Kim, and D.R. Chung. 2019. Development of label-free colorimetric assay for MERS-CoV Using Gold nanoparticles. ACS Sens 4(5):1306–1312. https://doi.org/10.1021/acssensors.9b00175

[68] Moitra, P., M. Alafeef, K. Dighe, M.B. Frieman, and D. Pan. 2020. Selective naked-eye detection of SARS-CoV-2mediated byN gene targeted antisense oligonucleotide capped plasmonic nanoparticles. ACS Nano 14(6):7617–7627. https://doi.org/10.1021/acsnano.0c03822

[69] Chou,W.C., W.P. Hu, Y.S. Yang, H.W.H. Chan, and W.Y. Chen. 2019. Neutralized chimeric DNA probe for the improvement of GCrich RNA detection specificity on the nanowire field-effect transistor. Scientific Reports 9(1):1–10. https://doi.org/10.1038/s41598-019-47522-9

[70] Uhm, M., J.M. Lee, J. Lee, et al. 2019. Ultrasensitive electrical detection of hemagglutinin for point-of-care detection of influenza virus based on a CMP-NANA probe and top-down processed silicon nanowire field-effect transistors. Sensors(Switzerland) 19(20):4502. https://doi.org/10.3390/s19204502

[71] Seo, G., G. Lee, M.J. Kim, et al. 2020. Rapid detection of COVID-19 causative virus (SARS-CoV-2) in human nasopharyngeal swab specimens using field-effect transistor-based biosensor. ACS Nano 14(9):12257–12258. https://doi.org/10.1021/acsnano.0c02823

[72] Wang, L.S., Y.R. Wang, D.W. Ye, and Q.Q. Liu. 2020a. A review of the 2019 Novel Coronavirus (COVID-19) based on current evidence. International Journal of Antimicrobial Agents 105948. https://doi.org/10.1016/j.ijantimicag.2020.105948

[73] Mahari, S., A. Roberts, D. Shahdeo, and S. Gandhi. 2020. eCovSensultrasensitive novel in-house built printed circuit board based electrochemical device for rapid detection of nCovid-19 antigen, a spike protein domain 1 of SARS-CoV-2. bioRxiv 4(24):059204. https://doi.org/10.1101/2020.04.24.059204

[74] Djaileb, A., B. Charron, M.H. Jodaylami, et al. 2020. A rapid and quantitative serumtest for SARS-CoV-2 antibodies with portable surface plasmon resonance sensing. ChemRxiv Cambridge open engage 2020. https://doi.org/10.26434/chemrxiv.12118914.v1

[75] Huang, J., Y. Jiang, Y. Ren, et al. 2020. Biomaterials and biosensors in intestinal organoid culture, a progress review. Journal of Biomedical Materials Research A 108(7):1501–1508. https://doi.org/10.1002/jbm.a.36921

[76] Elfiky, A.A. 2020. Anti-HCV, nucleotide inhibitors, repurposing against COVID-19. Life Sciences 117477. https://doi.org/10.1016/j.lfs.2020.117477

[77] Group, R.C. 2020. Dexamethasone in hospitalized patients with Covid-19—preliminary report. New England Journal of Medicine 384:693–704. https://doi.org/10.1056/NEJMoa2021436

[78] Cojocaru, F.D., D. Botezat, I. Gardikiotis, et al. 2020. Nanomaterials designed for antiviral drug delivery transport across biological barriers. Pharmaceutics 12(2):171. https://doi.org/10.3390/pharmaceutics12020171

[79] Wang, Y., L. Deng, S.M. Kang, and B. Z. Wang. 2018. Universal influenza vaccines: From viruses to nanoparticles. Expert Review of Vaccines 17(11):967–976. https://doi.org/10.1080/14760584.2018.1541408

[80] Vellingiri, B., K. Jayaramayya, M. Iyer, et al. 2020. COVID-19: A promising cure for the global panic. Science of the Total Environment 138277. https://doi.org/10.1016/j.scitotenv.2020.138277

[81] Thakur, A.K., D.K. Chellappan, K. Dua, M. Mehta, S. Satija, and I. Singh. 2020. Patented therapeutic drug delivery strategies for targeting pulmonary diseases. Expert Opinion on Therapeutic Patents 30(5):375–387. https://doi.org/10.1080/13543776.2020.1741547

[82] Pham, D.T., A. Chokamonsirikun, V. Phattaravorakarn, and W. Tiyaboonchai. 2021. Polymeric micelles for pulmonary drug delivery: A comprehensive review. Journal of Materials Science 56(3):2016–2036.

[83] Langellotto, F., B.T. Seiler, J. Yu, et al. 2020. A rapidly adaptable biomaterial vaccine for SARS-CoV-2. bioRxiv 7(7):192203. https://doi.org/10.1101/2020.07.07.192203

[84] Hajj, K.A., J.R. Melamed, N. Chaudhary, et al. 2020. A potent branched-tail lipid nanoparticle enables multiplexed mRNA delivery and gene editing in vivo. Nano Letters 20(7):5167–5175. https://doi.org/10.1021/acs.nanolett.0c00596

[85] Dai, L., T. Zheng, K. Xu, et al. 2020. A universal design of betacoronavirus vaccines against COVID-19, MERS, and SARS. Cell 182(3):722–733. e11 https://doi.org/10.1016/j.cell.2020.06.035

[86] Zhang, T., Y. He, W. Xu, A. Ma, Y. Yang, and K.F. Xu. 2020. Clinical trials for the treatment of Coronavirus disease 2019 (COVID-19): A rapid response to urgent need. Science China Life Science 63(5):774–776. https://doi.org/10.1007/s11427-020-1660-2

[87] Shan, B., Y.Y. Broza, W. Li, et al. 2020. Multiplexed nanomaterial-based sensor array for detection of COVID-19 in exhaled breath. ACS Nano 14(9):12125–12132. https://doi.org/10.1021/acsnano.0c05657

[88] Cho, N.J., and J.S. Glenn. 2020. Materials science approaches in the development of broad-spectrum antiviral therapies. Nature Materials 19(8):813–816. https://doi.org/10.1038/s41563-020-0698-4

[89] Ellah, N.H.A. 2020. Nanomedicine as a promising approach for diagnosis, treatment and prophylaxis against COVID-19. Nanomedicine 15(21):2085–2102. https://doi.org/10.2217/nnm-2020-0247

[90] Park, S.E., A. Georgescu, and D. Huh. 2019. Organoids-on-a-chip. Science 364(6444):960–965. https://doi.org/10.1126/science.aaw7894

12 An Intriguing, Versatile, and Sustainable Biomaterial

Nanocellulose

Shubhangi P. Patil
The Institute of Science, Mumbai

Rajendra R. Tayade
The Institute of Science, Nagpur

Aniruddha B. Patil
Maharshi Dayanand College, Mumbai

1 A NATURAL POLYMER–CELLULOSE

Every ecosystem in this world, including the human ecology, is beautifully shaped by nature. Nature employs the processes of creation and abolition to maintain the balance of each ecosystem. Consequently, luxury, comfort, and safety are by-products of human evolution. It was a remarkable trip for the human species from the Stone Age to the Age of Technology, and now we are here with practically all of our achievements, along with pollution, harmed ecosystems, and global warming [1]. The use of synthetic materials, which linger in the environment after years of use, is one of the primary reasons that nature's delicate equilibrium gets disrupted. Degradation of synthetic polymers is one of the largest issues since it directly or indirectly impacts all ecosystems and disturbs the natural order [2]. As a result, cellulose, a natural polymer, has caught the interest of numerous researchers recently. An abundant, widely available, renewable, and biodegradable natural polymer is cellulose. It is the element that generates the most output on earth. Therefore, it attracts the interest of numerous scholars [3].

1.1 STRUCTURE OF CELLULOSE

$C_6H_{10}O_5n$ is the chemical formula for cellulose, which is an organic molecule. Cellulose is made up of anhydroglucose units (AGUs), a semicrystalline polycarbohydrate that are linked together by 1,4-glycosidic connections as illustrated in Figure 12.1. There are three hydroxyl groups in the structure, with one of them being

DOI: 10.1201/9781003434313-12

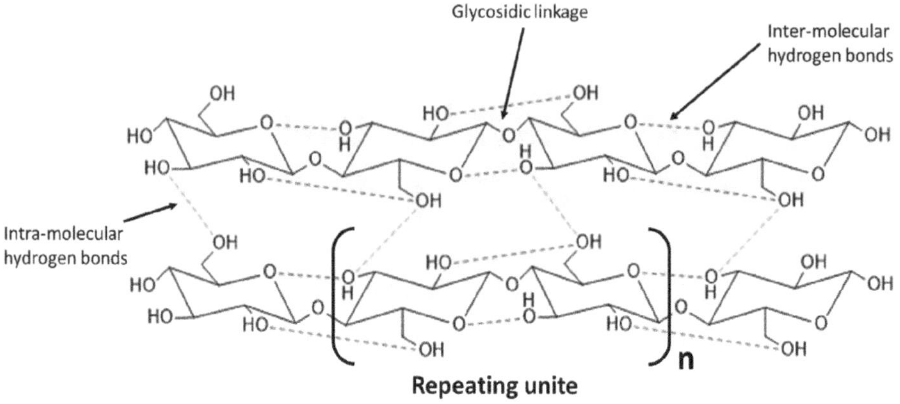

FIGURE 12.1 Chemical structure of cellulose.

a primary alcohol and the other two being secondary alcohols. Both intramolecular and intermolecular hydrogen bonds are caused by these hydroxyl groups. The equatorial orientation of the hydroxyls allows the AGU to create internal hydrogen connections, such as those between a unit's hydrogen atom and an oxygen atom in an adjacent unit's ring. Because of internal hydrogen bonds that prohibit the glucopyranose rings from freely revolving around the chemical glycoside bonds, cellulose chains are more rigid [4]. The hard, powerful, and highly organized cellulose components known as crystallites are not accessible to water or some chemical processes because of a strong network of intra- and intermolecular hydrogen connections. On the other hand, extremely weak hydrogen bonds in noncrystalline amorphous domains in cellulose materials are the reason for the higher hydrophilicity and accessibility of these materials. The structural chemistry of this cellulose makes it a changeable material [4, 5].

1.2 SOURCES

Typically, plant cell walls are composed of cellulose, hemicellulose, and lignin, with lignin making up between 10% and 25% of the total weight and serving as the material that provides the cell wall its strength and stiffness. The latter two components, cellulose and hemicellulose, make approximately 35–50% and 20–35%, respectively, of the material in cell walls [6]. A number of plant-based materials, such as sugarcane bagasse, rice straw, wheat straw, bamboo, rice husk, maize straw, sunflower stalks, cotton stalks, and many more, can be utilized as sources for cellulose extraction, depending on the quantity of cellulose present in a particular material [4, 6, 7]. You can use certain algae, including ulva lactuca, as a feedstock to extract cellulose. The Tunicata subphylum of marine invertebrates, which includes the group of organisms known as tunicates, is another source of cellulose. In terms of molecular weight, mechanical attributes, water-holding capacity, permeability, and thermal stability, tunicate cellulose outperforms plant-based cellulose. Tunicate cellulose has a high level of crystallinity (85–100%) but cannot be regarded as an acceptable source

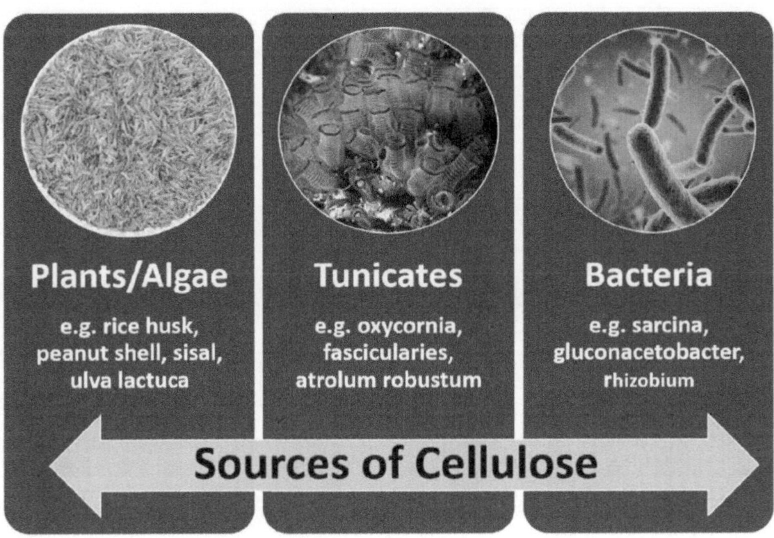

FIGURE 12.2 Various sources of cellulose.

because it comes from animals. In natural surroundings, the majority of bacteria produce extracellular polysaccharides like cellulose that wrap the cells in protective envelopes [8, 9]. Sources of cellulose include a wide variety of bacteria, such as *Acetobacter*, *Sarcina ventriculi*, and *Agrobacterium*. Bacterial cellulose stands out from other sources, thanks to its superior purity, strength, moldability, and water-holding capacity. Hence, broadly plants, tunicates, and bacteria are the primary sources of nanocellulose as shown in Figure 12.2.

1.3 NANOCELLULOSE AND ITS SUSTAINABILITY

Nanocellulose (NC) is a kind of cellulose with a diameter in one dimension of less than 100 nm and a length of a few micrometers. In addition to being nontoxic and biodegradable, this light-weight NC offers unique features due to its nanoform, including a high surface area to volume ratio, higher mechanical strength and stiffness, crystallinity, and changing surface chemistry. As a result, it is developing into a flexible, efficient, and sustainable material that will have a significant impact on the commercial sector [10].

1.4 TYPES OF NANOCELLULOSE

As shown in Figure 12.3 [11], nanocellulose may be classified into three different types: cellulose nanofiber (CNF), cellulose nanocrystal (CNC), and bacterial nanocellulose (BNC), depending on how it was made and its structural features [12–14]. CNF is characterized by a lengthy, threadlike structure that resembles a jigsaw puzzle made up of multiple threads based on Figure 12.3. To stabilize this structure, an intermolecular hydrogen bond is created between the hydroxyl group of one molecule

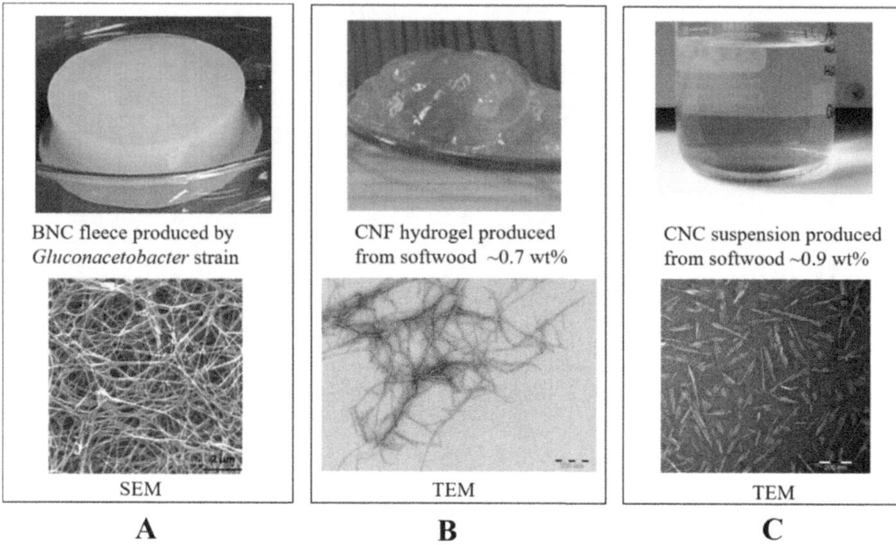

FIGURE 12.3 Classification of nanocellulose A. Bacterial nanocellulose (BNC) B. Cellulose nanofiber (CNF) C. Cellulose nanocrystal (CNC) Open access article [11].

and the oxygen atom of another molecule. As seen in the figure , CNC have a short, highly crystalline structure that resembles a bean with a rectangular cross section [12, 13]. The structure of CNF and CNC is influenced by the source of the material, the duration, and concentration needed for acid hydrolysis, as well as temperature. Top-to-bottom preparation is used for CNF, CNC, and BNC, whereas bottom-to-top preparation is used for CNF, CNC, and BNC [11]. Simple growth techniques are used by bacteria like *Acetobacter* to synthesize BNC, enabling low-cost, ecologically friendly small- and large-scale manufacturing. Because BNC is manufactured as a pure cellulose without the impurities of hemicellulose and lignin present in plants, it possesses remarkable purity, crystallinity, and a high degree of polymerization [15].

2 EXTRACTION OF NANOCELLULOSE

The many sources of nanocellulose have previously been discussed, now let's look at how CNC and CNF are produced utilizing a variety of processes, such as chemical and mechanical treatment for nanocellulose separation.

2.1 CHEMICAL TREATMENT

In this process, the hemicellulose and lignin-containing amorphous areas are eliminated to produce pure cellulose, which is then acid hydrolyzed to produce nanocellulose from cellulose. Rice husk, maize husk, and sugarcane bagasse are just a few of the selected cellulosic feedstocks that are washed in an alkaline solution containing 2–4% NaOH before being treated with acetic acid or HCl [16]. Delignification is the term used to describe this procedure, which eliminates amorphous lignin and hemicellulose.

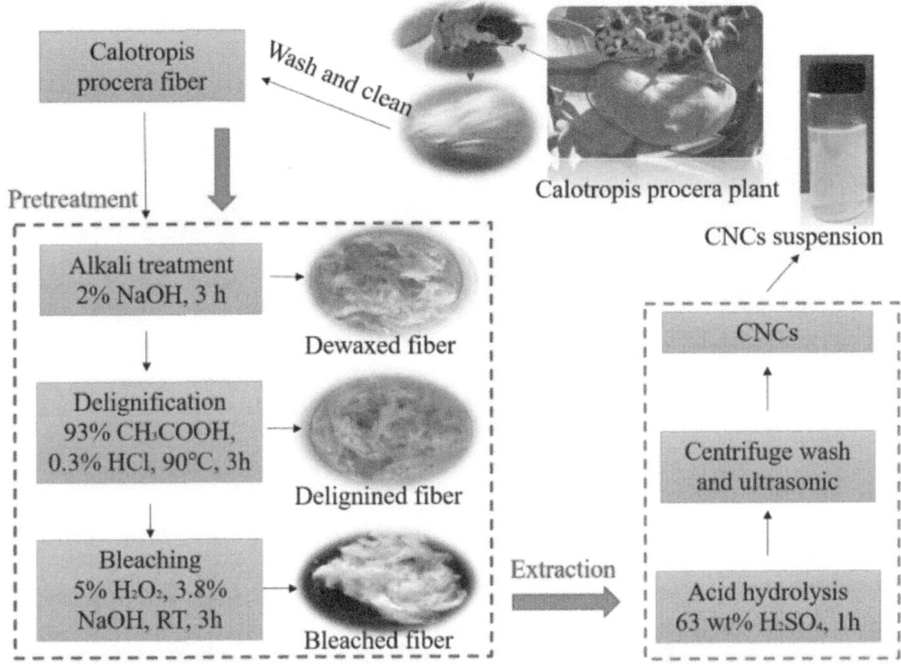

FIGURE 12.4 Chemical treatment for the isolation of nanocellulose. Adapted with permission from [18] Copyright © 2019, Springer Nature.

The material that has been alkali-treated is next bleached by being exposed to 2–6% H_2O_2 or sodium chlorite in an acidic atmosphere for three hours. After eliminating any remaining amorphous regions or other pollutants throughout this procedure, we receive pure, white, and crystalline cellulose. Then, 63 weight% H_2SO_4 is used to perform acid hydrolysis on this crystalline cellulose [6, 17, 18]. During this process, the cellulosic chain disintegrates and by performing a series of ultrasonications and several centrifugation cycles in water, we produce the nanocellulose solution seen in Figure 12.4. CNF can also infrequently be produced by acid hydrolysis; however CNC is often the product. Numerous studies claim that there are numerous variations to the chemical treatment process. Delignification can also be accomplished with the use of other alkali reagents, such as sodium carbonate, potassium hydroxide, calcium hydroxide, or $Ca(OH)_2$. After the acid hydrolysis procedure, further concentrated acids such HCl, HNO_3, and H_3PO_4 may be used. Additionally, it is known to employ maleic acid, oxalic acid, and citric acid for the same function [8, 19, 21].

2.2 MECHANICAL TREATMENT

Through the use of methods including microfluidization, high-pressure homogenization, ultrasonication, and micogrinding, nanocellulose may be mechanically removed from its source. Figure 12.5b illustrates the high-pressure homogenization (HPH) procedure, which was initially utilized to extract NC from wood pulp in 1983. In this

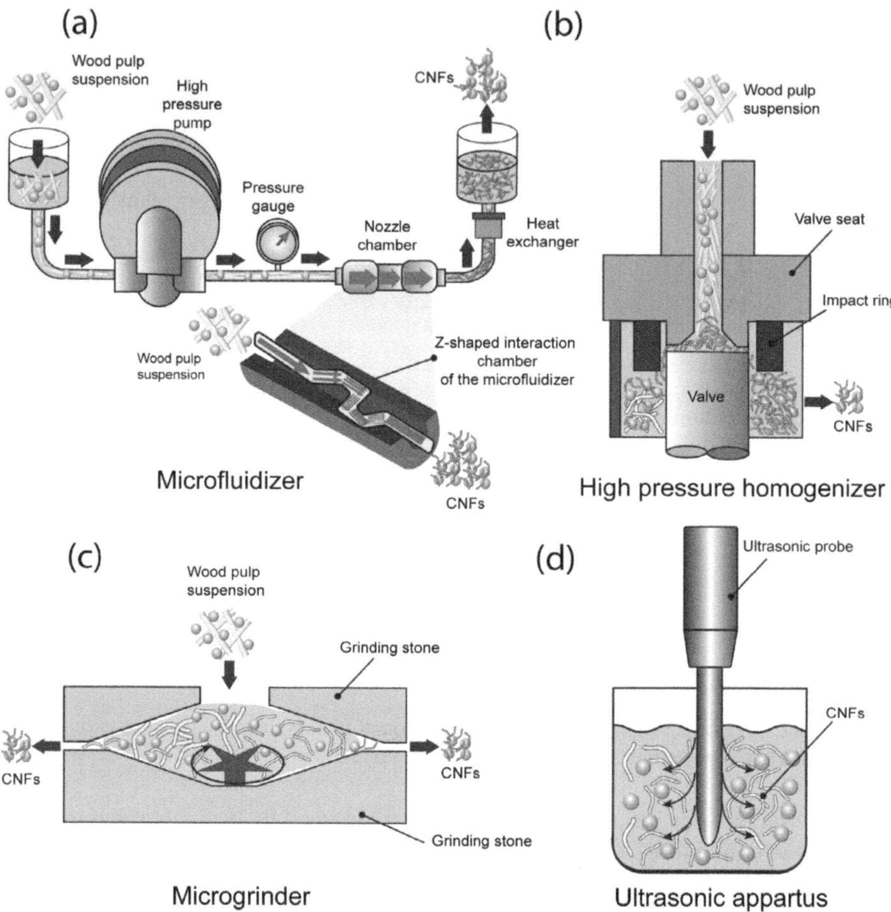

FIGURE 12.5 Different mechanical techniques for the isolation of nanocellulose (a). Microfluidizer (b). High-pressure homogenizer (c). Microgrinder (d) ultrasonic apparatus. Adapted with permission [25] © 2021 Elsevier.

procedure, pure cellulose is driven through a very small nozzle at high pressure [20]. High pressure and nozzle diameter are the two main factors that cause CNF to be generated. The process has the drawback that fibers might clog, thus the remedy is to microscopically cut the fibers before putting them through the HPH. Microfluidizers (Figure 12.5a) perform similarly to HPH for the separation of nanocellulose. The interaction chamber, which defibrillates fibers by imparting shear and impact forces to colliding streams, is used in microfluidizers to enhance pressure. The surface area of the fibers is increased and their size and structure are also affected by further runs through the microfluidizer [3, 20, 21].

High pressure and nozzle diameter are the two main factors that cause CNF to be generated. The process has the drawback that fibers might clog, thus the remedy is to microscopically cut the fibers before putting them through the HPH. Microfluidizers

(Figure 12.5a) perform similarly to HPH for the separation of nanocellulose. The interaction chamber, which defibrillates fibers by imparting shear and impact forces to colliding streams, is used in microfluidizers to enhance pressure. The surface area of the fibers is increased and their size and structure are also affected by further runs through the microfluidizer [2, 3, 20]. A combination of high power, time, and temperature produces the finest fabrillation. The biggest disadvantage of these mechanical procedures is that they consume a lot of energy. It is difficult to apply these technologies for large-scale synthesis as a result.

3 TAILORING OF NANOCELLULOSE

Nanocellulose, as we've already discussed, has special qualities, including high mechanical strength, a high surface area to volume ratio, crystallinity, and variable surface chemistry. Due to its active hydroxyl groups, large surface area, and hydrophilic makeup, nanocellulose is a versatile material whose qualities may be customized to alter in line with the needs of the primary use. This flexible material may be converted into the material needed for a particular application by converting the hydroxyl group into another functional group or essential component [21]. It may be a great reinforcing ingredient for many nanocomposites due to its strength and surface area. A naturally

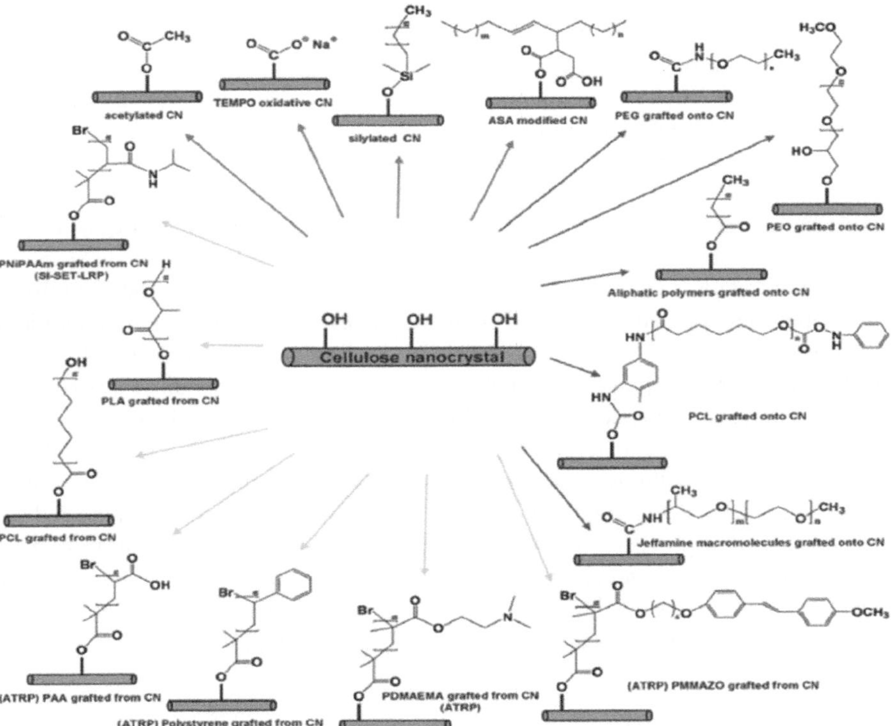

FIGURE 12.6 Tailoring of nanocellulose by different methods. Adapted with permission [21] @ RSC Pub.

occurring polymer called nanocellulose can graft with other polymers to improve its properties [5]. With the hydrophilic or hydrophobic character of the substance being changed in line with the demands of the application, nanocellulose may also be employed as a pickering agent to create stable emulsions or serums [22]. Numerous methods were reported for customizing nanocellulose as shown in Figure 12.6 [18].

3.1 Surface Modification through the Hydroxyl Group

Nanocellulose is mostly hydrophilic in nature due to the presence of the hydroxyl functional group. These nanoparticles are functionalized to create molecules with specialized chemical characteristics, such as hydrophobicity, active ionic surfaces, improved rheological nature, flexibility, and sorption capacity.

Concentrated H_2SO_4 (sulfuric acid) is utilized to hydrolyze nanocellulose during the extraction process, which produces half esters from the NC's hydroxyl groups and creates stable colloidal suspensions. The negatively charged sulfate groups on the surface produce a negative electrostatic layer that improves the environment for repelling particles, improves the dispersion of NC, and also raises the thermal stability of the substance. Additionally, it is said that the sulfonation procedure produces NC with a diameter range of 10–60 nm when $NaIO_4$ (periodate) and $NaHSO_3$ (bisulfite) are used. The TEMPO (2,2,6,6-tetramethylpiperidine-1-oxyl) mediated oxidation of nanocellulose is one way of transforming the main alcoholic group ($-CH_2OH$) of NC at the C_6 alcoholic position into carboxylic acid ($-COOH$), which imparts the material hydrophobicity. Carboxymethylation, a technique used to improve the negative charge repulsion on nanocellulosic surfaces, is another strategy that raises the possibility of generating more nanoscale materials with a diameter of 5–10 nm [5, 20]. During this process, the hydroxyl group ($-OH$) transforms into the carboxymethyl group ($-O-CH2OH$). This functionalization produced a hydrogel that most likely has oxygen barrier properties. The modification of nanocellulose by an inorganic ester, which results in phosphate and sulfate esters that liberate the phosphoric and sulfuric acids, favors the dehydration process, which enhances the crystallinity of the material. This functionalization of NC involves the introduction of additional substances including $POCl_3$ (phosphorus oxychloride), P_2O_5, H_3PO_4 (phosphoric acid), and organophosphates. Through the acetylation of nanocellulose utilizing acetic anhydride and acetic acid with perchloric acid as the catalyst, the hydrophobicity and crystallinity of the material were improved. The flexibility and rheological characteristics of nanocellulose were enhanced by silylation using $(CH_3)_2SiHCl$ (dimethylchlorosilane) and C_3H_8O (isopropyl) [5, 7]. During the urethanization method's functionalization process, the isocyanate (R–NCO) and hydroxyl ($-OH$) groups of nanocellulose engage and create covalent connections. Siqueira et al. described surface functionalization of CNCs and CNFs using the hydrophobicity of nanocellulose-increasing compound n-octadecyl isocyanate ($C_{19}H_{37}NO$) [23].

3.2 Nanocomposites

Nanocomposites are materials with many phases, at least one of which must be nanoscale (1–100 nm) in at least one dimension. Due to its unique properties, which make it an efficient reinforcing material for nanocomposites, nanocellulose is widely sought after for various industrial applications. In general, brittle composites fracture or shatter when subjected to stresses like impact, tensile, elongation, or compression [1, 24]. Since it can build a strong network of reinforcing nanofibers or nanocrystals throughout the composite material, absorbing stresses and increasing material strength, nanocellulose is a strong contender to solve this issue. Nanocellulose can thereby increase the material's critical stress tolerance threshold [17, 20]. When nanocellulose is added to a polymer matrix above the critical reinforcing concentration, a three-dimensional network is created; this network is assumed to be responsible for the mechanical stiffness of composites. Numerous polymers, such as polyethylene, polypropylene, and polystyrene, have made considerable use of nanocellulose as a reinforcing material [21, 25]. The rise in rubbery modulus coincided with an improvement in other mechanical properties, according to researchers. In addition to studying cellulose nanocomposite materials, several research teams have also explored natural rubber, synthetic rubbers, polyester, polyurethane, and epoxy resins [21].

Nanocomposites consisting of nanocellulose and metallic nanoparticles are one of the main areas of investigation. In bionanocomposites containing nanocellulose, a range of metal nanoparticles, including Ag, Au, Pt, Pd, Co, and Ni, have been reported as metallic dispersion phases [26]. These composites have a wide range of applications in the industries of packaging, biology, and electronics. Nanocellulose nanocomposites with various metal organic frameworks and 2D materials, such graphene, mxenes, and borophene, have also been claimed to have a wide range of uses [7, 27].

3.3 PICKERING EMULSION

Emulsions are made by sonicating, homogenizing, stirring, or simply shaking two or more immiscible liquids together with high energy. This results in the dispersion of droplets in a continuous phase, which is how emulsions are made. Oil-in-water (o/w) emulsions are produced when oil droplets are distributed in a continuous water phase. Water-in-oil (w/o) emulsions are produced when water droplets are dispersed in a continuous oil phase. Finally, water-in-water (w/w) emulsions are produced when immiscible salutes are present in water [28]. Emulsions are used in a number of

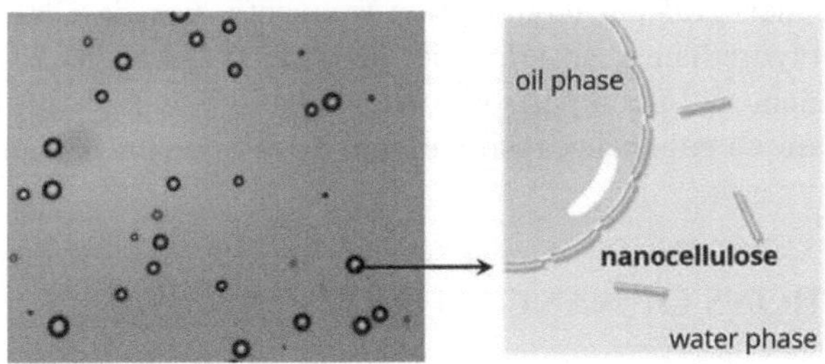

FIGURE 12.7 Pickering emulsion stabilized by nanocellulose. Adapted with permission [36] copyright @ 2022 Informa UK Limited.

items, including food, adhesives, firefighting agents, cosmetics, and medicines, giving them a wide range of applications. The stability of emulsions is important and a major aspect. An ideal emulsion does not vary significantly over time and is stable and consistent when measuring the droplet size of emulsions. An unstable emulsion experiences the consequences of phase separation, droplet coalescence, sedimentation, and creaming, all of which result in the emulsion's demise [28, 29].

As seen in Figure 12.7, pickering emulsions employ solid particles to stabilize the oil–water interface, which serves as a barrier to droplet coalescence [28]. The energy required to remove these solid particles is far higher than the thermal energy of the emulsion phase; hence, desorption of these solid particles from the interface is not possible, which leads to the formation of stable emulsions. Due to its structure, hydrophilic and green nature, and customizable surface chemistry, nanocellulose has attracted interest in the last few years for use as a pickering stabilizer [28]. Various nanocellulose alterations have reportedly led to the creation of stable pickering emulsions. For instance, polymer grafting can alter NC's hydrophilic character to hydrophobic, which helps stabilize the emulsions. An increase in CNC's capacity to adsorb on two-phase interfaces can be achieved by removing the anionic sulfate half ester group from the compound. Rapid oxidation of CNC and silylation of CNF with chlorodimethyl isopropylsilane result in the addition of a carboxylate group, which results in stable pickering agents [22, 30, 31].

3.4 POLYMER GRAFTING

The physico-chemical characteristics of the material may be improved by grafting method, which can expand the applicability of nanocellulosic material. The graphed nanocellulosic surfaces exhibit increased elasticity, heat resistance, ion exchange capacities, stability, and sorbent properties [20]. Due to its superior mechanical and biocompatible qualities and low degradability, grafted NC have been employed for surgical repair applications. There are three ways to graft with polymeric materials: to, from, and through. The "grafting to" method involves combining the nanocellulose's active hydroxyl group with another polymeric substance, such as polylactic acid or polystyrene [32], unlike grafting from route NC, which is initiated by a monomer produced by an initiator and then grafted via the surface. Grafting through, sometimes referred to as the macromonomer technique, is the connecting of two polymers through side chain linkage. Since CNF is a nanometer-wide long chain polymer, grafting modification frequently uses it. CNF grafting using PVA, starch, polyurethanes, acrylic acid, acrylonitrile, and glycidyl methacrylate has been documented. In comparison with synthetic polymers, grafted nanocellulose polymers are an enhanced, resilient, and environmentally friendly alternative [6, 20, 25, 32].

4 APPLICATIONS OF NANOCELLULOSE

A specific bio-based substance called nanocellulose has a characteristic surface chemistry that may be tuned. As seen in Figure 12.8, it also possesses favorable mechanical, thermal, and optical characteristics that make it appropriate for a variety

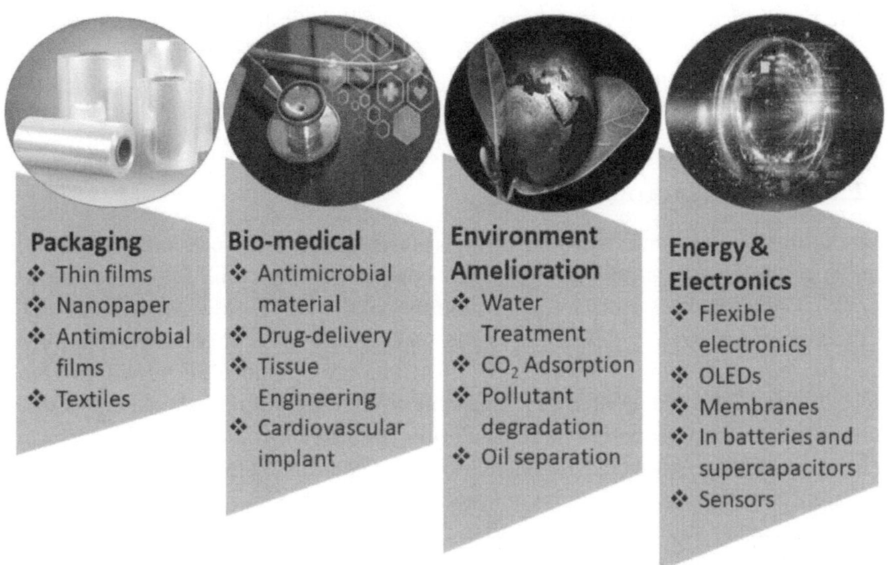

FIGURE 12.8 Applications of nanocellulose in various fields.

of uses in industries, including packaging, biomedicine, environmental improvement, energy harvesting, and electronics [17].

4.1 PACKAGING

Due to physical, biochemical, and microbiological factors, packing material can be justified for preventing the deterioration of food and drink, cosmetics, healthcare, and other consumer goods in addition to serving as packaging [6]. Packaging also has anticorrosive and fire-retardant characteristics. They should also provide an adequate barrier against air, water vapor, oil, and bacteria. Because of its nanoscale dimension, nanocellulose has the potential to replace synthetic polymers in the packaging field because it can behave as a barrier for water and gases [33]. With its appropriate modifications, nanocellulose gives various benefits to textiles, including reduced weight with high strength and increased filler content. Because of its multifunctionality and cross-linking nature, various studies have found that CNC's tensile strength improves dramatically, reaching nearly nine times that of steel. Several studies have been published in which nanocellulose was combined with other materials, such as polylactic acid, agar, carboxymethyl guar, polyvinyl alcohol, polyacrylamide, polypyrole, and polyaniline to form packaging films with good mechanical strength as well as water and gas barrier potential [19, 34]. Zwitterion films made of poly sulfobetaine methacrylate and nanocellulose were successfully produced as an antibacterial film for food packaging. Various researchers claim antimicrobial packaging films based on nanocellulose and illustrate their experiment for apples, oranges, bananas, and avocadoes; these films also extend the material's shelf life [35, 36]. Some cellulosic components, such as apple slices, restrict or lessen the enzymatic reactivity of food.

In the biomedical industry, antimicrobial nanocellulosic films are extensively used for treating and healing wounds. Additionally, this packaging is being explored for the packing of significant tools, materials, and, under exceptional circumstances, crucial organs [26, 37].

4.2 Biomedical Field

For the medical community, controlling infectious illnesses presents a continuing challenge. Everyone is on high alert and concerned about this following the pandemic COVID-19 breakout, not just in the medical and scientific industries but also at the grassroots level. It is possible to create antibacterial nanocellulose that may be used in a range of applications [6]. By functionalizing nanocellulose with different moieties like aldehyde, quaternary ammonium, and metallic nanoparticles (Zn, Ag, and Cu), nanocellulose is rendered bacteriostatic and biocompatible. As a pharmaceutical excipient, nanocellulose also compacts drug-loaded matrixes into tablets for oral administration [34, 38]. For some specific tablets, CNF are reported to provide a stronger encapsulation, and CNC with cyclodextrin works together to form hydrogel, which regulates the drug's release kinetics. For hydrophobic solid drug nanoparticles, nanocellulose also serves as a drug pool with socket [39].

The production of appropriate biomaterials for use in tissue engineering is possible because of the promising features of nanocellulose. BNC tubes for the replacement of blood arteries have been studied recently from a variety of perspectives, including cell attachment, hemodynamic analysis, microcirculatory evaluation, coagulation, and proliferation. It can be said that nanocellulose is a material that is on the rise in tissue engineering because it is being researched for a variety of tissue engineering applications, such as the replacement of soft tissues, the replacement of the nucleus pulposus, tissue regeneration and repair, wound healing, the restoration of skin tissue, and the renewal and repair of bone tissue [38, 40]. Crystalline NC has been investigated in relation to cardiovascular disease because of its strength and nanoscale dimensions. As a consequence, nanocrystalline cellulose has been used as a reinforcing material for a biocompatible matrix similar to fibrin to generate a new type of biomaterial for micro replacement vascular grafts [2, 19, 41].

4.3 Energy and Electronics

CNFs have the potential to be a contender for flexible electronics since they can be utilized to produce flexible, mechanically robust, transparent films with a poor charge transfer efficiency. According to some recent study, nanocellulose material is utilized to create a range of electrical devices. Nanocellulose and gallium arsenide combine to create more ecologically friendly and long-lasting microwave devices. Functional electronics and metal-oxide semiconductor circuits with operating frequencies up to 5 kHz may be constructed on a nanocellulose substrate. Another example of flexible electronics is the flexible organic field-effect transistors utilized in cellulose nanopaper. Printed electronics is another application where cellulose nanoparticles have shown to be effective substrates. CNFs have drawn interest from researchers and electronics producers over the past few years as substrates for screens and light-emitting diodes. It has been claimed that an organic light-emitting diode device was successfully constructed using a nanocomposite comprising cellulose nanofibers derived from wood.

Nanocellulose is also used as a separator membrane in lithium ion batteries, a solid-ion conducting electrolyte, for energy storage in battery electrodes, and for energy harvesting in solar cells [42]. According to several research findings, cellulosic materials perform better in energy storage devices than traditional materials. According to Jabbour et al., flexible electrodes made of graphite and cellulose microfibers had a conductivity of roughly 0.3 S/cm, which was much lower than that of reference electrodes [43, 44]. For Li-ion batteries, cellulose nanofibers might replace thermoplastic polymer sheets as the membrane separators. The potential use of cellulose nanoparticles as a component in supercapacitor devices has been demonstrated. The electrodes of electrochemical capacitors, which store energy by swiftly charging and discharging at the interface between an electrode and an electrolyte, have been employed in several research to include nanocellulose [42]. Another instance is the development of a very porous electrode structure by Liew et al. using cellulose nanocrystals covered with a thin polypyrrole layer [45]. They found that the porous three-dimensional structure of the cellulose nanocomposite was responsible for the composite's high capacitance (256 F/g) and rapid charge/discharge. It has been established that cellulose nanocomposites, including CNCs and polydimethylsiloxane, are capable of collecting mechanical energy through the triboelectric effect [42, 43, 46].

In the biomedical area, particularly, nanocellulose and its numerous constituents have been employed as sensors and stimuli-sensitive functional materials. The emission spectra of the CNC suspension altered as a function of pH, and there were clear transitions in the ratio of intensity at various wavelengths. Peptide-modified CNCs may be used as a biosensor wound dressing to identify a harmful protease when combined with a cellulose membrane [44]. In another instance, it was demonstrated that pyrene-modified CNCs could detect $Fe3+$ with significant discrimination between $Fe2+$ and $Fe3+$, indicating the possibility of using nanocellulosic materials as chemical sensors. As a result, there is no limit for nanocellulosic materials in the field of electronics since they touch practically every nook and cranny of the electronic sector, from energy storage to sensors [38, 46, 47].

4.4 ENVIRONMENT AMELIORATION

There have been reports on the usage of nanocellulose, as well as a number of its composites and surface-modified moieties, to remove oil contamination from the environment, as well as to purify the air and water and destroy dangerous pollutants. NC-based membranes are useful for the filtration of water because of their adaptability for the requisite pore size, mechanical strength, water flow, hydrophilicity, adsorption capacity, and operating pressure [21]. According to Jebali et al., the removal of humic and fulvic acids from waste water that are generated from biomass using amino group modified NC has been described [48]. Investigations on the breakdown of dye in modified nanocellulose and its various composites are conducted. The interaction between pollutants and the catalytic material is increased when nanocellulose membranes are changed with nanocatalysts like palladium or silver, automatically boosting the activity of the catalyst while removing impurities from the water. An 80–120 nm-sized filter made of CNF that is derived from cladophora algae is efficient in catching viruses like swine influenza. Nanocellulose functionalized with 1,2,3,4-butanetetracarboxylic acid has improved adsorption capacity for metallic contaminants like Cu2+ and Fe3+. Phosphorylation is another method to boost the sorption capacity of nanocellulose [10, 21, 37].

Nine out of ten people, according to a WHO assessment, breathe air that is more polluted than is safe, and the death curve linked to air pollution is continually rising. Hybrid membranes with superior CO_2 absorption are created when cellulose nanocomposite and metal organic frameworks (MOF) are combined. It has also been reported on bacterial cellulose with amino functionalization for CO_2 adsorption. Several zeolitic imidazolate frameworks (ZIFs) that are combined with nanocellulose perform better than stand-alone options for the adsorption of CO_2. Cello MOFs have a synergistic effect that improves the performance of both candidates as an aerogel and thin film membrane [10, 21, 37]. Oil spills can have a long-lasting negative impact on marine life and take months to clean up, which could have a devastating influence on the aquatic ecosystem. The NC material can be applied to the removal of oil spills with various modifications like silylation of nanocellulose makes it hydrophobic and possess sorption capacity for oils [49]. Nanocellulose composites based on ZIFs are also being investigated for the separation of oils, such as cyclohexane, *n*-decane, petroleum, *n*-octane, and heptane, which exhibit great flux and high oil purity [50, 51]. In light of this, it can be said that nanocellulose is a practical choice for the feeding of nature, which is exclusively derived from nature.

5   CONCLUSION

The importance of nanocellulose has grown over the past ten years, as seen by the scope, adaptability, and extent of contemporary nanocellulose research. Every facet of contemporary research trends, including those in the fields of energy, the environment, and biomedicine, is impacted by materials discoveries. The multifunctional nature of cellulose due to hydroxyl groups and hydrogen bonds was stressed in this research in order to draw attention to the cellulose's underlying chemistry. Additionally, we include the suppliers, varieties, extraction techniques, surface alterations, and uses of nanocellulose. Several challenges exist in terms of processability and performance, despite substantial advancements in turning nanocellulose into sustainable and practical materials. Initiatives in this field need a full comprehension of the interactions between cellulose and nanocellulose. Although it would be ideal, efforts to maintain the complex structure of nanocellulose after separation are unlikely to be successful. A new understanding of the natural assembly of nanocellulose is necessary to simplify separation and open up new possibilities. This calls for the creation of low-cost, green processes that fully use all the properties included in the fundamental elements of nanocellulose.

REFERENCES

[1] A. Kumar, H. Singh, P. Kumar, and B. AlMangour, *Handbook of Smart Manufacturing: Forecasting the Future of Industry 4.0*, 1st ed., Boca Raton: CRC Press, 2023. doi: 10.1201/9781003333760.

[2] K. Heise et al., "Nanocellulose: Recent Fundamental Advances and Emerging Biological and Biomimicking Applications," *Advanced Materials*, vol. 33, no. 3, p. 2004349, Jan. 2021, doi: 10.1002/adma.202004349.

[3] M. T. Islam, M. M. Alam, A. Patrucco, A. Montarsolo, and M. Zoccola, "Preparation of Nanocellulose: A Review," *AATCC Journal of Research*, vol. 1, no. 5, pp. 17–23, Sep. 2014, doi: 10.14504/ajr.1.5.3.

[4] Kargarzadeh, Hanieh, Ishak Ahmad, Sabu Thomas, and Alain Dufresne, eds. *Handbook of Nanocellulose and Cellulose Nanocomposites*. John Wiley & Sons, 2017. [5] K. Wieszczycka, K. Staszak, M. J. Woźniak-Budych, J. Litowczenko, B. M. Maciejewska, and S. Jurga, "Surface Functionalization – The Way for Advanced Applications of Smart Materials," *Coordination Chemistry Reviews*, vol. 436, p. 213846, Jun. 2021, doi: 10.1016/j.ccr.2021.213846.

[6] D. Tahir et al., "Sources, Chemical Functionalization, and Commercial Applications of Nanocellulose and Nanocellulose-Based Composites: A Review," *Polymers*, vol. 14, no. 21, p. 4468, Oct. 2022, doi: 10.3390/polym14214468.

[7] B. Thomas et al., "Nanocellulose, a Versatile Green Platform: From Biosources to Materials and Their Applications," *Chemical Reviews*, vol. 118, no. 24, pp. 11575–11625, Dec. 2018, doi: 10.1021/acs.chemrev.7b00627.

[8] P. Phanthong, P. Reubroycharoen, X. Hao, G. Xu, A. Abudula, and G. Guan, "Nanocellulose: Extraction and application," *Carbon Resources Conversion*, vol. 1, no. 1, pp. 32–43, Apr. 2018, doi: 10.1016/j.crcon.2018.05.004.

[9] M. Nasir, R. Hashim, O. Sulaiman, and M. Asim, "Nanocellulose," in *Cellulose-Reinforced Nanofibre Composites*, Elsevier, 2017, pp. 261–276. doi: 10.1016/B978-0-08-100957-4.00011-5.

[10] K. Dhali, M. Ghasemlou, F. Daver, P. Cass, and B. Adhikari, "A Review of Nanocellulose as a New Material towards Environmental Sustainability," *Science of the Total Environment*, vol. 775, p. 145871, Jun. 2021, doi: 10.1016/j.scitotenv.2021.145871.

[11] X. Wang, Q. Wang, and C. Xu, "Nanocellulose-Based Inks for 3D Bioprinting: Key Aspects in Research Development and Challenging Perspectives in Applications—A Mini Review," *Bioengineering*, vol. 7, no. 2, p. 40, Apr. 2020, doi: 10.3390/bioengineering 7020040.

[12] D. A. Gopakumar, V. Arumughan, D. Pasquini, S.-Y. (Ben) Leu, A. K. H.P.S., and S. Thomas, "Nanocellulose-Based Membranes for Water Purification," in *Nanoscale Materials in Water Purification*, Elsevier, 2019, pp. 59–85. doi: 10.1016/B978-0-12-813926-4.00004-5.

[13] Park, Ju Young, and In Hwa Lee. "Preparation of Electrospun Porous Ethyl Cellulose Fiber by THF/DMAc Binary Solvent System." *Journal of Industrial and Engineering Chemistry*, vol. 13, no. 6, pp. 1002–1008. 2007.

[14] F. Barja, "Bacterial Nanocellulose Production and Biomedical Applications," *Journal of Biomedical Research*, vol. 35, no. 4, p. 310, 2021, doi: 10.7555/JBR.35.20210036.

[15] N. Lin, J. Huang, and A. Dufresne, "Preparation, Properties and Applications of Polysaccharide Nanocrystals in Advanced Functional Nanomaterials: A Review," *Nanoscale*, vol. 4, no. 11, p. 3274, 2012, doi: 10.1039/c2nr30260h.

[16] K. Song, X. Zhu, W. Zhu, and X. Li, "Preparation and Characterization of Cellulose Nanocrystal Extracted from Calotropis Procera Biomass," *Bioresources and Bioprocessing*, vol. 6, no. 1, p. 45, Dec. 2019, doi: 10.1186/s40643-019-0279-z.

[17] A. Dufresne, *Nanocellulose: from Nature to High Performance Tailored Materials*, Berlin: De Gruyter, 2012.

[18] V. Thakur, A. Guleria, S. Kumar, S. Sharma, and K. Singh, "Recent Advances in Nanocellulose Processing, Functionalization and Applications: A Review," *Materials Advances*, vol. 2, no. 6, pp. 1872–1895, 2021, doi: 10.1039/D1MA00049G.

[19] D. Trache et al., "Nanocellulose: From Fundamentals to Advanced Applications," *Frontiers in Chemistry*, vol. 8, p. 392, May 2020, doi: 10.3389/fchem.2020.00392.

[20] M. Ghasemlou, F. Daver, E. P. Ivanova, Y. Habibi, and B. Adhikari, "Surface Modifications of Nanocellulose: From Synthesis to High-Performance Nanocomposites," *Progress in Polymer Science*, vol. 119, p. 101418, Aug. 2021, doi: 10.1016/j.progpolymsci.2021. 101418.

[21] H. N. Abdelhamid and A. P. Mathew, "Cellulose–metal Organic Frameworks (CelloMOFs) Hybrid Materials and their Multifaceted Applications: A Review," *Coordination Chemistry Reviews*, vol. 451, p. 214263, Jan. 2022, doi: 10.1016/j.ccr.2021.214263.

[22] A. G. de Souza, R. R. Ferreira, E. S. F. Aguilar, L. Zanata, and D. dos S. Rosa, "Cinnamon Essential Oil Nanocellulose-Based Pickering Emulsions: Processing Parameters Effect on Their Formation, Stabilization, and Antimicrobial Activity," *Polysaccharides*, vol. 2, no. 3, pp. 608–625, Aug. 2021, doi: 10.3390/polysaccharides2030037.

[23] G. Siqueira, H. Abdillahi, J. Bras, and A. Dufresne, "High Reinforcing Capability Cellulose Nanocrystals Extracted from Syngonanthus Nitens (Capim Dourado)," *Cellulose*, vol. 17, no. 2, pp. 289–298, Apr. 2010, doi: 10.1007/s10570-009-9384-z.

[24] A. Kumar, P. Kumar, A. K. Srivastava, and V. Goyat, Eds., Modeling, Characterization, and Processing of Smart Materials, in *Advances in Chemical and Materials Engineering*, IGI Global, 2023. doi: 10.4018/978-1-6684–9224–6.

[25] S. C. Agwuncha, C. G. Anusionwu, S. J. Owonubi, E. R. Sadiku, U. A. Busuguma, and I. D. Ibrahim, "Extraction of Cellulose Nanofibers and Their Eco/Friendly Polymer Composites," in *Sustainable Polymer Composites and Nanocomposites*, Inamuddin, S. Thomas, R. Kumar Mishra, and A. M. Asiri, Eds., Cham: Springer International Publishing, 2019, pp. 37–64. doi: 10.1007/978-3-030-05399-4_2.

[26] M. Oprea and D. M. Panaitescu, "Nanocellulose Hybrids with Metal Oxides Nanoparticles for Biomedical Applications," *Molecules*, vol. 25, no. 18, p. 4045, Sep. 2020, doi: 10.3390/molecules25184045.

[27] Y. Li et al., "Nanocellulose as Green Dispersant for Two-Dimensional Energy Materials," *Nano Energy*, vol. 13, pp. 346–354, Apr. 2015, doi: 10.1016/j.nanoen.2015.02.015.

[28] S. Fujisawa, E. Togawa, and K. Kuroda, "Nanocellulose-Stabilized Pickering Emulsions and their Applications," *Science and Technology of Advanced Materials*, vol. 18, no. 1, pp. 959–971, Dec. 2017, doi: 10.1080/14686996.2017.1401423.

[29] M. M. González, C. Blanco-Tirado, and M. Y. Combariza, "Nanocellulose as an Inhibitor of Water-in-Crude Oil Emulsion Formation," *Fuel*, vol. 264, p. 116830, Mar. 2020, doi: 10.1016/j.fuel.2019.116830.

[30] S. A. Kedzior, V. A. Gabriel, M. A. Dubé, and E. D. Cranston, "Nanocellulose in Emulsions and Heterogeneous Water-Based Polymer Systems: A Review," *Advanced Materials*, vol. 33, no. 28, p. 2002404, Jul. 2021, doi: 10.1002/adma.202002404.

[31] A. A. Wardana, A. Koga, F. Tanaka, and F. Tanaka, "Antifungal Features and Properties of Chitosan/Sandalwood Oil Pickering Emulsion Coating Stabilized by Appropriate Cellulose Nanofiber Dosage for Fresh Fruit Application," *Scientific Reports*, vol. 11, no. 1, p. 18412, Dec. 2021, doi: 10.1038/s41598-021-98074-w.

[32] K. Spence, Y. Habibi, and A. Dufresne, "Nanocellulose-Based Composites," in *Cellulose Fibers: Bio- and Nano-Polymer Composites*, S. Kalia, B. S. Kaith, and I. Kaur, Eds., Berlin, Heidelberg: Springer Berlin Heidelberg, 2011, pp. 179–213. doi: 10.1007/978-3-642-17370-7_7.

[33] C. G. Perdani and S. Gunawan, "A Short Review: Nanocellulose for Smart Biodegradable Packaging in the Food Industry," *IOP Conference Series: Earth and Environmental Science*, vol. 924, no. 1, p. 012032, Nov. 2021, doi: 10.1088/1755-1315/924/1/012032.

[34] V. B. Lunardi et al., "Nanocelluloses: Sources, Pretreatment, Isolations, Modification, and Its Application as the Drug Carriers," *Polymers*, vol. 13, no. 13, p. 2052, Jun. 2021, doi: 10.3390/polym13132052.

[35] M.-N. Efthymiou et al., "Development of Biodegradable Films Using Sunflower Protein Isolates and Bacterial Nanocellulose as Innovative Food Packaging Materials for Fresh Fruit Preservation," *Scientific Reports*, vol. 12, no. 1, p. 6935, Apr. 2022, doi: 10.1038/s41598-022-10913-6.

[36] A. Ferrer, L. Pal, and M. Hubbe, "Nanocellulose in Packaging: Advances in Barrier Layer Technologies," *Industrial Crops and Products*, vol. 95, pp. 574–582, Jan. 2017, doi: 10.1016/j.indcrop.2016.11.012.

[37] P. Thomas et al., "Comprehensive Review on Nanocellulose: Recent Developments, Challenges and Future Prospects," *Journal of the Mechanical Behavior of Biomedical Materials*, vol. 110, p. 103884, Oct. 2020, doi: 10.1016/j.jmbbm.2020.103884.

[38] S. Das, B. Ghosh, and K. Sarkar, "Nanocellulose as Sustainable Biomaterials for Drug Delivery," *Sensors International*, vol. 3, p. 100135, 2022, doi: 10.1016/j.sintl.2021.100135.

[39] H. N. Abdelhamid and A. P. Mathew, "Cellulose-Based Nanomaterials Advance Biomedicine: A Review," *IJMS*, vol. 23, no. 10, p. 5405, May 2022, doi: 10.3390/ijms23105405.

[40] A. Vikulina, D. Voronin, R. Fakhrullin, V. Vinokurov, and D. Volodkin, "Naturally Derived Nano- and Micro-Drug Delivery Vehicles: Halloysite, Vaterite and Nanocellulose," *New Journal of Chemistry*, vol. 44, no. 15, pp. 5638–5655, 2020, doi: 10.1039/C9NJ06470B.

[41] A. H. Bhat, Y. K. Dasan, I. Khan, H. Soleimani, and A. Usmani, "Application of Nanocrystalline Cellulose," in *Cellulose-Reinforced Nanofibre Composites*, Elsevier, 2017, pp. 215–240. doi: 10.1016/B978-0-08–100957-4.00009-7.

[42] O. A. T. Dias, S. Konar, A. L. Leão, W. Yang, J. Tjong, and M. Sain, "Current State of Applications of Nanocellulose in Flexible Energy and Electronic Devices," *Frontiers in Chemistry*, vol. 8, p. 420, May 2020, doi: 10.3389/fchem.2020.00420.

[43] B. Yao, J. Zhang, T. Kou, Y. Song, T. Liu, and Y. Li, "Paper-Based Electrodes for Flexible Energy Storage Devices," *Advanced Science*, vol. 4, no. 7, p. 1700107, Jul. 2017, doi: 10.1002/advs.201700107.

[44] Jabbour, Lara, Didier Chaussy, Benoit Eyraud, and Davide Beneventi. "Highly Conductive Graphite/Carbon Fiber/Cellulose Composite Papers." *Composites Science and Technology*, vol. 72, no. 5, pp. 616–623, 2012.

[45] Wu, Xinyun. "Conductive Cellulose Nanocrystals for Electrochemical Applications." PhD diss., University of Waterloo, 2016.

[46] R. Sabo, A. Yermakov, C. T. Law, and R. Elhajjar, "Nanocellulose-Enabled Electronics, Energy Harvesting Devices, Smart Materials and Sensors: A Review," *Journal of Renewable Materials*, vol. 4, no. 5, pp. 297–312, Oct. 2016, doi: 10.7569/JRM.2016.634114.

[47] V. Gabrielli and M. Frasconi, "Cellulose-Based Functional Materials for Sensing," *Chemosensors*, vol. 10, no. 9, p. 352, Aug. 2022, doi: 10.3390/chemosensors10090352.

[48] A. Jebali et al., "Adsorption of Humic Acid by Amine-Modified Nanocellulose: An Experimental and Simulation Study," *International Journal of Environmental Science and Technology*, vol. 12, no. 1, pp. 45–52, Jan. 2015, doi: 10.1007/s13762-014-0659-z.

[49] X. Wang et al., "Application of Nanocellulose in Oilfield Chemistry," *ACS Omega*, vol. 6, no. 32, pp. 20833–20845, Aug. 2021, doi: 10.1021/acsomega.1c02095.

[50] S. Fürtauer, M. Hassan, A. Elsherbiny, S. A. Gabal, S. Mehanny, and H. Abushammala, "Current Status of Cellulosic and Nanocellulosic Materials for Oil Spill Cleanup," *Polymers*, vol. 13, no. 16, p. 2739, Aug. 2021, doi: 10.3390/polym13162739.

[51] S. Mehanny, A. Kuzmin, and A. Elsherbiny, "Nanocellulosic Materials for Oil Spill," https://encyclopedia.pub/entry/14345

13 Sustainable Biomaterials for Pharmaceutical and Medical Applications

Bancha Yingngam
Ubon Ratchathani University

1 INTRODUCTION

In recent years, the focus on eco-friendly biomaterials for use in medical and pharmaceutical settings has increased, largely because of their distinctive characteristics and prospective advantages (Kumar, Kumar, Sharif, et al. 2023; Yang, Yao, and Zhang 2023). These biomaterials are sourced from renewable resources and have a minimal environmental impact during production and disposal (Kumar, Kumar, Srivastava, et al. 2023). They are also biocompatible and have the potential to improve patient outcomes. Sustainable biomaterials find applications in drug delivery (Syed et al. 2023), tissue engineering (Zarrintaj et al. 2023), and medical implants (Bansal et al. 2022). Additionally, they can be tailored to meet specific medical requirements, making them attractive to researchers and clinicians. The biopolymer market witnessed significant growth in 2022, reaching a value of US$12 billion. Analysts predict an impressive compound annual growth rate of 11% for the biopolymer market from 2023 to 2028. By 2028, the market is expected to expand significantly, reaching a value of US$23,910 million (Marketwatch, 2023).

The author analyzed trends in the utilization of biopolymers in pharmaceuticals and biomedicines. Data were obtained from two sources: the Scopus database (www. scopus.com) and the Lens Patent Search (https://www.lens.org/). The search used the keywords "biopolymer," "pharmaceutical," and "medicine" for the period between 2000 and April 19, 2023. The study revealed that the Scopus database contained 3,519 documents (Figure 13.1), including 1,717 research articles, 1,011 review articles, and others. Additionally, 16,824 granted patents were identified. Data from the first quarter of 2023 showed a relatively low number of publications (144) and granted patents (210). However, this limited time frame does not reflect the entire year's performance, and the numbers will likely increase over the full year. An analysis of these databases revealed a significant increase in publications and granted patents related to the utilization of biopolymers in pharmaceuticals and medicine. This trend can be attributed to the increased demand for sustainable and eco-friendly materials. Concerns about the environmental impact of synthetic materials have driven the need for sustainable alternatives, and biopolymers have emerged as a promising solution. Biopolymers, derived from renewable resources, can easily biodegrade after use.

DOI: 10.1201/9781003434313-13

Advances in materials science and biotechnology have improved our understanding of biopolymer properties, leading to new applications (Mehmood et al. 2023). Biopolymers can be engineered to exhibit specific properties, making them ideal for personalized medicine. Furthermore, regulatory agencies worldwide encourage the use of biopolymers in pharmaceuticals and biomedicine due to their safety and efficacy, resulting in increased funding and research in this field.

In the field of pharmaceutical and medical applications, recent advances encompass several key areas, reflecting up-to-date research findings. Ongoing research is targeting the development of new sustainable biomaterials, including biodegradable and compostable materials derived from agricultural waste, which have potential applications in medical implants, tissue engineering, and drug delivery (Wang, Xu et al. 2023). Concurrently, nanotechnology is leveraging sustainable biomaterials such as chitosan to create nanoparticles for targeted drug delivery, offering a more efficient approach that may reduce medication quantities and side effects (Syed et al. 2023). Three-dimensional (3D) printing is also making waves in tissue engineering by enabling the creation of intricate structures from sustainable biomaterials, with applications ranging from developing replacement tissues and organs to drug testing models (Moroni et al. 2022). Additionally, there is a concentrated research effort to find sustainable alternatives to conventional medical implants made from metals or plastics, which pose environmental and disposal challenges. Materials such as silk and collagen are being investigated as promising candidates for these sustainable alternatives (Bansal et al. 2022).

The aim of this chapter is to provide a comprehensive overview of the latest advancements in sustainable biomaterials for medical and pharmaceutical applications. It intends to discuss the properties and benefits of these biomaterials in comparison with traditional options and examine the associated challenges and opportunities. Overall, the chapter aims to equip researchers with a well-rounded understanding of the current landscape of sustainable biomaterials in medical and pharmaceutical contexts, highlighting both their potential advantages and challenges.

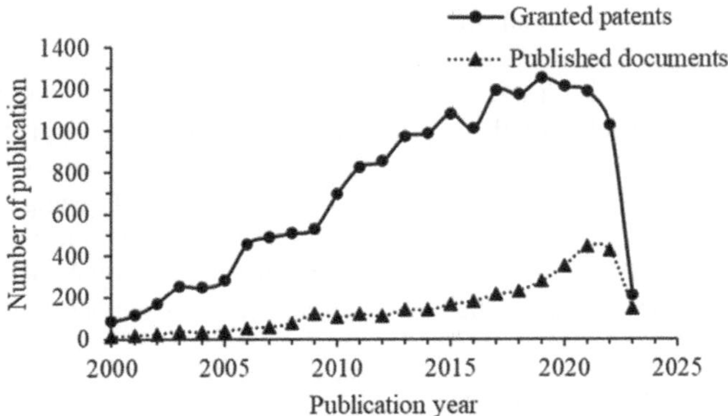

FIGURE 13.1 Annual trend of scientific publications and global patent searches conducted in the field of "biopolymer-based pharmaceuticals and medicines" over the past two decades.

2 DEFINITION OF BIOMATERIALS, BIOPOLYMERS, AND THEIR IMPORTANCE

Biomaterials are specifically designed to interact with biological systems for therapeutic or diagnostic purposes. This category includes various types of materials, such as metal alloys (Gudimetla et al. 2023; He et al. 2023; Tiwari et al. 2023), ceramics (Kumar, Kumar, Mittal, et al. 2023; Punj, Singh, and Singh 2021), polymers (Kumar, Singh, et al. 2023; Pires et al. 2023), and biocomposites (Carvalho et al. 2023). These are commonly referred to as "biomaterials" when used in biomedical applications (Hudecki, Kiryczyński, and Łos 2019; Kumar, Kumar, Dogra, et al. 2023). In this chapter, the focus will be on polymers as biomaterials. Sustainable biomaterials are gaining prominence in the pharmaceutical and medical fields due to their potential to reduce environmental impact, improve patient outcomes, and promote sustainability in healthcare. Utilizing these materials is crucial for several reasons. Conventional materials such as plastics and metals rely on nonrenewable resources, generate significant waste, and contribute to pollution. In contrast, sustainable biomaterials offer a more environmentally friendly alternative that mitigates the negative environmental impacts of healthcare (Kumar, Mittal, and Haleem 2022). Furthermore, many sustainable biomaterials possess unique properties, such as biocompatibility, biodegradability, and antimicrobial activity, which result in improved patient outcomes and enhanced safety (Aadil et al. 2023).

3 CLASSIFICATION OF POLYMER-BASED BIOPOLYMERS

Polymer-based biomaterials play a crucial role in pharmaceutical and medical applications. However, selecting the most suitable biopolymer for a specific application requires a thorough understanding of its classification based on origin. Figure 13.2 illustrates the two primary categories of biopolymers: natural and synthetic. Natural polymers are derived from a variety of sources, including plants, animals, marine organisms, and microbes. They have been used in pharmaceuticals and medicine for centuries due to their biocompatibility and biodegradability, making them attractive alternatives to synthetic polymers. On the other hand, synthetic polymers are chemically produced and are not naturally occurring. Nondegradable synthetic polymers, such as polyethylene glycol and silicone, are commonly used in medical devices owing to their durability and mechanical strength (Pires et al. 2023). However, they can cause long-term issues such as chronic inflammation and implant failure due to their non-degradability. Degradable synthetic polymers, such as polylactic acid (Swetha et al. 2023), polyglycolic acid (Mehmood et al. 2023), and polyhydroxyalkanoates (Yılmaz Nayır, Konuk, and Kara 2023), have the advantage of being metabolized by the body over a period of time, thus minimizing the potential for lasting negative impacts. To facilitate the choice of appropriate biopolymers for use in pharmaceutical and medical settings, a list of commonly used options is made available.

3.1 PLANT-DERIVED BIOPOLYMERS

3.1.1 Cellulose and Its Derivatives

Cellulose, originating from plants, is composed of glucose molecules connected by β-(1,4)-glycosidic bonds. Its unique robustness and stiffness are also due to hydrogen

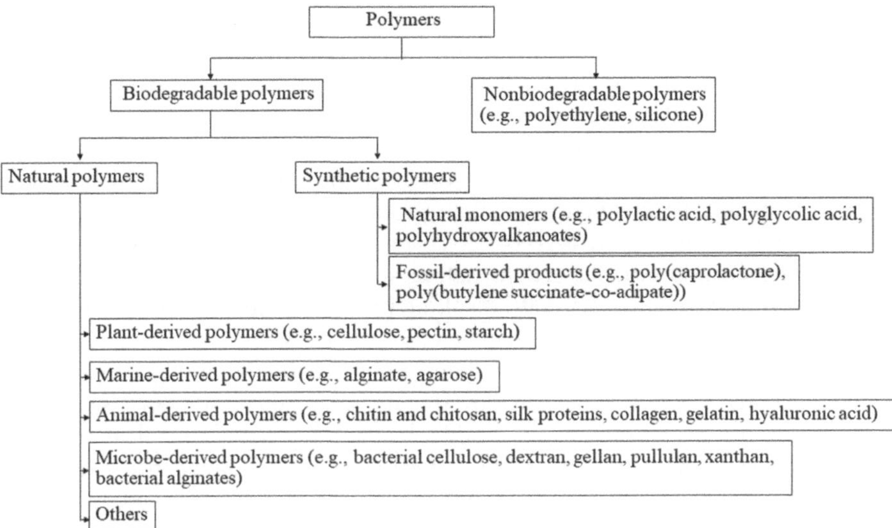

FIGURE 13.2 Diagram illustrating the categorization of biopolymers according to their biological or synthetic origins.

bonds within its structure. Its chief characteristics, such as being compatible with biological systems and having the ability to naturally decompose, make it valuable in medical contexts. It is used for applications such as covering wounds, administering medications, and providing structural support for tissue growth. It can absorb significant moisture and selectively transport small molecules. However, it also has limitations, such as poor microbial resistance and limited solubility, attributed to its hydroxyl groups. These challenges can be addressed through chemical modifications, such as substituting hydroxyl groups with functional groups such as acids or oxides. Researchers have developed various cellulose derivatives to optimize their biomedical uses. Common cellulose derivatives include ethyl cellulose, methyl cellulose, carboxymethyl cellulose (also known as cellulose ethers), cellulose acetate, and cellulose nitrate (also known as cellulose esters) (Otarola et al. 2020; Song et al. 2024).

3.1.2 Pectin

Pectin, a polysaccharide found in plant cell walls, is rich in fruits and vegetables such as apples and carrots (Zhu et al. 2024). Its chemical structure involves galacturonic acid monomers, with gelling properties influenced by the degree of esterification. Various extraction methods are available based on the plant source and desired pectin properties. Commonly used in food as a gelling agent, pectin also has potential biomedical applications, such as wound healing and drug delivery (Dang et al. 2023; Narala et al. 2023). Despite its promise, challenges exist. Variability in molecular weight affects its consistency, while solubility limitations restrict its use in alkaline conditions. Other factors, such as pH and temperature, can also affect its gelling properties, impacting product stability. Finally, its slow biodegradation and potential to trigger immune responses limit its broader medical applications (Dang et al. 2023).

3.1.3 Starch

Starch, a natural polymer, consists of two main types of glucose polymers: amylose and amylopectin. The proportions of these polymers vary depending on the plant source. Typically, starch contains 18–33% amylose and 72–82% amylopectin (Kunarbekova et al. 2023). Starch is made up of two types of glucose units: α-glucose and β-glucose. Both amylose and amylopectin, the components of starch, have α-(1→4) glycosidic bonds but differ in their structures. Amylose is a straightforward, linear polymer connected by α-(1→4) glycosidic bonds. On the other hand, amylopectin features a more complicated, branching structure, incorporating both α-(1→4) and α-(1→6) glycosidic connections, resulting in a multifaceted form with branches at its ends. Common sources of starch include corn, potatoes, and wheat. As a complex carbohydrate, starch can be broken down into glucose by digestive enzymes in the human body. Starch is widely employed in the food industry for purposes such as thickening, gelling, and stabilization. Additionally, it has uses in medical and pharmaceutical fields, including serving as a component in biodegradable systems for delivering medications (Vasquez-Martínez et al. 2023). However, the author's experience reveals limitations associated with starch use. Starch exhibits poor mechanical strength and high brittleness, which restricts its suitability for applications requiring materials with high tensile strength, such as sutures or implants. Environmental factors such as humidity and temperature can also adversely affect the properties and mechanical integrity of starch-based materials over time. Additionally, starch has limited solubility in both water and organic solvents, complicating its processing and shaping. When heated, starch tends to form gels, which hinder its homogeneous distribution within a product. Moreover, the biocompatibility of starch is compromised by its slow degradation rate and the potential for generating inflammatory byproducts.

3.2 Marine-Derived Biopolymers

Marine organisms provide a rich source of biomaterials with potential in pharmaceuticals and biomedicine. These materials are shaped by their unique marine environments, offering advantages such as biocompatibility, biodegradability, and therapeutic properties. Research supports their promise in drug development, drug delivery, and medical devices (Manna and Jana 2022). The following section will explore specific, well-known marine-derived biomaterials in detail.

3.2.1 Alginate

Alginate consists of a linear arrangement of carbohydrates, featuring a recurrent pattern of β-*D*-mannuronic acid (M) and α-*L*-guluronic acid (G), which are connected through 1,4-glycosidic bonds. The proportion of M to G can differ depending on its origin and the way it is extracted (Ahmad Raus, Wan Nawawi, and Nasaruddin 2021). Due to its polyanionic properties, alginate can create gel-like structures by connecting its carboxylic acid groups with doubly charged ions such as calcium or magnesium (Kruk and Winnicka 2023). Alginate is useful in wound dressings, drug delivery, and tissue engineering. Alginate is typically extracted from brown seaweed species such as *Laminaria* or *Macrosystis*. The seaweed is harvested, cleaned, chopped, and treated with mineral acids to dissolve cell wall calcium carbonate. After washing, the seaweed is boiled to extract alginate, which is then precipitated using water-miscible solvents such as ethanol and dried to a powder. Industrial extraction mimics the lab-scale process but incorporates machinery for harvesting and chopping. The seaweed undergoes treatment with a sodium carbonate and sodium hydroxide solution to extract alginate. This is precipitated, filtered, and purified before being dried and milled into a powder. The process can be optimized by adjusting the alkaline solution concentration and extraction time. Advanced purification techniques such as ion exchange chromatography or ultrafiltration can further improve efficiency.

3.2.2 Agarose

Agarose is a linear carbohydrate made up of recurring units of agarobiose, a two-sugar molecule consisting of *D*-galactose and 3,6-anhydro-*L*-galactose. These sugars are linked by a specific type of bond called a β-1,4-glycosidic bond. This two-sugar molecule serves as the main structural element of the polymer, and its linearity is maintained by these β-1,4-glycosidic bonds. The pattern that repeats within agarose consists of alternating β-*D*-galactopyranose and 3,6-anhydro-*L*-galactopyranose segments. The latter contains an ether linkage between carbons 3 and 6, increasing the rigidity of the molecule (Jung et al. 2023). In aqueous solutions, agarose forms a gel-like network through hydrogen bonding between the hydroxyl groups of its polymer chains. These hydrogen-bonded double helices aggregate to form a three-dimensional mesh, making agarose useful in DNA and protein electrophoresis, chromatography, and drug delivery.

3.3 ANIMAL-DERIVED BIOPOLYMERS

3.3.1 Chitin and Chitosan

Chitin and chitosan are naturally occurring polymers with increasing relevance in the pharmaceutical and medical fields. Chitin is a linear carbohydrate structure formed by *N*-acetylglucosamine (GlcNAc) units, which are linked together through β-1,4 glycosidic connections. It has a molecular formula of $(C_8H_{13}NO_5)_n$, where "n" indicates the degree of polymerization. Insoluble in water and most organic solvents, chitin has a structure similar to cellulose. Chitosan is obtained from chitin through a process called deacetylation, which removes acetyl groups from the GlcNAc units. This results in glucosamine units linked by β-1,4-glycosidic bonds. Although it shares the molecular formula of chitin, chitosan exhibits a lower degree of polymerization due to the deacetylation process. It is soluble in acidic solutions and can form gels under specific conditions (Khajavian et al. 2022). Both chitin and chitosan can be extracted from sources such as shrimp shells and fungal cell walls using chemical, enzymatic, or microbial methods. Chemical extraction often employs strong acids and bases, such as hydrochloric acid and sodium hydroxide, to remove calcium carbonate and proteins. The isolated chitin can then be deacetylated in an alkaline solution to yield chitosan. Enzymatic and microbial methods utilize specific enzymes or microbes to achieve comparable results. Industrial-scale extraction generally relies on acid–base methods and may include additional purification steps such as filtration and centrifugation. However, these chemical methods pose environmental risks due to high reagent consumption, prompting research into eco-friendly alternatives such as microbial fermentation and enzymatic hydrolysis. The conversion of chitin to chitosan requires a critical deacetylation step, generally performed using sodium hydroxide under specific conditions (60–100°C, 4–5 hours) to remove the acetyl groups (Novikov et al. 2023). Emerging technologies such as ionic liquids and electrochemical methods aim to improve both yield and quality.

Chitin's limited solubility in water and most organic solvents presents challenges for its large-scale application; various methods are being developed to overcome this limitation (Lv et al. 2023). The properties of chitin and chitosan vary based on their source, extraction method, and degree of deacetylation. While chitin is water-insoluble, chitosan is soluble in acidic solutions and can form gels. This makes it useful for applications such as wound healing and drug delivery. The degree of deacetylation also influences the solubility and charge density of chitosan. Nevertheless, current methods for isolating chitin and chitosan come with drawbacks, including environmental concerns related to the use of strong acids and bases, which generate substantial waste and consume significant energy. Therefore, developing sustainable, eco-friendly extraction methods is imperative. Additionally, the quality and purity of chitin and chitosan can differ depending on the source and extraction technique; impurities such as proteins and minerals may compromise their efficacy in pharmaceutical and medical applications.

3.3.2 Silk Proteins

Silk proteins, mainly from spiders and silkworms, are used in various biomedical applications. Comprising two key types—silk fibroin and silk sericin—these proteins offer distinct physical and mechanical benefits. Fibroin makes up 65–85% of natural silk and has a crystalline structure, providing strength and flexibility. Its primary components are amino acids, such as glycine, alanine, and serine. Sericin, accounting for 15–30%, is water-soluble and functions as an adhesive in silk fibers (Yang, Yao, and Zhang 2023). Fibroin is used for its strength, enabled by a β-sheet structure from hydrogen bonding. Its components have repeating units that contribute to water absorption and further flexibility. While sericin is typically eliminated in processing steps, it shows promise in fields such as skincare and medical treatments owing to its abilities to counter oxidative stress and inflammation. Both sericin and fibroin are able to naturally breakdown and are compatible with biological tissues, making them potentially useful in applications, such as controlled medication release, structural support for tissue growth, and the treatment of wounds.

3.3.3 Collagen and Gelatin

Collagen is the predominant protein in mammals, making up to 30% of the total protein content. It is commonly found in various tissues, such as skin, bones, and cartilage, and has a triple–helical configuration held together by hydrogen bonds. This arrangement consists of recurring segments of glycine, proline, and hydroxyproline (Puszkarska et al. 2022). On the other hand, gelatin is obtained from collagen through a process of partial hydrolysis, which breaks down the triple–helix formation into shorter, less-ordered chains of polypeptides. Like collagen, gelatin primarily consists of glycine, proline, and hydroxyproline but lacks the structured triple–helix (Rashid, Showva, and Hoque 2023). Both are valuable in pharmaceutical and medical applications. Collagen is often used in tissue engineering and wound healing, while gelatin is used in drug delivery and microcapsule production. Their properties can be modified further through chemical and enzymatic treatments.

3.4 MICROBE-DERIVED BIOPOLYMERS

3.4.1 Bacterial Cellulose

Bacterial cellulose, primarily produced by species such as *Gluconacetobacter xylinus*, is a highly crystalline and pure cellulose biopolymer. Unlike its plant-derived counterpart, it offers unique advantages, such as high tensile strength, water retention, and biocompatibility, making it ideal for biomedical applications (Ciecholewska-Juśko, Junka, and Fijałkowski 2022). Bacterial cellulose is generated through a simple, low-cost fermentation process using renewable sugars such as glucose. During this process, bacteria secrete cellulose fibrils that self-assemble into a 3D gel-like network. This can be shaped into sheets, tubes, or scaffolds. In biomedicine, it serves as an effective wound dressing due to its water retention and exudate absorption capabilities. It is also invaluable in tissue engineering, where its mechanical strength and biocompatibility are key attributes. Other bacterial strains, such as *Escherichia coli* and *Salmonella enterica*, can produce gels suitable for drug delivery and other biomedical applications (Chakrapani, Zare, and Ramakrishna 2022). Beyond medicine, its utility extends to the food and cosmetic industries, where it can act as a thickening agent or skincare ingredient. Consequently, bacterial cellulose presents a wide array of promising applications, most notably in biomedicine, where it has the potential to significantly impact areas such as wound care and tissue engineering.

3.4.2 Microbial Dextran

Produced primarily by lactic acid bacteria and *Leuconostoc mesenteroides*, microbial dextran is composed of glucose monomers, α-D-Glcp, connected by α-1,6-glycosidic bonds. The degrees of polymerization and branching are denoted by the subscripts m and n, respectively. α-1,3-linked glucose residues serve as branching points (Rahman, Pasupathi, and Karuppiah 2022). The unique branched architecture of the polymer opens up a wide array of uses, from delivering medications to engineering biological tissues. Altering its chemical structure, such as by introducing functional groups, can broaden its applicability even further. Dextran is versatile and is used across different sectors, such as food processing, pharmaceuticals, and biomedicine. In the food industry, it acts as a substance that thickens and stabilizes; in the pharmaceutical field, it plays roles in medication delivery and as an additional ingredient known as an excipient. In medical contexts, it serves as a blood plasma extender. In the field of biomedicine, dextran has proven useful in areas such as tissue construction and wound healing, especially when used in the form of dextran-based hydrogels. However, the high production cost and variability in molecular weight limit its commercial viability. Additionally, its susceptibility to enzymatic degradation can affect both its long-term stability and shelf life.

3.4.3   Polyhydroxyalkanoates (PHAs)

PHAs are microbial biomaterials composed of linear chains of hydroxyalkanoate monomers linked by ester bonds. The characteristics of these monomers influence the polymer's properties. These polymers are produced by culturing specific microorganisms in nutrient-rich media. Biopolymer formation is initiated by limiting nutrients such as nitrogen and phosphorus while supplying excess carbon sources. The fermentation process generally lasts 38–48 hours, after which PHAs are extracted using methods such as solvent extraction or enzymatic digestion. Optimization involves selecting the appropriate microorganism type, nutrient medium composition, and extraction methods. Genetic engineering can also enhance both yield and polymer properties (Martínez-Herrera et al. 2023). PHAs are highly valued for their ability to be compatible with biological systems, breakdown naturally, and withstand heat. These qualities make them particularly useful for medical uses, including stitching wounds and systems for administering medication. However, they face challenges, including potential inflammation upon implantation, slow degradation rates, variable mechanical properties, high production costs, and regulatory hurdles. Despite these limitations, ongoing research aims to improve their properties and cost-effectiveness for broader commercial use.

3.4.4   Pullulan

Pullulan, a hydrophilic polymer produced by *Aureobasidium pullulans*, consists of maltotriose units connected by α-(1→4) and α-(1→6) glycosidic bonds. Its molecular weight ranges from 10 to 400 kDa. These unique linkages confer high water solubility and enable it to form Newtonian flow solutions at concentrations above 10 wt% (Singh, Kaur, et al. 2023). Pullulan can be extracted from the fermentation liquid of *A. pullulans* through methods such as precipitation using ethanol or isopropanol. Subsequent purification can be achieved via techniques such as dialysis or ultrafiltration. In the pharmaceutical and medical fields, it serves as a coating material for tablets, aids in wound healing, and functions as a drug delivery system. Its biocompatibility and biodegradability make it well suited for tissue engineering. Approved by the Food and Drug Administration in 2002 and more recently by the European Union as a food additive, pullulan is also accepted for pharmaceutical use in several Asian countries. Toxicological studies confirm its low toxicity, lack of skin and eye irritation, and natural elimination from the body, supporting its broad, safe use (Singh, Kaur, et al. 2023). From the author's perspective, pullulan has both advantages and limitations as a biopolymer. While its water solubility is a benefit, its limited solubility in other solvents can be a challenge, especially for lipophilic drugs. It is generally more expensive than other biopolymers, such as chitosan or alginate, affecting its cost-effectiveness. Additionally, it is less extensively studied than these other biopolymers, leaving gaps in our understanding of their potential applications and properties.

3.4.5   Hyaluronic Acid

Hyaluronic acid is a type of glycosaminoglycan that lacks sulfate groups and is made up of recurring disaccharide segments. These segments are formed of *N*-acetylglucosamine and glucuronic acid, connected through β-1,3 and β-1,4 glycosidic linkages. The substance's chemical arrangement is represented as $-\beta$-D-glucuronic acid–(1→3)-β-D-N-acetylglucosamine–(1→4)–. Depending on the pH, the glucuronic acid residue can exist in either linear or ring forms. The molecule is highly hydrophilic, allowing it to bind large amounts of water and act as an effective lubricant in joints and tissues (Kikani, Dave, and Thakore 2023). It also plays roles in cell migration, wound healing, and inflammation. For industrial-scale production, hyaluronic acid is commonly sourced from *Streptococcus equi* sub. *zooepidemicus*. This method involves challenges, as the bacterium is a zoonotic pathogen. However, this limitation has spurred the exploration of recombinant production methods as safer alternatives (Torres-Acosta et al. 2021). Despite its benefits in the pharmaceutical and biomedical fields, hyaluronic acid has several drawbacks. High production costs can make it impractical for some applications. The material's performance varies based on its molecular weight and degree of cross-linking, complicating consistent production. It also poses potential risks of immunogenic reactions, particularly in high-molecular-weight forms. Its slow degradation rate may affect long-term performance and result in bodily accumulation. Furthermore, its limited mechanical strength restricts its utility in high tensile applications. Regulatory hurdles, including approval processes and safety regulations, can further restrict its use.

3.5 BIOPOLYMERS OBTAINED VIA BIOTECHNOLOGY

3.5.1 Polylactic Acid (PLA)

PLA is a heat-responsive, eco-friendly polymer made from sustainable plant-based materials such as cornstarch. Created by polymerizing lactic acid, PLA is mostly employed in the biomedical and pharmaceutical fields because it can be broken down naturally and is compatible with biological systems. It is particularly effective in systems that deliver medications, in the engineering of biological tissues, and in bio-compatible implants, providing regulated drug release and harmonious interaction with human cells. In tissue engineering, PLA is used to create scaffolds that support tissue regeneration, proving effective for various tissue types, such as bone, cartilage, and skin. Additionally, its mechanical strength and biocompatibility make it useful for producing medical implants such as screws, pins, and plates. However, PLA does have limitations, including relatively low mechanical strength and moisture sensitivity, which can restrict its use in certain applications. Additionally, the rate at which PLA breaks down can be affected by variables, such as heat levels, acidity or alkalinity, and water content. These factors make it complicated to precisely manage the release of medication in drug delivery systems that use PLA (Zhong et al. 2023).

3.5.2 Poly(Lactic-co-Glycolic Acid) (PLGA)

PLGA, a blend of lactic and glycolic acids, is both biodegradable and compatible with biological systems, making it popular in the pharmaceutical and biomedical sectors. By tweaking the proportions of the constituent acids, its degradation speed and structural robustness can be customized. PLGA is commonly employed in systems that deliver medications through micro- and nanoparticles. It is also used in the construction of biological tissues and various other medical applications. However, it has limitations such as a slow degradation rate, potential release of acidic byproducts, and high hydrophobicity, which can complicate its processing and use. Modifications, such as adding surfactants, can improve its properties. Although generally considered safe, some concerns exist about the potential toxicity of PLGA degradation products, necessitating thorough safety evaluations.

4 APPLICATIONS OF BIOMATERIALS IN PHARMACEUTICAL AND MEDICAL FIELDS

4.1 DRUG DELIVERY SYSTEMS

4.1.1 Beads

Hydrogel beads are typically made of hydrophilic polymer networks and are useful for drug delivery due to their water solubility and biocompatibility. While common materials such as polyethylene glycol have toxicity issues, biopolymers such as chitosan, alginate, and especially cellulose offer better biocompatibility and lower toxicity. Notably, cellulose stands out due to its abundant availability, with an annual production of 10^{11}–10^{12} tons, and its ability to naturally cycle back into the environment through decomposition processes (Nie et al. 2021). Considerable research, including a study by Nie et al. (2021), explores the use of biopolymers such as cellulose for hydrogel bead production. The authors dissolved cellulose in a $ZnCl_2$ solution and crosslinked it with calcium ions. Bead structure and crystallinity were influenced by calcium ion concentration and washing methods. These beads had a layered nanoporous structure and showed sustained loading of bovine serum albumin over time. Sodium alginate, a polysaccharide derived from brown seaweed, is commonly used in various applications, including the creation of hydrogel beads through ionotropic gelation. In unpublished research by the author, a microbead system encapsulating garlic oil was developed using sodium alginate. Garlic oil was chosen for its strong polysulfide odor, and encapsulation improved the product's palatability by masking the odor. The system could accommodate approximately 84% garlic oil and exhibited a spherical shape with a smooth surface, as shown in Figure 13.3.

FIGURE 13.3 Morphological features of calcium alginate beads with a diameter range of 10–20 mm: blank beads (left) and garlic oil-loaded beads (right). Photograph by the author.

4.1.2 Nanoparticles

Nanoparticles, colloidal structures ranging from 1 to 1,000 nm, are made from various natural and synthetic polymers, such as chitosan, alginate, and PLGA. They can encapsulate drugs, improve their stability and bioavailability, and reduce toxicity. Chitosan nanoparticles are especially notable for their biocompatibility, biodegradability, and low toxicity and have diverse applications in biomedicine and pharmaceuticals. Commercial products such as "oligonucleotide chitosan nanoparticles" and "nanochitosan" are used for delivering nucleic acids and a range of drugs, respectively (Jafarzadeh et al. 2023). These nanoparticles also have specialized uses in wound care; for example, "Celox Rapid" rapidly halts bleeding, while "MediPlus Advanced Wound Care Dressings" promote healing (Gegel et al. 2010). Additionally, in the cosmetics industry, "Nano Chitosan Cosmetic" uses chitosan nanoparticles to enhance skin hydration, reduce wrinkles, and offer ultraviolet protection (Guzmán, Ortega, and Rubio 2022).

4.1.3 Hydrogels

Hydrogels are three-dimensional networks made from water-attracting polymers, and they are used in a variety of fields, such as administering medications, constructing tissues, and promoting wound recovery. These gels are created using different types of cross-linking techniques, which can be physical, chemical, or enzyme-based. Physical cross-linking is reversible and sensitive to environmental factors, whereas chemical cross-linking is irreversible and creates a stable structure. Enzymatic cross-linking is biocompatible and frequently used in tissue engineering applications. Hydrogels can be formulated using synthetic polymers, such as polyethylene glycol, natural polymers such as collagen, or a combination of both in hybrid polymers. Self-assembling peptides are another option. Each material type offers unique mechanical properties, biocompatibility, and degradation rates. The versatility of hydrogel-based biomaterials is evidenced by their use in a wide range of commercial products, from topical treatments such as "hydrosal gel" to surgical adhesives such as "bioglue".

4.1.4 Biopolymer–Magnetite Composites

Composites of biopolymers and magnetite have a range of versatile uses, including but not limited to precise medication administration, treating cancer, and constructing biological tissues. These composites combine the biocompatibility of biopolymers with magnetite nanoparticles, enabling precise therapeutic delivery through external magnetic fields. This targeted approach has the potential to improve treatment efficacy while minimizing side effects. In the field of magnetic hyperthermia, magnetite nanoparticles generate heat when exposed to an alternating magnetic field, selectively destroying cancer cells while sparing healthy tissue. This offers a noninvasive, targeted option for cancer treatment. These composites are also promising in tissue engineering because they can form scaffolds that provide mechanical support, encourage cellular adhesion, and enable the controlled release of bioactive agents. The magnetic properties of these composites allow for precise manipulation of scaffolds, aiding in the optimal placement and orientation of cells for tissue regeneration. Additionally, these composites can enhance magnetic resonance imaging contrast, improving the visualization and detection of diseased tissue or specific anatomical features, thus aiding in diagnosis and monitoring.

4.2 Tissue Engineering

4.2.1 Scaffolds

Scaffolds serve as three-dimensional frameworks that aid in the proliferation and maturation of cells and tissues. Suitable materials for constructing these scaffolds range from natural polymers such as chitosan and cellulose to synthetic, biodegradable, and nontoxic polymers such as PLA and PGA. The use of sustainable scaffolds in tissue engineering can reduce waste and pollution associated with traditional scaffolds made from nonrenewable materials.

4.2.2 Hydrogels

Hydrogels made from natural polymers such as agarose, gelatin, and alginate are ideal for tissue engineering due to their ability to mimic natural tissues. This category includes silk fibroin, pullulan-polyethylene glycol, and hyaluronic acid–based hydrogels, whose properties can be tailored for specific applications. Singh et al. (2023) developed a silk fibroin hydrogel composite with vancomycin to address the slow healing and bacterial infections associated with diabetic wounds. The composite was comprehensively analyzed using methods including Fourier transform infrared spectroscopy, nuclear magnetic resonance spectroscopy, and scanning electron microscopy. It demonstrated controlled drug release and antibacterial properties in vitro. In vivo studies showed that the composite significantly outperformed traditional antiseptics, healing diabetic wounds at a rate of 86.66%, compared with 33% and 11.66%. Serum analysis further confirmed its wound-healing potential. The results suggest that these composites offer a promising antiseptic solution for diabetic wound care.

4.2.3 Nanofibers

Nanofibers, ultrafine fibers with nanoscale diameters, have shown promise as sustainable biomaterials for tissue engineering. These fibers can be made from natural polymers such as collagen and silk or from synthetic polymers such as PLA and polyethylene glycol. Nanofibers can serve as scaffolds that mimic the structure and function of natural tissues and can also deliver drugs and other therapeutic agents to specific sites in the body. While sustainable biomaterials offer many advantages for tissue engineering applications, there are some potential drawbacks and limitations. For instance, the mechanical properties of these biomaterials may not be as robust as those of nonrenewable materials, which could limit their applicability. Additionally, the production of sustainable biomaterials might require specialized equipment or processes, potentially making them more expensive than traditional options.

4.3 Wound Healing

4.3.1 Dressing

Wound dressings are vital for effective wound care, and various biomaterials, such as hydrogels, foams, and films, are used based on wound characteristics. Key features such as biocompatibility and exudate absorption make these materials beneficial for healing. For example, chitosan hydrogels have strong antibacterial properties and support cell growth. A study combined chitosan with aloe vera and copaiba oil to enhance wound healing in rats, showing promising results. Bacterial cellulose is another promising biomaterial known for its strength and water retention. However, there are challenges, such as cost and the potential for adverse reactions. For instance, Jokar et al. (2023) addressed these issues by creating a wound dressing from bacterial cellulose and dextran with pomegranate extract, which showed effective antibacterial activity and wound healing in mice. Despite their promise, biomaterial-based dressings have limitations, such as variable biodegradability, production costs, and inconsistent antibacterial efficacy, which need further research and evaluation.

4.3.2 Sutures

Biomaterials have been used in sutures to promote wound healing and facilitate closure. Tensile strength, elasticity, and biodegradability are key properties of the biomaterials used in sutures. Common materials for sutures include polyesters, polyglycolic acid, and polylactic acid. Sustainable biopolymers, such as silk, collagen, and PLGA, have been shown to effectively close wounds while reducing the risk of complications and infection. The benefits of using biomaterial-based sutures include their absorbable nature, eliminating the need for removal, and decreasing the risk of infection (Rakhmatullayeva et al. 2023). Some examples of sutures made from biopolymers include silk sutures, which are flexible, strong, and suitable for various surgeries, including ophthalmic and cardiovascular procedures (Li et al. 2023). Collagen sutures are commonly used in plastic surgery and dental procedures (Dasgupta et al. 2021). Chitosan sutures are also biodegradable, which eliminates the need for suture removal (Rakhmatullayeva et al. 2023). Polyhydroxyalkanoate sutures are degraded into natural components that can be easily metabolized by the body (Gregory et al. 2022). However, the drawbacks of biomaterial-based sutures are their higher cost and lower tensile strength compared with traditional sutures (Joseph et al. 2023).

4.3.3 Electrospun Nanofibers

Electrospun nanofibers are promising for wound dressings due to attributes such as a high porosity, surface area-to-volume ratio, and mechanical strength (Yao et al. 2023). Made from various natural and synthetic polymers, these fibers can be customized for different wound needs. Their high porosity aids in exudate management and gas exchange, which are crucial for healing, while the high surface area fosters cell adhesion and growth. Additionally, these nanofibers can be functionalized with bioactive agents to boost their therapeutic properties (Zhong et al. 2023). Dehghani et al. (2023) synthesized a nanocomposite wound dressing made of nitrogen-doped carbon quantum dots, α-tricalcium phosphate, chitosan, and silk fibroin. Tests showed that it had excellent antimicrobial and biocompatible qualities, promoted cell migration and proliferation, and effectively aided wound closure in animal models. Thus, this nanofiber dressing holds promise for wound healing and tissue regeneration.

4.4 DIAGNOSTIC DEVICES

4.4.1 Biosensors

Biosensors are instruments designed for analytical purposes, combining biological detection components such as enzymes, antibodies, or DNA strands with a transducer. This transducer is responsible for translating biological reactions into quantifiable signals, which can be electrical, optical, or mechanical in nature (Zhu et al. 2023). The development of biosensors relies on the utilization of biomaterials, which are crucial for achieving biocompatibility, stability, and functionality in these devices. One significant application of biomaterials in biosensor development is glucose monitoring for diabetes management. Glucose biosensors utilize immobilized glucose oxidase enzymes on conductive electrodes to detect glucose concentrations (Ozlu and Shim 2022). The choice of biomaterials significantly affects the stability, sensitivity, and specificity of these biosensors. For example, the incorporation of graphene-based materials has shown improved sensitivity and stability in glucose biosensors. In a recent study by Güler et al. (2023), an innovative method for detecting glucose electrochemically was introduced. This method relies on glucose oxidase that is fixed onto nanoparticles made of an Au@ Pd core-shell structure, which are further supported on carboxylated graphene oxide. To accomplish this immobilization, a glassy carbon electrode was modified with a cross-linking technique involving chitosan biopolymer, Au@Pd/carboxylated graphene oxide, and glutaraldehyde. Amperometry was employed to assess the performance of this biosensor, revealing a rapid reaction time of approximately 5.2 seconds, a consistent detection range from 2.0×10^{-5} to 4.2×10^{-3} M, and a detection limit of 10.4 µM. Moreover, the calculated apparent Michaelis–Menten constant was 3.04 mM. The biosensor showed high levels of repeatability, reliability, and long-term stability. It was also found to be resilient against interference from an array of substances, including dopamine, uric acid, ascorbic acid, paracetamol, folic acid, mannose, sucrose, and fructose. The utilization of carboxylated graphene oxide, with its large electroactive surface area, shows promise for sensor development. However, challenges related to reproducibility, scalability, and potential toxicity persist in the field of biomaterial-based glucose biosensors.

4.4.2 Imaging Agents

Imaging agents are used in medical imaging to highlight biological targets, such as tumors. Biomaterials enhance these agents by improving their solubility, biocompatibility, and targeting. One application is in MRI contrast agents, where biomaterial-based alternatives such as iron oxide nanoparticles offer safer and more effective options than traditional gadolinium-based agents. A study by King et al. (2020) approached the challenge of inconsistent molecular weights in polymers by using precision-engineered biopolymer mimics. These mimics stabilized iron oxide nanoparticles, resulting in superior MRI contrast and safe biological interactions. The study's findings suggest significant potential for biomaterial-based imaging agents in MRI diagnostics.

4.4.3 Microarrays

Microarrays are diagnostic tools that enable high-throughput detection of multiple biomolecules, including DNA, RNA, and proteins. These arrays depend on biomaterials to provide surface stability and biocompatibility, which are essential for the immobilization and detection of biomolecules (Fu and Nguyen 2021).

While they offer promise in areas such as protein-based diagnostics and drug discovery, challenges such as sensitivity and specificity remain. A recent study by Ferrara et al. (2020) introduced an innovative, environmentally friendly approach using aqueous-based inks for single-cell microarrays. These inks eliminate cytotoxic compounds, reducing the environmental impact and making the technology more scalable. By employing a green chemistry printing technique, the study successfully created polymeric platforms that enable precise cell localization. This opens up new avenues for disease research, drug discovery, and personalized medicine.

5 CASE STUDY: DISSOLVING MICRONEEDLE ARRAY PATCHES

This section presents a case study to deepen the readers' understanding of biopolymer applications in dissolving microneedle array patches (MAPs). Recently, MAPs have garnered attention as an innovative drug delivery system. They feature microscale needles that penetrate the skin's outermost layer, the stratum corneum, for effective drug delivery to underlying tissues. MAPs offer advantages over traditional methods, such as better patient compliance, reduced pain, and the potential for targeted, sustained drug release (Lin et al. 2023).

Zingiber montanum, a plant species common in Southeast Asia, has traditional uses in Thai medicine for various ailments (Navabhatra, Maniratanachote, and Yingngam 2022). The plant's rhizomes contain compounds with anti-inflammatory properties, which have been demonstrated in extensive preclinical and clinical studies (Yingngam and Brantner 2018). However, the limited water solubility and bioavailability of these phenylbutanoids pose challenges. MAPs containing these compounds are being explored to overcome these limitations. In an unpublished study by the author, MAPs were developed for transdermal delivery of a phenylbutanoid extract complexed with hydroxypropyl-β-cyclodextrin (HPβCD). To create the MAPs, plant extract was incorporated into the polymer matrix at varying concentrations: 0%, 10%, 20%, and 30% of the total polymer weight. The patches featured an array of microneedles (60 × 60 needles) composed of low molecular weight hyaluronic acid and loaded with prepared inclusion complexes. The molding method was selected as the preferred technique for preparing dissolving microneedle arrays, owing to its widespread use in incorporating drugs for transdermal delivery. This method employs molds to fabricate microneedle arrays with precise dimensions and shapes. The drug can be introduced either by mixing it with the polymer solution before casting or through postcasting drug loading. The preparation process consists of several steps. First, a hyaluronic acid solution containing the drug is prepared and poured into the mold. Next, the mold is dried, resulting in the formation of solid polymer needles. Finally, the needles are carefully detached from the mold and packaged for use (Figure 13.4). The molding method offers advantages, such as precise control over needle size, shape, and composition, and it also enables the incorporation of a variety of drugs for transdermal delivery. However, the method has limitations. The viscosity of the polymer solution can affect the reproducibility of needle formation, and removing the needles from the mold without damaging them can be challenging.

Additionally, the efficiency of drug loading into the needles may vary depending on the drug's properties and casting conditions.

Scanning electron microscopy images offer a detailed view of the microneedle structure and surface morphology for each formulation, emphasizing how the concentration of the active ingredient affects the physical characteristics of the microneedles. The results showed that MAPs prepared with 0%, 10%, 20%, and 30% phenylbutanoid extract–HPβCD complexes maintained good shape and exhibited dimensions of approximately 200 µm in base × 150 µm in height (Figure 13.5). These patches demonstrated excellent mechanical strength and effectively penetrated the stratum corneum layer of the skin. The dissolving process was assessed under a microscope using porcine skin samples at various time points: 0, 10, 20, and 30 minutes postapplication. All MAP formulations containing the plant extract dissolved in the skin within 30 minutes, as exemplified by patches with a 30% drug concentration (Figure 13.6). This indicates the efficient incorporation of the plant extract into the polymer matrix of the microneedles, which is released upon contact with the skin. The successful dissolution of the MAPs within 30 minutes highlights their potential for effective transdermal delivery. In vitro studies revealed higher transdermal delivery efficiency compared with traditional topical methods. However, further research is required to assess the pharmacokinetics, therapeutic efficacy, and safety profile of the plant extract when administered through these MAPs. In vivo studies, ideally using animal models followed by human trials, will provide valuable data on the systemic absorption and tissue distribution of the active compounds. These findings suggest that dissolving MAPs hold promise as an effective drug delivery system for the transdermal administration of anti-inflammatory phenylbutanoids.

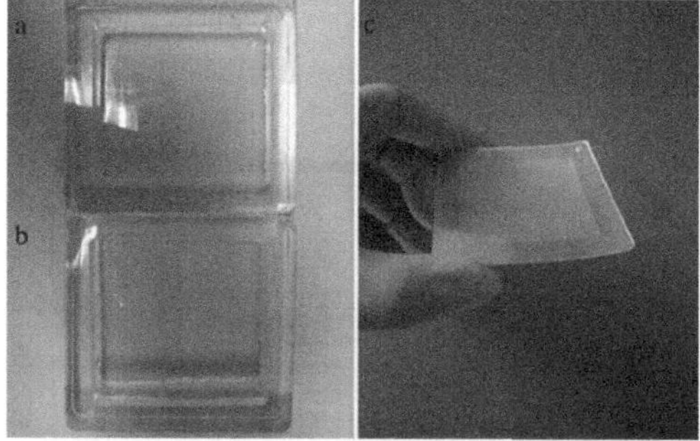

FIGURE 13.4 Preparation of microneedle array patches (MAPs) using the molding method: (a) represents a formulation containing a drug (phenylbutanoid extract–hydroxypropyl-β-cyclodextrin [HPβCD] complex), where the drug is incorporated into the polymer solution before casting; (b) depicts a blank formulation without the drug, serving as a control; and (c) shows the final appearance of the MAPs containing the phenylbutanoid extract–HPβCD complex, with solid polymer needles detached from the mold. Photograph by the author.

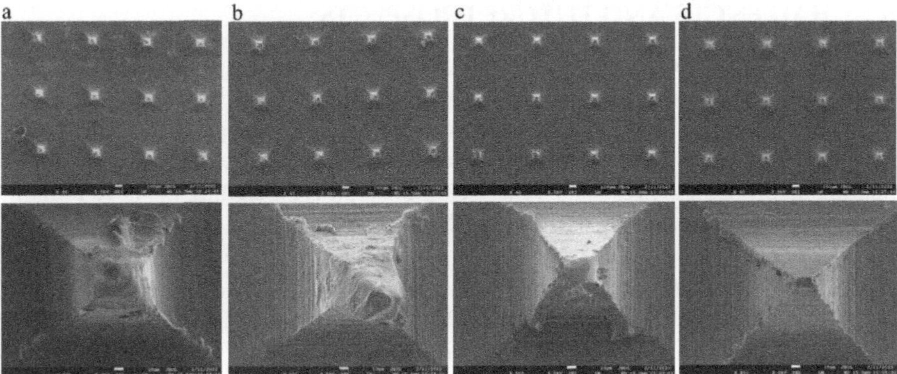

FIGURE 13.5 Scanning electron microscope images of microneedle array patches (MAPs) prepared with different concentrations of the phenylbutanoid extract–hydroxypropyl-β-cyclodextrin (HPβCD) complex. (a) MAPs prepared with 0% phenylbutanoid extract–HPβCD complex, (b) MAPs prepared with 10% phenylbutanoid extract–HPβCD complex, (c) MAPs prepared with 20% phenylbutanoid extract–HPβCD complex, and (d) MAPs prepared with 30% phenylbutanoid extract–HPβCD complex. Photograph by the author.

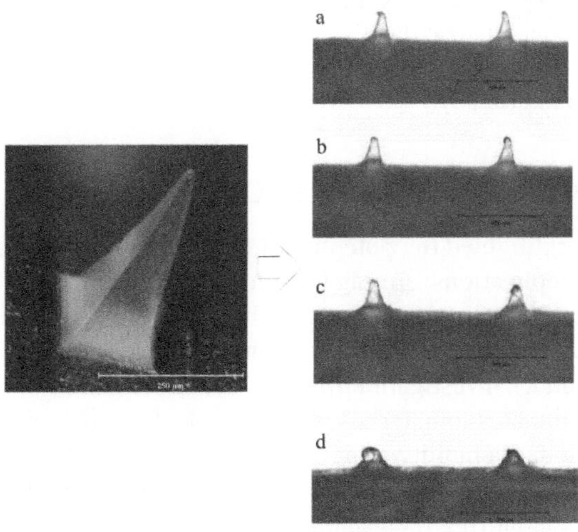

FIGURE 13.6 Dissolving properties of microneedle array patches (MAPs) containing a phenylbutanoid extract–hydroxypropyl-β-cyclodextrin (HPβCD) complex in porcine skin at different time points: (a) 0 minutes, (b) 10 minutes, (c) 20 minutes, and (d) 30 minutes. MAPs were prepared with the phenylbutanoid extract–HPβCD complex at a concentration of 30% relative to the weight of hyaluronic acid. Photograph by the author.

6 CHALLENGES AND FUTURE PROSPECTS

Regulatory challenges pose a significant barrier to the integration of biopolymers into the pharmaceutical and biomedical sectors. Agencies such as the FDA require time-consuming and costly tests that vary by jurisdiction, particularly affecting smaller entities struggling to navigate this complex landscape (Lemos et al. 2021). Scaling up production presents its own set of difficulties, including the need to maintain quality and consistency across large batches, as well as additional expenses and time for necessary modifications (Torres-Acosta et al. 2021). Furthermore, there are challenges in meeting regulatory criteria. Evaluating the cost-effectiveness of these biopolymers is complicated by higher upfront costs, a lack of standardized evaluation methods, and limited long-term data. Overall, the challenges span regulatory compliance, difficulties in scaling, and cost-effectiveness assessments. Concerted efforts for standardization, long-term studies, and more efficient manufacturing processes are required to truly unlock the potential of biopolymers in medical applications.

Despite the existing challenges, the future of biomaterials in pharmaceutical and medical applications looks bright, fueled by technological advancements such as additive manufacturing and 3D printing, sustainability concerns, and progress in tissue engineering. These technologies have democratized the production of biodegradable and nontoxic biomaterials that not only meet regulatory standards but also offer benefits such as reduced waste and fewer additional treatments. They also facilitate patient-specific customization in implants and drug delivery, improving patient outcomes while potentially reducing the need for further surgeries. However, these gains are tempered by challenges such as stringent regulatory hurdles and the high costs of new technologies, which could deter smaller companies. Tissue engineering, another promising field, particularly in creating scaffolds for tissue and organ regeneration, also faces obstacles, including material compatibility and concerns about long-term safety and efficacy.

7 CONCLUSION AND FUTURE SCOPE

This chapter has highlighted the potential of sustainable biomaterials in pharmaceutical and medical applications, aiming to address sustainability challenges within the industry by reducing waste, minimizing carbon footprints, and enhancing resource efficiency. Several newly discovered biopolymers, such as chitin, chitosan, pullulan, and alginate, are under investigation for their use in pharmaceutical and biomedical applications. Sustainable biomaterials offer several advantages over traditional materials, including biodegradability, renewability, and nontoxicity. The chapter covers various types of sustainable biomaterials, such as biopolymers, bioactive molecules, and nanocellulose, exploring their applications in drug delivery, tissue engineering, and medical devices. These materials have shown promising results in both pre-clinical and clinical studies, suggesting that they could replace traditional materials in the future. In summary, sustainable biomaterials have the potential to transform the pharmaceutical and medical industries by providing eco-friendly alternatives without sacrificing performance. Further research and development in this area are encouraged to expedite their incorporation into these fields.

REFERENCES

Aadil, Keshaw Ram, Sanu Awasthi, Raj Kumar, Sunil Dutt, and Harit Jha. 2023. "17-Advanced Functional Nanomaterials of Biopolymers: Structure, Properties, and Applications." In *Functional Materials from Carbon, Inorganic, and Organic Sources*, edited by Sanjay J. Dhoble, Amol Nande, N. Thejo Kalyani, Ashish Tiwari and Abdul Kariem Arof, 521–557. Woodhead Publishing. https://doi.org/10.1016/B978-0-323-85788-8.00015-X.

Ahmad Raus, Raha, Wan Mohd Fazli Wan Nawawi, and Ricca Rahman Nasaruddin. 2021. "Alginate and Alginate Composites for Biomedical Applications." *Asian Journal of Pharmaceutical Sciences* 16(3):280–306. https://doi.org/10.1016/j.ajps.2020.10.001.

Bansal, Priya, Deepti Katiyar, Surya Prakash, N. G. Raghavendra Rao, Vidhu Saxena, Vinay Kumar, and Abhishek Kumar. 2022. "Applications of Some Biopolymeric Materials as Medical Implants: An Overview." *Materials Today: Proceedings* 65:3377–3381. https://doi.org/10.1016/j.matpr.2022.05.480.

Carvalho, Tânia S. S., Nilza Ribeiro, Paula M. C. Torres, José C. Almeida, João H. Belo, J. P. Araújo, António Ramos, Mónica Oliveira, and Susana M. Olhero. 2023. "Magnetic Polylactic Acid-Calcium Phosphate-Based Biocomposite as a Potential Biomaterial for Tissue Engineering Applications." *Materials Chemistry and Physics* 296:127175. https://doi.org/10.1016/j.matchemphys.2022.127175.

Chakrapani, Gayathri, Mina Zare, and Seeram Ramakrishna. 2022. "Biomaterials from the Value-Added Food Wastes." *Bioresource Technology Reports* 19:101181. https://doi.org/10.1016/j.biteb.2022.101181.

Ciecholewska-Juśko, Daria, Adam Junka, and Karol Fijałkowski. 2022. "The Cross-Linked Bacterial Cellulose Impregnated with Octenidine Dihydrochloride-Based Antiseptic as an Antibacterial Dressing Material for Highly-Exuding, Infected Wounds." *Microbiological Research* 263:127125. https://doi.org/10.1016/j.micres.2022.127125.

Dang, G., X. Wen, R. Zhong, W. Wu, S. Tang, C. Li, B. Yi, L. Chen, H. Zhang, and M. Schroyen. 2023. "Pectin Modulates Intestinal Immunity in a Pig Model Via Regulating the Gut Microbiota-Derived Tryptophan Metabolite-AhR-IL22 Pathway." *Journal of Animal Science and Biotechnology* 14(1). https://doi.org/10.1186/s40104-023-00838-z.

Dasgupta, Amrita, Nardos Sori, Stella Petrova, Yas Maghdouri-White, Nick Thayer, Nathan Kemper, Seth Polk, Delaney Leathers, Kelly Coughenour, Jake Dascoli, Riya Palikonda, Connor Donahue, Anna A. Bulysheva, and Michael P. Francis. 2021. "Comprehensive Collagen Crosslinking Comparison of Microfluidic Wet-Extruded Microfibers for Bioactive Surgical Suture Development." *Acta Biomaterialia* 128:186–200. https://doi.org/10.1016/j.actbio.2021.04.028.

Dehghani, Niloofar, Fatemeh Haghiralsadat, Fatemeh Yazdian, Fatemeh Sadeghian-Nodoushan, Nasrin Ghasemi, Fahime Mazaheri, Mehrab Pourmadadi, and Seyed Morteza Naghib. 2023. "Chitosan/Silk Fibroin/Nitrogen-Doped Carbon Quantum Dot/α-Tricalcium Phosphate Nanocomposite Electrospinned as a Scaffold for Wound Healing Application: In Vitro and In Vivo Studies." *International Journal of Biological Macromolecules* 238:124078. https://doi.org/10.1016/j.ijbiomac.2023.124078.

Ferrara, Vittorio, Giovanni Zito, Giuseppe Arrabito, Sebastiano Cataldo, Michelangelo Scopelliti, Carlo Giordano, Valeria Vetri, and Bruno Pignataro. 2020. "Aqueous Processed Biopolymer Interfaces for Single-Cell Microarrays." *ACS Biomaterials Science and Engineering* 6(5):3174–3186. https://doi.org/10.1021/acsbiomaterials.9b01871.

Fu, Jinglin, and Kaitlyn Nguyen. 2021. "Reduction of Promiscuous Peptides-Enzyme Inhibition and Aggregation by Negatively Charged Biopolymers." *ACS Applied Bio Materials*. https://doi.org/10.1021/acsabm.1c01128.

Gegel, Brian, James Burgert, Brian Cooley, Jacob MacGregor, Jules Myers, Sean Calder, Ralph Luellen, Michael Loughren, and Don Johnson. 2010. "The Effects of Bleedarrest, Celox, and Traumadex on Hemorrhage Control in a Porcine Model." *Journal of Surgical Research* 164(1):e125–e129. https://doi.org/10.1016/j.jss.2010.07.060.

Gregory, David A., Caroline S. Taylor, Annabelle T. R. Fricker, Emmanuel Asare, Santosh S. V. Tetali, John W. Haycock, and Ipsita Roy. 2022. "Polyhydroxyalkanoates and Their Advances for Biomedical Applications." *Trends in Molecular Medicine* 28(4):331–342. https://doi.org/10.1016/j.molmed.2022.01.007.

Gudimetla, Avinash, Parveen Kumar, S. Sambhu Prasad, Satish Geeri, and V. V. N. Sarath. 2023. "Towards Smart Materials: Enhancing the Efficiency of the Materials." In *Modeling, Characterization, and Processing of Smart Materials*, edited by Ajay Kumar, Parveen Kumar, Ashish Kumar Srivastava and Vikas Goyat, 1–30. Hershey, PA: IGI Global. https://doi.org/10.4018/978-1-6684-9224-6.ch001.

Güler, M., A. Zengin, and M. Alay. 2023. "Fabrication of Glucose Bioelectrochemical Sensor Based on Au@Pd Core-Shell Supported by Carboxylated Graphene Oxide." *Analytical Biochemistry* 667. https://doi.org/10.1016/j.ab.2023.115091.

Guzmán, Eduardo, Francisco Ortega, and Ramón G. Rubio. 2022. "Chitosan: A Promising Multifunctional Cosmetic Ingredient for Skin and Hair Care." *Cosmetics* 9(5). https://doi.org/10.3390/cosmetics9050099.

He, Meifeng, Lvxin Chen, Meng Yin, Shengxiao Xu, and Zhenyu Liang. 2023. "Review on Magnesium and Magnesium-Based Alloys as Biomaterials for Bone Immobilization." *Journal of Materials Research and Technology* 23:4396–4419. https://doi.org/10.1016/j.jmrt.2023.02.037.

Hudecki, Andrzej, Gerard Kiryczyński, and Marek J. Łos. 2019. "Chapter 7- Biomaterials, Definition, Overview." In *Stem Cells and Biomaterials for Regenerative Medicine*, edited by Marek J. Łos, Andrzej Hudecki and Emilia Wiecheć, 85–98. Academic Press. https://doi.org/10.1016/B978-0-12-812258-7.00007-1.

Jafarzadeh, Hamid, Siavash Moushekhian, Narges Ghazi, Majdoddin Vahidi, Ali Bagherpour, Reyhaneh Shafieian, Shahin Moeini, Ali Kazemian, Amir Azarpazhooh, and Anil Kishen. 2023. "Bone Regeneration Effect of Nanochitosan with or without Temporally Controlled Release of Dexamethasone." *Journal of Endodontics* 49(5):496–503. https://doi.org/10.1016/j.joen.2023.03.001.

Jokar, Fatemeh, Somayeh Rahaiee, Mahboobeh Zare, Mehrab Nasiri Kenari, and Navideh Mirzakhani. 2023. "Bioactive Wound Dressing Using Bacterial Cellulose/Dextran Biopolymers Loaded with Pomegranate Peel Extract: Preparation, Characterization and Biological Properties."*Journal of Drug Delivery Science and Technology* 84:104461. https://doi.org/10.1016/j.jddst.2023.104461.

Joseph, Blessy, Jemy James, Nandakumar Kalarikkal, and Sabu Thomas. 2023. "Chapter 1- Advances in Biopolymer Based Surgical Sutures." In *Advanced Technologies and Polymer Materials for Surgical Sutures*, edited by Sabu Thomas, Phil Coates, Ben Whiteside, Blessy Joseph and Karthik Nair, 1–17. Woodhead Publishing. https://doi.org/10.1016/B978-0-12-819750-9.00008-5.

Jung, Hwabin, Lester C. Geonzon, Won Byong Yoon, and Shingo Matsukawa. 2023. "Change of Network Structure in Agarose Solution during Gelation Studied by Multiple Particle Tracking and NMR Measurements." *Food Hydrocolloids* 141:108740. https://doi.org/10.1016/j.foodhyd.2023.108740.

Khajavian, Mohammad, Vahid Vatanpour, Roberto Castro-Muñoz, and Grzegorz Boczkaj. 2022. "Chitin and Derivative Chitosan-Based Structures — Preparation Strategies Aided by Deep Eutectic Solvents: A Review." *Carbohydrate Polymers* 275:118702. https://doi.org/10.1016/j.carbpol.2021.118702.

Kikani, Twara, Sanskruti Dave, and Sonal Thakore. 2023. "Functionalization of Hyaluronic Acid for Development of Self-Healing Hydrogels for Biomedical Applications: A Review." *International Journal of Biological Macromolecules* 124950. https://doi. org/10.1016/j.ijbiomac.2023.124950.

King, A. M., C. Bray, S. C. L. Hall, J. C. Bear, L. K. Bogart, S. Perrier, and G. L. Davies. 2020. "Exploring Precision Polymers to Fine-Tune Magnetic Resonance Imaging Properties of Iron Oxide Nanoparticles." *Journal of Colloid and Interface Science* 579:401–411. https://doi.org/10.1016/j.jcis.2020.06.036.

Kruk, K., and K. Winnicka. 2023. "Alginates Combined with Natural Polymers as Valuable Drug Delivery Platforms." *Marine Drugs* 21(1). https://doi.org/10.3390/md21010011.

Kumar, Ajay, Parveen Kumar, Ahmad Sharif, Jyotsna Sharma, and Victor Gambhir, eds. 2023. *Handbook of Sustainable Materials, Modelling, Characterization, and Optimization.* CRC Press. https://doi.org/10.1201/9781003297772.

Kumar, Ajay, Parveen Kumar, Ashish Kumar Srivastava, and Vikas Goyat, eds. 2023. *Modeling, Characterization, and Processing of Smart Materials.* Hershey, PA: IGI Global. https:// doi.org/10.4018/978-1-6684-9224-6.

Kumar, Ajay, Parveen Kumar, Namrata Dogra, and Archana Jaglan. 2023. "Application of Incremental Sheet Forming (ISF) Toward Biomedical and Medical Implants." In *Handbook of Flexible and Smart Sheet Forming Techniques: Industry 4.0 Approaches,* edited by Vishal Gulati, Hari Singh, Parveen Kumar, Ajay Kumar and Pravin Kumar Singh, 247. Wiley.

Kumar, Ajay, Parveen Kumar, Ravi Kant Mittal, and Victor Gambhir. 2023. "Chapter 12-Materials Processed by Additive Manufacturing Techniques." In *Advances in Additive Manufacturing*, edited by Ajay Kumar, Ravi Kant Mittal and Abid Haleem, 217–233. Elsevier. https://doi.org/10.1016/B978-0-323-91834-3.00014-4.

Kumar, Ajay, Ravi Kant Mittal, and Abid Haleem. 2022. *Advances in Additive Manufacturing: Artificial Intelligence, Nature-Inspired, and Biomanufacturing.* Elsevier. https://doi. org/10.1016/C2020-0-03877-6.

Kumar, Ajay, Hari Singh, Parveen Kumar, and Bandar AlMangour. 2023. *Handbook of Smart Manufacturing: Forecasting the Future of Industry 4.0.* CRC Press. https://doi. org/10.1201/9781003333760.

Kunarbekova, M., K. Shynzhyrbai, M. Mataev, K. Bexeitova, K. Kudaibergenov, Ye Sailaukhanuly, S. Azat, K. Askaruly, Ye Tuleshov, S. U. Zhantikeyev, and D. Ybyraiymkul. 2023. "Biopolymers Synthesis and Application." *Materials Today: Proceedings.* https:// doi.org/10.1016/j.matpr.2023.02.367.

Lemos, P. V. F., H. R. Marcelino, L. G. Cardoso, C. O. D. Souza, and J. I. Druzian. 2021. "Starch Chemical Modifications Applied to Drug Delivery Systems: From Fundamentals to FDA-Approved Raw Materials." *International Journal of Biological Macromolecules* 184:218–234. https://doi.org/10.1016/j.ijbiomac.2021. 06.077.

Li, Xiaoxiao, Ying Luo, Fengbo Yang, Guoping Chu, Lingqiao Li, Ling Diao, Xiaoli Jia, Chunjing Yu, Xiaozhuo Wu, Wen Zhong, Malcolm Xing, and Guozhong Lyu. 2023. "In Situ-Formed Micro Silk Fibroin Composite Sutures for Pain Management and Anti-Infection." *Composites Part B: Engineering* 260:110729. https://doi.org/10.1016/j. compositesb.2023.110729.

Lin, Yixuan, Yang Chen, Ronghui Deng, Hao Qin, Nan Li, Yuting Qin, Hanqing Chen, Yaohua Wei, Zeming Wang, Qing Sun, Wenyi Qiu, Jian Shi, Long Chen, Yuguang Wang, Guangjun Nie, and Ruifang Zhao. 2023. "Delivery of Neutrophil Membrane Encapsulated Non-Steroidal Anti-Inflammatory Drugs by Degradable Biopolymer Microneedle Patch for Rheumatoid Arthritis Therapy." *Nano Today* 49:101791. https:// doi.org/10.1016/j.nantod.2023.101791.

Lv, Jiran, Xiaohui Lv, Meihu Ma, Deog-Hwan Oh, Zhengqiang Jiang, and Xing Fu. 2023. "Chitin and Chitin-Based Biomaterials: A Review of Advances in Processing and Food Applications." *Carbohydrate Polymers* 299:120142. https://doi.org/10.1016/j. carbpol.2022.120142.

Manna, S., and S. Jana. 2022. "Marine Polysaccharides in Tailor-made Drug Delivery." *Current Pharmaceutical Design* 28(13):1046–1066. https://doi.org/10.2174/13816128 28666220328122539.

Marketwatch, 2023. Biopolymers Market Recent Updates 2023–2028. Retrieved 24 April, 2023, from https://www.marketwatch.com/press-release/bcc-research-forecasts-industries-with-22-24-cagr-a-glimpse-into-rapid-growth-markets-98baa028

Martínez-Herrera, Raul E., María E. Alemán-Huerta, O. Miriam Rutiaga-Quiñones, Erick de J. de Luna-Santillana, and Temidayo O. Elufisan. 2023. "A Comprehensive View of Bacillus Cereus as a Polyhydroxyalkanoate (PHA) Producer: A Promising Alternative to Petroplastics." *Process Biochemistry* 129:281–292. https://doi.org/10.1016/j. procbio.2023.03.032.

Mehmood, Ahsan, Neelu Raina, Vanarat Phakeenuya, Benjamaporn Wonganu, and Kraipat Cheenkachorn. 2023. "The Current Status and Market Trend of Polylactic Acid as Biopolymer: Awareness and Needs for Sustainable Development." *Materials Today: Proceedings* 72:3049–3055. https://doi.org/10.1016/j.matpr.2022.08.387.

Moroni, Sofia, Shiva Khorshid, Annalisa Aluigi, Mattia Tiboni, and Luca Casettari. 2022. "Poly(3-hydroxybutyrate): A Potential Biodegradable Excipient for Direct 3D Printing of Pharmaceuticals." *International Journal of Pharmaceutics* 623:121960. https://doi. org/10.1016/j.ijpharm.2022.121960.

Narala, S., D. Nyavanandi, P. Mandati, A. A. A. Youssef, A. Alzahrani, P. Kolimi, F. Zhang, and M. Repka. 2023. "Preparation and In Vitro Evaluation of Hot-Melt Extruded Pectin-Based Pellets Containing Ketoprofen for Colon Targeting." *International Journal of Pharmaceutics: X* 5. https://doi.org/10.1016/j.ijpx.2022.100156.

Navabhatra, Abhiruj, Rawiwan Maniratanachote, and Bancha Yingngam. 2022. "Antiphotoaging Properties of Zingiber Montanum Essential Oil Isolated by Solvent-Free Microwave Extraction against Ultraviolet B-Irradiated Human Dermal Fibroblasts." *Toxicological Research* 38(2):235–248. https://doi.org/10.1007/s43188-021-00107-z.

Nie, Guangjun, Yipeng Zang, Wenjin Yue, Mengmeng Wang, Aravind Baride, Aliza Sigdel, and Srinivas Janaswamy. 2021. "Cellulose-Based Hydrogel Beads: Preparation and Characterization." *Carbohydrate Polymer Technologies and Applications* 2:100074. https://doi.org/10.1016/j.carpta.2021.100074.

Novikov, Vitaly Yu., Svetlana R. Derkach, Irina N. Konovalova, Natalya V. Dolgopyatova, and Yulya A. Kuchina. 2023. "Mechanism of Heterogeneous Alkaline Deacetylation of Chitin: A Review." *Polymers* 15(7):1729. https://doi.org/10.3390/polym15071729.

Otarola, J. J., A. K. Cobo Solis, M. E. Farias, M. Garrido, N. Mariano Correa, and P. G. Molina. 2020. "Piroxicam-loaded Nanostructured Lipid Carriers Gel: Design and Characterization by Square Wave Voltammetry." *Colloids and Surfaces A: Physicochemical and Engineering Aspects* 606. https://doi.org/10.1016/j.colsurfa. 2020.125396.

Ozlu, B., and B. S. Shim. 2022. "Highly Conductive Melanin-Like Polymer Composites for Nonenzymatic Glucose Biosensors with a Wide Detection Range." *ACS Applied Polymer Materials* 4(4):2527–2535. https://doi.org/10.1021/acsapm.1c01818.

Pires, Patrícia C., Filipa Mascarenhas-Melo, Kelly Pedrosa, Daniela Lopes, Joana Lopes, Ana Macário-Soares, Diana Peixoto, Prabhanjan S. Giram, Francisco Veiga, and Ana Cláudia Paiva-Santos. 2023. "Polymer-Based Biomaterials for Pharmaceutical and Biomedical Applications: A Focus on Topical Drug Administration." *European Polymer Journal* 187:111868. https://doi.org/10.1016/j.eurpolymj.2023. 111868.

Punj, Shivani, Jashandeep Singh, and K. Singh. 2021. "Ceramic Biomaterials: Properties, State of the Art and Future Prospectives." *Ceramics International* 47(20):28059–28074. https://doi.org/10.1016/j.ceramint.2021.06.238.

Puszkarska, Anna M., Daan Frenkel, Lucy J. Colwell, and Melinda J. Duer. 2022. "Using Sequence Data to Predict the Self-Assembly of Supramolecular Collagen Structures." *Biophysical Journal* 121(16):3023–3033. https://doi.org/10.1016/j.bpj.2022.07.019.

Rahman, Sameeha Syed Abdul, Saroja Pasupathi, and Sugumaran Karuppiah. 2022. "Conventional Optimization and Characterization of Microbial Dextran Using Treated Sugarcane Molasses." *International Journal of Biological Macromolecules* 220:775–787. https://doi.org/10.1016/j.ijbiomac.2022.08.094.

Rakhmatullayeva, Dilafruz, Aliya Ospanova, Zhanar Bekissanova, Ardak Jumagaziyeva, Balzhan Savdenbekova, Ayazhan Seidulayeva, and Aruzhan Sailau. 2023. "Development and Characterization of Antibacterial Coatings on Surgical Sutures Based on Sodium Carboxymethyl Cellulose/Chitosan/Chlorhexidine." *International Journal of Biological Macromolecules* 236:124024. https://doi.org/10.1016/j.ijbiomac.2023.124024.

Rashid, Adib Bin, Nazmir-Nur Showva, and Md Enamul Hoque. 2023. "Gelatin-Based Scaffolds: An Intuitive Support Structure for Regenerative Therapy." *Current Opinion in Biomedical Engineering* 26:100452. https://doi.org/10.1016/j.cobme.2023.100452.

Singh, Ram Sarup, Navpreet Kaur, Dhandeep Singh, Sukhvinder Singh Purewal, and John F. Kennedy. 2023. "Pullulan in Pharmaceutical and Cosmeceutical Formulations: A Review." *International Journal of Biological Macromolecules* 231:123353. https://doi.org/10.1016/j.ijbiomac.2023.123353.

Singh, Vandana, Deepak Kumar Tripathi, Vivek Kumar Sharma, Devika Srivastava, Umesh Kumar, Krishna Mohan Poluri, Brahma Nand Singh, Dinesh Kumar, and Venkatesh Kumar R. 2023. "Silk Fibroin Hydrogel: A Novel Biopolymer for Sustained Release of Vancomycin Drug for Diabetic Wound Healing." *Journal of Molecular Structure* 1286:135548. https://doi.org/10.1016/j.molstruc.2023.135548.

Song, J., X. Li, Y. Niu, L. Chen, Z. Wei, Y. Li, and Y. Wang. 2024. "pH-Sensitive KHA/ CMC-Fe3+@CS Hydrogel Loading and the Drug Release Properties of Riboflavin." *Particuology* 86:13–23. https://doi.org/10.1016/j.partic.2023.04.003.

Swetha, T. Angelin, Abhispa Bora, K. Mohanrasu, P. Balaji, Rathinam Raja, Kumar Ponnuchamy, Govarthanan Muthusamy, and A. Arun. 2023. "A Comprehensive Review on Polylactic Acid (PLA) – Synthesis, Processing and Application in Food Packaging." *International Journal of Biological Macromolecules* 234:123715. https://doi.org/10.1016/j.ijbiomac.2023.123715.

Syed, Murtaza Haider, Mior Ahmad Khushairi Mohd Zahari, Md Maksudur Rahman Khan, Mohammad Dalour Hossen Beg, and Norhayati Abdullah. 2023. "An Overview on Recent Biomedical Applications of Biopolymers: Their Role in Drug Delivery Systems and Comparison of Major Systems." *Journal of Drug Delivery Science and Technology* 80:104121. https://doi.org/10.1016/j.jddst.2022.104121.

Tiwari, Himanshu Kumar, Ashish Kumar Srivastava, Parveen Kumar, Manish Kumar Singh, Hritik Kumar, and Akshit Bhadauria. 2023. "Materials and Technology for Implant Manufacturing: Challenges and Opportunity." In *Modeling, Characterization, and Processing of Smart Materials*, edited by Ajay Kumar, Parveen Kumar, Ashish Kumar Srivastava and Vikas Goyat, 83–106. Hershey, PA: IGI Global. https://doi.org/10.4018/978-1-6684-9224-6.ch004.

Torres-Acosta, M. A., H. M. Castaneda-Aponte, L. M. Mora-Galvez, M. R. Gil-Garzon, M. P. Banda-Magaña, E. Marcellin, K. Mayolo-Deloisa, and C. Licona-Cassani. 2021. "Comparative Economic Analysis between Endogenous and Recombinant Production of Hyaluronic Acid." *Frontiers in Bioengineering and Biotechnology* 9. https://doi.org/10.3389/fbioe.2021.680278.

Vasquez-Martínez, N., D. Guillén, S. A. Moreno-Mendieta, P. Medina-Granados, R. G. Casañas-Pimentel, E. San Martín-Martínez, M. Á. Morales, S. Sanchez, and R. Rodríguez-Sanoja. 2023. "In Vivo Tracing of Immunostimulatory Raw Starch Microparticles after Mucosal Administration." *European Journal of Pharmaceutics and Biopharmaceutics* 187:96–106. https://doi.org/10.1016/j.ejpb.2023.04.013.

Wang, Lanqing, Zhenghong Xu, Han Zhang, and Cuiping Yao. 2023. "A Review on Chitosan-based Biomaterial as Carrier in Tissue Engineering and Medical Applications." *European Polymer Journal* 191:112059. https://doi.org/10.1016/j.eurpolymj.2023.112059.

Yang, Chao, Liang Yao, and Lei Zhang. 2023. "Silk Sericin-Based Biomaterials Shine in Food and Pharmaceutical Industries." *Smart Materials in Medicine* 4:447–459. https://doi.org/10.1016/j.smaim.2023.01.003.

Yao, Fei, Yanzhao Zheng, Yuhang Gao, Yan Du, and Fusheng Chen. 2023. "Electrospinning of Peanut Protein Isolate/Poly-L-Lactic Acid Nanofibers Containing Tetracycline Hydrochloride for Wound Healing." *Industrial Crops and Products* 194:116262. https://doi.org/10.1016/j.indcrop.2023.116262.

Yılmaz Nayır, Tülin, Selver Konuk, and Serdar Kara. 2023. "Extraction of Polyhydroxyalkanoate from Activated Sludge Using Supercritical Carbon Dioxide Process and Biopolymer Characterization." *Journal of Biotechnology* 364:50–57. https://doi.org/10.1016/j.jbiotec.2023.01.011.

Yingngam, Bancha, and Adelheid Brantner. 2018. "Boosting the Essential Oil Yield from the Rhizomes of Cassumunar Ginger by an Eco-Friendly Solvent-Free Microwave Extraction Combined with Central Composite Design." *Journal of Essential Oil Research* 30(6):409–420. https://doi.org/10.1080/10412905.2018.1503099.

Zarrintaj, Payam, Farzad Seidi, Mohamadreza Youssefi Azarfam, Mohsen Khodadadi Yazdi, Amir Erfani, Mahmood Barani, Narendra Pal Singh Chauhan, Navid Rabiee, Tairong Kuang, Justyna Kucinska-Lipka, Mohammad Reza Saeb, and Masoud Mozafari. 2023. "Biopolymer-Based Composites for Tissue Engineering Applications: A Basis for Future Opportunities." *Composites Part B: Engineering* 258:110701. https://doi.org/10.1016/j.compositesb.2023.110701.

Zhong, Guofeng, Mengyu Qiu, Junbo Zhang, Fuchen Jiang, Xuan Yue, Chi Huang, Shiyi Zhao, Rui Zeng, Chen Zhang, and Yan Qu. 2023. "Fabrication and Characterization of PVA@PLA Electrospinning Nanofibers Embedded with Bletilla Striata Polysaccharide and Rosmarinic Acid to Promote Wound Healing." *International Journal of Biological Macromolecules* 234:123693. https://doi.org/10.1016/j.ijbiomac.2023.123693.

Zhu, H., L. Zhang, F. Kou, J. Zhao, J. Lei, and J. He. 2024. "Targeted Therapeutic Effects of Oral Magnetically Driven Pectin Nanoparticles Containing Chlorogenic Acid on Colon Cancer." *Particuology* 84:53–59. https://doi.org/10.1016/j.partic.2023.02.021.

Zhu, Y., Q. Wei, Q. Jin, G. Li, Q. Zhang, H. Xiao, T. Li, F. Wei, and Y. Luo. 2023. "Polyethylene Glycol Functionalized Silicon Nanowire Field-Effect Transistor Biosensor for Glucose Detection." *Nanomaterials* 13(3). https://doi.org/10.3390/nano13030604.

14 Polymeric Biomaterials for Potential Pharmaceutical and Biomedical Applications

Shubhrat Maheshwari and Jagat Pal Yadav
Rama University and Sam Higginbottom University
of Agriculture, Technology and Sciences

Seema Yadav and Abhishek Singh
Amity University

Aditya Singh
Integral University

Bhupendra G. Prajapati
Ganpat University

Sudarshan Singh
Chiang Mai University

1 INTRODUCTION

Lignin is the second most common biopolymer after cellulose, accounting for approximately 10–30% (w/w) of total organic carbon in the biosphere (1). Lignocellulosic biomass (LCB) has sparked a lot of interest in energy generation due to its high cellulose content, particularly in the creation of bioethanol. However, lignin, a crucial component of LCB, lacks carbohydrate and is incompatible with bioethanol production. Historically, it has been regarded as an undesirable waste material and a low-value byproduct, frequently burnt to generate energy or heat. In recent years, world bioethanol production has expanded dramatically, driven by a desire to meet the world's energy needs while also lowering greenhouse gas emissions from fossil fuel consumption. The US biofuel production is estimated to exceed 60 billion gallons by 2030, while US refineries are expected to produce 0.225 billion tons of lignin (Halder et al. 2019). The chemical heterogeneity inherent in lignin has posed a significant obstacle to its utilization. Recent research has shown that lignin may be successfully extracted and even synthesized into valuable compounds; this chapter discusses

DOI: 10.1201/9781003434313-14

these findings in detail. The remarkable progress in science and technology has been instrumental in advancing the pharmaceutical industry, contributing to improved life expectancy and quality of life compared with a century ago. Additionally, the pharmaceutical sector plays a pivotal role in the global economy, with a substantial investment of nearly €35,200 in research and development in Europe alone in 2017. Despite these advancements, a considerable portion of the global population, estimated at around one-third, still lacks access to essential medicines. To address this disparity, reducing the costs associated with the raw materials and production of new medications is imperative. Lignocellulosic materials, therefore, hold significant potential for the facility. The substances that exist in nature or are artificially created, encompassing polymeric substances, metals, ceramic substances, and their combinations, which can be seamlessly integrated into the human body and materials that show an exceptional level of biocompatibility with cells and tissues are classified as biomaterials. Under the European Society for Biomaterials Consensus Conference-II, a biomaterial is a substance expected to interface within our biological systems, with the primary objective of assessing, enhancing, treating, or replacing any tissue, organ, or physiological function within the body (Biswal, BadJena, and Pradhan 2020). Biomaterials are generally categorized according to their intrinsic material characteristics; these include ceramic materials, polymer-based biomaterials, biopolymers, some active metals, and the composites that are composed of those materials. Additionally, these materials can be further differentiated according to their interaction or response to living tissue. Consequently, biomaterials are classified as biodegradable materials, bioactive materials, biomimetic materials, or biologically inactive materials. Specifically, materials that easily interact with bodily fluids and parts are essential for surgical operations involving tendons, knee or hip replacements, bones, and nerves. Materials that satisfy this requirement are referred to as biomaterials (Das, Ghosh, and Sarkar 2022; Pesode et al. 2023).

Backscattered Electron Microscopy (BSEM) is used to investigate the interaction of implanted biomaterials and bones. This technique allows for the visualization of interfaces between implanted biomaterials of varying densities and bone tissue at different maturation stages. Moreover, it facilitates the easy detection of cracks resulting from artifacts, rendering them visible. By directing electrons at a precise angle toward the surface of the biomaterial, BSEM enables the observation of interactions and reflections of varying energies. Subsequently, final energy is detected and represented graphically using gray tones through a photomicrograph. Based on this methodology, biomaterials can be classified into two categories (Abraham and Sridhar 2015; Aziz et al. 2022). Materials that promote a close and precise alignment of newly formed bone and materials that do not facilitate a close and precise alignment of newly formed bone.

Osteocytes are recognized within small micrometer distances from the material, exhibiting morphological characteristics of interlocking with the biomaterials. However, the term biomaterial refers to a substance engineered within a complex system. It may interact with living tissue in a controlled way when applied directly, which makes it appropriate for various types of therapeutic or diagnostic processes (Stoddart and Cleave 2009). Specific biomaterials play an important role in restoring function or completely replacing damaged or deteriorating organs and tissues. These materials include biodegradable polymers, ceramics, certain metals and alloys, bio-polymers, composites, biocomposite materials containing natural fibers, and other filler materials. They are chosen for their distinct biological, chemical, and physical characteristics (11).

1.1 Effective Biomaterials Display the Following Characteristics (12) (13)

- Biomaterials are typically nonviable or combined with other substances. Essentially, materials that lack the ability to grow, develop, or sustain life are categorized as biomaterials.
- These materials might be liquid or solid, synthesized or obtained from biological resources.
- Successfully employed in various applications, biomaterials play a crucial role in regenerating, replacing, repairing, or supplementing different parts of the body, organs, or tissues, contributing to both functional and structural aspects.
- Their application has a significant impact on maintaining a high quality of life for individuals.
- It is important to note that biomaterials are distinct from drugs and are not utilized in the manufacturing of pharmaceuticals.

2 SUSTAINABLE BIOMATERIALS

The idea of sustainable biomaterials is based on the synergistic relationship that exists between biomaterials and renewable natural resources. It holds very promising potential for the development of innovative and environmentally friendly solutions in the future (Tathe, Ghodke, and Nikalje 2010). Currently, significant efforts and attention are being given to the research and development of materials made from renewable resources, which will gradually replace conventional materials. In the contemporary era, there exists a critical need for sustainable energy, achievable through advancements in green technologies. The utilization of biomaterials is steadily gaining prominence, surpassing the usage of traditional materials (Kumar Gupta et al. 2015). Figure 14.1 depicts various biomedical applications of biomaterials. Several advanced techniques are employed for synthesizing diverse types of biomaterials tailored to specific properties. These methods include thermal and hydrothermal synthesis as well as the use of surfactant templates, which make it possible to create bionanomaterials precisely and in a controlled way (Pradhan, Panda, and Sukla 2018).

Various techniques are employed for the synthesis of nanomaterials for biological applications, with hydrothermal carbonization being widely utilized. This method uses a water medium with moderate heat and pressure. The resulting products from this method have versatile applications, including but not limited to drug delivery, CO_2 sequestration, acting as catalysts in various organic synthesis processes, serving as sensors, and playing a role in chemical vapor deposition (Rao et al. 2008). If biomaterials are developed and used in the following ways, they contribute to improved environmental sustainability:

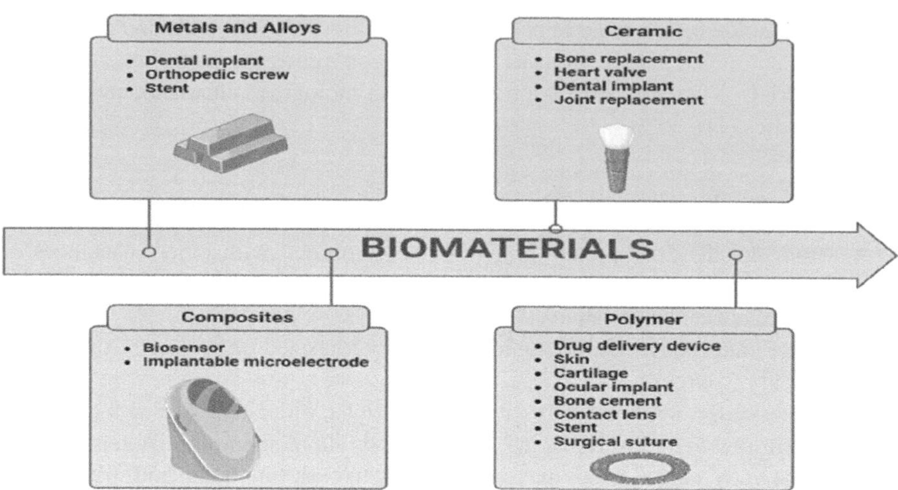

FIGURE 14.1 Various applications of biomaterials.

- Efforts are directed toward minimizing the volume of raw materials utilized, with a specific focus on the materials employed for packaging applications. Additionally, there is an increased emphasis on enhancing the quality of the manufactured products. This approach aims to optimize resource efficiency, reduce environmental impact, and ensure the production of high-quality goods.
- It is imperative to steer clear of single-use substances that cannot be recycled or composted. This approach aligns with sustainable practices by reducing environmental waste and promoting materials that can be reused or disposed of in an eco-friendly manner.
- A preference is given to products manufactured from renewable resources, actively avoiding materials derived from fossil fuels. This choice supports sustainability by encouraging the use of resources that can be replenished naturally, reducing dependence on finite and environmentally impactful fossil fuel-based materials.
- Sustainability must be upheld throughout the entire lifecycle of manufactured products. This includes considerations for the selection of materials, the responsible utilization of these materials during the product's life, efficient recovery of materials from waste, and the sustainable growth of feedstocks. This holistic approach ensures that environmental, economic, and social aspects are taken into account at every stage of a product's existence.
- Integrate sustainability into social, health, economic, and environmental concerns. This comprehensive approach ensures that sustainability principles are applied across multiple dimensions, fostering a balanced and responsible approach to development.
- Products are designed and manufactured with the aim of being recyclable, reusable, or compostable in nature. This strategy aligns with sustainable

- practices, promoting a circular economy by minimizing waste and encouraging responsible end-of-life disposal options.
- Both the state and central government should actively promote and support agricultural practices that are sustainable for farmers and other communities. This encouragement helps in fostering environmentally friendly and economically viable farming methods, ensuring the long-term well-being of agricultural communities and the broader environment.
- There is a recommendation to prohibit the use of genetically modified organisms (GMOs) in the production of feedstock. This suggestion aligns with concerns related to the environmental and health impacts associated with GMOs. Advocates for this position emphasize the importance of promoting natural and non-genetically modified alternatives in agricultural practices.
- Agricultural farms prefer to rely on chemicals that follow the 12 green chemistry principles.
- It is essential to impose restrictions on nanomaterial products and chemicals that pose environmental and public health hazards throughout their life cycle.
- The various stages of the production process should be decentralized.
- The field of sustainable biomaterials includes the use of renewable and natural resources to produce innovative and unique materials, and it also covers the sustainable production of bioenergy. It is a collaborative convergence of science, commerce, engineering, and technology (Lo et al. 2015; Miculescu et al. 2017). The prospective medical and pharmaceutical uses of lignin-based materials and compounds are exhaustively examined in this study. Significantly, the literature contains prior discussions regarding the pharmaceutical advantages of lignin and its derivatives (Vinardell and Mitjans 2017). This investigation encompasses data derived from articles published between 2010 and 2024, offering recent insights, particularly focusing on advancements in greener lignin extraction methods. The study systematically categorizes various types of lignins and delineates their pharmaceutical potential based on their distinct chemical structures.

Lignin-based hydrogels are explored in studies involving lignin–carbohydrate complexes, elucidating their diverse applications in the pharmaceutical realm. Notably, the potential of ligno-phenols is underscored throughout the discussion. The overarching purpose is to offer a comprehensive and up-to-date perspective, endorsing further research endeavors in the domain of lignin-based pharmaceutical production. The synthesis of recent findings, along with the systematic categorization of lignin types and their applications, is anticipated to contribute significantly to the ongoing progress in this field.

3   SOURCE AND STRUCTURE OF LIGNIN

Lignin, a 3-D amorphous polymer, is one of the essential components of LCB (Salehi Jouzani and Taherzadeh 2015). Agro-industrial activities produce significant amounts of lignocellulosic waste, which includes non-edible constituents from edible or non-edible plants. These waste items include straws, shells, leaves, bunches, and bagasse. During the early phases of the harvesting process, a part of these waste products is usually separated. Despite their high carbohydrate content, they are frequently left behind in the field, used as animal feed, or burned (Cardona and Sánchez 2007). The chemical composition of biomass and the amount of lignin it contains are determined by the particular source of agriculture or the current conditions of growth. Given the diverse structure of lignin, it becomes imperative to devise multiple methods for isolating the targeted chemical fractions from this complex polymer (Klugman 2015). Lignin represents a high molecular weight, intricate aromatic heteropolymer primarily constituted by three distinct monomers or monolignol units: (i) p-coumaryl alcohol-H-units, (ii) guaiacyl alcohol-G-units, and sinapyl alcohol. Lignin is a large, complex molecule comprising three distinct building blocks: p-coumaric alcohol, sinapyl alcohol, and guaiacyl alcohol. These building blocks represent the basic components that combine to produce lignin-S-units (Ponnusamy et al. 2019; Suseela 2019) (Figure 14.1). Hardwood lignin has lower molecular weight than softwood lignin. Softwood lignin is made up of approximately 90% guaiacyl alcohols (G-units), whereas hardwood lignin has a more balanced composition, with equal quantities of guaiacyl (G) and sinapyl (s) units (Suota et al. 2021). Given that the β-O-4 aryl ether bond constitutes the predominant linkage, comprising around half of ether bonds found in lignin, its degradation is a critical process in both softwood and hardwood lignin breakdown (Higuchi et al. 2019; Reiter et al. 2013).

4 TYPES OF LIGNIN AND CURRENT LIGNINS

4.1 Extraction Processes

The type and purity of lignin fractions generated are influenced by the biomass source as well as the extraction and isolation conditions utilized. Figure 14.2 provides a summary of current lignin types, categorized into four major groups, as outlined below.

5 SULFUR-FREE LIGNINS

Soda lignin is sulfur-free lignin derived from the soda pulping process. This method, developed by Watt and Burgess in 1853, stands as one of the oldest techniques for extracting cellulose from LCB (Hagiopol and Johnston 2011). In the soda pulping process, crop residues like bagasse, flax, and straw serve as raw materials (Sameni et al. 2020). This specific process involves treating the biomass at a temperature of 150–170°C with aqueous NaOH. The process results in lignin breakdown and the release of carbohydrates, which leads to a relatively low selectivity in lignin extraction (Macfarlane, Mai, and Kadla 2014). To address this limitation, certain additives or catalysts, such as anthraquinone, may be introduced. These additives play a role in mitigating carbohydrate degradation and facilitating lignin extraction specificity. Although the organosolv method is acknowledged for its comparatively high cost, the end product is regarded as being of superior quality, devoid of sulfates (although stays poorly water-soluble).

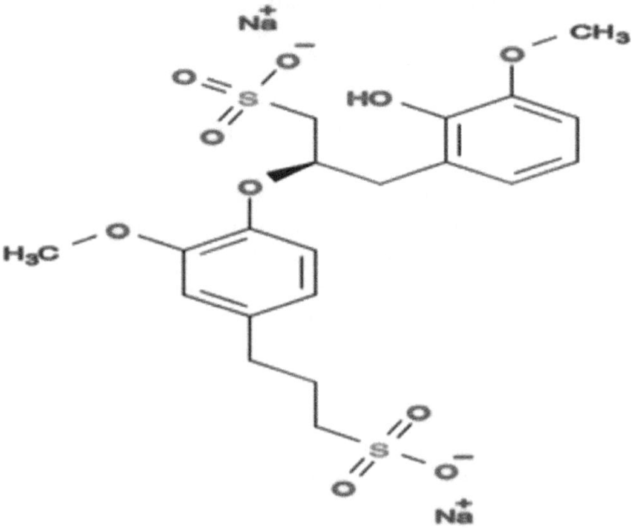

FIGURE 14.2 Macromolecular structure of lignin, including the main monolignol units. S-units (sinapyl alcohol) are shown in blue, H-units (p-coumaryl alcohol) in green, and G-units (guaiacyl alcohol) in red.

Organosolv lignin exhibits an elevated content of phenol hydroxyls and carbonyl groups (de la Torre et al. 2013), rendering it particularly promising for the production of epoxy resins and phenolics. The solvents that are used in the organosolv process, on the other hand, are usually higher in cost compared with the chemicals that are used in traditional pulping procedures. This is a key aspect to take into consideration (Domínguez et al. 2008). In a prior study, an organosolv technique combined with membrane filtering was investigated, revealing a production cost of approximately 52 €/ton for the obtained lignin. Considerably more expensive than Klason lignin, which was 33 €/ton (Alriols et al. 2010). Another comprehensive assessment of various lignin extraction techniques considered both economic and environmental impacts. This evaluation disclosed that the production costs and potential environmental effects associated with organosolv extraction surpassed those of alternative methods, such as soda extraction, lignosulfonate extraction, and Kraft extraction. For instance, steam explosion, a well-studied physicochemical pre-treatment method, involves the breakdown of LCB structural components using hot steam (180–240°C) under elevated pressure (1.0–3.5 MPa) (Bandyopadhyay-Ghosh, Ghosh, and Sain 2015).

6 SULFUR-CONTAINING LIGNINS

Sulfur-containing lignins, such as those found in Kraft lignin, can have significant impacts on their properties and applications. These lignins often contain sulfur in various forms like free sulfur, sulfides, and other sulfur-containing species. The presence of sulfur can inhibit the depolymerization of Kraft lignin, affecting its processability and potential uses. Kraft lignin, the most widely recognized type of lignin, is derived from commercial pulping processes.

Around 45 million tons of Kraft lignin are manufactured annually on a global scale, with the majority (98–99%) being incinerated to release energy. Only a small fraction, nearly 1–2% of this substance is utilized in chemical manufacturing (Demuner et al. 2019). A sulfuric acid precipitation is the conventional technique utilized in the recovery of lignin from Kraft liquor (Zhu and Theliander 2015). It is essential to maintain a chemical equilibrium between sulfur and sodium during the Kraft process to reduce production expenses. By separating the precipitated lignin from the liquid by filtration, one can obtain a product that comprises sulfur at a concentration of 1.5–3% (w/w) (Torres et al. 2020). As compared with Kraft lignin, lignosulfonates typically contain a large number of sulfur groups, making them anionic and water-soluble. Consequently, their precipitation by acidification is less straightforward (Li and Takkellapati 2018).

Ultrafiltration and amine extraction are frequent techniques employed to separate lignosulfonates. However, these methods face several challenges, including the need to eliminate amine, the formation of, NaCl, difficulties with foam and emulsion, and lengthy separation procedures (Pan et al. 2005). To isolate high molecular weight lignosulfonates and impurities, consecutive ultrafiltration processes using membranes with varying molecular weight cutoffs (MWCO) are carried out. The application of membrane-based filtration is considered an effective and sustainable method for recovering highly pure lignin (Domínguez et al. 2008). The incorporation of a fermentation step before membrane filtration is an alternative strategy that can be utilized to make use of residual sugars.

The global annual lignosulfonate production is anticipated to grow, reaching an estimated 1.75 million tons by 2025. Lignosulfonates hold a dominant position in the technical lignin market, constituting approximately 80% of the total production (Dessbesell et al. 2020). These substances are utilized in a wide range of products, including water-reducing concrete additives, coal briquettes, and plywood production (Niaounakis 2015). Redirecting a portion of lignin from lower-value applications to higher-value chemicals could significantly enhance the overall economics of the procedure.

7 NEXT-GENERATION "GREENER" LIGNINS

Ionic liquids have emerged as environmentally friendly alternatives to traditional organic solvents, given their capacity to dissolve significant amounts of biomass and their potential for multifunctionality. Various ionic liquids are used as a pretreatment agent to fractionate LCB, improving cellulose digestibility. The solubility of protonic ionic liquids, which are generated by an acid–base neutralization process, is associated with the cation and anion structures. These liquids can dissolve lignin at a concentration of up to 20% (w/w). They are economically viable and exhibit a high efficiency in extracting lignin (Lee et al. 2009). According to recent surveys, ionic liquids containing aromatic cations are considerably more soluble in lignin than those containing nonaromatic cations. However, the precise mechanism of lignin dissolution remains incompletely understood, necessitating further investigation into the potential applications and properties of the extracted lignin (Hossain and Aldous 2012). Another alternative to ionic liquids is deep eutectic solvents, characterized by their non-flammability, non-toxicity, and, at times, biodegradability. These solvents, composed of hydrogen-bond acceptors and donors, are less expensive than ionic liquids and can be easily produced from commonly accessible ingredients. Recent studies using deep eutectic solvents for the separation of lignin from rice straw showed promising results, with almost 60% of lignin separated at high purity (>90%) (Wang et al. 2017). Selective lignin extraction from loblolly pine, using various deep eutectic solvents, resulted in different lignin solubilities (range: 9–14%, w/w). The effectiveness of these solvents is especially dependent upon the chemical constituents of the biomass. Despite these advances, there is a lack of research on the characterization of deep eutectic lignin and its possible medicinal applications.

8 OPPORTUNITIES AND CHALLENGES IN THE PRODUCTION OF LIGNIN-BASED PRODUCTS

The economic viability and long-term sustainability of paper mills and biorefineries are dependent on their use and, ideally, valorization of the three primary biopolymers found in lignocellulosic biomass. As discussed previously, extracting particular compounds from lignin is difficult and time-consuming. Although lignin has the potential to yield numerous products, only a few of them are considered to be viable for large-scale industrial production (Ferri et al. 2020). To transform lignin into pharmaceutical compounds, numerous strategies have been developed. However, the extraction of the desirable compounds from lignin produced by biomass unveils a significant challenge from an engineering perspective. This problem arises from the complicated structure of lignin (see Figure 14.2) and often low concentrations of the target substances, which are typically less than 10%. Moreover, the majority of thermochemical procedures used in lignin fractionation lead to structural alterations, whereas (Lancefield et al. 2016) enzymatic treatments that are more selective are typically expensive (Wang et al. 2019). It has been found that conventional lignin degradation techniques are hindering the synthesis of lignin-based pharmaceuticals and chemicals. Alternative sources were studied for recent advances in the identification of bacterial enzymes that degrade lignin and their capacity to produce renewable compounds. On the other hand, biocatalytic, chemocatalytic, and non-catalytic thermochemical transformations are all viable methods for lignin conversion. Lancefield et al. (2016) offer the advantage of simultaneously releasing several pharmaceutical precursors having high selectivity and yield. Importantly, these strategies maintain the existing lower-value applications of the residual lignin, making them particularly attractive. The scavenging ability of phenolic structures against reactive oxygen species containing oxygen is widely recognized as the underlying mechanism by which lignin exerts its antioxidant influence. A previous study looked at how different lignin extraction procedures affected the antibacterial capabilities of several lignin-based products. Despite demonstrating the feasibility of pharmacological and therapeutic applications, the approaches addressed in these papers have yet to be commercialized.

9   PHARMACEUTICALS AND MEDICAL APPLICATIONS OF LIGNIN-DERIVED COMPOUNDS

The main monomeric phenolic compounds produced from lignin. Vanillin (4-hydroxy-3-methoxybenzaldehyde) is often found in the highest proportions, accounting for roughly 20% of the total phenolic content. At present, vanillin is as a single phenolic compound manufactured on a large scale from biomass, primarily through the Kraft process (Fache, Boutevin, and Caillol 2016). The annual production exceeds 9,000 tons. Just 5% of the world's vanilla production originates from the pods of the Vanilla orchid, with a staggering 95% of the vanillin being synthetically manufactured. Interestingly, 15% of this synthetic vanillin is obtained from lignin. Several methods have been developed to transform Kraft-produced lignin and ferulic acid into vanillin. Vanillin, the primary flavoring component of vanilla, finds broad applications in industries, such as food and beverages, fragrance, and the production of different pharmaceutical compounds (Karagoz et al. 2023). It is a part of several key drugs like Aldomet (used to control high blood pressure), L-dopa (helps in managing Parkinson's disease), and Trimethoprim (combats respiratory tract infections and some particular venereal diseases) (Gulsia 2020). Vanillin is also an anti-inflammatory substance that can be employed to develop medicinal products to address arthritis, allergies, and gout (Gulsia 2020). Vanillin possesses numerous pharmacological functions, including anti-cancer, anti-diabetic, antioxidant, antisickling, antibacterial, and anti-inflammatory properties (Olatunde et al. 2022). Several studies have explored the potential anti-cancer effects of vanillin and compounds derived from vanillin. One such study examined the antitumor effect of vanillin semicarbazone in vivo in Ehrlich ascite cancer cells in Swiss albino mice. The study revealed that administering vanillin semicarbazone (having a dose 10 mg kg^{-1}) significantly increased the lifespan of mice by more than 88.9%. This suggests that vanillin semicarbazone possesses potent anti-cancer properties (Ali et al. 2012). Elsherbiny et al. examined the combined effect of doxorubicin and vanillin, an anthracycline antibiotic commonly used in breast cancer chemotherapy. Their study demonstrated that administering vanillin with a dosage of 100 mg kg^{-1} protected against the toxic effects of doxorubicin. Moreover, the average survival time of mice that received doxorubicin alone (33.7 days) was considerably shorter than that of mice that had been administered with the combination of doxorubicin and vanillin (88.3 days) (Elsherbiny et al. 2016). Vanillic acid is an oxidized derivative of vanillin. Vanillic acid is known for its several pharmacological advantages, such as antioxidant, anti-inflammatory, immunostimulant, cardioprotective, hepatoprotective, neuroprotective, and antiapoptotic activities. Vanillic acid has been shown to reduce cognitive impairment and oxidative stress caused by Aβ1–42, making it a promising treatment for Alzheimer's disease (Sabzi and Jawad 2023). The study revealed that male albino rats administered with vanillic acid (at doses ranging from 50 to 100 mg kg^{-1}) experienced a restoration of elevated renal function levels and a reduction in antioxidant status, bringing it closer to normal levels compared with the animal group which has been treated solely with cisplatin. The findings suggested that both vanillin and vanillic acid could be employed as a combined regimen in cancer therapy (Table 14.1) and showed the lignin-derived bioactive molecules used in various diseases.

TABLE 14.1

Lignin-derived bioactive molecules used in various diseases

Bioactive molecule	Bioactivity	Model used	Activity/Finding	Reference
• Vanillin	• CNS Alzheimer's disease (AD)	• Interacts with the amyloid-β (Aβ)	• It has been found to disintegrate the Aβ plaques	• (Song et al. 2016)
• Ethyl vanillin	• AD	• PC12 cells treated with β-amyloid	• It alleviates brain damage by inhibiting oxidative stress and apoptosis	• (Zhong et al. 2019)
• Vanillin	• Anti-cancer	• Wild-type *E. coli* strain (NR9102)	• Antimutagenic, lowering the possibility of diseases linked to genetic damage or mutations	• (Shaughnessy et al. 2006)
• Anti-hyperlipidaemic and anti-hyperglycaemic	• Ethyl vanillin	• High glucose-triggered NRK-52E cells	• It Alleviates diabetic nephropathy	• (Tong et al. 2019)
	• Vanillin	• HepG2 cells	• Increases low-density lipoprotein receptor mRNA levels and inhibits the expression of the 3-hydroxy-3-methylglutaryl-coenzyme A reductase gene	• (Al-Naqeb et al. 2010)
• Anti-inflammatory, and Analgesic	• Vanillin	• Mesenchymal stem cells isolated from adipose tissue	• Lowers prostaglandin E2 levels	• (Lee et al. 2016)
• Anti-microbial	• Vanillin-coated gold nanoparticles	• *P. aeruginosa* strain	• Retards bacterial growth	• (Arya et al. 2019)
• Antioxidant	• Vanillin	• Inhibition of DPPH and ABTS radicals	• Suppresses oxidative damage	• (Tai, Sawano, and Yazama 2011)
• Anti-sickling	• Vanillin	• Suppression of SS cells	• Inhibits sickling of homozygous sickle red blood (SS) cells and enhances HbS affinity for oxygen	• (Safo et al. 2004)
• Cardioprotective	• Vanillin	• H9c2 cell line	• Decreases the production of reactive oxygen species and attenuates cell death	• (Sirangelo et al. 2020)
• Antioxidant	• Vanillic acid-coated chitosan microparticles	• H9c2 cardiomyocytes	• Reduces myocardial apoptosis and oxidative stress in the heart	• (Vishnu et al. 2018)
• Anti-cancer	• Vanillic acid	• Mice injected with B16BL6 melanoma and obesity induced by a high-fat diet	• Vanillic acid stimulates STAT3-mediated autophagy to prevent cancer progression and thermogenesis to alleviate obesity in cancer-obesity comorbidity	• (Park et al. 2020)

9.1 FERULIC ACID

Ferulic acid (4-hydroxy-3-methoxycinnamic acid) is a lignin-derived phenolic acid which can be converted into vanillin and vanillic acid. Ferulic acid (FA) is typically found in plant cell walls, cross-linked using hemicelluloses via ester linkages. Traditional Chinese medicine has utilized it to treat cerebrovascular and cardiovascular illnesses. It can be obtained using alkaline procedures, deep eutectic solvents, or hot water. It possesses numerous beneficial effects, including scavenging free radicals, serving as an antioxidant, reducing inflammation, preventing diabetes, and preventing cancer (Kumar and Pruthi 2014). A recent study found that FA reduces hepatic fibrosis by inhibiting the TGF-β1/Smad pathway both in vitro and also in vivo. These findings suggest that FA may help prevent liver fibrosis (Mu et al. 2018). FA may help lower pain sensitivity in rats suffering from chronic constriction injury, probably because of its capacity to combat inflammation and oxidative stress. It might work effectively in conjunction with other drugs. Additionally, FA is shown to protect from nerve pain (Aswar and Patil 2016). Johnson et al. (Johnson et al. 2020) developed an oral prodrug that selectively inhibits colorectal cancer cells by integrating an FA molecule along with fructo-oligosaccharide. In addition, scaffold materials for antioxidants have been reported to be biocompatible hydrogels composed of poly(N-isopropylacrylamide) (PNIPAM) as well as copolymers cross-linked with *N,N*-methylene bisacrylamide (BIS) (Casadey et al. 2020). A nanoformulation was developed by Hassanzadeh et al. (Hassanzadeh, Arbabi, and Rostami 2021) incorporating silk fibroin to serve as biomimetic material that was coated using FA, and modified with neutrophil membranes. The pharmacological profile of FA was enhanced, and targeted distribution throughout the inflammatory pancreatic lesion was more effectively facilitated by the synthesized nanoparticles. Praphakar et al. developed amphiphilic polymers leveraging chitosan, ε-caprolactone, and covalently attached FA to target antitubercular drug delivery (Praphakar, Munusamy, and Rajan 2017). To facilitate the concurrent delivery of resveratrol and FA, Poornima and Korrapati (Poornima and Korrapati 2017) designed novel nanofibers composed of chitosan grafted with polycaprolactone.

9.2 COUMARIC ACID

ρ-Coumaric acid is the most frequently found natural form of coumaric acid, also known as the hydroxy derivative of cinnamic acid. It exhibits many kinds of beneficial properties, including gastroprotective, anti-ulcer, cardioprotective, liver-protective, renoprotective, gout prevention, anti-cancer, anti-inflammatory, and anti-diabetic properties (Kaur and Kaur 2022). Taniguchi et al. reported that calocoumarin B, methoxyinophyllum P, and calophyllolide, all of which were extracted from the leaves of *Rhizophora mucronata*, exhibited anti-cancer properties against HeLa cells (cervical cancer) at IC50 values of 36.4, 3.8, and 29.9 µM, respectively (Taniguchi et al. 2018). Its antioxidant activity has been investigated in multiple experiments with phenolic substances like caffeic and FAs. A different study investigated at how ρ-coumaric acid prevented type 2 diabetic rats' hippocampus neurotoxicity (Abdel-Moneim et al. 2017). This study demonstrated that the strong anti-inflammatory, antioxidant, and anti-apoptotic characteristics of ρ-coumaric acid may prevent hippocampus neurodegeneration. As a result, investigators suggested that this chemical could be used as a potential adjuvant against diabetes neurodegeneration.

9.3 Syringic Acid

Syringic acid, also a phenolic with high antioxidant properties, is obtained through alkaline hydrolysis (4-hydroxy-3, 5-dimethoxybenzoic acid). It can be used to alleviate several conditions, including liver damage, diabetes, and cancer. It can change the dynamics of various biological targets, including growth factors and transcription factors (Srinivasulu et al. 2018). Syringic acid as well as other phenolic compounds have been extracted from leaves and barks of various Quercus species, which are small oak trees, to evaluate their biological activity. Quercus infectoria has been identified as a prominent traditional remedy in Asia, utilized for the treatment of dentition and wound infections (Basri et al. 2012). In DMBA-treated hamsters, oral syringic acid at 50 and 100 mg kg^{-1} bw may prevent adverse fluctuations in plasma or buccal mucosal tissue biochemical parameters and down-regulate molecular markers (PCNA, mutant p53, and Cyclin D1) (Velu et al. 2017). Oral syringic acid therapy (50 mg kg^{-1} body weight) restores to normal levels carbohydrate enzymes implicated in metabolism, as well as hepatic and renal marker enzymes, in male albino Wistar rats model diabetic rats induced with alloxan. Syringic acid mitigated alloxan-induced pancreatic damage in diabetic rodents and stimulated β-cell regeneration (Srinivasan et al. 2014). Advanced glycation end products produced internally are linked to endogenous diabetes, Alzheimer's disease, and heart attack complication. In vitro, syringic acid binds to serum albumin and inhibits the glycation pathway, preventing its structural change (Bhattacherjee and Datta 2015). Rats exposed to acetaminophen-induced hepatotoxicity have strong hepatoprotective and antihyperlipidaemic effects when compared with the standard hepatoprotective drug silymarin. After undergoing syringic acid treatment, the plasma of the treated rats had higher levels of HDL and lower levels of triglycerides, FFA, LDL, and VLDL (Ramachandran 2010).

9.4 Eugenol

Eugenol (4-allyl-2-methoxyphenol) is a compound produced from lignin found in woody biomass. It can be transformed into vanillin and FA by various metabolic processes (Ashengroph et al. 2011). Eugenol is commonly used in the food industry as an additive and flavoring. Additionally, some research has been conducted on its anti-microbial and antioxidant properties. It has been reported that eugenol-grafted chitosan hydrogel exhibits antioxidant properties. Low-density lipoprotein oxidation and lipid peroxidation can both be significantly suppressed by eugenol (Jung, Chung, and Lee 2006). Eugenol was combined with a standard antibiotic to determine the synergistic action against Gram-negative bacteria. A 50% decrease of membrane integrity was seen in the eugenol-treated cells, which increased the activity of the antibiotics under study. Eugenol combined with two antibiotics—vancomycin and a β-lactam—showed a synergistic effect by increasing membrane damage in bacteria.

Additionally, it has been shown that the combination of eugenol and vancomycin has improved penetration and produced greater antibacterial activity (Hemaiswarya and Doble 2009). Eugenol exhibits strong antihypercholesterolemic and anti-atherogenic properties. In intraperitoneally injected Wistar male rats that were hyperlipidemic due to injections of Triton WR-1339 (300 mg kg^{-1} per body weight), Venkadeswaran et al. (2014) reported the hypolipidemic effects of eugenol. Eugenol showed greater efficacy against elevated lipid and cholesterol levels when compared with a lipid-lowering medication such as lovastatin. Total cholesterol decreased by 55.88% as a result of it; LDL-C decreased by 79.48% and triglycerides decreased by 64.30%. In another study, Srinivasan et al. treated diabetic male rats with streptozotocin (40 mg kg^{-1} per body weight). To assess the effectiveness of eugenol, the activity of the major enzymes involved in glucose metabolism was investigated. However, in plasma (46.15%) samples, only 10 mg kg^{-1} per body weight dosage of eugenol significantly increased insulin and hepatic glycogen while lowering glycosylated hemoglobin (25.70%) as well as blood glucose (70%). Lethal impact on growth of several fungal strains, including *Aspergillus ochraceus*, *Penicillium citrinum*, *Candida tropicalis*, *Fusarium graminearum*, *F. moniliforme*, and *Tricophyton rubrum*, have been shown by eugenol. Eugenol either disrupts the integrity of the fungal cell membrane or induces cell cycle arrest (Zore et al. 2011). Furthermore, eugenol is reported to have analgesic, anti-pyretic, neuroprotection, dentistry, anti-viral, and cardioprotective activities (Nisar et al. 2021).

10   LIGNIN-DERIVED FRACTIONS AND COMPLEX WITH PHARMACEUTICAL APPLICATIONS

Technical lignins or depolymerized lignins can be conveniently fractionated through solvent fractionation, which produces a series of fractions with differing molecular weights and hydrophobicities that can be used in various ways. According to reports, these fractions exhibit increased antibacterial, antioxidant, and UV absorption capabilities—the latter of which might be beneficial for sunscreen application (Widsten 2020). If a single active species can be isolated, these fractions can be further purified or used as complex mixtures.

10.1   LIGNOPHENOLS

Lignophenol, a functional polymer derived from lignin, is isolated from lignin through phase separation using concentrated acid and phenol derivatives (Norikura et al. 2010). Despite having strong antioxidant characteristics not enough is known about the physiological function and possible medical uses of lignophenols. Many in vitro and in vivo have demonstrated the pharmacological properties of lignophenols in the literature. It has been observed that bamboo-derived lignophenols may inhibit the death of cells in vitro caused by hydrogen peroxide (Akao et al. 2004). In HepG2 cells, lignophenols inhibited sterol regulatory element binding protein-2 and apolipoprotein-B secretion produced by oleate. In the meantime, HepG2 cells' oleate-induced apo-B production was significantly and dose-dependently reduced by lignophenol administration. By reducing the rate of elimination through the rough endoplasmic reticulum, which is regulated by the microsomal triglyceride transfer protein, LPs may, at least in part, decrease apo-B production. Furthermore, in a dose-dependent manner, lignophenols lower the cholesterol levels in HepG2 (Norikura et al. 2010).

10.2 LIGNOSULFONIC ACID

Salts of lignosulfonic acid, or water-soluble lignosulfonates, are the primary byproduct generated during sulfite pulping. It has been shown that they are highly valuable raw materials for fine compounds like vanillin. The pulp and paper industry produces lignosulfonic acid (LSA), a low-cost polyanionic macromolecule based on lignin, as a byproduct (Bjørsvik and Liguori 2002). It studied how lignosulfonic acid affected the absorption of glucose in the intestine (Hasegawa et al. 2015). Lignosulfonic acid has been found to bind to enzymes as well as the enzyme–substrate complex due to its reversible nature and non-competitive inhibitory effect on α-glucosidase. The slow interaction of lignosulfonic acid with the enzyme during preincubation increased the inhibitory effect on α-glucosidase activity. When glucose with lignosulfonic acid was administered, there was a reduction in blood glucose levels and a delay in glucose uptake as compared with when glucose was administered alone. Based on findings obtained in vitro employing Caco-2 cells, where lignosulfonic acid dramatically decreased the uptake of 2-deoxyglucose, this supports the suppression of glucose transport by the colon. The authors proposed the potential application of lignins and lignosulfonic acid for the management of diabetes. The anti-microbial and anti-viral properties of lignosulfonic acid show its potential as an inexpensive pharmaceutical agent. Lignosulfonic acid (Laminaria saccharina) has been recently studied for its potential in controlled drug release. By cross-linking with glutaraldehyde, lignosulfonic acid was added to gelatin mix microspheres to enable the regulated release of an anti-malarial drug (Sekhar et al. 2014). The pH levels of these microspheres affected the drug release, which showed increased rates of release for up to ten hours.

10.3 LIGNIN–CARBOHYDRATE COMPLEXES

Lignin–carbohydrate complexes (LCCs) are made up of various types of polysaccharides that are covalently bonded to lignin in the cell walls of lignified plants. Eight different forms of lignin–carbohydrate bonds exist: ferulate ester, diferulate ester, hemiacetal linkages, glycosidic, phenyl glycosidic, benzyl ether, and benzyl ester bonds. The predominant forms of lignin–carbohydrate bonds in wood are ester linkages, phenyl glycoside, and benzyl ethers, whereas ferulate and diferulate esters are more commonly found in non-wood plants. Phenyl glycoside and benzoyl ether linkages in wood degrade readily in acidic environments (Yuan et al. 2011). Compared with other polyphenols, LCCs provided more effective cell protection against the cytopathic effects of UV radiation and HIV infection. The lignin moiety might be responsible for the significant anti-HIV action in limited digestion of LCCs, while the carbohydrate moiety might play a role in the immunopotentiating effect via a cell surface receptor (Sakagami 2014). LCCs have also been employed in studies on anti-cancer drugs (Fukuchi et al. 2020).

10.4 Lignin-Based Topical Hydrogels

Hydrogels are commonly referred to as hydrophilic polymers that form a three-dimensional network that can hold a significant amount of water. These characteristics make hydrogels potentially useful for regenerative medicine, wound healing dressings, drug release devices, and personal hygiene products (Zhang et al. 2019). Natural polymers have gained popularity in recent years for hydrogel formation. Numerous natural benefits of lignin include its antibacterial, antioxidant, and biodegradable qualities. As a result, lignin-based hydrogels have potential qualities for covering medicinal materials (Zhang and Too 2019). Using alkali lignin and chitosan, Ravishankar and coworkers reported creating biocompatible physic hydrogels (Ravishankar et al. 2019). These hydrogels were tested for cytotoxicity against zebrafish and mesenchymal stem cells in vivo/in vitro, respectively. The results of two experiments show that the hydrogels were safe. Since these hydrogels exhibit good NIH 3T3 murine fibroblast cell adhesion and proliferation, the authors also suggest a good potential relevance in wound healing. Furthermore, Zmejkoski et al. reported use of lignin-based hydrogels for wound healing. Utilizing coniferyl alcohol (DHP) as a model chemical for lignin, hydrogels containing bacterial cellulose were developed. To examine the antibacterial activity of these hydrogels, a variety of microorganisms, including *S. aureus* and *P. aeruginosa*, were tested. The release of DHP oligomers is linked to the antibacterial action of these gels, and it was found that the amounts of antibacterial compounds do not change over 72 hours. The release of oligomers may be very helpful in the healing of chronic wounds (Zmejkoski et al. 2018). Pinecones and pine nut shells were successively extracted alkaline and precipitated acidically, along with preparations of *Lentinus edodes* mycelia and *Sasa senanensis* Rehder leaves, to produce LCCs (Ali et al. 2012).

11 CONCLUSION

Lignin, a complex and abundant biopolymer, sourced primarily from plant cell walls, holds immense potential in various industries, particularly pharmaceuticals and medical applications. Understanding its source and structure is crucial for harnessing its benefits effectively. Lignin extraction processes vary, including sulfur-free methods, sulfur-containing techniques, and emerging greener approaches, each with its advantages and limitations. The diversity of lignin types presents both opportunities and challenges in production. While sulfur-free lignins offer environmental benefits, sulfur-containing lignins have unique properties. Next-generation greener lignins aim to mitigate environmental impact further. Exploring other types of lignins broadens the scope for applications but requires innovative extraction and processing techniques. Lignin-derived compounds exhibit promising pharmaceutical potentials. Vanillin, coumaric acid, FA, syringic acid, vanillic acid, and eugenol demonstrate various therapeutic properties, contributing to drug development and medical treatments. Additionally, lignin-derived fractions like lignophenols and lignosulfonic acid, along with lignin–carbohydrate complexes, offer novel avenues for pharmaceutical research. Overall, lignin's pharmaceutical and medical applications are diverse and promising, but challenges persist in extraction, purification, and formulation processes. With ongoing advancements in lignin science and technology, the pharmaceutical industry can harness lignin's full potential for the benefit of healthcare and beyond.

REFERENCES

Abdel-Moneim, Adel, Ahmed I. Yousef, Sanaa M. Abd El-Twab, Eman S. Abdel Reheim, and Mohamed B. Ashour. 2017. "Gallic Acid and P-Coumaric Acid Attenuate Type 2 Diabetes-Induced Neurodegeneration in Rats." *Metabolic Brain Disease* 32: 1279–186.

Abraham, Aby K., and V. G. Sridhar. 2015. "Materials, Design and Manufacturing Technologies for Orthopaedic Biomaterials: A Review." *International Journal of Applied Engineering Research* 10 (19): 40059–40062.

Akao, Yukihiro, Norio Seki, Yoshihito Nakagawa, Hong Yi, Kenji Matsumoto, Yukie Ito, Kuniyasu Ito, Masamitu Funaoka, Wakako Maruyama, and Makoto Naoi. 2004. "A Highly Bioactive Lignophenol Derivative from Bamboo Lignin Exhibits a Potent Activity to Suppress Apoptosis Induced by Oxidative Stress in Human Neuroblastoma SH-SY5Y Cells." *Bioorganic & Medicinal Chemistry* 12 (18): 4791–4801.

Al-Naqeb, Ghanya, Maznah Ismail, Gururaj Bagalkotkar, and Hadiza Altine Adamu. 2010. "Vanillin Rich Fraction Regulates LDLR and HMGCR Gene Expression in HepG2 Cells." *Food Research International* 43 (10): 2437–2443.

Ali, Shaikh M. Mohsin, M. Abul Kalam Azad, Mele Jesmin, Shamim Ahsan, M. Mijanur Rahman, Jahan Ara Khanam, M. Nazrul Islam, and Sha M.Shahan Shahriar. 2012. "In Vivo Anticancer Activity of Vanillin Semicarbazone." *Asian Pacific Journal of Tropical Biomedicine* 2 (6): 438–442. https://doi.org/10.1016/S2221-1691(12)60072-0.

Alriols, M. González, A. García, R. Llano-Ponte, and J. Labidi. 2010. "Combined Organosolv and Ultrafiltration Lignocellulosic Biorefinery Process." *Chemical Engineering Journal* 157 (1): 113–120.

Arya, Sagar S., Mansi M. Sharma, Ratul K. Das, James Rookes, David Cahill, and Sangram K. Lenka. 2019. "Vanillin Mediated Green Synthesis and Application of Gold Nanoparticles for Reversal of Antimicrobial Resistance in Pseudomonas Aeruginosa Clinical Isolates." *Heliyon* 5 (7), 1–11.

Ashengroph, Morahem, Iraj Nahvi, Hamid Zarkesh-Esfahani, and Fariborz Momenbeik. 2011. "Pseudomonas Resinovorans SPR1, a Newly Isolated Strain with Potential of Transforming Eugenol to Vanillin and Vanillic Acid." *New Biotechnology* 28 (6): 656–664.

Aswar, Manoj, and Vijay Patil. 2016. "Ferulic Acid Ameliorates Chronic Constriction Injury Induced Painful Neuropathy in Rats." *Inflammopharmacology* 24: 181–188.

Aziz, Tariq, Asmat Ullah, Amjad Ali, Muhammad Shabeer, Muhammad Naeem Shah, Fazal Haq, Mudassir Iqbal, Roh Ullah, and Farman Ullah Khan. 2022. "Manufactures of Bio-degradable and Bio-based Polymers for Bio-materials in the Pharmaceutical Field." *Journal of Applied Polymer Science* 139 (29): e52624.

Bandyopadhyay-Ghosh, S., S. B. Ghosh, and M. Sain. 2015. "The Use of Biobased Nanofibres in Composites." In *Biofiber Reinforcements in Composite Materials*, edited by Faruk, Omar, and Mohini Sain, 571–647. Elsevier.

Basri, Dayang Fredalina, Liy Si Tan, Zaleha Shafiei, and Noraziah Mohamad Zin. 2012. "In Vitro Antibacterial Activity of Galls of Quercus Infectoria Olivier against Oral Pathogens." *Evidence-Based Complementary and Alternative Medicine* 2012, 632796.

Bhattacherjee, Abhishek, and Abhijit Datta. 2015. "Mechanism of Antiglycating Properties of Syringic and Chlorogenic Acids in in Vitro Glycation System." *Food Research International* 77: 540–548.

Biswal, Trinath, Sushant Kumar BadJena, and Debabrata Pradhan. 2020. "Sustainable Biomaterials and Their Applications: A Short Review." *Materials Today: Proceedings* 30: 274–282.

Bjørsvik, Hans-René, and Lucia Liguori. 2002. "Organic Processes to Pharmaceutical Chemicals Based on Fine Chemicals from Lignosulfonates." *Organic Process Research & Development* 6 (3): 279–290.

Cardona, Carlos A., and Óscar J. Sánchez. 2007. "Fuel Ethanol Production: Process Design Trends and Integration Opportunities." *Bioresource Technology* 98 (12): 2415–2457.

Casadey, Rocio, Martin Broglia, Cesar Barbero, Susana Criado, and Claudia Rivarola. 2020. "Controlled Release Systems of Natural Phenolic Antioxidants Encapsulated inside Biocompatible Hydrogels." *Reactive and Functional Polymers* 156: 104729.

Das, Sudipta, Baishali Ghosh, and Keya Sarkar. 2022. "Nanocellulose as Sustainable Biomaterials for Drug Delivery." *Sensors International* 3: 100135.

de la Torre, M. Jesús, Ana Moral, M. Dolores Hernández, Elena Cabeza, and Antonio Tijero. 2013. "Organosolv Lignin for Biofuel." *Industrial Crops and Products* 45: 58–63.

Demuner, Iara Fontes, Jorge Luiz Colodette, Antonio Jacinto Demuner, and Carolina M Jardim. 2019. "Biorefinery Review: Wide-Reaching Products through Kraft Lignin." *BioResources* 14 (3): 7543–7581.

Dessbesell, Luana, Michael Paleologou, Mathew Leitch, Reino Pulkki, and Chunbao Charles Xu. 2020. "Global Lignin Supply Overview and Kraft Lignin Potential as an Alternative for Petroleum-Based Polymers." *Renewable and Sustainable Energy Reviews* 123: 109768.

Domínguez, J. C., M. Oliet, M. V. Alonso, M. A. Gilarranz, and F. Rodríguez. 2008. "Thermal Stability and Pyrolysis Kinetics of Organosolv Lignins Obtained from Eucalyptus Globulus." *Industrial Crops and Products* 27 (2): 150–156.

Elsherbiny, Nehal M., Nahla N. Younis, Mohamed A. Shaheen, and Mohamed M. Elseweidy. 2016. "The Synergistic Effect between Vanillin and Doxorubicin in Ehrlich Ascites Carcinoma Solid Tumor and MCF-7 Human Breast Cancer Cell Line." *Pathology Research and Practice* 212 (9): 767–777. https://doi.org/10.1016/j.prp.2016.06.004.

Fache, Maxence, Bernard Boutevin, and Sylvain Caillol. 2016. "Vanillin Production from Lignin and Its Use as a Renewable Chemical." *ACS Sustainable Chemistry & Engineering* 4 (1): 35–46.

Ferri, Maura, Anton Happel, Giulio Zanaroli, Marco Bertolini, Stefano Chiesa, Mauro Commisso, Flavia Guzzo, and Annalisa Tassoni. 2020. "Advances in Combined Enzymatic Extraction of Ferulic Acid from Wheat Bran." *New Biotechnology* 56: 38–45.

Fukuchi, Kunihiko, Hiroshi Sakagami, Yoshiaki Sugita, Koichi Takao, Daisuke Asai, Shigemi Terakubo, Hiromu Takemura, Hirokazu Ohno, Misaki Horiuchi, and Madoka Suguro. 2020. "Quantification of the Ability of Natural Products to Prevent Herpes Virus Infection." *Medicines* 7 (10): 64.

Gulsia, Oshin. 2020. "Vanillin: One Drug, Many Cures." *Resonance* 25 (7): 981–986.

Hagiopol, Cornel, and James W Johnston. 2011. *Chemistry of Modern Papermaking*. CRC Press.

Halder, Pobitra, Kalam Azad, Shaheen Shah, and Eity Sarker. 2019. "Prospects and Technological Advancement of Cellulosic Bioethanol Ecofuel Production." In *Advances in Eco-Fuels for a Sustainable Environment*, edited by Azad, Abul Kalam, 211–236. Elsevier.

Hasegawa, Yasushi, Yukiya Kadota, Chihiro Hasegawa, and Satoshi Kawaminami. 2015. "Lignosulfonic Acid-Induced Inhibition of Intestinal Glucose Absorption." *Journal of Nutritional Science and Vitaminology* 61 (6): 449–454.

Hassanzadeh, Parichehr, Elham Arbabi, and Fatemeh Rostami. 2021. "Coating of Ferulic Acid-Loaded Silk Fibroin Nanoparticles with Neutrophil Membranes: A Promising Strategy against the Acute Pancreatitis." *Life Sciences* 270: 119128.

Hemaiswarya, S., and M. Doble. 2009. "Synergistic Interaction of Eugenol with Antibiotics against Gram Negative Bacteria." *Phytomedicine* 16 (11): 997–1005.

Higuchi, Yudai, Ryo Kato, Koichiro Tsubota, Naofumi Kamimura, Nicholas J. Westwood, and Eiji Masai. 2019. "Discovery of Novel Enzyme Genes Involved in the Conversion of an Arylglycerol-β-Aryl Ether Metabolite and Their Use in Generating a Metabolic Pathway for Lignin Valorization." *Metabolic Engineering* 55: 258–267.

Hossain, Md Mokarrom, and Leigh Aldous. 2012. "Ionic Liquids for Lignin Processing: Dissolution, Isolation, and Conversion." *Australian Journal of Chemistry* 65 (11): 1465–1477.

Johnson, Eldin M., Hanki Lee, Rasu Jayabalan, and Joo-Won Suh. 2020. "Ferulic Acid Grafted Self-Assembled Fructo-Oligosaccharide Micro Particle for Targeted Delivery to Colon." *Carbohydrate Polymers* 247: 116550.

Jung, Byung-Ok, Suk-Jin Chung, and Sang Bong Lee. 2006. "Preparation and Characterization of Eugenol-grafted Chitosan Hydrogels and Their Antioxidant Activities." *Journal of Applied Polymer Science* 99 (6): 3500–3506.

Karagoz, Pinar, Sansanee Khiawjan, Marco P. C. Marques, Samir Santzouk, Timothy D. H. Bugg, and Gary J. Lye. 2023. "Pharmaceutical Applications of Lignin-Derived Chemicals and Lignin-Based Materials: Linking Lignin Source and Processing with Clinical Indication." *Biomass Conversion and Biorefinery*, 1–22.

Kaur, Jasleen, and Ramandeep Kaur. 2022. "P-Coumaric Acid: A Naturally Occurring Chemical with Potential Therapeutic Applications." *Current Organic Chemistry* 26 (14): 1333–1349.

Klugman, Anna. 2015. "Functionality Related Characterization of Pretreated Wood Lignin, Cellulose and Polyvinylpyrrolidone for Pharmaceutical Applications." *University of Tarti.*

Kumar Gupta, Girish, Sudipta De, Ana Franco, Alina Mariana Balu, and Rafael Luque. 2015. "Sustainable Biomaterials: Current Trends, Challenges and Applications." *Molecules* 21 (1): 48.

Kumar, Naresh, and Vikas Pruthi. 2014. "Potential Applications of Ferulic Acid from Natural Sources." *Biotechnology Reports* 4: 86–93.

Lancefield, Christopher S., Goran M. M. Rashid, Florent Bouxin, Agata Wasak, Wei-Chien Tu, Jason Hallett, Sharif Zein, Jaime Rodríguez, S. David Jackson, and Nicholas J. Westwood. 2016. "Investigation of the Chemocatalytic and Biocatalytic Valorization of a Range of Different Lignin Preparations: The Importance of β-O-4 Content." *ACS Sustainable Chemistry & Engineering* 4 (12): 6921–6930.

Lee, Sang Hyun, Thomas V. Doherty, Robert J. Linhardt, and Jonathan S. Dordick. 2009. "Ionic Liquid-mediated Selective Extraction of Lignin from Wood Leading to Enhanced Enzymatic Cellulose Hydrolysis." *Biotechnology and Bioengineering* 102 (5): 1368–1376.

Lee, Sang Yeol, See-Hyoung Park, Mi Ok Kim, Inhwan Lim, Mingyeong Kang, Sae Woong Oh, Kwangseon Jung, Dong Gyu Jo, Il-Hoon Cho, and Jongsung Lee. 2016. "Vanillin Attenuates Negative Effects of Ultraviolet A on the Stemness of Human Adipose Tissue-Derived Mesenchymal Stem Cells." *Food and Chemical Toxicology* 96: 62–69.

Li, Tao, and Sudhakar Takkellapati. 2018. "The Current and Emerging Sources of Technical Lignins and Their Applications." *Biofuels, Bioproducts and Biorefining* 12 (5): 756–787.

Lo, An-Ya, Chuan Wang, Wei Hsuan Hung, Anmin Zheng, and Biswarup Sen. 2015. "Nano-and Biomaterials for Sustainable Development." *Journal of Nanomaterials* 2015: 1.

Macfarlane, A. L., M. Mai, and J. F. Kadla. 2014. "Bio-Based Chemicals from Biorefining: Lignin Conversion and Utilisation." In *Advances in Biorefineries*, edited by Keith Waldron, 659–692. Elsevier.

Miculescu, F., A. Maidaniuc, Stefan Ioan Voicu, Vijay Kumar Thakur, G. E. Stan, and L. T. Ciocan. 2017. "Progress in Hydroxyapatite–Starch Based Sustainable Biomaterials for Biomedical Bone Substitution Applications." *ACS Sustainable Chemistry & Engineering* 5 (10): 8491–8512.

Mu, Mao, Shi Zuo, Rong-Min Wu, Kai-Sheng Deng, Shuang Lu, Juan-Juan Zhu, Gao-Liang Zou, Jing Yang, Ming-Liang Cheng, and Xue-Ke Zhao. 2018. "Ferulic Acid Attenuates Liver Fibrosis and Hepatic Stellate Cell Activation via Inhibition of TGF-β/Smad Signaling Pathway." *Drug Design, Development and Therapy*, 12, 4107–4115.

Niaounakis, Michael. 2015. "Building and Construction Applications." In *Biopolymers Application and Trends*, edited by Niaounakis, Michael, 445–505. Elsevier.

Nisar, Muhammad Farrukh, Mahnoor Khadim, Muhammad Rafiq, Jinyin Chen, Yali Yang, and Chunpeng Craig Wan. 2021. "Pharmacological Properties and Health Benefits of Eugenol: A Comprehensive Review." *Oxidative Medicine and Cellular Longevity* 2021. https://doi.org/10.1155/2021/2497354.

Norikura, Toshio, Yuuka Mukai, Shuzo Fujita, Keigo Mikame, Masamitsu Funaoka, and Shin Sato. 2010. "Lignophenols Decrease Oleate-Induced Apolipoprotein-B Secretion in HepG2 Cells." *Basic & Clinical Pharmacology & Toxicology* 107 (4): 813–817.

Olatunde, Ahmed, Aminu Mohammed, Mohammed Auwal Ibrahim, Nasir Tajuddeen, and Mohammed Nasir Shuaibu. 2022. "Vanillin: A Food Additive with Multiple Biological Activities." *European Journal of Medicinal Chemistry Reports* 5: 100055. https://doi.org/10.1016/j.ejmcr.2022.100055.

Pan, Xuejun, Claudio Arato, Neil Gilkes, David Gregg, Warren Mabee, Kendall Pye, Zhizhuang Xiao, Xiao Zhang, and John Saddler. 2005. "Biorefining of Softwoods Using Ethanol Organosolv Pulping: Preliminary Evaluation of Process Streams for Manufacture of Fuel-grade Ethanol and Co-products." *Biotechnology and Bioengineering* 90 (4): 473–481.

Park, Jinbong, Seon Yeon Cho, JongWook Kang, Woo Yong Park, Sujin Lee, Yunu Jung, Min-Woo Kang, Hyun Jeong Kwak, and Jae-Young Um. 2020. "Vanillic Acid Improves Comorbidity of Cancer and Obesity through Stat3 Regulation in High-Fat-Diet-Induced Obese and B16bl6 Melanoma-Injected Mice." *Biomolecules* 10 (8): 1098.

Pesode, Pralhad, Shivprakash Barve, Sagar V. Wankhede, and Akbar Ahmad. 2023. "Sustainable Materials and Technologies for Biomedical Applications." *Advances in Materials Science and Engineering* no. 1 (2023): 6682892.

Ponnusamy, Vinoth Kumar, Dinh Duc Nguyen, Jeyaprakash Dharmaraja, Sutha Shobana, J. Rajesh Banu, Rijuta Ganesh Saratale, Soon Woong Chang, and Gopalakrishnan Kumar. 2019. "A Review on Lignin Structure, Pretreatments, Fermentation Reactions and Biorefinery Potential." *Bioresource Technology* 271: 462–472.

Poornima, Balan, and Purna Sai Korrapati. 2017. "Fabrication of Chitosan-Polycaprolactone Composite Nanofibrous Scaffold for Simultaneous Delivery of Ferulic Acid and Resveratrol." *Carbohydrate Polymers* 157: 1741–1749.

Pradhan, Debabrata, Sandeep Panda, and Lala Behari Sukla. 2018. "Recent Advances in Indium Metallurgy: A Review." *Mineral Processing and Extractive Metallurgy Review* 39 (3): 167–180.

Praphakar, Rajendran Amarnath, Murugan A. Munusamy, and Mariappan Rajan. 2017. "Development of Extended-Voyaging Anti-Oxidant Linked Amphiphilic Polymeric Nanomicelles for Anti-Tuberculosis Drug Delivery." *International Journal of Pharmaceutics* 524 (1–2): 168–177.

Ramachandran, Vinayagam. 2010. "Preventive Effect of Syringic Acid on Hepatic Marker Enzymes and Lipid Profile against Acetaminophen-Induced Hepatotoxicity Rats." *International Journal of Pharmaceutical and Biological Science Archive* 1: 393–398.

Rao, Karanam Srinivasa, Amrita Mishra, Devbrata Pradhan, Gautam Roy Chaudhury, Birendra Kumar Mohapatra, Trupti Das, Lala Bihari Sukla, and Barada Kanta Mishra. 2008. "Percolation Bacterial Leaching of Low-Grade Chalcopyrite Using Acidophilic Microorganisms." *Korean Journal of Chemical Engineering* 25: 524–530.

Ravishankar, Kartik, Manigandan Venkatesan, Raj Preeth Desingh, Aparna Mahalingam, Balaji Sadhasivam, Rajalakshmi Subramaniyam, and Raghavachari Dhamodharan. 2019. "Biocompatible Hydrogels of Chitosan-Alkali Lignin for Potential Wound Healing Applications." *Materials Science and Engineering: C* 102: 447–457.

Reiter, Jochen, Harald Strittmatter, Lars O. Wiemann, Doris Schieder, and Volker Sieber. 2013. "Enzymatic Cleavage of Lignin β-O-4 Aryl Ether Bonds via Net Internal Hydrogen Transfer." *Green Chemistry* 15 (5): 1373–1381.

Sabzi, Noor Ali, and May Mohammed Jawad. 2023. "Synthesis, Characterization, and Preliminary Evaluation of Antimicrobial Activity of Imines Derived from Vanillic Acid Conjugated to Heterocyclic." *Iraqi Journal of Pharmaceutical Sciences (P-ISSN 1683–3597 E-ISSN 2521–3512)* 32 (Suppl.): 8–15.

Safo, Martin K., Osheiza Abdulmalik, Richmond Danso-Danquah, James C. Burnett, Samuel Nokuri, Gajanan S. Joshi, Faik N. Musayev, Toshio Asakura, and Donald.J Abraham. 2004. "Structural Basis for the Potent Antisickling Effect of a Novel Class of Five-Membered Heterocyclic Aldehydic Compounds." *Journal of Medicinal Chemistry* 47 (19): 4665–4676.

Sakagami, Hiroshi. 2014. "Biological Activities and Possible Dental Application of Three Major Groups of Polyphenols." *Journal of Pharmacological Sciences* 126 (2): 92–106.

Salehi Jouzani, Gholamreza, and Mohammad J. Taherzadeh. 2015. "Advances in Consolidated Bioprocessing Systems for Bioethanol and Butanol Production from Biomass: A Comprehensive Review." *Biofuel Research Journal* 2 (1): 152–195.

Sameni, Javad, Shaffiq A. Jaffer, Jimi Tjong, and Mohini Sain. 2020. "Advanced Applications for Lignin Micro-and Nano-Based Materials." *Current Forestry Reports* 6: 159–171.

Sekhar, E. Chandra, K. S. V. K. Rao, K. Madhu Sudana Rao, S. Eswaramma, and R. Ramesh Raju. 2014. "Development of Gelatin-Lignosulfonic Acid Blend Microspheres for Controlled Release of an Anti-Malarial Drug (Pyronaridine)." *Indian Journal of Advances in Chemical Science* 2: 228–237.

Shaughnessy, Daniel T., Roel M. Schaaper, David M. Umbach, and David M. DeMarini. 2006. "Inhibition of Spontaneous Mutagenesis by Vanillin and Cinnamaldehyde in Escherichia Coli: Dependence on Recombinational Repair." *Mutation Research/Fundamental and Molecular Mechanisms of Mutagenesis* 602 (1–2): 54–64.

Sirangelo, Ivana, Luigi Sapio, Angela Ragone, Silvio Naviglio, Clara Iannuzzi, Daniela Barone, Antonio Giordano, and Margherita Borriello. 2020. "Vanillin Prevents Doxorubicin-Induced Apoptosis and Oxidative Stress in Rat H9c2 Cardiomyocytes." *Nutrients* 12 (8): 2317.

Song, Shengmei, Xuewen Ma, Yehong Zhou, Maotian Xu, Shaomin Shuang, and Chuan Dong. 2016. "Studies on the Interaction between Vanillin and β-Amyloid Protein via Fluorescence Spectroscopy and Atomic Force Microscopy." *Chemical Research in Chinese Universities* 32 (2): 172–177.

Srinivasan, Subramani, Jayachandran Muthukumaran, Udaiyar Muruganathan, Rantham Subramaniyam Venkatesan, and Abdulkadhar Mohamed Jalaludeen. 2014. "Antihyperglycemic Effect of Syringic Acid on Attenuating the Key Enzymes of Carbohydrate Metabolism in Experimental Diabetic Rats." *Biomedicine & Preventive Nutrition* 4 (4): 595–602.

Srinivasulu, Cheemanapalli, Mopuri Ramgopal, Golla Ramanjaneyulu, C. M. Anuradha, and Chitta Suresh Kumar. 2018. "Syringic Acid (SA)–a Review of Its Occurrence, Biosynthesis, Pharmacological and Industrial Importance." *Biomedicine & Pharmacotherapy* 108: 547–557.

Stoddart, Alison, and Victoria Cleave. 2009. "The Evolution of Biomaterials." *Nature Materials* 8 (6): 444–445.

Suota, Maria Juliane, Thiago Alessandre da Silva, Sônia Faria Zawadzki, Guilherme Lanzi Sassaki, Fabricio Augusto Hansel, Michael Paleologou, and Luiz Pereira Ramos. 2021. "Chemical and Structural Characterization of Hardwood and Softwood LignoForce™ Lignins." *Industrial Crops and Products* 173: 114138.

Suseela, Vidya. 2019. "Potential Roles of Plant Biochemistry in Mediating Ecosystem Responses to Warming and Drought." In *Ecosystem Consequences of Soil Warming*, Mohan, Jacqueline E., eds., 103–124. Elsevier.

Tai, Akihiro, Takeshi Sawano, and Futoshi Yazama. 2011. "Antioxidant Properties of Ethyl Vanillin in Vitro and in Vivo." *Bioscience, Biotechnology, and Biochemistry* 75 (12): 2346–2350.

Taniguchi, Kaori, Mariko Funasaki, Akio Kishida, Samir K Sadhu, Firoj Ahmed, Masami Ishibashi, and Ayumi Ohsaki. 2018. "Two New Coumarins and a New Xanthone from the Leaves of Rhizophora Mucronata." *Bioorganic & Medicinal Chemistry Letters* 28 (6): 1063–1066.

Tathe, Amogh, Mangesh Ghodke, and Anna Pratima Nikalje. 2010. "A Brief Review: Biomaterials and Their Application." *International Journal of Pharmacy and Pharmaceutical Sciences* 2 (4): 19–23.

Tong, Yuna, Shan Liu, Rong Gong, Lei Zhong, Xingmei Duan, and Yuxuan Zhu. 2019. "Ethyl Vanillin Protects against Kidney Injury in Diabetic Nephropathy by Inhibiting Oxidative Stress and Apoptosis." *Oxidative Medicine and Cellular Longevity*, 2019: 2129350.

Torres, Luis Alberto Zevallos, Adenise Lorenci Woiciechowski, Valcineide Oliveira de Andrade Tanobe, Susan Grace Karp, Luiza Chemim Guimarães Lorenci, Craig Faulds, and Carlos Ricardo Soccol. 2020. "Lignin as a Potential Source of High-Added Value Compounds: A Review." *Journal of Cleaner Production* 263: 121499.

Velu, Periyannan, Veerasamy Vinothkumar, Sukumar Babukumar, and Duraisamy Ramachandhiran. 2017. "Chemopreventive Effect of Syringic Acid on 7, 12-Dimethylbenz (a) Anthracene Induced Hamster Buccal Pouch Carcinogenesis." *Toxicology Mechanisms and Methods* 27 (8): 631–640.

Venkadeswaran, Karuppasamy, Arumugam Ramachandran Muralidharan, Thangaraj Annadurai, Vasanthakumar Vasantha Ruban, Mahalingam Sundararajan, Ramalingam Anandhi, Philip A Thomas, and Pitchairaj Geraldine. 2014. "Antihypercholesterolemic and Antioxidative Potential of an Extract of the Plant, Piper Betle, and Its Active Constituent, Eugenol, in Triton WR-1339-Induced Hypercholesterolemia in Experimental Rats." *Evidence-Based Complementary and Alternative Medicine*, 2014: 478973.

Vinardell, Maria Pilar, and Montserrat Mitjans. 2017. "Lignins and Their Derivatives with Beneficial Effects on Human Health." *International Journal of Molecular Sciences* 18 (6): 1219.

Vishnu, K. V., K. K. Ajeesh Kumar, Niladri S. Chatterjee, R. G. K. Lekshmi, P. R. Sreerekha, Suseela Mathew, and C. N. Ravishankar. 2018. "Sardine Oil Loaded Vanillic Acid Grafted Chitosan Microparticles, a New Functional Food Ingredient: Attenuates Myocardial Oxidative Stress and Apoptosis in Cardiomyoblast Cell Lines (H9c2)." *Cell Stress and Chaperones* 23 (2): 213–222. https://doi.org/10.1007/s12192-017-0834-5.

Wang, Fu-Ling, Shuang Li, Yi-Xin Sun, Hui-Ying Han, Bi-Xian Zhang, Bao-Zhong Hu, Yun-Fei Gao, and Xiao-Mei Hu. 2017. "Ionic Liquids as Efficient Pretreatment Solvents for Lignocellulosic Biomass." *RSC Advances* 7 (76): 47990–47998.

Wang, Hongliang, Yunqiao Pu, Arthur Ragauskas, and Bin Yang. 2019. "From Lignin to Valuable Products–Strategies, Challenges, and Prospects." *Bioresource Technology* 271: 449–461.

Widsten, Petri. 2020. "Lignin-Based Sunscreens—State-of-the-Art, Prospects and Challenges." *Cosmetics* 7 (4): 85.

Yuan, Tong-Qi, Shao-Ni Sun, Feng Xu, and Run-Cang Sun. 2011. "Characterization of Lignin Structures and Lignin–Carbohydrate Complex (LCC) Linkages by Quantitative 13C and 2D HSQC NMR Spectroscopy." *Journal of Agricultural and Food Chemistry* 59 (19): 10604–10614.

Zhang, Congqiang, and Heng-Phon Too. 2019. "Revalorizing Lignocellulose for the Production of Natural Pharmaceuticals and Other High Value Bioproducts." *Current Medicinal Chemistry* 26 (14): 2475–2484.

Zhang, Lei, Xinwen Peng, Linxin Zhong, Weitian Chua, Zhihua Xiang, and Runcang Sun. 2019. "Lignocellulosic Biomass Derived Functional Materials: Synthesis and Applications in Biomedical Engineering." *Current Medicinal Chemistry* 26 (14): 2456–2474.

Zhong, Lei, Yuna Tong, Junlan Chuan, Lan Bai, Jianyou Shi, and Yuxuan Zhu. 2019. "Protective Effect of Ethyl Vanillin against Aβ-induced Neurotoxicity in PC12 Cells via the Reduction of Oxidative Stress and Apoptosis." *Experimental and Therapeutic Medicine* 17 (4): 2666–2674.

Zhu, Weizhen, and Hans Theliander. 2015. "Precipitation of Lignin from Softwood Black Liquor: An Investigation of the Equilibrium and Molecular Properties of Lignin." *BioResources* 10 (1): 1696–1714.

Zmejkoski, Danica, Dragica Spasojević, Irina Orlovska, Natalia Kozyrovska, Marina Soković, Jasmina Glamočlija, Svetlana Dmitrović, Branko Matović, Nikola Tasić, and Vuk Maksimović. 2018. "Bacterial Cellulose-Lignin Composite Hydrogel as a Promising Agent in Chronic Wound Healing." *International Journal of Biological Macromolecules* 118: 494–503.

Zore, Gajanan B., Archana D. Thakre, Sitaram Jadhav, and S. Mohan Karuppayil. 2011. "Terpenoids Inhibit Candida Albicans Growth by Affecting Membrane Integrity and Arrest of Cell Cycle." *Phytomedicine* 18 (13): 1181–1190.

15 The Advent of Biopolymers for Potential Applications in the Medical and Pharmaceutical Sector

Harsh Vardhan Singh, Sharon Nagpal,
and Alok Kumar Mishra
Lovely Professional University

Ajay Kumar
Department of Mechanical Engineering,
School of Engineering and Technology, JECRC
University, Jaipur, Rajasthan, India

Parveen Kumar
Department of Mechanical Engineering, Rawal Institute of
Engineering and Technology Faridabad, Haryana, India

1 INTRODUCTION

Biomaterials are specified materials which are designed and manufactured with the intent of interacting with biological systems to be used in the medical sector for therapeutic and diagnostic purposes. Biomaterials, according to the National Institute of Health, is defined as "any natural or synthetic substance or combination of substances, other than drugs, which can be used to augment or partially replace any tissue, organ or function of the body, to maintain or improve quality of life of the individual" [1]. The application of biomaterials to enhance medical care is growing in popularity. There are many biomaterials currently being developed and studied for use in the medical device market, which is valued at an estimated US$400 billion worldwide [2]. However, because biomaterials are foreign materials that are introduced into the body, they can elicit immune-related reactions that pose a significant challenge. This difficulty might result in a low standard of living for patients and is one of the factors that have hindered the clinical adoption of biomaterials. They have evolved over the past few decades and today find their utilization as a crucial aspect in various fields, particularly in pharmaceutical and medical research. Initially, biomaterials were used majorly in medical devices to treat or replace tissues and have expanded their applications beyond implants [3]. Because of their adaptable qualities

DOI: 10.1201/9781003434313-15

and environmental friendliness, sustainable biomaterials are essential to pharmaceutical and medical applications.

Since these materials are produced from renewable sources, such as plants, microorganisms, and animals, these materials are biodegradable, non-toxic, and bioresorbable [4]. They are frequently utilized in several medical domains, including tissue engineering, drug delivery, wound healing, and the production of human body parts. Biopolymers, a type of sustainable biomaterial, are particularly favored for their potential in implantable biomedical applications. Biomaterials and traditional materials are significantly different in terms of biodegradability and eco-friendliness, primarily due to their distinct properties and composition. Traditional materials, commonly sourced from non-renewable resources, tend to resist degradation in the environment, resulting in pollution and the accumulation of waste over time. In contrast, biomaterials are predominantly derived from renewable sources like plants or animals, which inherently makes them more sustainable and environmentally friendly. This difference in origin and composition allows biomaterials to naturally break down into harmless substances, reducing their environmental impact and promoting a greener approach to material usage and production [5].

2 EVOLUTION OF BIOMATERIALS

Biomaterials have been employed in healthcare for a long extensive duration of time, and yet tangible advancements in this field have been notable since the 1940s. Considerable progress has been made in the development of implants and therapeutic medical technology, especially in the previous 30 years [6, 7]. When the first cardiac resynchronization therapy or CRT devices were developed in 2001, sensor technology and software advancements allowed pacemakers to automatically modify their stimulation rates in response to a patient's changing degree of physical activity. This was a major turning point in the field [8]. The strategic evolution of biomaterials throughout the years included different types of biomaterials mainly divided into different groups.

2.1 INERT BIOMATERIALS (1960–1970)

First-generation biomaterials or inert biomaterials were first developed during the 20th century in the late 1960s and early 1970s [9]. It led to the development of materials that could be used within the human body, resulting in the creation of prostheses. Professor Bill Bonfield played a major role in the understanding of these inert biomaterials [10]. By 1980, the number of implants done had already exceeded more than 3 million annual goals, and over 50 prostheses, made from 40 different materials, were in clinical use worldwide. The principal objective of the initial biomaterials was to attain physical characteristics commensurate with the substituted tissues while reducing detrimental reactions inside the host. It's important to note that many materials utilized were biologically inert, and most were single-phase materials, having the same physical and chemical properties throughout. Implant materials were frequently modified versions of already available commercial materials that were purified further to reduce corrosion and stop harmful byproducts from leaking out.

2.2 BIOACTIVE MATERIALS (1980–1990)

Bioactive materials are still widely used in many different commercial goods. By applying coatings, these materials help to trigger a chemical–biological interaction at the material–host contact. This resulted in the enhanced effectiveness of medical devices. The goal of the biomaterials industry changed throughout the 1980s from producing bioinert tissue responses to producing bioactive materials. These bioactive substances are intended to start certain, regulated processes and reactions inside the body [11]. One example of bioactive materials is bioactive glasses, which were first established in 1971. Bioactive glasses interact with living tissue through an 11-reaction-step sequence, involving ion exchange followed by surface polycondensation reactions. Orthopedic and dental applications of bioactive materials were widely seen at the end of the 20th century during the 1980s. Clinical trials were being conducted on various compositions of bioactive glasses, components of ceramics, combined glass ceramics, and other composites to achieve bioactive fixation, synthetic hydroxyapatite (HA) ceramics have been used in metallic prostheses as coatings, powders, and porous implants. This encouraged the property of osteoconduction. Osteoconduction is the process where the bone forms along the covering to create a mechanically stable and robust interface. The original 45S5 Bio-glass formulation and other bioactive glasses and combined glass ceramics were utilized for the maintenance of endosseous ridge implants and middle ear prostheses. Bioactive composites, such as hydroxyapatite particles in a polyethylene matrix (Hapex), have been in use since the 1990s, mainly used for bone replacement and healing fractures, especially in the ear (middle ear).

2.3 BIODEGRADABLE BIOMATERIALS (2000–2010)

These materials can break down and then be absorbed by the body, which provides an important advantage against permanent implants. These biomaterials were developed to address the challenges of bioactive materials that caused infection and immunological reactions. This field of biodegradable biomaterials has seen tremendous advancements in the 21st century in both development and use. There are many benefits to using these materials, including less need for follow-up surgery, better patient comfort, and environmental sustainability. For example, biodegradable materials reduce environmental impacts because they decompose into safe byproducts. These materials can be manipulated to copy the mechanical characteristics of the tissues surrounding them, which helps in improving performance and comfort when implanted into the human body [12]. The research on biodegradable biomaterials is to optimize their mechanical strength, biocompatibility, and rate of degradation for particular uses in the medical field. Biodegradable materials have also been used as carriers for targeted drug delivery, which would enable the precise release of medication at a specific location. These biomaterials also have another property of scaffolds that promote cell proliferation and direct tissue regeneration made from these materials is especially helpful for tissue engineering.

2.4 SMART BIOMATERIALS (2010–NOW)

These biomaterials can replicate natural structures and processes by encouraging specific cellular responses, aiding in the regeneration and repair of injured tissues. Biomimetic biomaterials can serve as temporary tools like screws or plates or take on a permanent role like prosthetics. Their functions are still being studied as their uses are widespread [13, 14]. Synthetic polymers are becoming increasingly popular due to their uniformity, well-understood structures and properties, reduced immunogenicity, and dependable source materials. Among them, materials like polytetrafluoroethylene (PTFE), polyurethane foams (PU), and ultra-high molecular weight polyethylene (UHMWPE), exhibit diverse properties, ranging from brittleness to toughness, and from softness to viscosity, which can also be molecularly adjusted to achieve desired properties. Smart biomaterials are preferred over traditional materials due to their cost-effectiveness and easy processing, making them ideal materials for developing structures at the nano to macro scales in the biomedical and bioengineering sectors.

3 HISTORICAL PERSPECTIVE AND EVIDENCE

The utilization of biomaterials can be traced back to ancient civilizations, alongside the earliest human societies. Over the past 200 years, evidence has suggested the use of these biomaterials as implants or prostheses. This is mainly due to the discovery of human skeletons or skulls at sites associated with ancient civilizations like the Etruscans, Romans, Egyptians, and Greeks. Notably, historical literature extensively discusses the "presumed" applications of biomaterials within dental practices by ancient civilizations and subsequent eras. The emphasis on dental applications likely stems from the wealth of evidence available since antiquity, including the use of wire for dental binding, teeth implants, and various filling materials [15]. The replacement of damaged or amputated limbs with wooden prostheses is a common topic in historical narratives. However, a need of certain evidence that can point to their specific use in medical applications exists so that the materials today may consider them biomaterials, particularly in orthopedic, vascular, and ophthalmological surgeries.

One of the earliest instances of biomaterials can potentially be traced back to a hybrid material crafted during the Etruscan civilization approximately 2,600 years ago. This prosthesis was of dental use and supposedly consisted of an animal tooth, probably from a cow, attached with a gold wire to the patient's natural teeth. With the prosthesis firmly affixed to the nearby teeth with gold wire, this novel technique was specifically utilized to replace the upper incisors. This artifact illuminates early attempts to use biomaterials for dental prosthetic applications and demonstrates the inventiveness and resourcefulness of ancient societies in resolving dental issues [16]. A brief idea can be understood from Figure 15.1.

4 CLASSIFICATION OF BIOMATERIALS

Thousands of biomedical devices and diagnostic products have been invented and discovered that are used to support the regeneration ability of human tissues and organs after damage and to ensure better functioning of the body [17]. The count of medical devices listed in the medical device registry of the U.S. Food and Drug Administration Center for Medical Devices and Radiological Health is about 6,000; the registry is done to ensure both the safety and the performance of biomaterials. Many devices and types of equipment including pacemakers, bone plates, heart valves, neurostimulators, and artificial knees are now used to treat different diseases or injuries [18].

There are several ways biomaterials have been classified; these include the basis of:

4.1 SPECIFICITY OF UTILIZATION

The basic classification of biomaterials can be done based on their use in the body. The main aspect includes the use in the case of a specific organ and whole of the body systems.

4.1.1 Organ Specificity

This includes biomaterials that are being employed in different organs of the body for their property that helps in regeneration, healing, and restoring functioning. Some of the materials include:

a Intraocular lenses (eyes), work as an artificial lens for the eye [19]
b Catheters and stents (bladder), are used to drain the bladder to save the kidneys from the pressure [20]
c Kidney dialysis machine (kidneys), is used as an artificial filtration system in case of kidney failures [21]
d Artificial stapes and cochlear implants (ears), work as a hearing aid and also improve hearing in people with hearing problems.

4.1.2 Different Body Systems

In this type of classification, different biomaterials are used to repair the body systems including nervous and circulatory systems, etc. Some examples of the type of biomaterials and the specific implants include:

a. **Nerve stimulators (nervous system):** These types of biomaterials are used to provide targeted stimulations to a nerve in case of compromise in the working of the nerve [22].
b. **Joint replacement and bone plates (skeletal system):** These types of biomaterials are used to repair broken bones and joints in the body [23].
c. **Tracheal stents (respiratory):** These types of biomaterials are hollow materials that prevent the collapse of the tracheal lumen and stabilization and reconstruction of the larynx or trachea [24].

4.2 Type of Reaction of the Biomaterials

This type of classification is based on the different reactions that are seen toward these biomaterials after implementation in the human body. These include:

4.2.1 Bioinert Biomaterials

Bioinert materials include materials like stainless steel, titanium, alumina, and zirconia that are safe for the body after being placed inside it. These materials have less contact with the surrounding tissue. Non-bioinert implants have sometimes been found with a fibrous capsule around them [25].

4.2.2 Bioactive Materials

Bioactive materials can react with the body by exchange of ions between the implant and the body fluid, which produces a layer made of carbonate apatite (CHAp), one of the active carbonate compounds with a composition very similar to the chemical components of the bones of the human body [26].

4.2.3 Bioresorbable Materials

Bioabsorbable materials are transient structures that dissolve and are removed from the body's environment. The first recorded application of a bioabsorbable substance dates to 175 AD, when catgut suture—a naturally occurring polymer—was used [27]; other examples include tricalcium phosphate [Ca3(PO4)2] and many more.

4.3 Materials Used in Biomaterials

The material used to create the biomaterial is very important due to the different factors associated with different materials that influence the specificity and usage. Biomaterials can be made of different types of materials; the broad classification of biomaterials includes:

4.3.1 Bioceramic Materials

Bioceramic biomaterials consist of metallic and non-metallic elements that may contain covalent bonds or ionic bonds. Some of the examples include oxides of aluminum i.e., Al_2O_3 or aluminum oxide, silicon i.e., Si_2O_3 or silicon dioxide, and magnesium i.e., MgO or magnesium oxide that have both non-metallic and metallic components. Other examples include the diamond and carbonaceous structures. The strong covalent and ionic bonds of the ceramic elements are what contribute to their hardness, brittleness, and stiffness [28].

4.3.2 Polymeric Biomaterials

In the 1950s, a different approach was taken to make biomaterials, including using polymers and associated materials; this led to the use of many different types of polymers being used and the new materials formed were termed polymeric materials. Their uses were in non-load-bearing applications, like ophthalmic prostheses, facial prostheses, skin/cartilage prostheses, catheters, drug-delivery devices, vascular prostheses, and orthopedic and dental applications in conjunction with metals. The

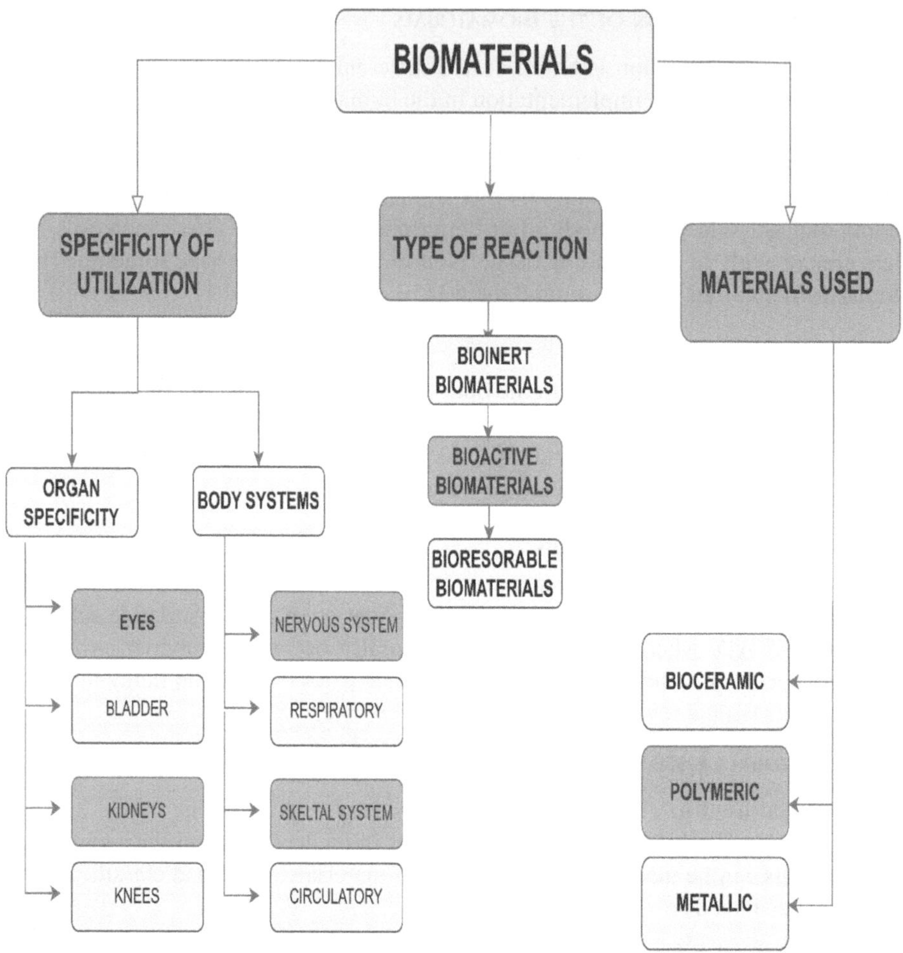

FIGURE 15.1 Classification of different biomaterials based on specificity, type of reaction, and material used.

contemporary polymeric biomaterial classes additionally consist of liquid, film, and gel polymeric compounds used as fillers, coatings, medical adhesives, and sealants [29]. In specifics of medical applications, different types of synthetic polymeric materials are used, including polyethylene, polyethylene terephthalate, polyurethane, polypropylene, and polymethyl methacrylate.

4.3.3 Metallic Biomaterials

Metal has many properties that make it one of the best materials to make biomaterials. This includes their property of thermal, conductivity, and metallic biomaterials that are being employed in various body parts, such as heart valves, stents, and hip replacements. To tackle the corrosive property of certain matter, certain metals are being employed as alloys.

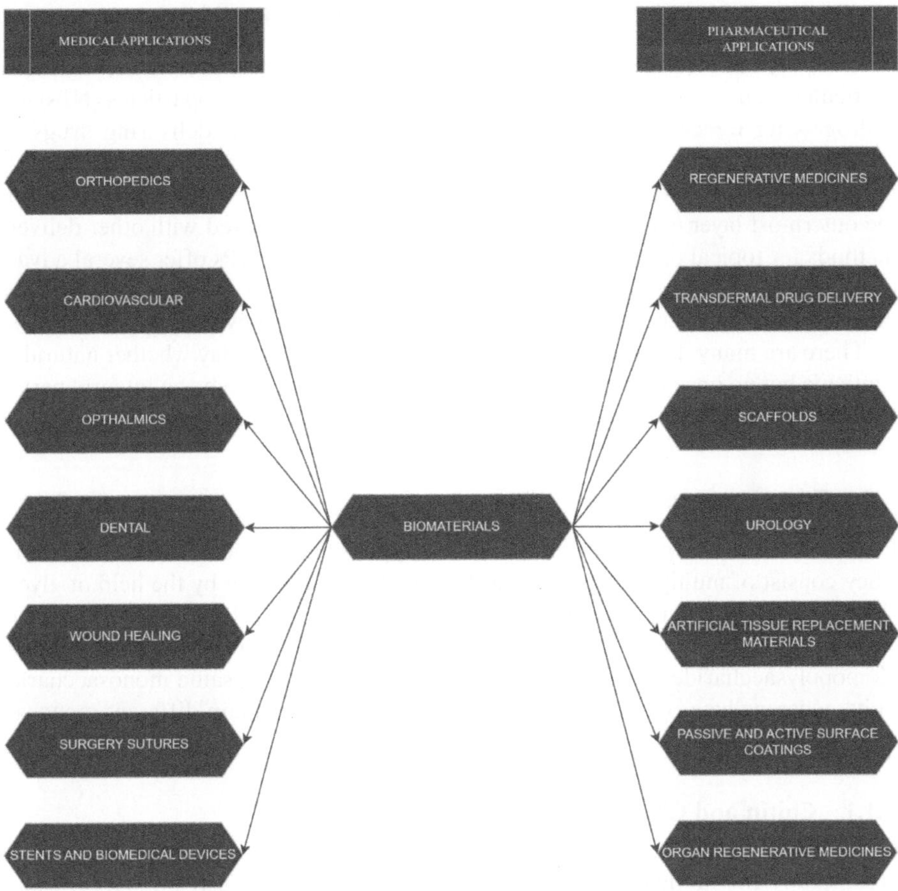

FIGURE 15.2 Application of biomaterials in medical and medical sectors.

The alloy material, mixing two or other metals to provide the efficiency of all the metals included. This makes that alloy corrosive, free, and stable to use in the body [30]. Currently, the major materials that dominate metallic biomaterials include:

1. Pure titanium (Ti)
2. Titanium alloys (Ti-6Al-4V)
3. Stainless steel
4. Cobalt–chromium alloys (Co–Cr)

The ability to withstand corrosion, biocompatibility, affordability, and suitable physical and mechanical characteristics are some of the properties that make the use of metallic metals and alloys as biomaterials in medical applications [31]. However, there are some disadvantages to using the metals, such as their high modulus, certain level of cytotoxicity, and metal ion sensitivity, which can cause blood-related issues and body defects. Figure 15.2 shows applications of biomaterials in different sectors.

5 BIOMATERIALS FOR PHARMACEUTICAL APPLICATIONS

Natural polymers are considered important in pharmaceutical and cosmetics research. Particular focus is on using these natural polymers to create nanoparticles (NPs) and hydrogels for topical applications. These applications include delivering drugs, as cosmetic ingredients, in promoting skin regeneration and tissue regeneration [32]. Polymeric NPs are great at getting drugs to different skin layers, all while protecting the outermost layer (stratum corneum) from damage. Compared with other delivery methods for topical or transdermal applications, polymeric NPs offer several advantages. These include precise control over how quickly the drug is released, efficient drug encapsulation, and increased resistance to break down by enzymes [33].

There are many different types of biomaterials available today whether natural or synthetic but no one can deny that these biomaterials have become an integral part of human life and all pharmaceutical as well as the biomedical sectors.

5.1 POLYSACCHARIDES

Polysaccharides are one of the most common types of carbohydrates found in nature. They consist of multiple monosaccharide units linked together by the help of glycosidic bonds. These units can be either sugar residues themselves linked together or can be covalently bonded with structures, such as peptides, amino acids, and lipids. Homopolysaccharides or homoglycans, are composed of the same monosaccharide units, whereas heteropolysaccharides, or heteroglycans, contain different monosaccharide units [34].

5.1.1 Chitin and Chitosan

Chitin and chitosan are natural polysaccharides that are extracted from the shells of crustaceans and mollusks as well as from insects and fungal cell walls [35]. Chitin is primarily extracted from shrimp and crab residue leftovers from the seafood industry. Chitin is primarily composed of N-acetyl glucosamine subunits that are linked by β-1,4 glycosidic bonds. It transforms into its active form, chitosan, through deacetylation [36, 37]. Unlike natural polysaccharides, chitin and chitosan are alkaline owing to their high nitrogen content. Chitin is hydrophobic and is insoluble in water and many organic solvents, whereas chitosan being an active form is soluble in aqueous and acidic solutions, which makes it useful in various applications. This behavior of chitosan in solution depends on its degree of deacetylation which is the removal of acetyl groups, the arrangement of acetyl groups along the chain, and its molecular weight. Chitosan is unique in its ability to create polymeric nanoparticles, making it mucoadhesive and enhancing its ability of skin permeability by interacting with junctions between the cells. It also exhibits the bioactive property of antimicrobial effect by engaging bacterial membranes and promoting wound healing by interacting with macrophages [38]. Along with its role in polymeric nanoparticle preparation, chitosan is also used as a surface stabilizer when coating other nanoparticles to mitigate their toxicity. Nanoparticles have a small particle size and large surface area, which makes them advantageous; however, they can accidentally penetrate human cells and pose a cytotoxic threat to skin cells [39].

5.1.2 Alginate

Alginate is a type of natural polysaccharide that is derived from algae, having basic properties that make it desirable, including its biocompatibility, affordability, and abundance in nature. Its molecular structure consists of alternating M-blocks and G-blocks. In the presence of divalent cations like calcium, barium, or strontium, its alginate forms gels, which is one of the important properties to make biomaterials [40]. Alginate gels are majorly used as drug carriers in wound dressings and controlled drug-delivery systems. Nowadays, they are also being used in tissue engineering.

Alginate chains can be broken down by replacing the divalent ions with monovalent cations like sodium ions. Although these chains remain stable under normal physiological conditions, there is a possibility of gradual cleavage of glycosidic bonds under acidic or basic conditions. There is still a lack of evidence that confirms the existence of human enzymes that can degrade alginate [41, 42]. To enhance alginate's in vivo biodegradation, strategies have been devised, such as partially oxidizing the polymer chains. Weakly oxidized chains readily degrade in aqueous media which makes alginate a viable option as a biodegradable carrier in controlled drug-delivery applications [43].

5.1.3 Agarose

Agarose is a linear polysaccharide that is extracted mainly from red algae found in marine environments. It is known for its hydrophilicity, neutrality, and non-pH sensitivity. The structure of agarose is composed of repeating units of D-galactose and 3,6-anhydro-L-galactose that are linked together by α (1 $\rightarrow$ 4) and β (1 $\rightarrow$ 3) bonds. What's interesting is that agarose can form stable gels in water at low concentrations which are thermoreversible, when cooled below 40°C, without using any kind of cross-linking agents [44, 45].

The process of physical cross-linking has a significant advantage over chemical methods because chemical methods can be toxic due to the cross-linking agents used. Unlike chemical cross-linking, physical cross-linking uses weak interactions such as ionic and hydrogen bonds to form agarose helices that aggregate into a flexible structure with a pore size of about 100–300 nm (average). These bonds are established upon cooling and do not involve covalent bond formation between polymer chains. The exceptional ability of agarose to form gels, along with its mechanical strength, biocompatibility, availability, affordability, and non-toxicity, has made it a popular choice in the fields of biomedicine and cosmetics, specifically in the development of drug-delivery systems and tissue engineering [46]. Additionally, agarose's anti-adhesive characteristics make it favorable for hydrogel-based wound dressings, where low cell adhesion and proliferation prevent tissue ingrowth, minimizing injury upon removal. Furthermore, agarose's high mechanical strength fulfills the durability, stress resistance, softness, and elasticity requirements of wound dressings, accommodating body movement stresses. Notably, agarose's physicochemical properties can be tailored by introducing functional groups, enhancing drug-delivery efficiency, such as incorporating pH-responsive properties via carboxymethyl group introduction.

Other biomaterials that can be utilized include

a. Starch
b. Hyaluronic acid
c. Dextan [47].

5.2 PROTEINS

Protein-based biomaterials are used in biomedical fields like tissue engineering, medical science, and regenerative medicine. Some of the protein-based biomaterials include:

5.2.1 Silk Fibrinogen

Silk fibrinogen is a type of a natural protein derived from the silkworm (*Bombyx mori*), having many important properties such as the property of biocompatibility, controlled biodegradation with non-inflammatory byproducts, and adjustable mechanical strength [47]. Biocompatibility is one of the properties that make it safe for use in the human body and in the medical sector. The controlled biodegradation property of silk fibroin produces non-inflammatory byproducts, reducing the risk of inflammation and adverse reactions in the body. It also has its tunable mechanical strength, which can be adjusted to specific applications making it an important biomaterial that makes it possible to be used in a wide variety of medical devices, from soft tissue implants to hard bone replacements [48]. Silk fibroin can act as a drug-delivery system, stabilizing drugs and releasing them at a controlled rate, improving the effectiveness of treatments [49]. These characteristics make fibroin an important biomaterial that can be used for various applications, ranging from sustained drug-delivery systems and tissue regeneration. Silk fibroin-based hydrogels due to their unique molecular structure and high water content show gelation behavior which means having adhesion properties; these properties together are best for wound healing done by maintaining skin hydration and preventing cracks formation. The structure of many silks' fibroin-based hydrogels is sponge-like which facilitates the controlled release of therapeutic nanocarriers and act as a direct carrier for drug molecules. These silk hydrogels are versatile and can be very effective making them useful for a variety of wound management scenarios.

5.2.2 Collagen and Gelatin

Collagen and gelatin are two closely related naturally derived protein materials that are derived from animal connective tissues and have many different properties which make them important biomaterials to be used mostly in tissue engineering. Their unique properties are brought to an end in the form of gels, scaffolds, microspheres, and films. These biomaterials are important in the development of bioengineered tissues, offering reconstruction of bodily structures like blood vessels, heart valves, and ligaments [50] collagen's important hemostatic property actively promotes blood clotting, playing a crucial role in tissue repair. Collagen sponges or gels initiate the process which results in adhesion and aggregation of platelets, ultimately forming a clot (thrombus) at the site of injury.

The molecular structure of collagen plays an important role in interaction with platelets. Monomeric collagen lacks the organized molecular arrangement needed to activate platelets while polymeric collagen has a regular molecular structure which effectively triggers platelet aggregation. The specific amino acid side chains in collagen, particularly arginine, are responsible for this interaction [51]. Apart from its role in blood clotting, collagen also acts as a valuable scaffold for cell organization. The development of provisional extracellular support systems using type I collagen lattices (three-dimensional structure in vitro), enables the organization and growth of cells. The use of collagen has been extended for tissue engineering which shows the property of potential integration into the host cells as it has been used in creating small-diameter grafts constructed from type I bovine collagen, which have demonstrated the property of the potential integration into host tissue. These grafts act as a temporary scaffold, allowing for remodeling and ultimately forming functional blood vessels [52]. The versatility and biocompatibility of collagen and gelatin are revolutionizing the field of tissue engineering.

6 BIOPOLYMERS FOR MEDICAL APPLICATION

When discussing biomaterials used in medical applications, the thing that comes to mind is the use of various devices in the human body. These devices can either be used to treat injuries or to support the structure of the body. Some of the medical devices that have been used in the body include:

6.1 METALLIC BIOMATERIALS

There are different intrinsic properties of metal materials, such as electrical conductivity, good corrosion resistance, reliability, elasticity, high strength, mechanical reliability, and electrical conductivity. In the early 20th century, the first metal implants were made of stainless steel and chromium. In about 1960, titanium and nickel were designed to be used as biomaterials. Today, about 70–80% of the implants that are used in medicine practice are metallic [53]. Some examples are given in Table 15.1.

6.1.1 Titanium and Titanium Alloys

Titanium is an allotrophic element, meaning its crystalline structure changes with temperature. At a temperature of 885°C, the structure of Ti changes from a compact hexagonal to a cubic body-centered lattice. Titanium alloys are generally classified based on their microstructure, which can be α, almost α, α–β, or β. To enhance mechanical properties and lower the β phase transition temperature in α-phase alloys, stabilizers such as aluminum, gallium, and germanium are used. Meanwhile, vanadium, molybdenum, chromium, iron, niobium, silicon, and tantalum are used as stabilizers for β-phase alloys. However, vanadium's toxicity has resulted in replacement, using both iron or niobium in some cases. Pure titanium is classified into four categories based mainly on its oxygen and iron content. Titanium and its alloys are the preferred materials for manufacturing medical components, particularly implants, due to their low density for lightweight properties, high mechanical strength, corrosion resistance, low Young's modulus that mirrors cortical bone, and

TABLE 15.1

Common metallic, ceramic, and polymeric biomaterials and their mechanical properties

Type	Material	Modulus of elasticity (E) (Gpa)	Yield strength (Mpa)	Ultimate tensile strength (Mpa)	Density (g/cm³)	ρ umts (MPa)
Metallic Biomaterials	316 L Stainless steel	200–215	160–760	460–960	8.4	—
	Titanium alloys	105–125	840–1,100	590–1,024	4.5	—
	Ni–Ti alloys	35	240–241	895	6.45	—
	Pt–Ir alloys	168–528	380	380–620	21.56	—
	Cr–Co alloys	220–210	275–1,585	600–1,785	8	—
	Cortical bone	10–30	45–55	45–50	1–2	—
Ceramic Biomaterials	Alumina	350–400	—	3,800–4,000	—	—
	Zirconia	200–230	—	1,990–2,100	—	—
	Bio-glass	70–75	—	1,000	—	—
	Calcium phosphate	1–27	—	38–48	3.14	—
	Carbon (graphite)	23–26	—	130–140	—	—
	Carbon (vitreous)	30–33	—	170–175	—	—
	Carbon (pyrolytic)	20–30	—	800–900	—	—
	Hydroxyapapatite (HA)	117	—	550–650	—	—
	Vitroceramic	118	—	1,070–1,080	—	—
Polymeric biomaterials	Polyethylene	0.88	—	—	—	35
	Polyurethane	002	—	—	—	35
	Polytetraflourethylene	0.5	—	—	—	27.5
	Polyamide	2.1	—	—	—	67
	Polymethylmethacrylate	2.55	—	—	—	59
	Polyethylene terephthalate	2.85	—	—	—	61
	Polyrtherketone	8.3	—	—	—	139
	Synthetic rubber	0.008	—	—	—	7.6
	Polystyrene	2.65	—	—	—	75

superb biocompatibility. These characteristics position them as better materials than stainless steel and chromium–cobalt alloys [54].

Titanium alloys, commonly used in the medical industry as implant devices, are for the replacement of damaged or failed hard tissue. These devices include the majority of devices, such as artificial hip joints (ball and socket), knee joints (hinge joint), bone plates, screws for fracture fixation, cardiac valve prostheses, and pacemakers. Ti-6Al-4V is one of the important titanium alloys used for medical implants. There are various vanadium and aluminum-free titanium alloys have been introduced for implant applications. These new alloys are all based on the Ti-6Al-4V implants, including all Ti-6Al-7Nb (ASTM F1295), the Ti-13Nb-13Zr (ASTM F1713), and Ti-12Mo-6Zr (ASTM F1813).

6.1.2   Stainless Steel

Stainless steel is an alloy mainly composed of iron but highly resistant to corrosion due to the presence of chromium in concentrations. Due to 12% chromium in composition, there is the formation of a protective layer of chromium oxide. This protective layer is formed on the surface of the alloy, preventing the diffusion of oxygen, and making the steel corrosion resistant. The presence of elements such as molybdenum (2–3%) and nickel (12–15%) also enhances the corrosion resistance of stainless steel, making it durable in various environments. Based on their microstructure, stainless steel can be classified into four types [55].

A. **Austenitic stainless** steel has about 8% nickel in its composition and has low carbon content, resulting in good mechanical properties, weldability, corrosion resistance, and nonmagnetic properties. It is generally used in orthopedics as 316L austenitic stainless steel, denoting low carbon content. Austenitic stainless steel constitutes 70% production of worldwide steel production.
B. **Ferritic stainless steel** makes up around 25% of global production. This type of steel consists of magnetic ferro–chromium alloys and has about 12–14% chromium and minimal nickel, resulting in lower corrosion resistance compared with austenitic but altogether better.
C. **Martensitic stainless steel** is produced in lower volumes and has high carbon content and a low chromium concentration of 12–14%. Sometimes, nitrogen is added to improve its resistance, while sulfur helps to enhance its machinability. Unlike other types of stainless steel, it contains little to no nickel. Due to its hardness, which is due to the high carbon content, it is widely used in the production of surgical instruments.
D. **Duplex stainless steel** combines both austenitic and martensitic structures that offer enhanced resistance compared with its parent steels. With high chromium content (20–25%) and reduced nickel content (1.4–7%), it exhibits the property of fracture resistance due to corrosion.

6.1.3   Chromium–Cobalt Alloy

First originating at the start of the 20th century as "Stellite" chromium (Cr) and cobalt (Co) alloys due to their bioinert characteristics were used as biomaterials.

A refined version of stellite was called vitallium, introduced in 1932, having a composition of 60.6% Co, 31.5% Cr, and 6% molybdenum, tailored majorly for medical use [56]. Nonmagnetic Cr–Co alloys are still widely employed in dental procedures, artificial joints, and implants, due to their resilience against wear, heat, and corrosion, which makes them ideal for long-term implantation. This property of hardness and resistance poses machining challenges, leading to diminished tool lifespan, reduced productivity, and elevated manufacturing costs. Consequently, manufacturing methods like smelting or hot forging are preferred to mitigate expenses. Precise adjustment of cutting parameters is necessary for machining operations. ASTM identifies four recommended Cr–Co alloys for surgical implants [57].

6.1.4 Noble Metals

Noble metals utilization can be traced back to ancient times due to the evidence in human remains. Gold, silver, and platinum having their high corrosion resistance, have historical roots in medical applications [58]. However, their high cost, density, and limited mechanical strength often make them secondary options for medical components. Gold historically has been a popular choice for dental implants due to its properties of bacterial resistance, biocompatibility, and high malleability. It has also been used in pacemaker wiring and dental crowns. Silver is often used in surgical equipment as the coating. Silver compounds also help in burn treatment, infection prevention, stethoscope diaphragms, and germicidal and analgesic wound dressing fibers. Platinum, known is used for pacemaker electrodes to stabilize heart rhythms and in hearing aids. This is due to its corrosion resistance and stable electrical properties. Commonly used as platinum coils are utilized in the clearance of aneurysms [59].

6.2 Ceramic Biomaterials

Ceramics are inorganic materials that have become important in the field of biomaterials in recent years. This can be due to their various properties that include biocompatibility, non-corrosion, low compression, low electrical conductivity, and heat stability. Two different types of ceramics show presence in the composition of bone, these are HA or hydroxyapatite and TCP or tricalcium phosphate. These ceramics initially were only available and used as powders and then their use was also expanded as fillers in bone defects. It was later discovered that ceramics can also up to a certain extent, promote the regeneration of bone tissue as well. Compared with metals, ceramics are less dense, yet still exhibit a high level of hardness, brittleness, and minimal plastic deformation. The biocompatibility property of ceramics is due to their inertness and not causing and eliciting any immune response in the body. In addition, they can mimic the properties of natural bone, which makes them an ideal choice for implant materials. Ceramics such as alumina, zirconia, and glass ceramics have been used as biomaterials in various biomedical applications in the form of dental implants, artificial joints, bone plates, and screws. Ceramics having a similar chemical composition to natural bone can serve as a scaffold for new bone growth; this property makes it one of the most used materials in the skeletal system of the body. Ceramics can also release calcium and phosphate ions, which further are essential for bone growth and repair.

6.2.1 Alumina

Alumina is a material that has shown a high degree of compatibility with the human body and is commonly used in various medical applications such as devices and stents. This is due to their property of biocompatibility with the host body, immunocompatibility which makes it safe for use in the body, and resistance to corrosion, ensuring a long life of the material. It is especially useful for joint prostheses, like the acetabulum and femoral heads in hip arthroplasty, because it can be polished and is highly resistant to wear and tear [60]. To minimize the chances of joint components experiencing friction and harmful residue formation, manufacturers often polish two pieces together during the production process to ensure an optimal fit. The alumina femoral head is commonly combined with UHMWPE acetabular components. Alumina is also fragile and susceptible to fracture with the possibility of long-term implant relaxation, which can pose significant challenges [61].

6.2.2 Zirconia

Zirconia was used as a biomaterial in the late 1960s when it was first examined for use in femoral heads. After extensive research on various chemical compositions, the most commonly used combination used today is zirconia with yttrium. This combination of zirconia with yttrium produces fine-grain microstructures that are known as polycrystalline tetragonal zirconia or simply TZP. TZP with its high mechanical and chemical properties, is considered to be a prominent ceramic material in prosthetic applications [62]. Upon exposure to oxygen, zirconia forms zirconia oxide (ZrO_2), which has enhanced biocompatibility. Zirconia is a highly stable material used for hip arthroplasty due to its exceptional chemical and dimensional stability. ZrO_2 implants have high mechanical strength with Young's modulus similar to that of stainless steel (metallic biomaterial). ZrO_2 also provides remarkable resistance to wear and corrosion. They are also highly biocompatible and possess high flexural strength, and can resist fractures [63].

6.2.3 Carbon

Carbon is used as a biomaterial. Due to the property of being both inert and biocompatible, carbon fibers are strong under tension and are majorly used carbon materials. Graphene is a nootropic form of carbon that is dense, biocompatible, and has high flexural strength. Carbon nanotubes (CNTs), have the properties of optical and mechanical conductivity, good chemical stability, biocompatibility and high surface-to-volume ratio, good chemical stability, biocompatibility, and easy functionality. Fullerene is a class of carbon allotrope with a hollow core due to single or double-bonded carbons and a molecular structure that resembles a cylindrical tube, a sphere, or a cube nanodiamond. Carbon nanomaterials have good mechanical, thermal, and optical properties along with a high surface area and tunable surface structure.

Some others include:

1. Graphene
2. Bio-glass
3. Hydroxyapatite.

6.3 POLYMERIC BIOMATERIALS

In medical field applications, polymers are utilized in the majority of the components whether in the manufacturing of devices or in forming the coating to be used on them. Some polymeric biomaterials comprise PE or polyethylene PU or polyurethane, PA or polyamide, PTFE or polytetrafluoroethylene, and many more. These materials are considered highly reliable for long-term use. Additionally, composite materials, such as glass or carbon-reinforced PEEK, and variants like UHMWPE or ultra-high-molecular-weight polyethylene are also employed.

These polymers are classified into

1. natural (e.g., latex, cellulose, gums, starch),
2. synthetic,
3. semi-synthetic types.

These materials are further categorized as biodegradable or non-biodegradable. Biodegradable polymers contain all the natural materials, including polysaccharides, such as starch, and hyaluronic acid derivatives along with proteins, such as soy, collagen, and silk; it may also contain certain synthetic polymers reinforced or mixed with various bio-fibers [64]. These biodegradable synthetic polymers have emerged as one of the best in biomaterials due to their property of post-surgical self-degradation, excluding the need for material removal. This emphasizes the need for developing new polymers and processing methods to improve efficiency, such as cross-linking different types of polymers to enhance biomaterial quality for various medical uses.

There are many advantages that polymers offer, such as reduced density, making the implants lighter, and more, due to this there has been a trend seen in the replacement of metal devices with these polymers. Polymers allow for complex shape fabrication through injection molding on a large scale, which reduces production costs. Polymers are now utilized in various medical practices, including disposable equipment, orthopedic interventions, ocular prostheses, dental reconstruction materials, cardiovascular stents and interventions, membranes, sutures, and adhesive applications. In selecting the appropriate polymer for medical use, there is one factor that plays an important role called the degree of crystallinity, with polymers existing in amorphous, crystalline, and semi-crystalline states.

7 NATURAL POLYMERS

Polymers are substances found in nature, such as proteins and polysaccharides. Natural polymers are mainly derived from plants, animals, and microorganisms that gives them the property of being biodegradable, safe, and easily available. However, to enhance their materialistic properties, they may be mixed with some synthetic polymers [65]. The extraction and purification of natural polymers can be expensive and complex, which makes their use in demanding fields like medical components very difficult. However, recent advancements in chemical and enzymatic synthesis have improved the purity of natural polysaccharides and reduced their harmfulness. By combining natural and synthetic polymers, new materials with superior biological, chemical, and mechanical properties can be created. For instance, using autogenous bone marrow aspirated from the iliac bone to combine collagen with hyaluronic acid and tricalcium phosphate has been successful in treating fractures and pseudarthroses [66].

8 SYNTHETIC POLYMERS

Synthetic polymers have significant advantages, including mass production capability, intricate shape formation, and chemical manipulability to enhance mechanical properties. These encompass polyamides (PA), polyethylenes (PE), polypropylenes (PP), polyurethanes (PU), and others. Their limited biocompatibility may elicit inflammatory responses in patients, necessitating meticulous usage to eliminate such effects [67].

8.1 POLYOLEFINS

Polyolefins, derived from propylene (PP) or ethylene (PE), is a type of inert and hydrophobic polymer which does not degrade in the body. PE is generally used in medical equipment, such as drainage tubes and catheters; polyolefins are available in various forms such as LDPE or low-density polyethylene, HDPE or high-density polyethylene, LLDPE or linear low-density polyethylene, UHMWPE or ultra-high-molecular-weight polyethylene, and ULDPE or ultra-low-density polyethylene. UHMWPE is employed on artificial joint sliding surfaces due to its high-impact strength, biocompatibility, and high chemical stability [68]. However, friction-induced particle release may trigger intolerance reactions and mechanical prosthetic failure. To address this, highly crosslinked polyethylenes, such as UHMWPE, undergo gamma radiation to enhance wear resistance and biocompatibility. Second-generation crosslinked UHMWPE incorporates antioxidants like vitamin E. UHMWPE are extensively used in the medical sector specifically in orthopedic, cardiovascular, and neurological applications, in the making of surgical cables, catheters, and stent-grafts.

8.2 POLYTETRAFLUOROETHYLENE (PTFE)

PTFE, commonly known by its commercial name "Teflon" with its low coefficient of friction makes it an ideal biomaterial to be used for vascular stapling and various medical applications. Despite its hydrophobic nature and non-degradable properties, PTFE may elicit a mild inflammatory response in patients. ePTFE or expanded PTFE is a derived form of polytetrafluoroethylene that has to some extent the property of malleability, i.e., can be shaped into different membranes, rods, or other forms, suitable best for diverse functions, such as vascular grafts and soft tissue fillers post-surface modification.

8.3 POLYVINYL CHLORIDE (PVC)

PVC is used in matters of careful handling in medical applications due to potential toxicity induced by stabilizing and plasticizing agents. Its uses include the materials used to make tubes, blood bags, and catheters [69].

8.4 SILICONE

Silicone is a substance that is characterized by its ability to repel water and its stability in biological environments without the use of plasticizers [70]. Its effects on the body vary depending on how it is used. It is widely used in ophthalmology, breast implantation, and intra-articular implants. However, it can cause inflammation of the synovial membrane in the joints. There have been cases observed where liver tumors were seen in the presence of unused silicone oil.

8.5 POLYMETHYL METHACRYLATE (PMMA)

PMMA, or acrylic is extensively applied in stemmatological and orthopedics as bone cement for load transfer between implants and bones. PMMA polymerization entails significant heat generation, requiring cautious handling to prevent tissue damage and subsequent aseptic loosening. Despite its inherent biological inertness, reactions to its monomers may occur. PMMA is also being used to make contact lenses and in preparation for other ophthalmic applications. Polypropylene (PP) exhibits biological inertness and finds utility in sutures and nets.

9 COMPOSITE BIOMATERIALS

Sometimes two or more materials can be combined to make a material with better qualities than its constituents, forming a new form of biomaterial in this case called a composite material. A foundation material and a reinforcement material, such as carbon fibers added to a polymer, are combined, resulting in the formation of composite biomaterial having enhanced characteristics like less fatigue, high mechanical strength, and hardness is the goal. In composite materials, bioactivity and mechanical strength are enhanced using bioactive and bioenergetic ceramic materials e.g. HA and zirconia together can strengthen HA's mechanical characteristics while preserving its adherence to bone tissue.

Ceramic coatings are utilized in metal-based composite implants to enhance stability and biocompatibility. This promotes bone formation on the surface of the implant. Ceramic reinforcements, such as calcium phosphates, HA, or bio-glass, enhance the base material's elastic modulus and biocompatibility in polymer matrix implants. This results in mechanical properties like bone, reducing the possibility of stress-shielding. The use of polymer-based composite materials as opposed to metal implants provides numerous benefits. These materials do not rust and are biodegradable i.e., they don't produce toxic effects by releasing metal ions like nickel or chromium, which may cause allergic responses in patients. They also do not provide any hindrance to the accuracy of radiographs and thus are well-suited for diagnostic procedures, such as magnetic resonance imaging and computed tomography [71]. Table 15.1 represents mechanical properties of various biomaterials.

10 CONCLUSION AND FUTURE PROSPECTS

Biomaterials have always been a part of human evolution whether the usage was pure invention or necessity. This evolution has led to all those biomaterials becoming better and better. Today we have thousands of such biomaterials, improving the lives of humans every day. The importance of these materials has only risen in the last two centuries and with time seems to be only increasing. This has led scientists all over the world to make and explore all different types of biomaterials that are useful, not only in the pharmaceutical and medical fields but also in the industrial sectors. The biomaterials that are derived from renewable resources like animals and plants, offer numerous benefits i.e., non-toxicity, biodegradability, and tunable biodegradability. The use of sustainable biomaterials can lead to a greener approach to material usage and production, reducing the environmental impact. Traditional materials, commonly sourced from non-renewable resources, tend to be non-biodegradable, resulting in pollution and the accumulation of waste over time. These materials can offer a healthy alternative.

REFERENCES

[1] Farag, Mohammad M. 2023. "Recent Trends on Biomaterials for Tissue Regeneration Applications: Review." *Journal of Materials Science* 58 (2): 527–558. https://doi.org/10.1007/S10853-022-08102-X.

[2] Salthouse, Daniel, Katarina Novakovic, Catharien M. U. Hilkens, and Ana Marina Ferreira. 2023. "Interplay between Biomaterials and the Immune System: Challenges and Opportunities in Regenerative Medicine." *Acta Biomaterialia* 155 (January): 1–18. https://doi.org/10.1016/J.ACTBIO.2022.11.003.

[3] Naidu, Nikita A., Kamlesh Wadher, and Milind Umekar. 2021. "An Overview on Biomaterials: Pharmaceutical and Biomedical Applications." *Journal of Drug Delivery and Therapeutics* 11 (1-s): 154–161. https://doi.org/10.22270/jddt.v11i1-s.4723.

[4] Arif, Zia Ullah, Muhammad Yasir Khalid, Reza Noroozi, Mokarram Hossain, Hao Tian Harvey Shi, Ali Tariq, Seeram Ramakrishna, and Rehan Umer. 2023. "Additive Manufacturing of Sustainable Biomaterials for Biomedical Applications." *Asian Journal of Pharmaceutical Sciences*. Shenyang Pharmaceutical University. https://doi.org/10.1016/j.ajps.2023.100812.

[5] Biswal, Trinath, Sushant Kumar BadJena, and Debabrata Pradhan. 2020. "Sustainable Biomaterials and Their Applications: A Short Review." *Materials Today: Proceedings* 30: 274–282. https://doi.org/10.1016/j.matpr.2020.01.437.

[6] Vallet-Regí, Maria. 2022. "Evolution of Biomaterials." *Frontiers in Materials* 9 (March): 864016. https://doi.org/10.3389/fmats.2022.864016

[7] Bhat, Sumrita, and Ashok Kumar. 2013. "Biomaterials and Bioengineering Tomorrow's Healthcare." *Biomatter* 3 (3). https://doi.org/10.4161/BIOM.24717.

[8] Young, James B., William T. Abraham, Andrew L. Smith, Angel R. Leon, Randy Lieberman, Bruce Wilkoff, Robert C. Canby, et al. 2003. "Combined Cardiac Resynchronization and Implantable Cardioversion Defibrillation in Advanced Chronic Heart Failure: The MIRACLE ICD Trial." *JAMA* 289 (20): 2685–2694. https://doi.org/10.1001/JAMA.289.20.2685.

[9] Hench, Larry L., and Ian Thompson. 2010. "Twenty-First Century Challenges for Biomaterials." *Journal of the Royal Society Interface* 7 (Suppl 4): S379. https://doi.org/10.1098/RSIF.2010.0151.FOCUS.

[10] Murali Rao, H., K. Rajkumar, Vanamala Narayana, M. Akash, Jeeba Sabu, and Ramya Ramadoss. 2022. "Bioinert and Bioactive Materials – Narrative Review." *Journal of Pharmaceutical Negative Results* 13 (December): 2251–2255.

[11] Zhou, Lisheng, Liming Zhou, Chengxiu Wei, and Rui Guo. 2022. "A Bioactive Dextran-Based Hydrogel Promote the Healing of Infected Wounds via Antibacterial and Immunomodulatory." *Carbohydrate Polymers* 291 (September). https://doi. org/10.1016/J.CARBPOL.2022.119558.

[12] Nair, Lakshmi S., and Cato T. Laurencin. 2007. "Biodegradable Polymers as Biomaterials." *Progress in Polymer Science (Oxford)* 32 (8–9): 762–798. https://doi. org/10.1016/J.PROGPOLYMSCI.2007.05.017.

[13] Festas, A. J., A. Ramos, and J. P. Davim. 2019. "Medical Devices Biomaterials – A Review." *Proceedings of the Institution of Mechanical Engineers, Part L: Journal of Materials: Design and Applications* 234 (1): 218–228. https://doi.org/10.1177/1464420719882458.

[14] Özel, Tuğgrul, Paolo Jorge Bártolo, Elisabetta Ceretti, Joaquim D. de Ciurana Gay, Ciro Angel Rodríguez, and Jorge Vicente Lopes da Silva. 2016. "Biomedical Devices: Design, Prototyping, and Manufacturing." *Biomedical Devices: Design, Prototyping, and Manufacturing*, October, 1–190. https://doi.org/10.1002/9781119267034.

[15] Migonney, Véronique. 2014. "History of Biomaterials." *Biomaterials*, November, 1–10. https://doi.org/10.1002/9781119043553.CH1.

[16] Corruccini, Robert S., and Elsa Pacciani. 1989. "'Orthodontistry' and Dental Occlusion in Etruscans." *Angle Orthod.* Accessed March 22, 2024. https://meridian.allenpress. com/angle-orthodontist/article-abstract/59/1/61/57142.

[17] Chen, Fa Ming, and Xiaohua Liu. 2016. "Advancing Biomaterials of Human Origin for Tissue Engineering." *Progress in Polymer Science* 53 (February): 86–168. https://doi. org/10.1016/J.PROGPOLYMSCI.2015.02.004.

[18] "Products and Medical Procedures | FDA." n.d. Accessed March 23, 2024. https://www. fda.gov/medical-devices/products-and-medical-procedures.

[19] Uwaezuoke, Onyinye J., Pradeep Kumar, Viness Pillay, and Yahya E. Choonara. 2021. "Fouling in Ocular Devices: Implications for Drug Delivery, Bioactive Surface Immobilization, and Biomaterial Design." *Drug Delivery and Translational Research* 11 (5): 1903–1923. https://doi.org/10.1007/S13346-020-00879-1.

[20] Zare, Mina, Erfan Rezvani Ghomi, Prabhuraj D. Venkatraman, and Seeram Ramakrishna. 2021. "Silicone-Based Biomaterials for Biomedical Applications: Antimicrobial Strategies and 3D Printing Technologies." *Journal of Applied Polymer Science* 138 (38): 50969. https://doi.org/10.1002/APP.50969.

[21] Eswari, J. Satya, and Sweta Naik. 2020. "A Critical Analysis on Various Technologies and Functionalized Materials for Manufacturing Dialysis Membranes." *Materials Science for Energy Technologies* 3 (January): 116–126. https://doi.org/10.1016/J. MSET.2019.10.011.

[22] Farokhi, Mehdi, Fatemeh Mottaghitalab, Mohammad Reza Saeb, Shahrokh Shojaei, Negin Khaneh Zarrin, Sabu Thomas, and Seeram Ramakrishna. 2021. "Conductive Biomaterials as Substrates for Neural Stem Cells Differentiation towards Neuronal Lineage Cells." *Macromolecular Bioscience* 21 (1): 2000123. https://doi.org/10.1002/MABI.202000123.

[23] Singh, Ranjeet Kumar, and Swati Gangwar. 2021. "An Assessment of Biomaterials for Hip Joint Replacement." *International Journal of Engineering, Science and Technology* 13 (1): 25–31. https://doi.org/10.4314/IJEST.V13I1.4S.

[24] O'Leary, Cian, Luis Soriano, Aidan Fagan-Murphy, Ivana Ivankovic, Brenton Cavanagh, Fergal J. O'Brien, and Sally Ann Cryan. 2020. "The Fabrication and in Vitro Evaluation of Retinoic Acid-Loaded Electrospun Composite Biomaterials for Tracheal Tissue Regeneration." *Frontiers in Bioengineering and Biotechnology* 8 (March): 504885.

[25] Bharadwaz, Angshuman, and Ambalangodage C. Jayasuriya. 2020. "Recent Trends in the Application of Widely Used Natural and Synthetic Polymer Nanocomposites in Bone Tissue Regeneration." *Materials Science and Engineering: C* 110 (May): 110698. https://doi.org/10.1016/J.MSEC.2020.110698.

[26] Agrawal, Reeya, Anjan Kumar, Mustafa K. A. Mohammed, and Sangeeta Singh. 2004. "Biomaterial Types, Properties, Medical Applications, and Other Factors: A Recent Review." *Journal of Zhejiang University-SCIENCE A (Applied Physics & Engineering)* *2023* 24 (11): 1027–1042. https://doi.org/10.1631/jzus.A2200403.

[27] Khan, Hiba, Ben Barkham, and Alex Trompeter. 2021. "The Use of Bioabsorbable Materials in Orthopaedics." *Orthopaedics and Trauma* 35 (5): 289–296. https://doi.org/10.1016/J.MPORTH.2021.07.005.

[28] Subedi, Mitra Mani. 2013. "Ceramics and Its Importance." *Himalayan Physics* 4 (December): 80–82. https://doi.org/10.3126/HJ.V4I0.9433.

[29] Hasegawa, Masahiro, Shine Tone, Yohei Naito, and Akihiro Sudo. 2021. "Reconstruction of Patellar Tendon Rupture after Total Knee Arthroplasty Using Polyethylene Cable." *Knee* 29 (March): 63–67. https://doi.org/10.1016/j.knee.2021.01.008.

[30] Pandey, Anamika, Ankita Awasthi, and Kuldeep K. Saxena. 2020. "Metallic Implants with Properties and Latest Production Techniques: A Review." *Advances in Materials and Processing Technologies* 6 (2): 167–202. https://doi.org/10.1080/2374068X.2020.1731236.

[31] Prasad, Karthika, Olha Bazaka, Ming Chua, Madison Rochford, Liam Fedrick, Jordan Spoor, Richard Symes, et al. 2017. "Metallic Biomaterials: Current Challenges and Opportunities." *Materials* 10 (8): 884. https://doi.org/10.3390/MA10080884.

[32] Caldas, Mariana, Ana Cláudia Santos, Francisco Veiga, Rita Rebelo, Rui L. Reis, and Vitor M. Correlo. 2020. "Melanin Nanoparticles as a Promising Tool for Biomedical Applications – A Review." *Acta Biomaterialia* 105 (March): 26–43. https://doi.org/10.1016/J.ACTBIO.2020.01.044.

[33] Hussain, Zahid, Haliza Katas, Mohd Cairul Iqbal Mohd Amin, Endang Kumulosasi, and Shariza Sahudin. 2013. "Antidermatitic Perspective of Hydrocortisone as Chitosan Nanocarriers: An Ex Vivo and in Vivo Assessment Using an NC/Nga Mouse Model." *Journal of Pharmaceutical Sciences* 102 (3): 1063–1075. https://doi.org/10.1002/JPS.23446.

[34] Mohammed, Aiman Saleh A., Muhammad Naveed, and Norbert Jost. 2021. "Polysaccharides; Classification, Chemical Properties, and Future Perspective Applications in Fields of Pharmacology and Biological Medicine (A Review of Current Applications and Upcoming Potentialities)." *Journal of Polymers and the Environment* 29 (8): 2359–2371. https://doi.org/10.1007/S10924-021-02052-2.

[35] Jacob, Joby, Józef T. Haponiuk, Sabu Thomas, and Sreeraj Gopi. 2018. "Biopolymer Based Nanomaterials in Drug Delivery Systems: A Review." *Materials Today Chemistry* 9 (September): 43–55. https://doi.org/10.1016/J.MTCHEM.2018.05.002.

[36] Torres, Fernando G., Omar P. Troncoso, Anissa Pisani, Francesca Gatto, and Giuseppe Bardi. 2019. "Natural Polysaccharide Nanomaterials: An Overview of Their Immunological Properties." *International Journal of Molecular Sciences* 20 (20). https://doi.org/10.3390/IJMS20205092.

[37] Sahana, T. G., and P. D. Rekha. 2018. "Biopolymers: Applications in Wound Healing and Skin Tissue Engineering." *Molecular Biology Reports* 45 (6): 2857–2867. https://doi.org/10.1007/S11033-018-4296-3.

[38] Park, Sung bin, Eugene Lih, Kwang Sook Park, Yoon Ki Joung, and Dong Keun Han. 2017. "Biopolymer-Based Functional Composites for Medical Applications." *Progress in Polymer Science* 68 (May): 77–105. https://doi.org/10.1016/J.PROGPOLYMSCI.2016.12.003.

[39] Krzyszczyk, Paulina, Rene Schloss, Andre Palmer, and François Berthiaume. 2018. "The Role of Macrophages in Acute and Chronic Wound Healing and Interventions to Promote Pro-Wound Healing Phenotypes." *Frontiers in Physiology* 9 (May). https://doi.org/10.3389/FPHYS.2018.00419.

[40] Dadashzadeh, Arezoo, Rana Imani, Saeid Moghassemi, Kobra Omidfar, and Nabiollah Abolfathi. 2020. "Study of Hybrid Alginate/Gelatin Hydrogel-Incorporated Niosomal Aloe Vera Capable of Sustained Release of Aloe Vera as Potential Skin Wound Dressing." *Polymer Bulletin* 77 (1): 387–403. https://doi.org/10.1007/S00289-019-02753-8.

[41] Lee, Kuen Yong, and David J. Mooney. 2012. "Alginate: Properties and Biomedical Applications." *Progress in Polymer Science (Oxford)* 37 (1): 106–126. https://doi.org/10.1016/J.PROGPOLYMSCI.2011.06.003.

[42] Salehi, Majid, Arian Ehterami, Saeed Farzamfar, Ahmad Vaez, and Somayeh Ebrahimi-Barough. 2021. "Accelerating Healing of Excisional Wound with Alginate Hydrogel Containing Naringenin in Rat Model." *Drug Delivery and Translational Research* 11 (1): 142–153. https://doi.org/10.1007/S13346-020-00731-6.

[43] Kurczewska, Joanna, Paulina Pecyna, Magdalena Ratajczak, Marzena Gajęcka, and Grzegorz Schroeder. 2017. "Halloysite Nanotubes as Carriers of Vancomycin in Alginate-Based Wound Dressing." *Saudi Pharmaceutical Journal* 25 (6): 911–920. https://doi.org/10.1016/J.JSPS.2017.02.007.

[44] Singh, Yogendra Pratap, Nandana Bhardwaj, and Biman B. Mandal. 2016. "Potential of Agarose/Silk Fibroin Blended Hydrogel for in Vitro Cartilage Tissue Engineering." *ACS Applied Materials and Interfaces* 8 (33): 21236–21249. https://doi.org/10.1021/ACSAMI.6B08285.

[45] Singh, Tejwant, Tushar J. Trivedi, and Arvind Kumar. 2010. "Dissolution, Regeneration and Ion-Gel Formation of Agarose in Room-Temperature Ionic Liquids." *Green Chemistry* 12 (6): 1029–1035. https://doi.org/10.1039/B927589D.

[46] Grolman, Joshua M., Mansher Singh, David J. Mooney, Elof Eriksson, and Kristo Nuutila. 2019. "Antibiotic-Containing Agarose Hydrogel for Wound and Burn Care." *Journal of Burn Care and Research* 40 (6): 900–906. https://doi.org/10.1093/JBCR/IRZ113.

[47] Pires, Patrícia C., Filipa Mascarenhas-Melo, Kelly Pedrosa, Daniela Lopes, Joana Lopes, Ana Macário-Soares, Diana Peixoto, Prabhanjan S. Giram, Francisco Veiga, and Ana Cláudia Paiva-Santos. 2023. "Polymer-Based Biomaterials for Pharmaceutical and Biomedical Applications: A Focus on Topical Drug Administration." *European Polymer Journal* 187 (April): 111868. https://doi.org/10.1016/J.EURPOLYMJ.2023.111868.

[48] Li, Zheng, Jiangbo Song, Jianfei Zhang, Kaige Hao, Lian Liu, Baiqing Wu, Xinyue Zheng, Bo Xiao, Xiaoling Tong, and Fangyin Dai. 2020. "Topical Application of Silk Fibroin-Based Hydrogel in Preventing Hypertrophic Scars." *Colloids and Surfaces B: Biointerfaces* 186 (February). https://doi.org/10.1016/J.COLSURFB.2019.110735.

[49] Mao, Kai Li, Zi-Liang Fan, Jian-Dong Yuan, Pian-Pian Chen, Jing-Jing Yang, Jie Xu, De-Li ZhuGe, et al. 2017. "Skin-Penetrating Polymeric Nanoparticles Incorporated in Silk Fibroin Hydrogel for Topical Delivery of Curcumin to Improve Its Therapeutic Effect on Psoriasis Mouse Model." *Colloids and Surfaces B: Biointerfaces* 160 (December): 704–714. https://doi.org/10.1016/J.COLSURFB.2017.10.029.

[50] Shi, Rui, Dafu Chen, Quanyong Liu, Yan Wu, Xiaochuan Xu, Liqun Zhang, and Wei Tian. 2009. "Recent Advances in Synthetic Bioelastomers." *International Journal of Molecular Sciences* 10 (10): 4223–4256. https://doi.org/10.3390/IJMS10104223.

[51] Ulrich, Theresa A., Amit Jain, Kandice Tanner, Joanna L. MacKay, and Sanjay Kumar. 2010. "Probing Cellular Mechanobiology in Three-Dimensional Culture with Collagen-Agarose Matrices." *Biomaterials* 31 (7): 1875–1884. https://doi.org/10.1016/J.BIOMATERIALS.2009.10.047.

[52] Lin, Ange, Yanan Liu, Xufeng Zhu, Xu Chen, Jiawei Liu, Yanhui Zhou, Xiuying Qin, and Jie Liu. 2019. "Bacteria-Responsive Biomimetic Selenium Nanosystem for Multidrug-Resistant Bacterial Infection Detection and Inhibition." *ACS Nano* 13 (12): 13965–13984. https://doi.org/10.1021/ACSNANO.9B05766.

[53] Niinomi, Mitsuo, Masaaki Nakai, and Junko Hieda. 2012. "Development of New Metallic Alloys for Biomedical Applications." *Acta Biomaterialia* 8 (11): 3888–3903. https://doi.org/10.1016/J.ACTBIO.2012.06.037.

[54] Hussein, Abdelrahman H., Mohamed A. H. Gepreel, Mohamed K. Gouda, Ahmad M. Hefnawy, and Sherif H. Kandil. 2016. "Biocompatibility of New Ti-Nb-Ta Base Alloys." *Materials Science and Engineering C* 61 (April): 574–578. https://doi.org/10.1016/J.MSEC.2015.12.071.

[55] Gomes, L. S. M. 2010. "Biomateriais Em Artroplastia de Quadril: Propriedades, Estrutura e Composição." São Paulo: Atheneu. Accessed March 25, 2024. https://www.researchgate.net/profile/Luiz-Gomes-17/publication/309788298_Biomateriais_em_Artoplastia_de_Quadril_PropriedadesEstrutura_e_Composicao/links/582ae65508ae004f74af1ba8/Biomateriais-em-Artoplastia-de-Quadril-Propriedades-Estrutura-e-Composicao.pdf.

[56] Markatos, K., G. Tsoucalas, and M. Sgantzos. 2016. "Hallmarks in the History of Orthopaedic Implants for Trauma and Joint Replacement." *Acta Medico-Historica Adriatica: AMHA* 14 (1): 161–176. https://hrcak.srce.hr/ojs/index.php/amha/article/view/19235.

[57] Shokrani, A., V. Dhokia, and S. T. Newman. 2016. "Cryogenic High Speed Machining of Cobalt Chromium Alloy." *Procedia CIRP*. Accessed March 25, 2024. https://www.sciencedirect.com/science/article/pii/S2212827116302001.

[58] Medici, Serenella, Massimiliano Peana, Valeria Marina Nurchi, Joanna I. Lachowicz, Guido Crisponi, and Maria Antonietta Zoroddu. 2014. "Noble Metals in Medicine: Latest Advances." *Coordination Chemistry Reviews* 284: 329–350. https://doi.org/10.1016/j.ccr.2014.08.002.

[59] Jeong, Seok. 2016. "Basic Knowledge about Metal Stent Development." *Clinical Endoscopy* 49 (2): 108–112. Accessed March 25, 2024. https://www.ncbi.nlm.nih.gov/pmc/articles/PMC4821512/.

[60] Judas, Fernando, Helena Figueiredo, Rui Dias, and Professora Associada Da Fmuc. 2009. "Biomateriais Em Cirurgia Ortopédica Reconstrutiva." https://core.ac.uk/download/pdf/61496975.pdf.

[61] Katti, Kalpana S. 2004. "Biomaterials in Total Joint Replacement." *Colloids and Surfaces. B, Biointerfaces* 39 (3): 133–142. https://doi.org/10.1016/J.COLSURFB.2003.12.002.

[62] Chevalier, J., and L. Gremillard. 2011. "Zirconia as a Biomaterial." *Comprehensive Biomaterials* 1 (October): 95–108. https://doi.org/10.1016/B978-0-08-055294-1.00017-9.

[63] Sollazzo, Vincenzo, Furio Pezzetti, Antonio Scarano, Adriano Piattelli, Carlo Alberto Bignozzi, Leo Massari, Giorgio Brunelli, and Francesco Carinci. 2008. "Zirconium Oxide Coating Improves Implant Osseointegration *in Vivo*." *Dental Materials* 24 (3): 357–361. Accessed March 25, 2024. https://www.sciencedirect.com/science/article/pii/S0109564107001285.

[64] Rezwan, K., Q. Z. Chen, J. J. Blaker, and Aldo Roberto Boccaccini. 2006. "Biodegradable and Bioactive Porous Polymer/Inorganic Composite Scaffolds for Bone Tissue Engineering." *Biomaterials* 27 (18): 3413–3431. https://doi.org/10.1016/J.BIOMATERIALS.2006.01.039.

[65] Guo, Linqi, Zhihui Liang, Liang Yang, Wenyan Du, Tao Yu, Huayu Tang, Changde Li, and Hongbin Qiu 2021. "The Role of Natural Polymers in Bone Tissue Engineering." *Elsevier*. Accessed April 12, 2024. https://www.sciencedirect.com/science/article/pii/S0168365921004697.

[66] Benavente, R., and Csic Madrid. 1997. "Polímeros Amorfos, Semicristalinos, Polímeros Cristales Líquidos y Orientación." https://ruc.udc.es/dspace/bitstream/handle/2183/9633/CC_32_art_2.pdf.

[67] Singh, C., C. S. Wong, and X. Wang. 2015. "Medical Textiles as Vascular Implants and Their Success to Mimic Natural Arteries." *Journal of Functional Biomaterials* 6 (3): 500–525. Accessed March 28, 2024. https://www.mdpi.com/2079-4983/6/3/500.

[68] Jasinska-Walc, Lidia, Miloud Bouyahyi, and Rob Duchateau. 2022. "Potential of Functionalized Polyolefins in a Sustainable Polymer Economy: Synthetic Strategies and Applications." *ACS Publications* 55 (15): 1985–1996. https://doi.org/10.1021/acs. accounts.2c00195.

[69] "2 Applications of PVC-P as a Blood-Contacting Biomaterial | Download Scientific Diagram." n.d. Accessed April 19, 2024. https://www.researchgate.net/figure/ Applications-of-PVC-P-as-a-blood-contacting-biomaterial_fig10_265359394.

[70] Jung, Yuna, Youngbuhm Huh, and Dokyoung Kim. 2021. "Recent Advances in Surface Engineering of Porous Silicon Nanomaterials for Biomedical Applications." *Microporous and Mesoporous Materials* 310 (January): 110673. https://doi.org/10.1016/J. MICROMESO.2020.110673.

[71] Kargozar, Saeid, Seeram Ramakrishna, and Masoud Mozafari. 2019. "Chemistry of Biomaterials: Future Prospects." *Current Opinion in Biomedical Engineering* 10 (June): 181–190. https://doi.org/10.1016/J.COBME.2019.07.003.

16 Applications of Biopolymers in the Agriculture and Food Industry

Avani S, Isha Sharma, and Sharon Nagpal
Lovely Professional University

Pawan Saini
CSB-Central Sericultural Research &
Training Institute (CSR & TI)

Parveen Kumar
Department of Mechanical Engineering, Rawal Institute of
Engineering and Technology, Faridabad, Haryana, India

Ajay Kumar
Department of Mechanical Engineering,
School of Engineering and Technology, JECRC
University, Jaipur, Rajasthan, India

1 INTRODUCTION

The Greek phrases "bio" and "polymer," which signify nature and living things, respectively, are the origin of the term "biopolymer." Biopolymers are large, repeating macromolecules composed of multiple components. A single molecule is referred to as a macromolecule in International Union of Pure and Applied Chemistry (IUPAC). Biopolymers are excellent resources produced by living organisms, possessing properties, such as biocompatibility, recyclability, and degradability. The increasingly prevalent idea that biopolymers may compete with fossil-based polymers raises many environmental concerns. The use of bio-based polymers is growing in some sectors, such as electronics, healthcare, energy, food packaging, and power. Biological polymers are derived from bacteria, plants, animals, and agricultural waste. The qualities that make a material biocompatible and biodegradable allow a wide range of uses, like emulsions and edible films, drug transport supplies, packaging for the food industry, medical implants, tissue scaffolds, wound care, and pharmaceutical and medical industry dressing materials [1]. Reducing the carbon footprint is another

DOI: 10.1201/9781003434313-16

advantage of biopolymers for the environment and living organisms. Instead of forming any harmful compounds when exposed to mold and bacteria, they break down into their monomers to form CO_2, CH_4, and water. Because of the world's population growth and increased use of technology, biopolymers are finding applications both in industry and in everyday life.

The most well-studied biopolymers in the current development of biocomposites are starch-based plastics, cellulose esters, polylactic acid (PLA), and polyhydroxy-alkanoates (PHAs) [2]. Renewable polymers, including PLA, may be reabsorbed by the human body and broken down while composting plants. It has a biocidal impact because it tends to produce lactic acid through surface hydrolysis. Additionally, it is among the best substitutes for petroleum-derived polymers in areas of agriculture, packing, tissue engineering, and personal hygiene. To expand its processability and range of applications, several studies have been carried out utilizing a variety of nanofillers, including metal oxide chitin nanofibers, clays, and cellulose nanocrystals [3]. Chitin nanofibers and PLA in particular offer a significant chance to build bioplastic composites with enhanced structural and functional properties because of their complementing effects. At the nanoscale, it is difficult to scatter these nanofibers uniformly throughout the PLA matrix. Compatibility may be achieved by substituting certain cellulose microfibers.

Potato pulp powder, a useful byproduct of starch extraction and manufacture, is an intriguing PLA nanofiller [4]. Its primary constituents are lignocellulosic fibers, proteins, and carbohydrates. Its low cost as a raw material makes it a particularly appealing commodity for industrial use. Injection molding and extrusion were used to create the PLA/potato pulp biocomposites, and in addition to their processability, their mechanical, rheological, and thermal characteristics were assessed. Citric acid is the source of acetyl tributyl citrate, a naturally occurring plasticizer that was used to make processing easier. To aid in the detachment of the injection-molded specimens, inert filler calcium carbonate was also introduced in small amounts, which will increase the stiffness, improve biodegradability, and reduce the final cost of the composite. Chitosan, a biopolymer of interest, is often derived from the crustacean exoskeleton. Its antimicrobial activity, biodegradability, and biocompatibility are some of its most valuable properties. Because chitosan is ionic in solution, forms networks in three dimensions, and has high hydrogen bonding, making it insoluble in organic solvents, producing chitosan products is challenging [5]. However, there have been considerable advancements in the method of producing chitosan fiber. Still, there have been significant improvements in the chitosan fiber production process. Anionic biodegradable polyvinyl alcohol may produce more fiber by using electrospinning technology. Chitosan concentrations of 0.5, 1, 2, and 3 weight percent were investigated using the electrospinning process at different voltages, syringe/collector distances, and solution flow parameters. To provide the fibers with chemical stability, they were also treated with an ethanolic NaOH solution. Conversely, combining chitosan, α-tocopherol succinate, and the glycol skeleton leads to the creation of an amphiphilic polymer [6]. The powerful anti-tumor medication paclitaxel, which is frequently used to treat cancer, may be transported by this polymer.

Biopolymers are revolutionizing food and agriculture packaging, waste management, and production processes. Functional properties, including gelling, emulsification, and film formation, make biopolymers like proteins, cellulose, and starch useful in food applications. For instance, edible coatings composed of biopolymers are used to extend the shelf life of vegetables and fruits, prevent deterioration, and reduce the need for pesticides. Moreover, packaging made of biopolymers that degrade naturally offers a sustainable alternative to traditional plastics by reducing their negative environmental consequences and promoting the use of circular economic ideas. Biopolymers are used in agriculture to improve crop yield and sustainability through soil stability, water retention, and controlled-release fertilization [7]. Reduced soil contamination is achieved with biodegradable mulches composed of biopolymers, which are occasionally mistaken for traditional plastic mulches. They achieve this by keeping the soil moist, preventing weed development, and breaking down organically after application. Furthermore, especially in nutrient-poor or dry soils, hydrogels based on biopolymers are utilized to coat seedlings, encourage germination, and guarantee efficient nutrient absorption. Biopolymers are frequently employed in the agricultural and food sectors to address important issues, including resource conservation, environmental pollution, and food waste, while fostering sustainability and innovation in these crucial fields [8]. They also represent a move toward ecologically friendly practices.

2 BIOPOLYMERS TYPES

They exist in a multitude of forms and may be divided into many groups according to their characteristics, origin, and composition. These are only a few of the most common kinds of biopolymers.

2.1 NATURAL BIOPOLYMERS

Biopolymers obtained from nature can be used as an alternative to nonbiodegradable plastics made from petroleum. They are made up of gelatin and its derivatives, cellulose, starch, chitosan, and carrageenan. Motion images are made possible by biopolymers, which are easily accessible, cheaply priced, renewable, biodegradable, and biocompatible materials.

2.1.1 Silk Fibroin

Silks are, by definition, a unique class of biopolymers derived from the "spinning" secretions of numerous arthropod species. A few lineages that employ silks to build their webs, nests, and cocoons include silkworms, wasps, and honey bees. Together with more than 30,000 recognized spider species, the majority of Lepidoptera have 113,000 recognized species that are also capable of making silk. In addition, it has been documented that glowworms, fungus gnats, caddisfly larvae, wasps, fleas, lacewings, and crickets may also produce silk. Silk fibroin is a biomaterial that has been studied in great detail because of its potential applications in the photonic, electrical, medicinal, and textile areas. Similar to collagen, silk fibroin is a protein with an organized structure that is produced by a living, complex creature by the solution of amino-acid extrusion [9]. Collagen, on the other hand, is synthesized inside the extracellular area by the cell-produced monomer self-assembly. Moreover, due to its amino-acid nature, silk fibroin can include a range of side-chain chemistries that are advantageous for drug administration or cellular instructions [10].

The intrinsic properties of silk fibroin have been leveraged by the regenerated silk materials. The coupling of intra- and intermolecular hydrogen bonding with beta-sheet topologies not only yields extraordinary flexibility in natural fiber but also ensures the highest level of conformity in the regenerated film format [11]. Silk fibroin sheets are used as palatable, biocompatible, and biodegradable bases for electrodes that are implanted into the brain, as well as palatable, sticky devices to detect fruit ripening and the aging of cheese. Silk fibroin suspension is a consumable coating material for the preservation of perishable fruit. Two of the most appealing characteristics of silk fibroin that may be used in cuisine coating and packaging applications include flavor and odorlessness. Fresh strawberries were covered with a suspension of silk fibroin using the dip method. A regulated boiling duration of 30 minutes was used to extract silk fibroin with a molecular weight that was designed to be between 170 and 90 kDa [12]. The final solution protein concentration in the dipping suspension was also modified to 1 weight percent to achieve surface tension and viscosity equivalent to earlier biopolymer-based coatings.

2.1.2 Chitin and Chitosan

An important polysaccharide found in nature, chitin is sometimes referred to as poly (β (1→4) N-acetyl-D-glucosamine). The word "chitin" comes from the Greek word "chiton," which signifies a coat of mail. Chitin ($C_8H_{13}O_5N$) structure is similar to that of cellulose; furthermore, chitin has monomer units of 2-acetamido-2-deoxy-β-D-glucose (NAG [N-acetylglucosamine]) connected by β (1→4) links. Chitosan, a derivative of chitin that is deacetylated, has different potencies and can be challenging to dissolve in acidic environments. Chitin can be hydrolyzed chemically or enzymatically to create chitosan. It is anticipated that 1,011 tons of chitin will naturally occur worldwide each year. Chitin is a stiff, inelastic, and white nitrogenous polymer. Furthermore, evidence has shown that it is the main factor contributing to beach pollution in coastal areas. Chitin may be found in both the structural components of arthropod exoskeletons and the crystalline microfibrils present in the cell walls of fungi. Many species in the lower plant and animal kingdoms also synthesize chitin, but this substance has more potent reinforcing and protective qualities. Microorganisms are assumed to be the main mediators of chitin degradation in the natural world. The water and soil systems both offer proof of their operation. The rate of chitin hydrolysis in soil systems is associated with the quantity and number of bacteria present. Here, factors like pH and temperature affect the deterioration. Both bacteria and fungi are considered to be numerically significant participants in the process of chitin breakdown. The principal mediators of chitin breakdown are bacteria, as the findings of plating tests conducted in aquatic settings adequately revealed [13].

2.1.3 Cellulose

One of the most prevalent polymers on Earth, cellulose, is easily separated from plant cell walls. Anselme Payene made the initial discovery of cellulose. He found the chemical formula for both starch and this plant component [14]. Cellulose and the attention that its derivatives are receiving are believed to have significant effects since they are biocompatible and ecologically sustainable [14]. Affordable cellulose is highly sought after globally, especially in the food and textile sectors, because it possesses sufficient mechanical properties. But because of its supramolecular properties and ease of usage as a carrier, cellulose can readily encapsulate a large variety of natural and manmade polymers. In the past, cellulose-based paper was used for printing. Because of their special properties, biopolymers made of natural and renewable resources are crucial to the food-packaging sector. Cellulose is used extensively in the food-packaging sector because of its distinct qualities, composition, and attributes. On a micro- and nanoscale, cellulose is an ingredient in both widely used and freshly launched goods. Bio-composite cellulose, or nano cellulose, is used in a variety of industrial applications, including paper packaging. Furthermore, because cellulose-based molecules are biodegradable, they can satisfy the criteria for sustainable packaging. These molecules are also referred to as cellulose fibers or microfibrillated cellulose. Moreover, because of its barrier and preservation properties, it can help increase the length of time that fruits and vegetables are stored.

2.2 Synthetic Biopolymers

Monomers are repeating units that combine to form massive compounds known as synthetic polymers. Chemical bonding is the mechanism that polymerizes these monomers into lengthy polymer strands. There are several methods by which the polymerization process might take place, including addition polymerization and condensation polymerization.

2.2.1 Polyethylene (PE)

PE is the most widely used polymer worldwide. Ethylene is the principal monomer used in synthesis, much like polystyrene (PS) and polyvinyl chloride (PVC). It is typically derived by distillation from petroleum feedstock. Though there is increasing interest in generating this polymer from biological sources, the present process involves drying bio-ethanol, which is derived from glucose. This process creates the ethylene monomer. To produce glucose, various biological feedstocks can be used, including lignocellulosic materials, sugar cane, sugar beet, and starch crops generated from cereals like wheat, maize, or other grains. For example, bagasse, or sugar cane fiber, is a byproduct of processing sugar cane, which is first used to make sugar cane juice. These operations consist of cutting, slicing, cleaning, and shredding. The juice, which has a 12–13% sugar content, is fermented anaerobically to produce ethanol. An azeotropic solution of hydrous ethanol at 95.5% vol.% and vinasse, a byproduct, is produced by distilling ethanol to remove water. In terms of mechanical recycling processes, the corresponding biopolymer has the same mechanical, chemical, and physical characteristics as fossil-based PE. This type of additional ethylene monomer polymerization is identical to that which occurs when ethylene obtained from petroleum is utilize.

Dow and Crystalsev established a partnership focused on producing bio-PE for various packaging uses. Despite Dow's global ranking as the second-largest chemical producer, Crystalsev is Brazil's leading ethanol producer. Other significant players in the bio-polyethylene industry are Petrobras, Nova, Solvay, and Chemicals, which account for 20% of global bio-based plastics production. Projections indicate that Braskem could secure a 10% share of the global plastics market. In the past, there was a common belief that producing ethylene from petrochemicals was more cost-effective than from biomass.

2.2.2 Polyethylene Terephthalate (Bio-PET)

Biological feedstocks have the potential to be used to produce polyesters, an extensive polymer family [15]. The materials most relevant to a transition to bio-based chemicals comprise the copolymers poly butylene succinate adipate, polybutylene adipate terephthalate (PBAT), polybutylene succinate, PET poly butylene succinate terephthalate, and PBAT. Additional materials include thermoplastic polyester elastomer, polyacrylates, poly (trimethylene naphthalate), polyvinyl acetate, and poly (trimethylene isophthalate). While diesters and acids can come from petrochemicals or bio-based sources like adipic acid or succinic acid (like purified terephthalic acid and/or dimethyl terephthalate [DMT]), these polymers are generated from a bio-based diol. PET is the polyester that is used most commonly among themselves because of its mechanical and physical properties; it may be utilized in packaging (35%) and fibers (65%). The final application is related to containers (11%), bottles (76%), and films (13%). It is a prominent participant in the plastics sector but is employed in food packaging. It offers serious environmental problems due to its poor sustainability (it dissolves much more slowly). Its primary applications at the time of its 1940 release were in motion pictures and synthetic textiles. The Coca-Cola Company unveiled the "PlantBottle," a beverage container made of 30% petroleum-derived TA and 100% bio-based EG. The companies that make Bio-PET include Japan Future Polyesters, PepsiCo Venture, Coca-Cola Venture, and Toyota Tsusho Corporation [16]. This process yields bio-based PET that may be formed by blow molding, injection molding, or extrusion, just like petrochemical PET. PET is nonbiodegradable; however, a fascinating study [17] revealed that it may be broken down microbially by a substance called microbial polyester hydrolase. This is seen as a substantial substitute for recycling PET [17].

2.2.3 PLA

When bacteria break down, the carbohydrates found in crops like potatoes, cassava, and maize, poly (lactic acid), or PET, is produced. It is a naturally degradable polyester polymer made of lactic acid (LA) or 2-hydroxypropionic acid. The chemical conversion of bio-based monomers, specifically LA, results in the production of PLA. PLA is the preferred material for medical applications since it is compatible with the human body [18]. With its renewable, biodegradable, recyclable, and compostable properties, PLA is highly investigated as one of the most promising biopolymers produced from non-toxic renewable feedstock. Biopolymer is a possible substitute for traditional petroleum-based polymers. Furthermore, PLA offers a wide range of production opportunities due to its versatile treatment options. Various techniques such as film extrusion, blow molding, thermoforming, fiber spinning, and film-forming are used for synthesizing PLA. PLA materials have several industrial uses in the biomedical, automotive, packaging, textile, and structural engineering industries. PLA is still not widely utilized in many areas due to several issues, even if it is becoming increasingly useful and accessible. The surface of PLA is not as hard as that of polycarbonate (PC) or PET, despite being harder than that of PS. This might restrict the use of PLA as a structural element. Furthermore, PLA is often not recommended for use in biomedical settings due to its hydrophobicity and slow rate of biodegradation. Other techniques, including copolymerization, polymer mixing, and polymer compositing, have been employed to get around the drawbacks of PLA materials [19]. Among these processes, polymer mixing has garnered a lot of attention as a simple and affordable way to make polymers for a variety of uses.

2.3 HYBRID BIOPOLYMERS

Hybrid biopolymers derived from natural resources are renewable, eco-friendly, biodegradable, and compostable; they present an excellent substitute for petroleum-based latex and synthetic polymers. However, biopolymers might not be able to offer every needed property in its native form for a given application.

2.3.1 Bio-Based Polyesters

Polyester is one of the most versatile polymers and finds extensive application in plasticizers, coatings, elastomers, textiles, and plastics. Because petrochemical products pose a significant environmental risk and the biomass industry is growing quickly, scientists are focusing their efforts on developing polyesters made from bio-based monomers. The synthesis of polyesters derived from biomass is intimately related to the development of bio-based monomers [20]. In the polyester synthesis process, bio-based monomers and monomers with distinct structures and properties have been employed in place of the majority of conventional monomers. In addition to providing environmentally friendly substitutes for petroleum-based polyesters like PET, PTT or PPT, and PBS, the development and use of these bio-based monomers also provide insights into the synthesis of novel polyesters like PLA, bio-based engineering polyester elastomer, and poly (ethylene 2,5-furandicarboxylate) (PEF) [20]. Additionally, a variety of fully and partly bio-based polyesters, produced via modification of copolymerization from bio-based monomers, provide additional possibilities for creating and utilizing bio-based polyesters. Based on the characteristics of the monomer structure, bio-based polyesters with aromatic units can be further separated into bio-based aromatic and bio-based aliphatic polyesters. Polyesters developed using biotechnological strategies are now going through an unprecedentedly rapid development phase. Market-oriented polyester generated via bio-based monomers as well as novel bio-based polyesters with distinctive features derived from monomers or combinations have been shown to be useful in the industrial sector [21].

2.3.2 Nanocellulose Composites

A linear biopolymer with nanofibrils makes up the large surface area of cellulose. This naturally occurring polymer, including tunicates, microbes, and plants, has a high specific surface area, low cost, low density, and is easy to process. Due to its exceptional qualities, like low cytotoxicity, excellent mechanical properties, great chemical stability, biocompatibility, and affordability, cellulose is an excellent material for biomedical applications. The distinct physicochemical characteristics of nanocellulose are the outcome of cellulose nanoscale modification [22]. An intriguing field is cellulose nanocomposite, which has potential applications in biological domains like bone tissue production, medicine delivery, wound healing, and printing in three dimensions.

Plant- and microbe-derived nanocellulose fibers differ from one another. Plant-derived cellulose has a semi-crystalline structure and a hierarchical organization, but bacterial nanocellulose has a greater critical surface tension and a higher breakdown temperature. While bacterial cellulose is relatively pure and has a higher degree of crystallinity, plant-based cellulose is more widely available. Bacterial cellulose is relatively uncommon, though. Depending on how long plants take to grow, cellulose harvesting takes a longer time. In addition to these variables, food, climate, plant and soil types, and susceptibility to insect pest infestation all have an impact on cellulose fibers derived from bacteria and plants. The components of cellulose that originate from plants include cellulose, ash, hemicellulose, and lignin [23].

3   AGRICULTURAL APPLICATIONS

Pests and infections that produce harmful diseases are to blame for the global losses in agricultural output since they have an impact on seedling stands and productivity. The seedlings and plant stand will be shielded from these enemies by maintaining the quality of the seed. With the help of modern technology and different chemical seed treatments, it is now possible to combat diseases and pests, and with them, economic loss. Nowadays, applications ranging from plant protection, gel plantings, seed coatings, elastomers, coatings, plastic, fibers, and water-soluble polymers are the main categories of polymers used in soil conditioning to regulate the release of nutrients and insecticides. Degradable polymers are nonetheless appealing for use as planting pots and mulch in agriculture. It might be possible to mix degradable plastics with other biodegradable materials to create beneficial soil-improving elements if they have ultimate biodegradability, as in composting. Mulches made of biopolymers encourage plant development, which in turn causes the mulch to photodegrade in the fields and reduces the need for expensive removal. These mulches expedite the growth of plants because they may inhibit weeds, retain moisture, and increase soil temperature. It was demonstrated that using black polyethylene mulch increased melon yield by a ratio of two to three [24]. Mulch keeps the soil from being compacted and eliminates weeds, so there is less need to cultivate and less chance of root damage and plant death. The need for fertilizer and water is also reduced. Natural compounds called chitin and chitosan can be used in agriculture to control plant diseases. It has been reported that they are effective in combating bacteria, viruses, fungi, and other pests. Using them to chelate minerals and nutrients to keep pathogens out has also enhanced plant inherent defenses [25]. Microorganisms in the soil may readily degrade poly-lactone and poly (vinyl alcohol) films, although polyethylene degrades more quickly in the presence of iron or calcium. Biodegradable mulches should break down into small, fragile pieces that are readily broken up by harvesting machinery and do not get in the way of new planting. Fumigant mulches are useless if their porous films do not become thinner. This is because reduced porous films, which also prevent volatile chemicals from escaping, allow for lower application rates of nematicides, insecticides, herbicides, and other agents [26].

3.1 Hydrogels-Based Biopolymers and Their Use in Agriculture

Three-dimensional networks of very absorbent hydrophilic polymers are known as hydrogels. Hydrogels can also be precisely employed as active ingredient delivery methods. Synthetic polymers made from natural sources are called biopolymers. Biopolymer-derived hydrogels are believed to be biodegradable, affordable, non-toxic, and biocompatible. As a result, biopolymeric hydrogels are developed and widely utilized worldwide. Because biopolymer-based hydrogels enhance the capacity of the soil to retain water and serve as a conduit for additional agrochemicals and nutrients, they oftentimes serve as soil-conditioning agents in agriculture [7]. Hydrogels are also used to provide fertilizer to plants that are under control. Polymer derivatives leverage their interactions with other artificial polymers to create useful polysaccharide hydrogel composites that might be applied in agriculture as transporters of fertilizer and soil conditioners. Crop viability and total food production are enhanced by agricultural polysaccharide hydrogels, or APHs, which also improve soil porosity and water-holding capacity, all of which help to create a flora-friendly environment [27].

Numerous scholarly works exist that specify the use of soil conditioners made from hybrid formulations containing both synthetic hydrogels and polysaccharides. Cross-linked hydrogels with high stability, needed for usage as a soil conditioner and for nutrient management, are difficult to make using just polysaccharides, using cellulose starch derivatives in a dual coating method as a substitute to enhance slow-release urea effectiveness. Research aimed at covering the urea granules with ethyl cellulose on the inside and starch-polyacrylamide (PAAm) on the outside to produce fertilizer with two layers that release urea slowly for more than 96 hours. Radiation is another technique used for polymerization, where radiation-exposed natural polymers undergo a chain reaction. They are using hydrogels based on natural polymers for agricultural (degradation) processes as an alternative to chain-constructive (cross-linked) methods. Stable, cross-linked soil conditioners require a combination of synthetic vinyl monomers and natural polymers. They are composed of artificial monomers and polysaccharides that may be modified to produce soil conditioners [7, 28].

Management of urea fertilizer for crops was mediated by a range of hydrogels made of calcic montmorillonite (MMt), methylcellulose (MC), and polyacrylamide (PAAM) [29]. Two agents that cross-link, such as N'-N-methylene bisacrylamide and N, N, N', N' tetramethyl ethylenediamine catalysts, were combined with a sodium persulfate initiator agent to perform the radical polymerization. Montmorillonite clay optimizes the hydrogel formation, its mechanical resistance, and water absorption, among other things. In general, the transport of nutrients is slowed down by the combination of hydrogel and clay minerals. The regulated release of urea during the hydrogel hydrolysis was studied before and during the hydrolysis of the obtained sodium acetate hydroxide of polyacrylamide (PAAm) and methylcellulose (MC) using (sodium hydroxide) NaOH [7].

The application of starch-based superabsorbent hydrogel in agricultural settings has been covered in several research studies [28]. Slow-release fertilizer, or SRF, is crucial for lowering fertilizer leaching loss. Xiao et al. (2017) synthesized a superabsorbent hydrogel based on starch and filled it with urea to investigate the regulated release of urea. After a day, less than 15% of the urea was released; after 30 days, more than 80% was released. Sulfonated maize starch and poly (acrylic acid) to form an embedded phosphate rock that contained water and delivered controlled-release fertilizer. Urea was encapsulated by using two layers of acrylamide, starch, and AA copolymers, respectively. For the next step of the polymerization process, ammonium persulfate was employed as a radical polymerization activator. On the second, fifth, and thirtieth days, respectively, the area was progressively released into the soil through the increased network at a rate of around 10%, 15%, and 61%.

The quantity of soil that is permeable to air and water is increased by hydrogels. The texture of this soil is ideal for planting. Water will quickly move from the inside polymer network to the exterior of the hydrogel when the soil dries up. The water was stored in a hydrogel network until it was needed, and then it was returned to the soil. The interactions between soil and water as well as the integrity of the soil structure were enhanced by swollen hydrogel [28]. Such nutrient and fertilizer seepage from agricultural soils has also been reported. To reduce fertilizer leaching, hydrogel is added to fertilizer formulations. Fertilized hydrogel behavior allows for the progressive release and maintenance of fertilizer. Because there is less leaching, this increases crop output while reducing the need for fertilizer. Fertilizers injected into hydrogel have been shown to release half as quickly as those dissolved in water alone. Hydrogel is one technique for gradually releasing dissolved nutrients and water into the soil. Fertilizers are often released via the diffusion technique. Deswelling is the initial reaction phase in which water diffuses and moves fertilizer outside of the porous network.

3.2 BIOPOLYMERS FROM WASTE FOR SUSTAINABLE AGRICULTURE AND THE PRODUCTION OF SAFE FOOD

Biopolymers, which provide a useful way to support safe food production and environmentally sustainable agriculture, may be made from waste. Biopolymers are better for the environment than traditional petroleum-based polymers since their composition consists of sustainable materials, including plants, animals, and microbes. Waste biopolymers can improve food safety and enduring agriculture in numerous ways when used properly. Sustainable agriculture, often referred to as eco-friendly agriculture, is a comprehensive method of producing food that emphasizes striking a balance between environmental health, economic viability, and social justice. With this approach, the adverse effects of farming on the surroundings are mitigated while maintaining a steady and nutritious supply of food. The goal of eco-friendly agriculture is to establish a robust and sustainable food production system by utilizing techniques that improve soil health, minimize resource use, and foster biodiversity. Higher crop yields are achieved by applying minimal fertilizer dosages than what the soil and plants can absorb to meet the requirement for food. A surplus of nutrients builds up in the soil and seeps into the groundwater, causing eutrophication. The best method for preserving fruits and vegetables is to treat them with wax. Wax is occasionally mixed with carcinogenic substances; currently, many edible coatings have been developed using natural polymers to increase the shelf life of food. The edible coatings prevent bacteria and gases from coming into direct contact with the food by acting as a barrier between the food and its surroundings. This lowers the respiration rate and lengthens the shelf life of food. Due to their high cost, edible biofilms are only available at the industrial level; small-scale fruit and vegetable merchants cannot afford to purchase them.

3.3 BACTERIAL POLYMERS AND THEIR USE AS BIOFERTILIZERS AND BIOSTIMULANTS

A wide range of biopolymers, including polyesters, polysaccharides, and polyamides, are generated spontaneously by a multitude of bacteria. Microbial biopolymers are produced by a variety of microbes grown under different growing and feeding conditions. These polymers, which are frequently lipids, congregate in the cell cytoplasm as granules that act as mobile, amorphous, liquid storage materials intended to promote microbe survival in hostile conditions. Lemoigne (1926) identified polyhydroxybutyrate (PHB) granules, a kind of biopolymer, in the Gram-positive bacterium *Bacillus megaterium*. Since then, it has been shown that certain archaebacteria, in addition to both Gram-positive and Gram-negative bacteria, may elaborate on different polyhydroxyalkanoates (PHAs). *Bacillus mycoides*, *Rhodococcus ruber*, and *Alcanivorax borkumensi*, along with other *Pseudomonas species*, *Pseudomonas putida*, and other *Ralstonia* spp. and *Ralstonia eutropha*, are a few examples. Numerous polyhydroxyalkanoates with short and medium chains are of bacterial origin, such as those produced by *Ralstonia eutropha* and *Pseudomonas*. It has been demonstrated that using microbial systems to generate biopolymers more efficiently is a low-cost and environmentally safe option. Biopolymers synthesized by *Bacillus species*, *Leuconostoc mesenteroides*, and *Pseudomonas pseudomallei* have attracted a lot of attention because of how easily they may be modified using genetic engineering methods [30]. Certain microorganisms extracted from plant roots and root zones are added to organic preparations known as biofertilizers. Plant productivity and growth are enhanced by 10–40%. Certain bacteria, including rhizobacteria, promote plant growth and fix and solubilize insoluble minerals (P, K, and Zn), organic acids, and siderophores to acquire nitrogen [31]. These bacteria also solubilize phosphorus, which makes it possible for plants to absorb it. They help make potassium, zinc, and other essential minerals soluble for plant growth. With the help of bacterial biofertilizers, plants can withstand oxidative damage, heavy metals, extreme temperatures, drought, and high soil salt. Biostimulants based on plant growth-promoting rhizobacteria also increase crop plant growth [32]. Bacterial polymers improve soil fertility, crop productivity, and nutrient availability when they are used as biostimulation and fertilizers, contributing to sustainable agriculture as routes toward efficient and ecologically friendly farming practices [33].

3.4 BIOPOLYMERS IN CROP PROTECTION

Farmers now have two challenges: battling plant infections and pests that often destroy crops and boosting agricultural output to feed the world's growing population.

TABLE 16.1

Biopolymer coatings on fruits and vegetables

Vegetables/Fruits	Phytopathogen	Biopolymer used	References
Orange	*Penicillium digitatum*	Chitosan	[35]
	Penicillium italicum	Hydroxypropyl methylcellulose	[36]
Apple	*Penicillium expansum*	Chitosan	[37]
	Listeria innocua	Alginate	[38]
Cherry tomato	*Aspergillus niger, Penicillium expansum, Rhizopus stolonifera*	Chitosan	[39]
	Alternaria alternata	Hydroxypropyl methylcellulose	[39]
Grapes	*Botrytis cinereal*	Chitosan, nanoparticles	[40]
	Botrytis cinereal	Chitosan	[41]
	Botrytis cinereal	Alginate	[42]
Blueberry	*Aspergillus niger*	Alginate	[43]
Papaya	*Not specific*	Alginate	[44]
Pineapple	*Yeast and molds*	Alginate	[45]
Raspberry	*Yeast and molds*	Alginate	[46]
Peach	*Penicillium expansum*	Alginate	[47]
Shiitake mushrooms	*Bacteria, yeasts*, and *molds*	Alginate	[48]
Strawberry	*Not specified*	Alginate	[49]
Kiwifruit	*Botrytis cinerea*	Alginate oligosaccharide	[50]
Guava	*Not specified*	Carboxymethyl cellulose	[51]
Mandarin	*Penicillium italicum*	Carboxymethyl cellulose	[52]
Tangerine	*Penicillium* spp.	Carboxymethyl cellulose	[53]
Avocado	*Lasiodiplodia theobromae*	Carboxymethyl cellulose	[54]

Pests and diseases have been projected to cause worldwide productivity losses of 17.2–30.0% in terms of losses for five main crops for food: rice, wheat, potatoes, soybeans, and maize. Most destruction is caused by pathogens, fungi, and oomycetes. From an epidemiological standpoint, the wide host ranges of these phytopathogens, their high pathogenicity, high capability for reproduction, and capacity to survive outside of their plant hosts as persistent spores or saprophytes are essential components of their lives [34]. There were not many different strategies put out to deal with the problem of a large drop in pesticide use. Based on their meta-analysis, Lázaro et al. (2021) proposed that using decision support systems, which assist farmers in scheduling fungicide treatment based on a documented or forecasted possibility of fungus-related illness, can lead to a 50% reduction in fungicide consumption.

Carbohydrate biopolymer-established plant protection agents might be developed as a substitute for conventional fungicides. Large amounts of these biopolymers are available from a diversity of organic materials. As they are biodegradable and non-toxic, they may be utilized while farming organically. They can interact in increasingly complicated formulations with a variety of hydrophobic and hydrophilic molecules. Three strategies exist for how carbohydrate biopolymers could shield plants against damaging fungi. Firstly, they might interact directly with the fungus by mycelial growth and blocking spore germination, as was seen in the chitosan instan. Second, they could act as potent elicitors, strengthening the plant's defenses against harmful organisms. Thirdly, they might serve as carriers in mixtures that include active agents with regulated releases, such as agrochemicals. Table 16.1 shows the biopolymers coatings used on vegetables/fruits.

4 BIOPOLYMER APPLICATIONS IN THE FOOD INDUSTRY

Over the last 50 years, a growing number of improved as well as substitute foods have been produced with the aid of various scientific, engineering, and biotechnology tools. Because of this, the majority of food that is available on a commercial level today has been improved and changed by some means to enhance its taste, appearance, and economic viability [55]. The food industry has seen remarkable advancements in bio-technologies and applied engineering. Numerous advancements, such as enhanced preservatives, genetic engineering, and advanced technologies in innovative components, product quality assurance, and wrapping, are aimed at reducing global food scarcity in a world where the population is expanding [1]. Through the promotion of an ecologically friendly food engineering approach, their application is revolutionizing the design, production, and packaging of food. Human and environmental health are at risk due to the growing usage of synthetic polymers. Growing environmental concerns have led to an increase in the popularity of materials comprising natural polymers [56]. Nowadays, biopolymers are extensively utilized in food processing as components of specialized structures because of their ability to interact with other food ingredients to enhance their stability, and physical and chemical properties [1]. Polysaccharides and proteins are the two types of food biopolymers.

Plant and animal tissues contain polymers of amino acids called proteins. The structural stability of proteins in meals is influenced by pH, shear, and heat. One of the main sources of energy is branched polysaccharides, which are monomers of sugars joined by glycosidic links and found in nature. Hydrated compounds called polysaccharides are present in the exo along with endo core matrixes of plants and marine animals. Biopolymers have been proposed as an appropriate medium for the synthesis and stability of silver nanoparticles [56, 57]. The method of polymer-assisted synthesis facilitates the dispersion of nanoparticles within a matrix of polymer, which ultimately influences the rigidity of the structure and consistency of the nanocomposite film. This ultimately leads to the retention of notable antibacterial properties in nanocomposite films [58]. The primary advantages of nanoparticles are improved food protection and durability that can be achieved by enhancing the physical and functional characteristics of film-forming materials. Different types of biopolymers and their applications in the food sector are shown in Figure 16.1 that are discussed in this book chapter.

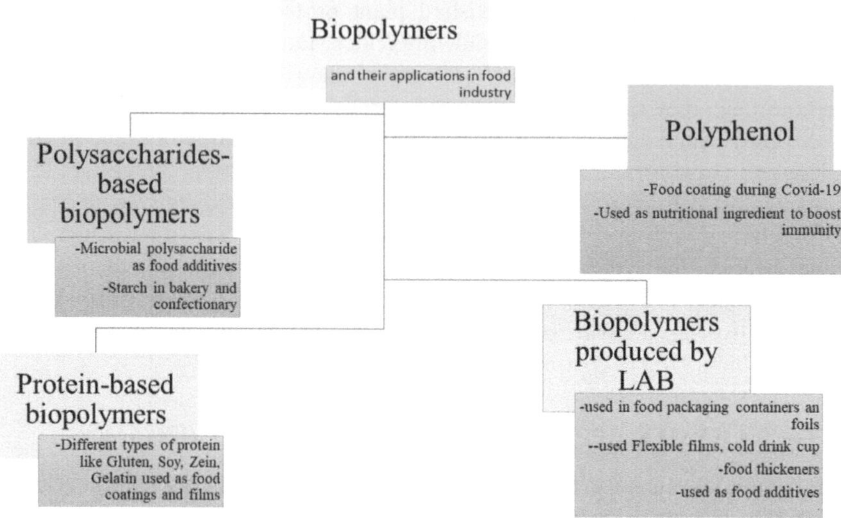

FIGURE 16.1   Types of biopolymers and their applications.

4.1   POLYPHENOL APPLICATION IN THE FOOD INDUSTRY DURING COVID-19

The emergence of the COVID-19 pandemic in late 2019 originated from a new coronavirus in Wuhan, China. The global impact of COVID-19 included millions of confirmed cases and fatalities by October 2020 [59]. The pandemic not only affected public health but also disrupted supply chains, leading to challenges in fruit preservation and waste due to limited shelf life. To address these issues, the development of nanostructured coatings with antimicrobial properties for perishable fruits is crucial for extending shelf life and ensuring nutritional security. Fruit coatings that are edible to prolong fruit shelf life and limit viral transmission—a critical factor in the recovery of COVID-19 patients [59]. Because of their antiviral, antioxidant, and antibacterial qualities, polyphenols—found in fruits like citrus, cranberries, and apples—offer several health advantages. Certain polyphenols, such as quercetin and epigallocatechin-gallate, have demonstrated antiviral and immune-stimulating properties against coronaviruses without causing adverse consequences. These substances can aid those who are infected with lung conditions such as cytokine storms [60]. By extending the shelf life of polyphenolic fruits and preventing virus pre-infection, the use of edible antimicrobial coatings on fruits can assist COVID-19 patients in their recuperation [60].

4.2   PROTEIN-BASED BIOPOLYMERS USED AS FOOD COATING

Food packaging created from proteins—such as gluten, zein, gelatin, and casein—obtained from both plant and animal sources is known as protein-based packaging. Several specialists are concentrating on creating food-packaging films with proteins since they are readily available, easily processed, and biodegradable. Strict health and safety regulations must be met by these protein-based films, and their characteristics must be customized to the particular needs of the food being packaged [53]. By lowering the amount of moisture and gas exchanged between the environment and packaged food, biopolymers made of proteins can increase the shelf life of vegetables and fruits. Protein-based films are selected based on their planned use, required longevity, and preferred qualities of the wrapped product. Fruits and vegetables are frequently preserved with protein-based films or coatings.

4.3 GLUTEN-BASED COATINGS AND FILMS

Food packaging known as "gluten-based coating/films" is derived from the protein gluten, which is present in wheat, barley, and rye. Gliadin and glutenin are the two main categories of gluten proteins, owing to their varying solubility in hydroalcoholic solutions [61]. Usually, wheat flour is used to extract gluten for use in gluten-based coating/films. To make a film, water along with additional components, like plasticizers, are mixed with gluten. After the dough has been rolled out and flattened, a thin layer is formed. To make the film stronger, it is then dried and heated. It has been discovered that these films have good barrier qualities, which means they can shield food from elements like moisture and oxygen that can lead to deterioration or a loss of quality [62]. Furthermore, this film has excellent mechanical qualities that make it a good choice for food packaging, including a high degree of impermeability, high rate of water-vapor transmission, and adaptability. Nevertheless, the utilization of films composed of gluten for food packaging is limited by their complexity and opacity in comparison with other biopolymeric coatings and films.

4.4 SOY-BASED COATINGS AND FILMS

Soy protein isolate (SPI) is used to make soy protein-based coating/films [56]. These films can replace synthetic packaging materials since they are biodegradable, eco-friendly, and environmentally favorable [63]. Additionally, they have strong oxygen barrier qualities, which may contribute to the shelf life of goods in packaging. For soy-based films, a plasticizer such as polyethylene oxide or low-density polyethylene (LDPE) can effectively regulate many characteristics, including emulsification, texture, cohesion, adhesion, fat retention and fiber alignment, moisture, as well as texturizing proficiency [63]. Films made of soy protein, however, cannot be devoid of limitations. They may not be as heat-resistant or as clear as synthetic packing films, which could restrict their use in some situations. SPI and gluten coating prolong shelf life by preventing peanut fat from deteriorating [53].

4.5 ZEIN-BASED COATINGS AND FILMS

About 60–70% of the protein in corn is composed of prolamin proteins, such as zein protein, which is an isolate from maize. Zein is a protein that is widely utilized to create films and coatings since it is thought to be edible and non-toxic. Zein solutions are poured into ethanol, and allowed to evaporate, and the protein is subsequently cross-linked to make the films [64]. It has been discovered that these films have strong barrier qualities, which means they may shield food from elements like oxygen and moisture as well as other external elements that may cause degradation or loss of quality. To increase shelf life and maintain food freshness, these films are applied as coatings to certain food varieties, including vegetables and fresh fruits, applying a zein coating on tomatoes softens and slows down their color change, whereas the respiration rate of apples and pears decreases due to the impact of zein coatings [65]. Zein-based films, on the other hand, crumble in high humidity and are water-sensitive.

4.6 GELATIN-BASED COATINGS AND FILMS

Food, medicine, and pharmaceuticals are just a few of the industries that use gelatin-based films and coatings [66]. Gelatin-based films have several uses, including vegetable and fruit packing, confectionery, and packaging of meat. They are translucent, flexible, and have strong gas barrier qualities [66]. Its longevity as well as proper film formation, biocompatibility, biodegradability, and affordability make it an essential raw material for the formation of films that degrade naturally [67]. These film/coating biopolymers are ideal for edible films because they are translucent, possess a melting coefficient, minimal exchange of gases, and have semi-permeability, and are antioxidant and antibacterial. Gelatin films or coatings are used to preserve a variety of vegetables and fruits, like cherry tomatoes, carrots, as well as stalks [53, 68]. Research revealed that the gelatin/GSE/TiO$_2$ composite films exhibited antibacterial and antioxidant properties. Additionally, there may be ethical and environmental issues with the procurement and manufacture of animal-based proteins. Further casein and whey protein-based coatings and films have also been used but are not discussed in this chapter.

5 MICROBIAL POLYSACCHARIDES-BASED BIOPOLYMERS AS FOOD ADDITIVES

Different microorganisms with unique and creative properties produce gums that dissolve in water, known as polysaccharides. Not only are these biopolymers inexpensive, but they have developed into distinctive and significant polymeric materials for industry [57]. Despite their structural and physicochemical diversity, the food industry can employ microbial polysaccharides for a variety of purposes [69]. They serve as coagulants, gelling agents, suspending agents, binders, stabilizers, and emulsifiers. Due to their high purity and regular structure, these biopolymers have distinctive rheological properties that make them perfect for the food industry. Microbial polysaccharides remain active at high temperatures, pH levels, and salinities and are non-toxic, biodegradable, and harmless to the environment. These are suitable replacements for synthetics and naturally occurring water-soluble gums. Due to their superior properties, they may prove to be essential as suspending, gelling agents, and thickening in the food industry [70]. By using genetically modified bacteria in under-regulated fermentation environments, new exopolysaccharide rides with enhanced properties may be synthesized, opening up new industrial applications. Polysaccharides can contain linked non-sugar molecules and/or anionic (uronic acid) sugars or neutral (pentoses and hexoses) ones, and they can be heteropolymers or homopolymers. Microbial polysaccharides, as, thickeners, stabilizers, texturizers, and gelling agents, enhance the flavor, food texture, and quality [1].

6 BIOPOLYMERS DERIVED FROM STARCH IN BAKERY AND CONFECTIONERY

One kind of polysaccharide is starch, which is a long chain of glucose molecules. Starch contains two distinct kinds of glucose chains. First, there is a simple chain known as amylose, and then there is a complex branch form known as amylopectin [71]. It is a highly helpful element in the food industry due to its adaptability. Starch is the primary ingredient in bread and cookies, particularly gluten-free varieties, as it provides the final product with structure and texture. Conversely, starches are mostly utilized as dusting and gelling agents in the confectionery industry [1]. Under various processing conditions, native starches can enhance their functional properties and adaptability. Among other things, starch derivatives were used in the baking and confectionery industries to substitute for fat in chocolate fillings, replace gluten in gluten-free products, and stop bread from sticking together [72]. Interest in using starch to create new foods will likely increase because it is a cheap, accessible, and non-toxic ingredient. Among the most common carbohydrates in nature, starch is second only to cellulose. In recent times, modified starches have been utilized more often in convenience foods like biscuits, bakeries, noodles, and confections [73]. Because people want food products that are healthier and last longer, the market for modified starches has expanded. Modified starches are used to increase the shelf life of bread loaves, increase the volume of cakes and bread, and replace saturated fats in cookies and confections. Modified starches are used to replace confectionery fillings and saturated fats in cookies, increase the content of fiber, delay the staling of loaves of bread, and improve the volume of cakes and bread [53]. Beyond their conventional use in baked goods and confections, modified starches used in food manufacturing are expected to become more varied as research into double and multi-modifier starches to increase starch effectiveness proceeds [53].

7 BIOPOLYMERS PRODUCING LACTIC ACID BACTERIA AND THEIR APPLICATIONS IN THE FOOD-MANUFACTURING INDUSTRY

7.1 DIFFERENT BIOPOLYMERS PRODUCE BY LACTIC ACID BACTERIA AND THEIR STRAINS

Lactic acid bacteria (LAB) were first used in industrial settings thousands of years ago as starters and biocontrol agents. More recently, LAB has been used as an ingredient in some nutritious foods (such as probiotics) [74, 75]. These bacteria are known for their ability to cause uncontrolled fermentation, which produces food products that meet people's needs for survival [74]. Based on their qualities and innovative, creative technological approaches, LABs are considered as promising microbes for biological refineries converting the waste biomasses into high-value-added outcomes [75]. LAB produce a wide range of macromolecules known as biopolymers, which are categorized into groups such as PLA, polyhydroxyalkanoates (PHAs) [76], and exopolysaccharides (EPSs). Based on their chemical makeup, EPSs—a type of biopolymer—can be further divided into homopolysaccharides (HoPSs) and heteropolysaccharides (HePSs) [77]. HoPSs are made up of glucose, fructose, and galactose repeat units, with subclasses like α- and β-glucans resulting from glucose polymerization. Their synthesis is aided by particular enzymes such as glucansucrase [77]. HePSs, on the other hand, can have varying chain lengths and structures and comprise repeating units of distinct monosaccharides [78]. These biopolymers, which differ in composition and structure, are essential to many biological processes. LAB, particularly the *Lactobacillus* genus, are highlighted for their ability to synthesize these biopolymers with significant applications in the food industry [78, 79]. Lactobacilli, being Gram-positive bacteria, are known for their diversity, with a reclassification into 25 genera proposed in 2020. They are primarily mesophiles and are commonly recognized as generally recognized as safe. Lactobacilli are strict fermenters, capable of producing LA predominantly, with some strains being homofermentative and others heterofermentative [78]. Different groups of lactobacilli exhibit varying fermentation characteristics, influencing the types of acids they produce. Additionally, LAB like *Pediococcus* [80], *Leuconostoc* [76], *Weisella* [81], *Lactococcus* [82], *Streptococcus* [83], and *Enterococcus* [84] are discussed for their roles in biopolymer production, highlighting their unique characteristics and applications in the food industry.

7.2 APPLICATIONS IN FOOD INDUSTRY

These LAB genera contribute to synthesizing various biopolymers, showcasing their metabolic diversity and significance in biopolymer biosynthesis for food-related purposes. PLA and polyhydroxyalkanoates, such as poly(3-hydroxybutyrate) (PHB) and poly(3-hydroxyvalerate) (PHV), are examples of biodegradable polymers that have been used to reduce environmental damage to a certain extent [85].

7.2.1 PLA Applications in the Food Industry Synthesis by LAB

LA monomers that are produced by fermenting starch as a feedstock are polymerized to create compostable bioplastics, such as PLA. PLA is an enticing replacement for materials derived from petroleum because it can be used to create a wide range of polymer products [86]. The primary application of PLA, an economically beneficial and biodegradable natural product, is in the production of packaging foils and containers for both perishable and dry goods, including vegetables and fruits. Additionally, it is promoted for usage in applications involving disposable packaging, including flexible films, blister packaging, overwrap packaging, thermoformed trays and lids, bottles, and cold drink cups. The fact that PLA is made from renewable resources, biocompatible, recyclable, compostable in industrial settings, and can replace traditional plastic components are just a few of the advantages of using it for food-packaging applications [86]. The Food and Drug Administration of the United States has also certified PLA as safe enough for ingestion. The benefits of utilizing

PLA in the packaging of food can be seen in its superior transparency, convenience of being processed, and ecological qualities [87]. Its low mechanical performance, sensitivity to heat deterioration, and weak barrier qualities to different gases and oxygen that depend on the degree of crystallinity are some of its drawbacks, which prevent it from being used in industrial settings [19]. Currently, vegetables, fruit, dairy products, fresh juices, meat, as well as other products with shorter shelf lives and low requirements for resistance to water-vapor transmission or oxygen are the main uses for bio-based food packaging. Due to the aforementioned restrictions, PLA is not as widely used in food packaging. However, research is being done to investigate how to improve PLA's properties by adding plasticizers and nanoparticles, as well as by using sophisticated processing methods or treatments [88].

7.2.2 PHA Applications in the Food Industry Synthesis by LAB

PHA are microbial polyesters produced by fermenting pure microbial cultures on various renewable sources through bacterial fermentation. Because of their enhanced biodegradability and comparable physicochemical characteristics, these could serve as viable alternatives to traditional plastics. The high cost of production of PHA still prevents its widespread use, even though its qualities are similar to those of regular plastics. The most widely used polyhydroxybutyrates (PHAs) are poly-3-hydroxybutyrate (PHV), poly-4-hydroxybutyrate (PHB), poly(3-hydroxybutyrate-co-3-hydroxyvalerate) (PHBV), and poly (3-hydroxybutyrate-co-3 hydroxyhexanoate) [74]. Their mechanical and physical characteristics are strikingly similar to those of conventional plastics. Microbial polyesters known as PHAs can be cut or molded into foils, diaphragms, and films with superior oxygen and moisture absorption capabilities. Consequently, they serve as a substitute for aluminum in the construction of cardboard boxes with water-resistant ability. They can be utilized to make adaptable packaging for foods with high oil content, such as cheese, nuts, and marinated olives [74] as well as reusable containers and utensils for food. They are hydrophobic, flexible, elastomers, crystalline, brittle, and thermoplastic polymers [89]. These biopolymers can be utilized to create bioplastics that have a high melting strength and are appropriate for thermal forming with little heat deformation. They work well for a variety of packaging uses, such as single-serve food packaging, yogurt containers, tubs, trays, hot and cold cups, and cup lids. PHA films have extremely high water-vapor barrier qualities, comparable to LDPE because polyesters are hydrophobic. Heat-shaping and producing adaptable plastic bags for the food industry are suitable uses for PHBV.

7.2.3 EPS Synthesis by LAB and Their Application in Food Industry

Lactic acid bacterial EPSs are considered natural thickeners and natural functional food ingredients because they can enhance the rheological qualities of food, particularly fermented foods [78, 89]. This has attracted particular attention from the food industry. The food industry uses EPSs produced by LAB for a variety of purposes, particularly in fermented goods where they are formed in nature. Furthermore, when produced ex situ, they can be added as additives or ingredients. Foods like low-fat cheese, cheese, yogurt, kefir, and curd, as well as bakery goods like bread [90], sourdough, and gluten-free products [91], as well as fermented vegetable products like pureed food

[92], yogurt made from cereals, fermented beverages made from cereals, protein concentrate [43], and vegetable doughs, are produced using EPSs. An interesting topic is the use of EPSs to create food packaging [93]. "Food films" or "food coatings" [94] that are resistant to UV rays [95], oxygen moisture [93], and microbes can be used to preserve food. The thin film known as the "food film" is deposited to safeguard the food item. Application techniques for "food coatings" include spraying or dipping suspensions or emulsions directly onto food surfaces. The aroma, texture, taste, and food appearance can be preserved as they solidify and create a film that is edible with good mechanical qualities and low water-vapor permeability [69, 93]. Kefiran, levan, and dextran can all form nanocomposite films when certain ingredients are present. To obtain edible or food packaging, these films are used in the food industry [94].To stop the deterioration of light-sensitive nutrients, food discoloration, lipid photo-oxidation, as well as flavor quality loss, [95] have developed a packaging film with biodegradable nanocomposite material made of kefiran, starch, and zinc oxide with UV protection

TABLE 16.2

Lactic acid bacteria producing biopolymers and their applications

S.no	Lactic acid-producing bacteria	Type of biopolymers	Applications	References
1.	*Lactococcus lactis, Cupriavidus necator*	Polyhydroxybutyrates	Edible films and coatings	[76]
2.	*Pediococcus and Streptococcus*	Poly-4-hydroxybutyrate	Edible films production	[75]
3.	*Leuconostoc mesenteroides*	Polyhydroxybutyrates	Microencapsulation	[76]
4.	*Enterococcus species*	Poly-4-hydroxybutyrate	Manufacturing of throwaway food containers and cutlery; manufacturing of yogurt containers, tubs, trays, hot and cold cups, and cup lids; manufacturing of single-serve food packaging	[96]
5.	*Lactobacillus acidophilus, Lactobacillus casei*	Polylactic acid	Cutlery, packaging, and food items	[75]
6.	*Lactobacillus bulgaricus, Lactobacillus leichmannii*	Polylactic acid	Dairy product packaging, yogurt pots made of PLA, and translucent films	[97]
7.	*Leuconostoc mesenteroides*	Exopolysaccharides	Edible coating for packaging food	[98]
8.	*Lactobacillus kefiranofaciens*	Exopolysaccharides	Packaging of food items	[99]

[95]. To replace packaging that harms the environment, new food coatings can be made using EPSs generated by LAB [94]. EPSs have limited uses in food packaging and are not extensively researched since LAB produces them in small amounts. Types of biopolymers synthesized by different LAB and their applications are given in Table 16.2.

8   APPLICATIONS OF BIOPOLYMERS IN AGRICULTURE AND THE FOOD INDUSTRY: BENEFITS AND DRAWBACKS AS WELL AS CHALLENGES

Biopolymers, or polymers made from renewable resources like plants, animals, or microbes, have drawn a lot of interest from the food industry and agriculture because of their many uses and potential environmental advantages. The use of biopolymers in these industries has the following advantages and disadvantages:

8.1   ADVANTAGES OF BIOPOLYMERS IN THE FOOD INDUSTRY

Biopolymers are more environmentally friendly than traditional plastics derived from petroleum because they are frequently compostable or biodegradable. By using renewable resources like plants (like cellulose and starch) or microorganisms (like polylactic acid-producing bacteria), biopolymers can reduce our reliance on fossil fuels. A portion of biopolymers can be utilized to make materials for edible packaging that are both waste-free and nutritionally enhanced [100]. Certain biopolymers exhibit strong barrier properties against moisture and oxygen, enhancing the shelf life of food products [89]. Production of biopolymers generally emits lower greenhouse gas emissions compared with traditional plastics, contributing to a reduced carbon footprint.

8.2   DISADVANTAGES OF BIOPOLYMERS IN THE FOOD INDUSTRY

Biopolymers may have inferior mechanical and thermal properties compared with conventional plastics, limiting their applications in certain food packaging or processing. Biopolymer production can be more expensive than traditional plastic production due to factors such as raw material costs and processing methods. The properties of biopolymers can vary based on the source material and manufacturing process, leading to inconsistency in performance [57]. Some biopolymers require specialized processing equipment or conditions, which may increase production costs and complexity. Biopolymers may not be compatible with all food types or processing methods, restricting their use in certain applications [85].

8.3   ADVANTAGES OF BIOPOLYMERS IN AGRICULTURE

Biopolymers used in agriculture can degrade naturally in the soil, reducing environmental pollution and soil contamination [7]. Biopolymers can improve soil structure, water retention, and nutrient availability, promoting healthier plant growth and higher crop yields [101]. Some biopolymers can be formulated for the controlled release of fertilizers, pesticides, or growth regulators, enhancing their efficacy and reducing environmental impact [102]. Using biopolymers in agriculture aligns with sustainable farming practices, reducing reliance on synthetic chemicals and promoting eco-friendly alternatives [7]. Biopolymers can be used as seed coatings to improve germination rates, protect against pests and diseases, and enhance seedling vigor [101].

8.4 Disadvantages of Biopolymers in Agriculture

Biopolymer-based agricultural products may be more expensive than conventional alternatives, limiting adoption by farmers, especially in resource-constrained regions [101]. The degradation rate of biopolymers in agricultural applications can vary depending on environmental conditions, potentially affecting their long-term effectiveness [7]. Some biopolymers may not be compatible with certain agricultural practices or soil types, limiting their applicability in certain regions or cropping systems [101]. Regulatory approval processes for biopolymer-based agricultural products may be complex and time-consuming, delaying their market introduction [102]. The performance of biopolymer-based agricultural products may vary depending on factors such as formulation, application method, and environmental conditions, leading to inconsistent results [101].

8.5 Challenges Faced by Biopolymers in the Agriculture and Food Industry

Biopolymers face several challenges when used in the food industry and agriculture sectors, including scaling up biopolymer production processes to industrial levels while maintaining cost-effectiveness and quality standards can be challenging. Achieving consistent and desirable properties in biopolymers for specific applications requires continuous research and development efforts [101]. Biopolymers face competition from conventional polymers and other alternative materials in the market, requiring innovation and differentiation to gain market acceptance. Biopolymers may not always offer the same level of performance as synthetic polymers [85]. They may have limitations in terms of mechanical strength, barrier properties, thermal stability, and shelf life, which can affect their suitability for certain food-packaging applications and agricultural uses. Meeting regulatory requirements for food safety, quality, and labeling can be challenging for biopolymer manufacturers and users. Compliance with standards related to food contact materials, biodegradability, and recyclability adds complexity and may require additional testing and certification.

The availability and consistency of raw materials for biopolymer production can be influenced by factors such as seasonal variations, agricultural practices, and land-use changes. Ensuring a reliable and sustainable supply chain is essential for the consistent production of biopolymer. Biopolymers may require specialized processing equipment and techniques that differ from those used for synthetic polymers. Adapting existing manufacturing processes or developing new ones to accommodate biopolymers can be costly and time-consuming [7]. Consumer acceptance of biopolymer-based products may vary due to factors, such as unfamiliarity, perceived performance, and concerns about safety and quality. Educating consumers about the benefits and advantages of biopolymers, as well as addressing any misconceptions or doubts, is important for promoting their adoption in the food industry and agriculture [101, 102].

Effective management of biopolymer waste streams, including recycling, composting, or disposal, is essential to minimize environmental impact and maximize resource utilization [102]. Increasing awareness and understanding among stakeholders, including consumers, policymakers, and industry professionals, about the benefits and limitations of biopolymers is crucial for their widespread adoption and acceptance [1]. Biopolymers designed to be biodegradable or compostable require appropriate end-of-life management to ensure they break down efficiently and do not contribute to environmental pollution. Establishing infrastructure for composting facilities or recycling programs for biopolymer waste can be challenging and costly. Biopolymer-based packaging materials and agricultural products must be compatible with existing processing, packaging, and distribution infrastructure to ensure seamless integration into the supply chain. Compatibility issues may arise with equipment, labeling systems, transportation methods, and recycling facilities [77].

Addressing these challenges will require collaboration among stakeholders across the food industry and agriculture sectors, including manufacturers, suppliers, policymakers, research institutions, and consumers. Investing in research and development, improving processing technologies, optimizing supply chains, and raising awareness about the benefits of biopolymers are essential steps toward overcoming these challenges and promoting the sustainable use of biopolymers in food and agriculture [101].

9 CONCLUSION AND FUTURE PERSPECTIVES

Biopolymers such as hydrogels and bacterial polymers have diverse agricultural applications, from soil conditioning to crop protection. They enhance water retention, nutrient delivery, and plant growth while combating pests and diseases. Biopolymers derived from waste contribute to sustainable agriculture, improving food safety and reducing environmental impact. Bacterial polymers act as biofertilizers and biostimulants, supporting soil fertility, crop productivity, and nutrient availability. Carbohydrate biopolymer-based plant protection agents offer a biodegradable alternative to fungicides. Overall, biopolymers are pivotal in advancing agricultural sustainability and global food production. The use of biopolymers in the food industry, as described in this chapter, demonstrates a notable move in the direction of environmentally friendly and sustainable food-packaging options. Enhancing food preservation, cutting waste, and addressing environmental issues related to conventional synthetic packaging materials can be accomplished through the use of biopolymers, such as protein-based coatings, gluten-based films, soy-based coatings, zein-based coatings, and gelatin-based films. Due to the addition of antioxidant and antimicrobial qualities, these biopolymers have strong barrier qualities, biodegradability, and possible health benefits.

The need for sustainable solutions in the face of escalating environmental challenges highlights the growing significance of biopolymers in transforming food-packaging and preservation techniques. The outlook for the future calls for a sustained emphasis on R&D to improve the characteristics of biopolymers, address limitations like heat resistance and opacity, and explore novel uses in food packaging. Furthermore, developments in LAB and their ability to synthesize biopolymers create new avenues for the development of functional foods and the enhancement of food quality through the use of natural thickeners and additives. All things considered, the use of biopolymers in the food industry points to a promising path toward more environmentally friendly, health-conscious, and sustainable food-packaging options.

REFERENCES

1. Baranwal, J., Barse, B., Fais, A., Delogu, G., and Kumar, A. "Biopolymer: A Sustainable Material for Food and Medical Applications." *Polymers*, February 28, 2022. https://doi.org/10.3390/polym14050983.
2. Díez-Pascual, A. M., and Díez-Vicente, A. L. "ZnO-Reinforced Poly(3-Hydroxybutyrate-Co-3-Hydroxyvalerate) Bionanocomposites with Antimicrobial Function for Food Packaging." *ACS Applied Materials & Interfaces*, June 3, 2014. https://doi.org/10.1021/am502261e.
3. Nakagaito, A. N., Kanzawa, S., and Takagi, H. "Polylactic Acid Reinforced with Mixed Cellulose and Chitin Nanofibers—Effect of Mixture Ratio on the Mechanical Properties of Composites." *Journal of Composites Science*, June 19, 2018. https://doi.org/10.3390/jcs2020036.
4. Righetti, M. C., Cinelli, P., Mallegni, N., Massa, C. A., Bronco, S., Stäbler, A. et al. "Thermal, Mechanical, and Rheological Properties of Biocomposites Made of Poly(Lactic Acid) and Potato Pulp Powder." *International Journal of Molecular Sciences*, February 5, 2019. https://doi.org/10.3390/ijms20030675.
5. Homayoni, H., Ravandi, S. A. H., and Valizadeh, M. "Electrospinning of Chitosan Nanofibers: Processing Optimization." *Carbohydrate Polymers*, July 1, 2009. https://doi.org/10.1016/j.carbpol.2009.02.008.
6. Liang, N., Sun, S. Y., Gong, X., Li, Q., Yan, P., and Cui, F. "Polymeric Micelles Based on Modified Glycol Chitosan for Paclitaxel Delivery: Preparation, Characterization and Evaluation." *International Journal of Molecular Sciences*, May 23, 2018. https://doi.org/10.3390/ijms19061550.
7. Tariq, Z., Iqbal, D. N., Rizwan, M., Ahmad, M., Faheem, M., and Ahmed, M. "Significance of Biopolymer-based Hydrogels and their Applications in Agriculture: A Review in Perspective of Synthesis and their Degree of Swelling for Water Holding." *RSC Advances* 13, no. 35 (2023): 24731–24754. https://doi.org/10.1039/d3ra03472k.
8. Verma, K., Sarkar, C., and Saha, S. "Biodegradable Polymers for Agriculture." *Materials Horizons*, January 1, 2023. https://doi.org/10.1007/978-981-99-3307-5_9.
9. Zheng, H., and Zuo, B. "Functional Silk Fibroin Hydrogels: Preparation, Properties and Applications." *Journal of Materials Chemistry B*, January 1, 2021. https://doi.org/10.1039/d0tb02099k.
10. Altman, G. H., Diaz, F., Jakuba, C. M., Calabro, T., Horan, R. L., Chen, J., et al. "Silk-Based Biomaterials." *Biomaterials*, February 1, 2003. https://doi.org/10.1016/s0142-9612(02)00353-8.
11. Jin, H.-J., and Kaplan, D. L. "Mechanism of Silk Processing in Insects and Spiders." *Nature*, August 1, 2003. https://doi.org/10.1038/nature01809.

12. Matsumoto, A., Lindsay, A., Abedian, B., and Kaplan, D. L. "Silk Fibroin Solution Properties Related to Assembly and Structure." *Macromolecular Bioscience* 8, no. 11 (2008): 1006–1018.

13. Beier, S., and Bertilsson, S. "Bacterial Chitin Degradation—Mechanisms and Ecophysiological Strategies." *Frontiers in Microbiology*, January 1, 2013. https://doi.org/10.3389/fmicb.2013.00149.

14. Seddiqi, H., Oliaei, E., Honarkar, H., Jin, J., Geonzon, L. C., Bacabac, R. G., et al. "Cellulose and Its Derivatives: Towards Biomedical Applications." *Cellulose*, January 27, 2021. https://doi.org/10.1007/s10570-020-03674-w.

15. Rabnawaz, M., Wyman, I., Auras, R., and Cheng, S. "A Roadmap towards Green Packaging: The Current Status and Future Outlook for Polyesters in the Packaging Industry." *Green Chemistry*, January 1, 2017. https://doi.org/10.1039/c7gc02521a.

16. Reddy, M. M., Vivekanandhan, S., Misra, M., Bhatia, S. K., and Mohanty, A. K. "Biobased Plastics and Bionanocomposites: Current Status and Future Opportunities." *Progress in Polymer Science*, October 1, 2013. https://doi.org/10.1016/j.progpolymsci.2013.05.006.

17. Salvador, M., Abdulmutalib, U., Salicio González, J. M., Kim, J.-H., Smith, A., Faulon, J.-L., et al. "Microbial Genes for a Circular and Sustainable Bio-PET Economy." *Genes*, May 16, 2019. https://doi.org/10.3390/genes10050373.

18. Dhandayuthapani, B., Yoshida, Y., Maekawa, T., and Kumar, D. S. "Polymeric Scaffolds in Tissue Engineering Application: A Review." *International Journal of Polymer Science*, January 1, 2011. https://doi.org/10.1155/2011/290602.

19. Farah, S., Anderson, D. G., and Langer, R. "Physical and Mechanical Properties of PLA, and their Functions in Widespread Applications — A Comprehensive Review." *Advanced Drug Delivery Reviews*, December 1, 2016. https://doi.org/10.1016/j.addr.2016.06.012.

20. Zhang, Q., Song, M., Xu, Y., Wang, W., Wang, Z., and Zhang, L. "Bio-Based Polyesters: Recent Progress and Future Prospects." *Progress in Polymer Science*, September 1, 2021. https://doi.org/10.1016/j.progpolymsci.2021.101430.

21. Cai, S., Cheng, Z., Djouonkep, L. D. W., Wang, L., Wang, H., and Gauthier, M. "Synthesis and Properties of Bio-Based Copolyesters Based on Phydroxyphenylpropionic Acid." *Journal of Polymers and the Environment*, December 20, 2022. https://doi.org/10.1007/s10924-022-02732-7.

22. Joseph, B., Sagarika, V. K., Sabu, C., Kalarikkal, N., and Thomas, S. "Cellulose Nanocomposites: Fabrication and Biomedical Applications." *Journal of Bioresources and Bioproducts*, November 1, 2020. https://doi.org/10.1016/j.jobab.2020.10.001

23. Meftahi, A., Heravi, M. E. M., Barhoum, A., Samyn, P., Najarzadeh, H., and Alibakhshi, S. "Cellulose Nanofibers." *Springer eBooks*, January 1, 2022. https://doi.org/10.1007/978-3-030-89621-8_13.

24. Ahmad, A., Yaseen, M., Hussain, H., Tahir, M. N., Gondal, A. H., Iqba, M., et al. (2022). Effects of Mulching on Crop Growth, Productivity and Yield. In: Akhtar, K., Arif, M., Riaz, M., Wang, H. (eds) *Mulching in Agroecosystems*. Springer, Singapore. https://doi.org/10.1007/978-981-19-6410-7_14.

25. Hadrami, A. E., Adam, L. R., El Hadrami, I., and Daayf, F. "Chitosan in Plant Protection." *Marine Drugs*, March 30, 2010. https://doi.org/10.3390/md8040968.

26. Abd-Elgawad, M. M. M., and Askary, T. H. "Fungal and Bacterial Nematicides in Integrated Nematode Management Strategies." *Egyptian Journal of Biological Pest Control*, September 24, 2018. https://doi.org/10.1186/s41938-018-0080-x.

27. Patra, S. K., Poddar, R., Brestič, M., Acharjee, P. U., Bhattacharya, P., Sengupta, S., et al. "Prospects of Hydrogels in Agriculture for Enhancing Crop and Water Productivity under Water Deficit Condition." *International Journal of Polymer Science*, June 22, 2022. https://doi.org/10.1155/2022/4914836.

28. Kaur, Prabhpreet, Agrawal, R., Pfeffer, F. M., Williams, R. J., and Bohidar, H. B. "Hydrogels in Agriculture: Prospects and Challenges." *Journal of Polymers and the Environment*, April 7, 2023. https://doi.org/10.1007/s10924-023-02859-1.

29. Bortolin, A., Aouada, F. A., Mattoso, L. H. C., and Ribeiro. C. "Nanocomposite PAAm/ Methyl Cellulose/Montmorillonite Hydrogel: Evidence of Synergistic Effects for the Slow Release of Fertilizers." *Journal of Agricultural and Food Chemistry* 61, no. 31 (2013): 7431–7439.

30. Zikmanis, P., Brants, K., Koļesovs, S., and Semjonovs, P. "Extracellular Polysaccharides Produced by Bacteria of the Leuconostoc Genus." *World Journal of Microbiology & Biotechnology*, September 29, 2020. https://doi.org/10.1007/s11274-020-02937-9.

31. Saeed, Q., Wang, X., Haider, F. U., Kučerik, J., Mumtaz, M. Z., Holátko, J., et al. "Rhizosphere Bacteria in Plant Growth Promotion, Biocontrol, and Bioremediation of Contaminated Sites: A Comprehensive Review of Effects and Mechanisms." *International Journal of Molecular Sciences*, September 29, 2021. https://doi.org/10.3390/ ijms221910529.

32. Kour, D., Rana, K. L., Yadav, A. N., Yadav, N., Kumar, M., Kumar, V., et al. "Microbial biofertilizers: Bioresources and Eco-friendly Technologies for Agricultural and Environmental Sustainability." *Biocatalysis and Agricultural Biotechnology*, January 1, 2020. https://doi.org/10.1016/j.bcab.2019.101487

33. Bhardwaj, D. K., Ansari, M. W., Sahoo, R. K., and Tuteja, N. "Biofertilizers Function as Key Player in Sustainable Agriculture by Improving Soil Fertility, Plant Tolerance and Crop Productivity." *Microbial cell Factories*, May 8, 2014. https://doi. org/10.1186/1475-2859-13-66.

34. Savary, S., Willocquet, L., Pethybridge, S. J., Esker, P. D., McRoberts, N., and Nelson, A. "The Global Burden of Pathogens and Pests on Major Food Crops." *Nature Ecology & Evolution*, February 4, 2019. https://doi.org/10.1038/s41559-018-0793-y.

35. Kharchoufi, S., Parafati, L., Licciardello, F., Muratore, G., Hamdi, M., Cirvilleri, G., et al. "Edible Coatings Incorporating Pomegranate Peel Extract and Biocontrol Yeast to Reduce Penicillium Digitatum Postharvest Decay of Oranges." *Food Microbiology*, September 1, 2018. https://doi.org/10.1016/j.fm.2018.03.011.

36. Valencia-Chamorro, S., Pérez-Gago, M. B., Del Río, M. A., and Palou, L. "Effect of Antifungal Hydroxypropyl Methylcellulose (HPMC)–Lipid Edible Composite Coatings on Postharvest Decay Development and Quality Attributes of Cold-Stored 'Valencia' Oranges." *Postharvest Biology and Technology*, November 1, 2009. https:// doi.org/10.1016/j.postharvbio.2009.06.001.

37. Madanipour, S., Alimohammadi, M., Rezaie, S., Nabizadeh, R., Khaniki, G. J., Mirfazaelian, H., et al. "Influence of Postharvest Application of Chitosan Combined with Ethanolic Extract of Liquorice on Shelflife of Apple Fruit." *Journal of Environmental Health Science & Engineering*, February 14, 2019. https://doi.org/10.1007/s40201-019- 00351-4.

38. Rojas-Graü, M. A., Raybaudi-Massilia, R. M., Soliva-Fortuny, R., Avena-Bustillos, R. J., McHugh, T. H., and Martín-Belloso, O. "Apple Puree-Alginate Edible Coating as Carrier of Antimicrobial Agents to Prolong Shelf-Life of Fresh-Cut Apples." *Postharvest Biology and Technology*, August 1, 2007. https://doi.org/10.1016/j. postharvbio.2007.01.017.

39. Guerra, I. C. D., Oliveira, P., De Souza Pontes, A. L., Carneiro Lúcio, A. S. S., Tavares, H. F., Barbosa-Filho, J. M., et al. "Coatings Comprising Chitosan and Mentha Piperita L. or Menthaxvillosa Huds Essential Oils to Prevent Common Postharvest Mold Infections and Maintain the Quality of Cherry Tomato Fruit." *International Journal of Food Microbiology*, December 1, 2015. https://doi.org/10.1016/j.ijfoodmicro.2015. 08.009.

40. Youssef, K., De Oliveira, A. G., Tischer, C. A., Hussain, I., and Roberto, S. R. "Synergistic Effect of a Novel Chitosan/Silica Nanocomposites-Based Formulation against Gray Mold of Table Grapes and Its Possible Mode of Action." *International Journal of Biological Macromolecules*, December 1, 2019. https://doi.org/10.1016/j.ijbiomac.2019.08.249.

41. Shen, Y., and Yang, H. "Effect of Preharvest Chitosan- g -Salicylic Acid Treatment on Postharvest Table Grape Quality, Shelf Life, and Resistance to Botrytis Cinerea -Induced Spoilage." *Scientia Horticulturae*, October 1, 2017. https://doi.org/10.1016/j.scienta.2017.06.046.

42. Takma, D. K., and Korel, F. "Impact of Preharvest and Postharvest Alginate Treatments Enriched with Vanillin on Postharvest Decay, Biochemical Properties, Quality and Sensory Attributes of Table Grapes." *Food Chemistry*, April 1, 2017. https://doi.org/10.1016/j.foodchem.2016.09.195.

43. Xu, Y., Coda, R., Holopainen-Mantila, U., Laitila, A., Katina, K., and Tenkanen, M. "Impact of In Situ Produced Exopolysaccharides on Rheology and Texture of Fava Bean Protein Concentrate." *Food Research International* 115 (2019): 191–199. https://doi.org/10.1016/j.foodres.2018.08.054.

44. Tabassum, N., and Ali Khan, M. "Modified Atmosphere Packaging of Fresh-Cut Papaya Using Alginate Based Edible Coating: Quality Evaluation and Shelf Life Study." *Scientia Horticulturae*, January 1, 2020. https://doi.org/10.1016/j.scienta.2019.108853.

45. Azarakhsh, N., Osman, A., Ghazali, H. M., Tan, C. P., and Adzahan, N. M. "Lemongrass Essential Oil Incorporated into Alginate-Based Edible Coating for Shelf-Life Extension and Quality Retention of Fresh-Cut Pineapple." *Postharvest Biology and Technology*, February 1, 2014. https://doi.org/10.1016/j.postharvbio.2013.09.004.

46. Guerreiro, A., Gago, C., Miguel, M. G., and Faleiro, M. L. "The Influence of Edible Coatings Enriched with Citral and Eugenol on the Raspberry Storage Ability, Nutritional and Sensory Quality." *Food Packaging and Shelf Life*, September 1, 2016. https://doi.org/10.1016/j.fpsl.2016.05.004.

47. Li, X., Du, X., Liu, Y., Tong, L., Wang, Q., and Li, J. "Rhubarb Extract Incorporated into an Alginate-Based Edible Coating for Peach Preservation." *Scientia Horticulturae*, November 1, 2019. https://doi.org/10.1016/j.scienta.2019.108685.

48. Jiang, T., Feng, L., and Li, H. "Effect of Alginate/Nano-Ag Coating on Microbial and Physicochemical Characteristics of Shiitake Mushroom (Lentinus Edodes) during Cold Storage." *Food Chemistry*, November 1, 2013. https://doi.org/10.1016/j.foodchem.2013.03.093.

49. Emamifar, A., and Bavaisi, S. "Nanocomposite Coating Based on Sodium Alginate and Nano-ZnO for Extending the Storage Life of Fresh Strawberries (Fragaria × Ananassa Duch.)." *Journal of Food Measurement and Characterization*, January 2, 2020. https://doi.org/10.1007/s11694-019-00350-x.

50. Zhuo, R., Li, B., and Tian, S. "Alginate Oligosaccharide Improves Resistance to Postharvest Decay and Quality in Kiwifruit (Actinidia Deliciosa Cv. Bruno)." *Horticultural Plant Journal*, January 1, 2022. https://doi.org/10.1016/j.hpj.2021.07.003.

51. Godara, S. K., Baswal, A. K., Ramezanian, A., Gill, K. S., and Mirza, A. "Impact of Carboxymethyl Cellulose Based Edible Coating on Storage Life and Quality of Guava Fruit Cv. 'Allahabad Safeda' under Ambient Storage Conditions." *Journal of Food Measurement and Characterization*, July 9, 2021. https://doi.org/10.1007/s11694-021-01057-8.

52. Baswal, A. K., Dhaliwal, H. S., Singh, Z., Mahajan, B. V. C., Kalia, A., and Gill, K. S. "Influence of Carboxy Methylcellulose, Chitosan and Beeswax Coatings on Cold Storage Life and Quality of Kinnow Mandarin Fruit." *Scientia Horticulturae*, January 1, 2020. https://doi.org/10.1016/j.scienta.2019.108887.

53. Chen, H., Wang, J., Cheng, Y., Wang, C., Liu, H., Bian, H., Pan, Y., Sun, J., and Han, W. "Application of Protein-Based Films and Coatings for Food Packaging: A Review." *Polymers* 11, no. 12 (2019): 2039. https://doi.org/10.3390/polym11122039.

54. Tesfay, S. Z., Magwaza, L. S., Mbili, N. C., and Mditshwa, A. "Carboxyl Methylcellulose (CMC) Containing Moringa Plant Extracts as New Postharvest Organic Edible Coating for Avocado (Persea Americana Mill.) Fruit." *Scientia Horticulturae*, December 1, 2017. https://doi.org/10.1016/j.scienta.2017.08.047.

55. Padoan, E., Montoneri, E., Bordiglia, G., Boero, V., Ginepro, M., Evon, P., et al. "Waste Biopolymers for Eco-Friendly Agriculture and Safe Food Production." *Coatings* 12, no. 2 (2022). https://doi.org/10.3390/coatings12020239.

56. Bhaskar, R., Zo, S. M., Narayanan, K. B., Purohit, S. D., Gupta, M. K., and Han, S. S. "Recent Development of Protein-based Biopolymers in Food Packaging Applications: A Review." *Polymer Testing* 124 (2023). https://doi.org/10.1016/j.polymertesting.2023.108097.

57. Nicolescu, C. M., Bumbac, M., Buruleanu, C. L., Popescu, E. C., Stanescu, S. G., Georgescu, A. A., et al. "Biopolymers Produced by Lactic Acid Bacteria: Characterization and Food Application." *Polymers* 15, no. 6 (2023). https://doi.org/10.3390/polym15061539.

58. Udayakumar, G. P., Muthusamy, S., Selvaganesh, B., Sivarajasekar, N., Rambabu, K., Banat, F., et al. "Biopolymers and Composites: Properties, Characterization and their Applications in Food, Medical and Pharmaceutical Industries." *Journal of Environmental Chemical Engineering* 9, no. 4 (2021). https://doi.org/10.1016/j.jece.2021.105322.

59. Chitrakar, B., Zhang, M., and Bhandari, B. "Improvement Strategies of Food Supply Chain through Novel Food Processing Technologies during COVID-19 Pandemic." *Food Control* 125 (2021): 108010. https://doi.org/10.1016/j.foodcont.2021.108010.

60. Paraiso, I. L., Revel, J. S., and Stevens, J. F. "Potential use of polyphenols in the battle against COVID-19." *Current Opinion in Food Science* 32 (2020): 149–155. https://doi.org/10.1016/j.cofs.2020.08.004.

61. Guillaume, C., Pinte, J., Gontard, N., and Gastaldi, E. "Wheat Gluten-Coated Papers for Bio-Based Food Packaging: Structure, Surface and Transfer Properties." *Food Research International* 43, no. 5 (2010): 1395–1401. https://doi.org/10.1016/j.foodres.2010.04.014.

62. Bibi, F., Guillaume, C., Vena, A., Gontard, N., and Sorli, B. "Wheat Gluten, a Bio-Polymer Layer to Monitor Relative Humidity in Food Packaging: Electric and Dielectric Characterization." *Sensors and Actuators A: Physical* 247 (2016): 355–367. https://doi.org/10.1016/j.sna.2016.06.017.

63. Tian, H., Guo, G., Fu, X., Yao, Y., Yuan, L., and Xiang, A. "Fabrication, Properties and Applications of Soy-Protein-Based Materials: A Review." *International Journal of Biological Macromolecules* 120 (2018): 475–490. https://doi.org/10.1016/j.ijbiomac.2018.08.110.

64. Moreno, M. A., Orqueda, M. E., Gómez-Mascaraque, L. G., Isla, M. I., and López-Rubio, A. "Crosslinked Electrospun Zein-Based Food Packaging Coatings Containing Bioactive Chilto Fruit Extracts." *Food Hydrocolloids* 95 (2019): 496–505. https://doi.org/10.1016/j.foodhyd.2019.05.001.

65. Pavlátková, L., Sedlaříková, J., Pleva, P., Peer, P., Uysal-Unalan, I., and Janalíková, M. "Bioactive Zein/Chitosan Systems Loaded with Essential Oils for Food-Packaging Applications." *Journal of the Science of Food and Agriculture* 103, no. 3 (2023): 1097–1104. https://doi.org/10.1002/jsfa.11978.

66. Lu, Y., Luo, Q., Chu, Y., Tao, N., Deng, S., Wang, L., and Li, L. "Application of Gelatin in Food Packaging: A Review." *Polymers* 14, no. 3 (2022): 436. https://doi.org/10.3390/polym14030436.

67. Riahi, Z., Priyadarshi, R., Rhim, J.-W., and Bagheri, R. "Gelatin-based Functional Films Integrated with Grapefruit Seed Extract and TiO2 for Active Food Packaging Applications." *Food Hydrocolloids* 112 (2021): 106314. https://doi.org/10.1016/j.foodhyd.2020.106314.

68. Farhan, A., and Hani, N. M. "Active Edible Films Based on Semi-Refined κ-carrageenan: Antioxidant and Color Properties and Application in Chicken Breast Packaging." *Food Packaging and Shelf Life* 24 (2020): 100476. https://doi.org/10.1016/j.fpsl.2020.100476.

69. Mohamed, S. A. A., El-Sakhawy, M., and El-Sakhawy, M. A.-M. "Polysaccharides, Protein and Lipid -Based Natural Edible Films in Food Packaging: A Review." *Carbohydrate Polymers* 238 (2020): 116178. https://doi.org/10.1016/j.carbpol.2020.116178.

70. Mahmoud, Y. A.-G., El-Naggar, M. E., Abdel-Megeed, A., and El-Newehy, M. "Recent Advancements in Microbial Polysaccharides: Synthesis and Applications." *Polymers* 13, no. 23 (2021): 4136. https://doi.org/10.3390/polym13234136.

71. Ludwicka, K., Kaczmarek, M., and Białkowska, A. "Bacterial Nanocellulose— A Biobased Polymer for Active and Intelligent Food Packaging Applications: Recent Advances and Developments." *Polymers* 12, no. 10 (2020): 2209. https://doi.org/10.3390/polym12102209.

72. Eerlingen, R. C., and Delcour, J. A. "Formation, Analysis, Structure and Properties of Type III Enzyme Resistant Starch." *Journal of Cereal Science* 22, no. 2 (1995): 129–138. https://doi.org/10.1016/0733-5210(95)90042-X.

73. Özcan, E., and Öner, E. T. (2015). Microbial Production of Extracellular Polysaccharides from Biomass Sources. In *Polysaccharides* (pp. 161–184). Springer International Publishing. https://doi.org/10.1007/978-3-319-16298-0_51.

74. Monilola, W. S., and Makinde, O. E. "Production and Characterization of Polyhydroxyalkanoates from Lactic Acid Bacteria Isolated from Dairy Wastewater, Fermented Cow Milk and 'Ogi.'" *Journal of Advances in Microbiology* (2020): 31–46. https://doi.org/10.9734/jamb/2020/v20i930279.

75. Mazzoli, R., Bosco, F., Mizrahi, I., Bayer, E. A., and Pessione, E. "Towards Lactic Acid Bacteria-based Biorefineries." *Biotechnology Advances* 32, no. 7 (2014): 1216–1236. https://doi.org/10.1016/j.biotechadv.2014.07.005.

76. Bosco, F., Cirrincione, S., Carletto, R., Marmo, L., Chiesa, F., Mazzoli, R., et al. "PHA Production from Cheese Whey and 'Scotta': Comparison between a Consortium and a Pure Culture of Leuconostoc Mesenteroides." *Microorganisms* 9, no. 12 (2021): 2426. https://doi.org/10.3390/microorganisms9122426.

77. Ahmed, Z., and Ahmad, A. (2017). Biopolymer Produced by the Lactic Acid Bacteria: Production and Practical Application. In *Microbial Production of Food Ingredients and Additives* (pp. 217–257). Elsevier. https://doi.org/10.1016/B978-0-12-811520-6.00008-8.

78. Jurášková, D., Ribeiro, S. C., and Silva, C. C. G. "Exopolysaccharides Produced by Lactic Acid Bacteria: From Biosynthesis to Health-Promoting Properties." *Foods* 11, no. 2 (2022): 156. https://doi.org/10.3390/foods11020156.

79. Coelho, M. C., Malcata, F. X., and Silva, C. C. G. "Lactic Acid Bacteria in Raw-Milk Cheeses: From Starter Cultures to Probiotic Functions." *Foods* 11, no. 15 (2022): 2276. https://doi.org/10.3390/foods11152276.

80. Qiu, Z., Fang, C., Gao, Q., and Bao, J. "A Short-Chain Dehydrogenase Plays a Key Role in Cellulosic D-Lactic Acid Fermentability of Pediococcus Acidilactici." *Bioresource Technology* 297 (2020): 122473. https://doi.org/10.1016/j.biortech.2019.122473.

81. Bancalari, E., Alinovi, M., Bottari, B., Caligiani, A., Mucchetti, G., and Gatti, M. "Ability of a Wild Weissella Strain to Modify Viscosity of Fermented Milk." *Frontiers in Microbiology* 10 (2020). https://doi.org/10.3389/fmicb.2019.03086.

82. Bintsis, T. "Lactic Acid Bacteria as Starter Cultures: An Update in their Metabolism and Genetics." *AIMS Microbiology* 4, no. 4 (2018): 665–684. https://doi.org/10.3934/microbiol.2018.4.665.

83. George, F., Daniel, C., Thomas, M., Singer, E., Guilbaud, A., Tessier, F. J., et al. "Occurrence and Dynamism of Lactic Acid Bacteria in Distinct Ecological Niches: A Multifaceted Functional Health Perspective." *Frontiers in Microbiology* 9 (2018). https://doi.org/10.3389/fmicb.2018.02899.

84. Mende, S., Peter, M., Bartels, K., Rohm, H., and Jaros, D. "Addition of Purified Exopolysaccharide Isolates from S. Thermophilus to Milk and their Impact on the Rheology of Acid Gels." *Food Hydrocolloids* 32, no. 1 (2013): 178–185. https://doi.org/10.1016/j.foodhyd.2012.12.011.

85. Vinod, A., Sanjay, M. R., Suchart, S., and Jyotishkumar, P. "Renewable and Sustainable Biobased Materials: An Assessment on Biofibers, Biofilms, Biopolymers and Biocomposites." *Journal of Cleaner Production* 258 (2020): 120978. https://doi.org/10.1016/j.jclepro.2020.120978.

86. Ahmad, A., Banat, F., Alsafar, H., and Hasan, S. W. "An Overview of Biodegradable Poly (Lactic Acid) Production from Fermentative Lactic Acid for Biomedical and Bioplastic Applications." *Biomass Conversion and Biorefinery* 14, no. 3 (2024): 3057–3076. https://doi.org/10.1007/s13399-022-02581-3.

87. Moradi, M., Guimarães, J. T., and Sahin, S. "Current Applications of Exopolysaccharides from Lactic Acid Bacteria in the Development of Food Active Edible Packaging." *Current Opinion in Food Science* 40 (2021): 33–39. https://doi.org/10.1016/j.cofs.2020.06.001.

88. Yildirim, S., Röcker, B., Pettersen, M. K., Nilsen-Nygaard, J., Ayhan, Z., Rutkaite, R., et al. "Active Packaging Applications for Food." *Comprehensive Reviews in Food Science and Food Safety* 17, no. 1 (2018): 165–199. https://doi.org/10.1111/1541-4337.12322.

89. Torino, M. I., Font de Valdez, G., and Mozzi, F. (2015). "Biopolymers from Lactic Acid Bacteria. Novel Applications in Foods and Beverages." *Frontiers in Microbiology*, 6. https://doi.org/10.3389/fmicb.2015.00834.

90. Torrieri, E., Pepe, O., Ventorino, V., Masi, P., and Cavella, S. "Effect of Sourdough at Different Concentrations on Quality and Shelf Life of Bread." *LWT - Food Science and Technology* 56, no. 2 (2014): 508–516. https://doi.org/10.1016/j.lwt.2013.12.005.

91. Wolter, A., Hager, A.-S., Zannini, E., Galle, S., Gänzle, M. G., Waters, D. M., et al. "Evaluation of Exopolysaccharide Producing Weissella Cibaria MG1 Strain for the Production of Sourdough from Various Flours." *Food Microbiology* 37 (2014): 44–50. https://doi.org/10.1016/j.fm.2013.06.009.

92. Juvonen, R., Honkapää, K., Maina, N. H., Shi, Q., Viljanen, K., Maaheimo, H., et al. "The Impact of Fermentation with Exopolysaccharide Producing Lactic Acid Bacteria on Rheological, Chemical and Sensory Properties of Pureed Carrots (Daucus carota L.)." *International Journal of Food Microbiology* 207 (2015): 109–118. https://doi.org/10.1016/j.ijfoodmicro.2015.04.031.

93. Tan, K.-X., Chamundeswari, V. N., and Loo, S. C. J. "Prospects of Kefiran as a Food-Derived Biopolymer for Agri-Food and Biomedical Applications." *RSC Advances* 10, no. 42 (2020): 25339–25351. https://doi.org/10.1039/D0RA02810J.

94. Sakr, E. A. E., Massoud, M. I., and Ragaee, S. "Food Wastes as Natural Sources of Lactic Acid Bacterial Exopolysaccharides for the Functional Food Industry: A Review." *International Journal of Biological Macromolecules* 189 (2021): 232–241. https://doi.org/10.1016/j.ijbiomac.2021.08.135.

95. Babaei-Ghazvini, A., Shahabi-Ghahfarrokhi, I., and Goudarzi, V. "Preparation of UV-protective starch/kefiran/ZnO nanocomposite as a packaging film: Characterization." *Food Packaging and Shelf Life* 16 (2018): 103–111. https://doi.org/10.1016/j.fpsl.2018.01.008.

96. Nilsen-Nygaard, J., Fernández, E. N., Radusin, T., Rotabakk, B. T., Sarfraz, J., Sharmin, N., et al. "Current Status of Biobased and Biodegradable Food Packaging Materials: Impact on Food Quality and Effect of Innovative Processing Technologies." *Comprehensive Reviews in Food Science and Food Safety* 20, no. 2 (2021): 1333–1380. https://doi.org/10.1111/1541-4337.12715.

97. Vijayendra, S. V. N., and Shamala, T. R. "Film Forming Microbial Biopolymers for Commercial Applications—A Review." *Critical Reviews in Biotechnology* 34, no. 4 (2014): 338–357. https://doi.org/10.3109/07388551.2013.798254.

98. Davidović, S., Miljković, M., Tomić, M., Gordić, M., Nešić, A., and Dimitrijević, S. "Response Surface Methodology for Optimisation of Edible Coatings Based on Dextran from Leuconostoc Mesenteroides T3." *Carbohydrate Polymers* 184 (2018): 207–213. https://doi.org/10.1016/j.carbpol.2017.12.061.

99. Hasheminya, S.-M., Mokarram, R. R., Ghanbarzadeh, B., Hamishekar, H., Kafil, H. S., and Dehghannya, J. "Development and Characterization of Biocomposite Films made from Kefiran, Carboxymethyl Cellulose and Satureja Khuzestanica Essential Oil." *Food Chemistry* 289 (2019): 443–452. https://doi.org/10.1016/j.foodchem.2019.03.076.

100. Guérin, M., Silva, C. R.-D., Garcia, C., and Remize, F. "Lactic Acid Bacterial Production of Exopolysaccharides from Fruit and Vegetables and Associated Benefits." *Fermentation* 6, no. 4 (2020): 115. https://doi.org/10.3390/fermentation6040115.

101. Gamage, A., Liyanapathiranage, A., Manamperi, A., Gunathilake, C., Mani, S., Merah, O., et al. "Applications of Starch Biopolymers for a Sustainable Modern Agriculture." *Sustainability (Switzerland)* 14, no. 10 (2022). https://doi.org/10.3390/su14106085.

102. Rkhaila, A., Chtouki, T., Erguig, H., El Haloui, N., and Ounine, K. "Chemical Proprieties of Biopolymers (Chitin/chitosan) and their Synergic Effects with Endophytic Bacillus Species: Unlimited Applications in Agriculture." *Molecules* 26, no. 4 (2021). https://doi.org/10.3390/molecules26041117.

Index

Note: Page references in *italics* and **bold** refer to figures and tables respectively.